Applications of
Electronic
Structure
Theory

MODERN THEORETICAL CHEMISTRY

Editors: **William H. Miller**, *University of California, Berkeley*
Henry F. Schaefer III, *University of California, Berkeley*
Bruce J. Berne, *Columbia University, New York*
Gerald A. Segal, *University of Southern California, Los Angeles*

Applications of
Electronic Structure Theory

Edited by
Henry F. Schaefer III
University of California, Berkeley

PLENUM PRESS · NEW YORK AND LONDON

Library of Congress Cataloging in Publication Data

Main entry under title:

Applications of electronic structure theory.

(Modern theoretical chemistry; 4)
Includes bibliographical references and index.
1. Molecular theory. I. Schaefer, Henry F. II. Series.
QD461.A66   541'.28   77-349
ISBN 0-306-33504-2 (v. 4)

© 1977 Plenum Press, New York
A Division of Plenum Publishing Corporation
227 West 17th Street, New York, N.Y. 10011

Printed in the United States of America

Contributors

Leland C. Allen, Department of Chemistry, Princeton University, Princeton, New Jersey

J. Demuynck, Université L. Pasteur, Strasbourg, France

Warren J. Hehre, Department of Chemistry, University of California, Irvine, California

C. William Kern, Battelle Columbus Laboratories, Columbus, Ohio, and The Ohio State University, Department of Chemistry, Columbus, Ohio

Peter A. Kollman, Department of Pharmaceutical Chemistry, School of Pharmacy, University of California, San Francisco, California

Stephen R. Langhoff, Battelle Columbus Laboratories, Columbus, Ohio

Marshall D. Newton, Department of Chemistry, Brookhaven National Laboratory, Upton, New York

Philip W. Payne, Department of Chemistry, Princeton University, Princeton, New Jersey

J. A. Pople, Department of Chemistry, Carnegie-Mellon University, Pittsburgh, Pennsylvania

Péter Pulay, Eötvös L. University, Department of General and Inorganic Chemistry, Budapest, Hungary

Leo Radom, Research School of Chemistry, Australian National University, Canberra, Australia

Maurice E. Schwartz, Department of Chemistry and Radiation Laboratory, University of Notre Dame, Notre Dame, Indiana

A. Veillard, C.N.R.S., Strasbourg, France

Preface

These two volumes deal with the quantum theory of the electronic structure of molecules. Implicit in the term *ab initio* is the notion that approximate solutions of Schrödinger's equation are sought "from the beginning," i.e., without recourse to experimental data. From a more pragmatic viewpoint, the distinguishing feature of *ab initio* theory is usually the fact that no approximations are involved in the evaluation of the required molecular integrals. Consistent with current activity in the field, the first of these two volumes contains chapters dealing with methods *per se*, while the second concerns the application of these methods to problems of chemical interest. In a sense, the motivation for these volumes has been the spectacular recent success of *ab initio* theory in resolving important chemical questions. However, these applications have only become possible through the less visible but equally important efforts of those developing new theoretical and computational methods and models.

Henry F. Schaefer

Contents

Chapter 3.   Hydrogen Bonding and Donor–Acceptor
            Interactions

Peter A. Kollman

Chapter 6.   Strained Organic Molecules
             Marshall D. Newton

Chapter 7.   Carbonium Ions: Structural and Energetic
             Investigations
             Warren J. Hehre

Chapter 8.   Molecular Anions
Leo Radom

Chapter 9.   Electron Spectroscopy
Maurice E. Schwartz

Chapter 10.   Molecular Fine Structure
Stephen R. Langhoff and C. William Kern

Contents of Volume 3

A Priori Geometry Predictions

J. A. Pople

1. Introduction

Early quantum mechanical computations on the electronic structure of molecules were generally concerned with the determination of a wave function for a single assumed nuclear geometry (usually an experimentally determined equilibrium structure). As techniques have improved, however, it has increasingly become possible to make explorations of potential surfaces (total energies for heavy nuclei in the Born–Oppenheimer approximation) and hence to use theory directly to locate the minima in such surfaces and the corresponding equilibrium structural parameters. Such explorations can either be carried out partially, that is assuming some parameters and varying others (as in the study of "rigid" internal rotation with fixed bond lengths and angles) or, more desirably, by complete minimization of the energy with respect to all variables. Given the quantum mechanical procedure, the latter leads to *a priori* predictions of structure making no appeal to experimental data other than using the values of fundamental constants. Theoretical structures of this sort have been used for two main purposes. The first is to assess how well experimentally known structures are reproduced at a given level of theory and hence evaluate the limitations of the theory in a systematic manner. Second, the theory has been increasingly used to investigate structures of molecules for which experimental data are insufficient. Many such predictions have been made and there is an increasing number of examples of subsequent experimental verification.

In this chapter most attention will be devoted to structural prediction at the Hartree–Fock level, that is using a single-determinant molecular orbital wave function with a basis-set expansion. Most of the applications carried out to date have been at this level. If the basis set is sufficiently large the results

J. A. Pople • Department of Chemistry, Carnegie-Mellon University, Pittsburgh, Pennsylvania

should be independent of detail and characteristic of a geometry at the Hartree–Fock limit. However, for basis sets of limited size, as is presently necessary for large molecules, the predicted structures will depend on the basis used. Under these circumstances it is desirable to select a well-defined small basis set and then use it systematically for extensive exploration. This is the concept of a "theoretical model chemistry" in which a uniform level of approximation is used throughout.[1] It is a valuable approach in that it permits discussion of relative structural features and general trends even though absolute agreement with experimental data is lacking. We shall review some Hartree–Fock model chemistries of this type. Probably the most widely used models are those including a minimal basis of Slater-type orbitals.[2]

For some small molecules, geometrical studies have been carried out with sufficiently large basis sets for the Hartree–Fock limiting structure to be well established. These show a number of systematic deviations from experimental data. To correct these it is necessary to go beyond single-determinant wave functions and investigate modifications of equilibrium geometry due to corrections for the correlation of the motion of electrons with different spin. This involves configuration interaction or some other equivalent higher level of approximation. If carried out to sufficient extent, such treatments should generally lead to highly accurate structural predictions since the remaining relativistic energy corrections are mostly associated with inner-shell electrons and, hence, are likely to be largely independent of nuclear positions. The number of structural studies of polyatomic molecules using correlated wave functions is still relatively small, but some interesting trends are already becoming evident. These will be reviewed in the latter part of this chapter.

2. Equilibrium Geometries by Hartree–Fock Theory

2.1. Restricted and Unrestricted Hartree–Fock Theories

The fundamental feature of Hartree–Fock theory is the use of a single-determinant wave function Ψ which is formed from a set of occupied spin-orbitals $\chi_1, \chi_2, \ldots, \chi_n$,

$$\Psi = A \det\{\chi_1(1)\chi_2(2) \cdots \chi_n(n)\} \tag{1}$$

where n is the number of electrons, A is a normalization constant, and the diagonal element product of the determinant is specified explicitly in Eq. (1). The spin-orbitals are, in general, one-electron functions in the space of the Cartesian and spin coordinates. Before proceeding to the discussion of geometry predictions using Hartree–Fock theory it is desirable to indentify constraints which are sometimes imposed on the spin-orbitals. These constraints are of some importance since the self-consistent, or Hartree–Fock, orbitals χ_i are derived by variation to minimize the energy expectation,

$$E = \int \cdots \int \Psi^* \mathcal{H} \Psi \, d\tau_1 \cdots d\tau_n \tag{2}$$

$\mathcal{H}$ being the total Hamiltonian. Imposition of constraints must raise (or possibly leave unchanged) the value of the energy (2) for any nuclear arrangement.

In a completely unconstrained Hartree–Fock theory no limitations are placed on the functions χ_i other than orthonormality, which may be imposed without restriction of the full wave function Ψ. This version of the theory is rarely used. A common assumption is to use products of ordinary molecular orbitals ψ (functions of only the Cartesian coordinates x, y, and z) and spin functions α and β. The spin-orbitals are then

$$\psi_i^\alpha \alpha, \qquad \psi_i^\beta \beta, \qquad i = 1, 2, \ldots \tag{3}$$

If the molecular orbitals ψ_i^α associated with α spin are varied, completely independently of those associated with β spin (ψ_i^β), the method is described as "spin-*un*restricted *H*artree–*F*ock" (SUHF or UHF) or sometimes as "different orbitals for different spins." If, on the other hand, the molecular orbitals for the α set are constrained to be identical with the β set where both are occupied, they are then described as "doubly or singly occupied" and the corresponding theory is "*s*pin-*r*estricted *H*artree–*F*ock" (SRHF or RHF). In some situations the additional flexibility in UHF theory does not lead to any energy lowering relative to RHF, but for most systems with unpaired electrons there is such a difference.

The relation between RHF and UHF theories is of some importance in geometry predictions for the following reason. It can be shown that RHF wave functions are strictly eigenfunctions of the total spin operator $\mathbf{S}^2$ and so correspond to pure singlet, doublet, triplet, etc., electronic states. UHF functions, on the other hand, are not always eigenfunctions of $\mathbf{S}^2$ and may lead to "contaminated" wave functions where a primarily double wave function, for example, may also have some quartet character. If this contamination becomes large, the search for a minimum, corresponding to a doublet state, may produce a structure partly characteristic of a neighboring higher-multiplicity quartet state.

Another constraint that is sometimes imposed is the requirement that the molecular orbitals ψ_i belong to irreducible representations of the point group of the nuclear framework. Again, this restriction does not necessarily raise the energy, but if it does, the constraint is undesirable since small distortions which break the symmetry would lead to discontinuity in the potential surface.

In practice, the molecular orbitals are normally expanded as linear combinations of a set of basis functions $\varphi_\mu(x, y, z)$. For UHF theory,

$$\psi_i^\alpha = \sum_\mu c_{\mu i}^\alpha \varphi_\mu$$
$$\psi_i^\beta = \sum_\mu c_{\mu i}^\beta \varphi_\mu \tag{4}$$

In RHF theory, the c^α and c^β matrices are constrained to be identical. Algebraic equations for RHF coefficients, if the orbitals are doubly occupied (or empty), were given by Roothaan.[3] The corresponding UHF equations are due to Pople and Nesbet.[4] The RHF equations with both doubly and singly occupied orbitals are rather more complicated, but several algebraic procedures are available.[5–8] Once the optimized self-consistent energy is obtained, as a function of geometrical parameters, the subsequent minimization to locate the potential surface minimum can be carried out by any of a wide range of general optimization techniques.[9]

2.2. Basis Sets for Hartree–Fock Studies

In published Hartree–Fock calculations a wide variety of basis sets φ_μ are used. Since certain features of predicted structures depend on the nature of the basis set, some survey of the bases commonly used is desirable.

The earliest wave function computations were generally based on the use of Slater-type functions[2] as approximations to atomic orbitals. These are functions with exponential radial parts such as

$$\varphi_{1s} = (\zeta_1^3/\pi)^{1/2} \exp(-\xi_1 r)$$
$$\varphi_{2s} = (\zeta_2^5/3\pi)^{1/2} r \exp(-\xi_2 r) \tag{5}$$
$$\varphi_{2p} = (\zeta_2^5/\pi)(x, y, z) \exp(-\xi_2 r)$$

and similar forms for $3s$, $3p$, $3d$, ... If the Slater-type functions correspond just to those shells which are populated (or partly populated) in the atomic ground states, the basis is described as *minimal* (e.g., $1s$ only for hydrogen, $1s$, $2s$, $2p$ for carbon, etc.).

The direct use of the Slater-type basis functions (5) leads to difficult two-electron repulsion integrals. Such integrals can be evaluated by methods such as the Gaussian transform technique proposed by Shavitt and Karplus.[10] A number of structural studies have been made by such methods. However, it turns out that a more efficient procedure, leading to equivalent results, uses the replacement of Slater-type functions with linear combinations of Gaussian functions obtained by least-squares fitting.[11–14] All integrals can then be explicitly evaluated. In the fits proposed by Hehre, Stewart, and Pople,[4] each Slater-type function φ_μ is replaced by a linear combination of K Gaussian functions (contracted Gaussian functions)

$$\varphi'_\mu(\xi, \mathbf{r}) = \xi^{3/2} \varphi'_\mu(1, \xi\mathbf{r}) \tag{6}$$

where

$$\varphi'_{ns}(1, \mathbf{r}) = \sum_{k=1}^{K} d_{ns,k} g_{1s}(\alpha_{nk}, \mathbf{r})$$

$$\varphi'_{np}(1, \mathbf{r}) = \sum_{k=1}^{K} d_{np,k} g_{2p}(\alpha_{nk}, \mathbf{r}) \tag{7}$$

g_{1s} and g_{2p} being normalized Gaussian functions

$$g_{1s}(\alpha, \mathbf{r}) = (2\alpha/\pi)^{3/4} \exp(-\alpha r^2)$$

$$g_{2p}(\alpha, \mathbf{r}) = (128\alpha^5/\pi^3)^{1/4}(x, y, z) \exp(-\alpha r^2) \tag{8}$$

The constants d and α are chosen to give best least-squares fits to the Slater orbitals (with $\zeta = 1$). A particular feature is the sharing of Gaussian exponents α_{2k} between $2s$ and $2p$ functions, leading to an increase in computational efficiency.

The minimal basis sets (6)–(8) are termed STO-KG (*Slater-type orbital* at the *K-G*aussian level) and are documented[14] for $K = 2$ through 6. As K becomes larger, the results for STO-KG must approach those for the Slater-type basis itself. Comparative studies of STO-3G and STO-4G have given very similar theoretical structures[15,16] which are also in agreement with some obtained directly from the full STO basis.[17,18]

Complete specification of a Slater minimal basis requires values for the exponents ζ_1 and ζ_2 in (5) or (6). Ideally, these should be optimized in each calculation, but in practice it is necessary to use some set of average values. ζ values appropriate to isolated atoms are generally rather different from optimum molecular values. The set of average molecular ζ values shown in Table 1 (for atoms up to fluorine), in combination with the STO-3G basis, has been most widely used for minimal basis studies of structure.[14,19] The full theoretical models may be described as RHF/STO-3G or UHF/STO-3G

Table 1. *Standard ζ-Values for Slater Minimal Bases*

Atom	ζ_1	ζ_2
H	1.24	—
Li	2.69	0.80
Be	3.68	1.15
B	4.68	1.50
C	5.67	1.72
N	6.67	1.95
O	7.66	2.25
F	8.65	2.55

depending on whether restricted or unrestricted Hartree–Fock theory is used. These techniques are quite inexpensive to use and have had fair success in predicting structure. We shall review the results in some detail.

An alternative minimal basis set is one in which the φ_μ are just the Hartree–Fock atomic orbitals, rather than Slater-functions. This is a "*l*inear *c*ombination of *a*tomic *o*rbitals," an LCAO approach in a strict sense. In practice, the atomic orbitals again have to be approximated by linear combinations of Gaussian functions but the limiting procedure can be explored by making the number of these functions sufficiently large. A systematic study of such a basis was made (least-energy minimal atomic orbitals) with up to six Gaussians per atomic orbital,[16] but the overall results were less successful than those with the Slater-type basis. In particular, some of the bond lengths in such a model were found to be much too long.

The next step, beyond the minimal basis, involves the use of additional functions to provide extra flexibility for the molecular orbitals. A common procedure is to replace each minimal basis function by two functions (an inner and an outer part). Such basis sets are described as "double zeta." A number of such bases have been used. The double-zeta contracted Gaussian basis of Snyder and Basch[20] has been widely used, but not for many different geometry studies. A slightly simpler type is a "split-valence basis" in which the number of functions is only doubled in the valence shell, the splitting of the inner shell being probably less important for the description of chemical bonds. The basis described as 4-31G was introduced[21] as a relatively simple theoretical model at this level and this has been used for some systematic studies which will be reviewed.

Beyond the split-valence or double-zeta levels, completion of a basis set requires the addition of "polarization functions" or basis functions with higher angular quantum numbers. Thus d functions should be added to carbon atoms and p functions to hydrogen. Early work, particularly by Rauk, Allen, and Clementi,[22] indicated that such refinements were important in structure prediction and this is now becoming increasingly clear. The basis set 6-31G* was introduced[23] to give, as simply as possible, a theoretical model containing d polarization functions and some geometrical studies have been carried out at this level. Finally, for certain very small molecules studies have been made with very large bases (multiple splits of s and p functions and more than one set of d functions on nonhydrogen atoms) to give energies and geometries which must approach the Hartree–Fock limit quite closely.

2.3. Hartree–Fock Structures for Small Molecules

In this section we shall review the application of Hartree–Fock theory to the determination of structures of molecules containing up to two nonhydrogen atoms. This class of molecules incorporates those which characterize the basic types of chemical bond (single, double, and triple) and the simplest stereochemical arrangements as specified by bond angles and dihedral angles. Emphasis is laid on systems for which the experimental structure is well documented, so that the level of success of Hartree–Fock theory at the various levels of the basis set can be well established. Predictive structures, where experimental data are incomplete, will be discussed later in the chapter.

Before presenting comparisons between theoretical and experimental structures, it is advisable to make a few remarks on error sources. In theoretical computations the structural parameters obtained are subject to some errors other than those implicit in the underlying theory and basis set. Such errors arise from rounding and incomplete accuracy in integral evaluation. Also, parameters obtained by energy minimization are subject to considerable uncertainty as the potential surface is flat in the immediate vicinity of the minimum. Errors of this type would, of course, be removed by carrying out the computations at a higher level of precision, but this is usually not worthwhile because of cost. Most of the theoretical results given in the following tables are subject to such rounding errors in the third decimal place (Å) for bond lengths and in the first decimal place (degrees) for angles. For dihedral angles, with flatter potential curves, the errors may be larger.

The experimental structure data are also subject to various sources of error which should be taken into account in comparisons with theory.[24,25] The theoretical structures refer to the true equilibrium form, that is the actual potential minimum (bond lengths r_e and angles θ_e). Experimental data often refer to averages over zero-point vibrational motion (r_0, θ_0) and may differ slightly. The force field data necessary to obtain (r_e, θ_e) from (r_0, θ_0) are often unavailable. Differences between r_e and r_0 may range up to 0.01 Å. Another limitation is that, for some of the larger molecules, there is insufficient data for complete structure determination and values for some parameters are often obtained by assuming values for others. We have not attempted to identify these limitations in individual cases, but they do represent some error. Overall, it is clear that the combined effects of theoretical and experimental uncertainties render comparisons dubious below the level of 0.01 Å for lengths and 1° for angles, except for the very smallest systems.

We begin with the series of molecules H_2 and AH_n where A is a first row atom in the series Li–F. All these systems which have been studied experimentally have been found to have symmetric structures with equivalent AH bonds ($n > 1$) and this constraint has been imposed in most of the theoretical work.[23,26] The bond lengths are listed in Table 2 which compares the values for various basis sets and the experimental results. The "best" theoretical results in this and other tables arise from varied basis sets which lead to total energies lower than STO-3G, 4-31G and 6-31G*. This column gives the best approximations available to the structural parameter at the Hartree–Fock

Table 2. Theoretical (Hartree–Fock) and Experimental Bond Lengths for H_2 and AH_n Molecules[a]

Molecule[b]	Bond length[c]				
	STO-3G	4-31G	6-31G*	Best	Experimental
H_2	0.172	0.730	0.730	0.737[d]	0.742
LiH	—	—	—	1.605[e]	1.595
BeH	—	—	—	1.338[e]	1.343
BeH_2	—	—	—	1.333[f]	—
BH	1.213[g]	—	—	1.220[e]	1.236
BH_2	1.161[g]	—	—	—	1.18
BH_3	1.160[g]	—	—	1.190[h]	—
CH	1.143	1.118	1.108	1.104[e]	1.120
CH_2 ($^3B_1$)	1.082	1.069	1.071	1.069[i]	1.078[j]
CH_2 ($^1A_1$)	1.123	1.100	1.096	1.095[i]	1.110[j]
CH_3	1.080	1.070	1.073	1.072[k]	1.079
CH_4	1.083	1.081	1.084	1.084[l]	1.085[l]
NH	1.082	1.033	1.024	1.018[e]	1.038
NH_2	1.058	1.015	1.013	1.019[m]	1.024
NH_3	1.033	0.991	1.004	1.000[n]	1.012[o]
OH	1.014	0.968	0.958	0.950[e]	0.971
OH_2	0.990	0.951	0.948	0.941[p]	0.957[q]
FH	0.956	0.922	0.911	0.897	0.917

[a] Values are given in angstroms.
[b] Electronic ground state except for CH_2, for which the lowest singlet and triplet states are listed. All polyatomic structures assumed to have equal AH bond lengths.
[c] Unless otherwise specified, STO-3G and 4-31G results are from Ref. 26, 6-31G* results are from Ref. 23, and experimental values are from Refs. 24 or 25.
[d] Obtained with a $(4s, 1p)$ uncontracted Gaussian basis.
[e] Ref. 27.
[f] J. J. Kaufman, L. M. Sachs, and M. Geller, J. Chem. Phys. 49, 4369 (1968).
[g] Ref. 31.
[h] M. Gelus and W. Kutzelnigg, Theor. Chim. Acta 28, 103 (1973).
[i] V. Staemmler, Theor. Chim. Acta 31, 49 (1973).
[j] r_0, θ_0 values from G. Herzberg and J. W. C. Johns, J. Chem. Phys. 54, 2276 (1971).
[k] F. Driessler, R. Ahlrichs, V. Staemmler, and W. Kutzelnigg, Theor. Chim. Acta 30, 315 (1973).
[l] Theoretical value from W. Meyer, J. Chem. Phys. 58, 1017 (1973). Experimental values from K. Kuchitsu and L. S. Bartell, J. Chem. Phys. 36, 2470 (1962). Meyer suggests that the experimental value (r_e) is based on an incorrect anharmonicity and should really be close to 1.090 Å.
[m] C. F. Bender and H. F. Schaefer III, J. Chem. Phys. 55, 4798 (1972).
[n] Ref. 22.
[o] From electron diffraction results (r_e, θ_e) of K. Kuchitsu, J. P. Guillory, and L. S. Bartell, J. Chem. Phys. 49, 2488 (1968) and from spectroscopic results of W. S. Benedict and E. K. Plyler, Can. J. Phys. 35, 1235 (1957).
[p] Ref. 29.
[q] W. S. Benedict, N. Gailar, and E. K. Plyler, J. Chem. Phys. 24, 1139 (1956).

limit. The comparison with experimental values indicates that these deviations are quite systematic. The STO-3G basis gives bond lengths which are mostly too long (H_2 being an exception), the mean absolute deviation from experimental values being 0.023 Å for the entries listed. The relative values for a given type of bond are well reproduced (the five types of CH bonds, for example). Proceeding to the split-valence 4-31G basis, some improvement may be noted, the mean absolute deviation from experiment being reduced to 0.008 Å. However, the theoretical values are now too small (with the exception of HF). The next step to the polarized basis 6-31G* leads to only small changes, the theoretical values still being too small (mean absolute deviation 0.010 Å). Further refinement of the basis leads to additional shortening. The results of Cade and Huo[27] for the AH diatomics are close to the limit and all lengths are too short except for LiH. The mean absolute deviation from experiment of the best Hartree–Fock lengths is 0.012.Å (15 comparisons). This underestimation seems well established and is more notable for bonds to N, O, and F. Nesbet[28] has suggested that this systematic effect is primarily due to the neglect of interaction with an electron configuration in which two electrons are raised from the highest-occupied bonding σ orbital to an antibonding σ^* orbital. As the molecule dissociates, the energy gap between these orbitals decreases and the energy lowering due to configuration interaction increases, thereby increasing the bond length.

Table 3 gives a similar listing of theoretical and experimental bond angles in AH_n molecules. Again the deviations of the theoretical models are quite systematic. The STO-3G basis gives bond angles which are too small for all the states listed (mean deviation of 4.4°), whereas the 4-31G values are mostly too large (mean absolute deviation of 4.6°). The split-valence results reflect a general tendency of double-zeta type basis sets to give excessively large bond angles at atoms with lone-pair electrons. It is only when polarization functions are added (particularly d functions on the heavy atoms) that the theoretical angles become reasonably close to the experimental values. This general dependence on basis set was first clearly noted by Rauk, Allen, and Clementi[22] in a detailed study of ammonia. They found that if a very large basis of only s and p functions was used, the HNH angle opened right out to 120°, giving a planar structure, and a pyramidal structure then only followed if d functions were also added. Similar effects were also found in studies of the water molecule.[29] These effects can reasonably be interpreted in terms of the

Table 3. Theoretical (Hartree–Fock) and Experimental Bond Angles for AH_n Molecules

Molecule[a]	Bond angle[b]				
	STO-3G	4-31G	6-31G*	Best	Experimental
BeH_2	—	—	—	180.0	—
BH_2	123.5	—	—	—	131
CH_2 ($^3B_1$)	125.5	132.0	—	—	136
CH_2 ($^1A_1$)	100.5	105.4	—	—	102.4
CH_3	118.3	120.0	120.0	120.0	120
NH_2	100.2	108.3	104.3	105.4	103.4
NH_3	104.2	115.8	107.5	107.2	106.7
CH_2	100.0	111.2	105.5	106.6	104.5

[a] Electronic ground state unless otherwise specified.
[b] For sources, see Table 2.

local symmetry around the heavy atom. In planar ammonia, for example, the lone-pair electrons occupy an a_2'' molecular orbital (point group D_{3h}) and this cannot utilize any d basis functions on the nitrogen atom. If on the other hand the molecule is pyramidal, the lone-pair orbital has a_1 symmetry (point group C_{3v}) and such mixing can take place. Thus the d functions participate more effectively in the less symmetrical form and hence favor a pyramidal structure.

The next set of molecules for which extensive geometrical studies have been carried out is $H_m ABH_n$ with two nonhydrogenic atoms A and B. If A, B are carbon, nitrogen, oxygen, or fluorine, the complete set of such neutral

Table 4. *Theoretical (Hartree–Fock) and Experimental Bond Lengths for AB Bonds in $H_m ABH_n$ Molecules[a]*

		Bond length			
Bond	Molecule	STO-3G[b]	4-31G	Best	Experimental[b]
C≡C	C_2H_2	1.168	1.190[c]	1.205[d]	1.203
C=C	C_2H_4	1.306	1.316[c]	—	1.330
C—C	C_2H_6	1.538	1.529[c]	1.551[e]	1.531
C≡N	HCN	1.153	1.140[c]	—	1.154
C=N	H_2CNH	1.273	1.257[f]	—	—
C—N	H_3CNH_2	1.486	—	—	1.474
$\overset{-}{C}{\equiv}\overset{+}{O}$	CO	1.146	1.128[g]	1.101[h]	1.128
C=O	H_2CO	1.217	1.206[c]	1.178[i]	1.203
C—O	H_3COH	1.433	1.437[j]	—	1.427
C—F	H_3CF	1.384	1.412[c]	—	1.385
N≡N	N_2	1.134	1.085[g]	1.065[k]	1.094
N=N	N_2H_2	1.267	—	—	1.238
N—N	N_2H_4	1.459	—	—	1.453
N=O	NO	1.184	—	—	1.150
N=O	HNO	1.231	—	—	1.211
N—O	H_2NOH	1.420	—	—	1.46
N—F	H_2NF	1.387	—	—	—
O=O	O_2	1.217	1.19[l]	—	1.207
O—O	HOOH	1.396	1.468[l]	1.475[m]	1.475
O—F	HOF	1.355	—	—	1.442
F—F	F_2	1.315	—	1.323[n]	1.418

[a]Values are given in Angstroms.
[b]STO-3G and references to experimental lengths are given in Ref. 31.
[c]Ref. 21.
[d]R. J. Buenker, S. D. Peyerimhoff, and J. L. Whitten, *J. Chem. Phys.* **46**, 2029 (1967).
[e]E. Clementi and H. Popkie, *J. Chem. Phys.* **57**, 4870 (1972).
[f]R. Macauley, L. A. Burnelle, and C. Sandorfy, *Theor. Chim. Acta* **29**, 1 (1973).
[g]W. A. Lathan, unpublished.
[h]W. M. Huo, *J. Chem. Phys.* **43**, 624 (1965).
[i]W. Meyer and P. Pulay, *Theor. Chim. Acta* **32**, 253 (1974).
[j]Partial optimization by G. A. Jeffrey, L. Radom, and J. A. Pople, *Carbohydrate Res.* **25**, 117 (1972).
[k]P. E. Cade, K. D. Sales, and A. C. Wahl, *J. Chem. Phys.* **44**, 1973 (1966).
[l]R. J. Blint and M. D. Newton, *J. Chem. Phys.* **59**, 6220 (1973).
[m]Ref. 33.
[n]G. Das and A. C. Wahl, *J. Chem. Phys.* **44**, 87 (1966).

Table 5. Theoretical (Hartree–Fock) and Experimental Bond Angles for H_mABH_n Molecules

| Angle | Molecule | Angle[a] | | | |
		STO-3G	4-31G	Best	Experimental
HCH	C_2H_4	115.4	116.0	—	116.6
	H_2CO	114.5	116.4	115.9	116.5
	C_2H_6	108.2	107.7	—	107.8
	H_3CF	108.3	110.7	—	109.9
COH	H_3COH	103.8	—	—	105.9
OOH	H_2O_2	101.1	100.8	101.3	94.8

[a] For sources, see Table 4.

molecules and cations have been considered at the STO-3G level.[30,31] These studies have recently been extended to boron compounds.[32] Table 4 lists the theoretical AB bond lengths in the subset which characterize the principal single and multiple bonds between the heavy atoms C, N, O, and F. Most of these have structures which are experimentally characterized. The complete set of AB lengths at the STO-3G level shows a mean absolute deviation from experiment of 0.030 Å (19 comparisons). STO-3G CC lengths are somewhat shorter, but other bonds involving N, O, and F are mostly too long. The worst results are those for the single bonds O—O, O—F, and F—F, where the theoretical values are of the order of 0.1 Å too small. Again some improvement is found for the split-valence 4-31G basis, where the mean absolute deviation from experiment is 0.011 Å (11 comparisons). For those molecules for which there are geometry studies with lower computed energies, the bond lengths, close to the Hartree–Fock limit, are mostly too small.

Table 5 shows some bond angles for the same set of H_mABH_n molecules. Certain significant features are satisfactorily reproduced such as the considerable deviations of HCH angles in ethylene and formaldehyde from the ideal value of 120°, expected for sp^2 hybridization. The very small OOH bond angle for hydrogen peroxide is not well reproduced, even if d functions are included in the basis set.[33]

The dihedral angles of the H_mABH_n compounds, specifying the internal rotation conformation about the AB bond are most satisfactorily given by Hartree–Fock theory.[31] For single bonds to carbon (ethane, methylamine, and methanol), a staggered conformation (bonds to hydrogen *trans*) is always preferred in agreement with observation. All those molecules with double bonds give planar structures (ethylene, formaldimine, formaldehyde, and azene) and those with triple bonds give linear structures (acetylene, hydrogen cyanide). Azene is predicted to have *trans*-NH bonds. Hydrogen peroxide is predicted to have a twisted structure (C_2 symmetry) at STO-3G, but the barrier to rotation to the trans (C_{2h}) arrangement is so low (~0.1 kcal/mole) that the result is rather inconclusive. This is consistent with a study by Stevens[18] using an STO basis directly. Full optimization of the geometry using the 4-31G basis leads to a *trans* structure and it is not until *d* functions are added on oxygen that a twisted structure is clearly indicated. This was first demonstrated by Veillard,[33] who found a dihedral angle of 123° as compared with an experimental value of 111°. Polarization functions apparently play a role here similar to that in the nonplanarity of ammonia; preferentially stabilizing conformations with lower local symmetry at the oxygen atom. Hydrazine shows a twisted structure at STO-3G and hydroxylamine gives a C_s structure with the OH bond trans to the direction of the lone pair on nitrogen. For all these molecules, these predicted dihedral conformations are consistent with experimental observations.

2.4. Hartree–Fock Structures for Larger Molecules

For molecules with more than two nonhydrogen atoms, fewer full theoretical structure determinations have been carried out and most of these have been hydrocarbons with the minimal STO-3G basis.[34,35] Some work with larger bases has involved partial structures, that is variation of certain geometrical parameters while others are held fixed at assumed standard experimental values. We shall briefly review some aspects of work on these larger molecules. The first is the variation of the length of a given type of bond from one molecule to another, the second is rotational isomerism as determined by dihedral angle; and the third is the nature and magnitude of the puckering or nonplanarity of ring structures.

Table 6 gives a comparative study of CH bond lengths in molecules for which complete geometrical optimization has been carried out. It is evident that certain general trends are well reproduced, even at the minimal basis level. CH bonds are significantly shorter when attached to triple bonds and exceptionally long when the carbon is bonded to oxygen (in formaldehyde or methanol).

Table 7 gives a comparison of STO-3G and experimental CC bond lengths in a series of hydrocarbons. All of the compounds in this table have been completely optimized (subject to some symmetry constraints) and represent the largest group of related molecules with which a theoretical model can be compared with available experimental data. The correlation here between theoretical and experimental lengths is very good, as illustrated in Fig. 1, the principal deviation being an underestimation of multiple carbon-carbon bonds by 0.02–0.03 Å. Many observed trends are well reproduced. Among single bonds we may note the short CC length in cyclopropane, the long bonds in cyclobutane, the exceptionally long C_3-C_4 bond in cyclobutene, the short bridge bond in bicyclobutane, and the very short partially conjugated single

Table 6. Comparative Study (Hartree–Fock) of CH Bond Lengths

Bond type	Molecule	Bond length			
		STO-3G[a]	4-31G[b]	Best	Experimental[c]
$\equiv$C—H	Acetylene	1.065	1.051	—	1.061
	Propyne	1.064[d]	—	—	1.056
	Hydrogen cyanide	1.070	1.051	—	1.063
⟩C—H	Ethylene	1.082	1.073	—	1.076
	Allene	1.083[d]	—	—	1.087
	Cyclopropene	1.075[d]	—	—	1.070
	Benzene	1.08[e]	—	—	1.084
	Formaldehyde	1.101	1.081	1.092[f]	1.101
—C—H	Methane	1.083	1.081	1.083[g]	1.090
	Ethane	1.086	1.083	—	1.096
	Propyne	1.088[d]	—	—	1.105
	Cyclopropane	1.081[d]	—	—	1.089
	Cyclopropene	1.087[d]	—	—	1.087
	Methyl fluoride	1.097	1.076	—	—
	Methanol	1.094	—	—	1.096

[a]STO-3G values from Ref. 31 unless otherwise indicated.
[b]4-31G values from Ref. 21.
[c]For experimental sources, see theoretical references.
[d]Ref. 34.
[e]Ref. 15.
[f]W. Meyer and P. Pulay, *Theor. Chim. Acta* **32**, 253 (1974).
[g]Theoretical value from W. Meyer, *J. Chem. Phys.* **58**, 1017 (1973). Experimental values from K. Kuchitsu and L. S. Bartell, *J. Chem. Phys.* **36**, 2470 (1962). Meyer suggests that the experimental value (r_e) is based on an incorrect anharmonicity and should really be close to 1.090 Å.

bonds in methylenecyclopropane, 1,3-butadiene, propyne, and but-1-yne-3-ene. The variation in double bonds is also well reproduced, apart from systematic underestimation. The shortest double bond is $C_2=C_3$ in butatriene and the longest is in but-1-yne-3-ene in agreement with observation. This is consistent with expectations in terms of conjugation, the bond in butatriene acquiring some triple-bond character and that in but-1-yne-3-ene some single-bond character. Finally, the difference between CC lengths in 1-3-cyclobutadiene (triplet) and benzene, although not yet tested experimentally, illustrates the influence of aromatic character on bond lengths. A 4-31G study of triplet 1,3-cyclobutadiene[36] also gives a bond length of 1.434 Å.

A further series of bonds which show large variations in length are the CF bonds in fluorocarbons. Here it is found that the bonds are substantially shortened if several fluorines are attached to the same atom, a phenomenon sometimes ascribed to resonance of the type,

$$F—C—F \leftrightarrow F^-C=F^+$$

Table 7. Comparative Study of (Hartree–Fock) CC Bond Lengths in Hydrocarbons

Bond type	Molecule	Bond length[a]	
		STO-3G	Experimental
$\searrow$C—C$\swarrow$ Ethane	Ethane	1.538	1.531
	Propane	1.541	1.526
	Cyclopropane	1.502	1.510
	Cyclobutane	1.554	1.548
	Cyclobutene	1.565	1.566
	Bicyclobutane (bridge)	1.501	1.498
	Methylenecyclopropane	1.522	1.542
$\rightthreetimes$C—C$\swarrow$	Propene	1.520	1.501
	Cyclopropene	1.493	1.515
	Cyclobutene	1.526	1.517
	Methylenecyclopropane	1.474	1.457
$\rightthreetimes$C—C$\nwarrow$	1,3 Butadiene	1.488	1.483
≡C—C$\swarrow$	Propyne	1.484	1.459
≡C—C$\diagdown$	But-1-yne-3-ene	1.459	1.431
≡C—C≡	Butadiyne	1.408	—
$\diagup$C=C$\diagdown$	Ethylene	1.306	1.330
	Propene	1.308	1.336
	Cyclopropene	1.277	1.300
	Cyclobutene	1.314	1.342
	Methylenecyclopropene	1.298	1.332
	But-1-yne-3-ene	1.320	1.341
	1,3-Butadiene	1.313	1.337
=C=C$\diagdown$	Allene	1.288	1.308
	Butatriene	1.296	1.318
=C=C=	Butatriene	1.257	1.283
—C≡C—	Acetylene	1.168	1.203
	Propyne	1.170	1.206
	But-1-yne-3-ene	1.171	1.208
	Butadiyne	1.175	—
$\rightthreetimes$C⋯C$\swarrow$	1,3-Cyclobutadiene[b]	1.431	—
	Benzene	1.39	1.397

[a]STO-3G values from Refs. 15, 31, 34, and 35. For experimental references, see the theoretical papers.
[b]Triplet state.

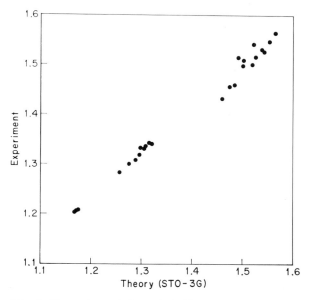

Fig. 1. Comparison of theoretical (Hartree–Fock/STO-3G) and experimental carbon–carbon bond lengths.

Available theoretical data on fluoromethanes are summarized in Table 8. A shortening along the series is found with the STO-3G basis but the magnitude of the effect is much too small. This is perhaps another example of the rather poor description of fluorine with a minimal Slater-type basis. Rather limited studies with the 4-31G basis, also listed in Table 8, indicate that the full extent of the shortening is probably reproduced at the split-valence level.

Another molecule involving fluorine with anomalous bond lengths is FOOF, where the observed OO distance (1.22 Å) is much shorter than in hydrogen peroxide (1.48 Å) and the OF bond is very long (1.58 Å). An STO-3G study correctly gives a twisted nonplanar structure, but the theoretical lengths ($R_{OO} = 1.392$ Å and $R_{OF} = 1.358$ Å) fail to reproduce the observed

Table 8. Comparative Study of CF Bond Lengths

Molecule	Bond length		
	STO-3G[a]	4-31G[b]	Experimental[a]
CH_3F	1.384	1.412	1.385
CH_2F_2	1.378	1.375[c]	1.358
CHF_3	1.371	—	1.332
CF_4	1.366	—	1.317

[a] Ref. 15.
[b] Ref. 21.
[c] S. Visheswera, unpublished result with partial optimization.

effects. A more recent study of Newton[37] at the 4-31G level ($R_{OO} = 1.39$ Å, $R_{OF} = 1.43$ Å, $\theta_{OOF} = 105°$, and dihedral angle FOOF = 84°) still fails. This molecule is apparently an exception to the usual success of Hartree–Fock theory in predicting bond lengths. Since the observed structure may be rationalized in terms of resonance of the type,

$$F—O—O—F \leftrightarrow F^-O^+=O—F$$

it is possible that the difficulty of reproducing the electron affinity of fluorine at the Hartree–Fock level is related to this deficiency of the theory for this molecule.

We next consider theoretical results for dihedral angles in moderate-sized organic molecules and related problems in stereochemistry. These topics are related, of course, to internal rotation potentials. Most studies in this field have only involved limited geometrical variation and many have been based on a "rigid rotation model" in which only the dihedral angle has been varied. Nevertheless, the level of success of Hartree–Fock theory has been high.

Most of the C_3 and C_4 neutral hydrocarbons have been thoroughly studied. Early work involved only limited geometrical variation,[38] but the majority of these compounds have now been examined with full geometry optimizations at the STO-3G level.[34,35] In these studies, all known structural features are reproduced, at least qualitatively. In the C_3 systems, for example, allene is found to have D_{2d} symmetry with perpendicular methylene groups and propene is found to be most stable in a form I with single bonds *trans* about

I

the carbon–carbon single bond. The reasons underlying this preference have been discussed by Hehre and Salem.[39] The analogous structure II for *cis*-2-butene is also found to be the most stable structure in spite of the steric

II

crowding. However, this result is only achieved theoretically if flexibility in the C—C=C angles are taken into account.[38] The most stable form for *n*-butane is found to be with a *trans* carbon skeleton (partial geometry optimization). A

second minimum is found in the internal rotation potential curve corresponding to a *gauche* structure.[38] The fullest study carried out[40] (at the 4-31G level
with some angle optimization) gives a C—C—C—C dihedral angle of 68.5°
(experimental value, 67.5°) for this *gauche* structure indicating some opening
about an ideal "bond-staggered" value of 60°.

For other molecules with three–five nonhydrogenic atoms, a number
of studies of dihedral angle geometries have been carried out,[38,40–42] but
almost all with rigid rotation and usually with standard values for bond lengths
and angles. Most of this work has used the 4-31G basis. Among the theoretical
predictions that are consistent with experimental results are the trans structure
III for ethanol and the CCCO, CCCF *gauche* structures for 1-propanol and

III

1-fluoropropane. One system in which there is a possible failure of Hartree–
Fock theory is 1,2-difluoroethane. Here the STO-3G basis leads to a *gauche*
FCCF structure,[43] but the 4-31G basis (without length and angle optimization) gives the *trans* FCCF structure a lower energy by about 1 kcal/mole.[38]
Experimental evidence slightly favors the gauche form.[44] It is possible that the
4-31G basis gives too much polarity to CF bonds (dipole moments are
generally too large) and this may lead to excessive stabilization of the trans
structure in which the negatively charged fluorine atoms are kept far apart.

The remaining structural feature which is being tested at the Hartree–
Fock level is the shape of the ring skeletons of cyclic compounds. An extensive
set of three-membered ring compounds has been examined at the STO-3G
level,[45] but the question of nonplanarity does not arise until the four-
membered rings. The puckering of cycloalkanes is apparently underestimated
at the Hartree–Fock level, at least by the simpler basis sets. The first *ab initio*
study of cyclobutane was performed by Wright and Salem,[46] using an STO
basis. They found a slight puckering to a D_{2d} structure with an angle of 15°
between the bisectors of opposite CCC angles. Full optimization with the
STO-3G basis gives only 7° for this angle. Use of the 4-31G basis gives some
improvement, but only if d functions are added is a reasonable description of
the inversion barrier achieved.[47] It may be noted that a larger puckering
amplitude was obtained by Nelson and Frost[48] using floating spherical Gaussian functions (away from the nuclear centers).

The nonplanarity of five- and six-membered rings has only been quite
recently studied by these *ab initio* molecular orbital techniques. It has been
recognized for many years that the cyclopentane ring is puckered and that
interconversion among the various possible forms takes place very easily by a
pseudorotational path.[49] Using a partially constrained geometrical model,

Cremer and Pople[50] found that the STO-3G basis underestimated the puckering, but the agreement with experiment improved at the 4-31G level. This parallels results for cyclobutane. It was also found that a satisfactory description of the small barriers in the pseudorotational path in less symmetrical five-membered rings was not achieved until the 6-31G* level was reached. A comparable treatment of cyclohexane at the STO-3G level shows good agreement with the experimental structure.[51]

3. Equilibrium Geometries with Correlation

Following the extensive exploration of potential surface minima at the Hartree–Fock level, the next step is to test whether more sophisticated quantum mechanical methods, taking account of electron correlation, can account for the deficiencies that are evident at the simpler level. Although the significance of correlation has been recognized for many years, detailed exploration of potential surfaces taking it into account have only become possible very recently. Most of the work published to date has been concerned with the smallest H_2 and AH_n molecules which have already been discussed.

A wide variety of techniques has been proposed for obtaining wave functions beyond the Hartree–Fock level and several have now been used for equilibrium studies. The most straightforward is configuration interaction (CI) following a Hartree–Fock calculation. Here, determinantal functions corresponding to new electron configurations are constructed utilizing some of the virtual orbitals (eigenfunctions of the Fock Hamiltonian which are unoccupied at the Hartree–Fock level) and the full wave function is chosen variationally as a linear combination of such functions together with the Hartree–Fock determinant itself. The difficulty with such a technique is the large number of possible configurations unless some restrictive and rather subjective selection is made.

Some attempts have been made to utilize CI methods which avoid large secular equations. One is the independent electron-pair approximation (IEPA) which has been used extensively by Kutzelnigg and co-workers. The steps in this method[52] are first to transform the occupied Hartree–Fock orbitals to a localized set, second to solve the configuration interaction problem separately for each electron pair (by considering all configurations in which only this one pair are excited into the virtual manifold), and finally to add the correlation energy corrections so obtained. This technique neglects higher-order correlation effects and so involves some approximation. A further development along these lines is the pseudonatural orbital configuration interaction (PNO-CI) method of Meyer.[53] This technique uses the pseudonatural orbitals of the pair approach[54] in a limited CI treatment (single and double substitutions) and leads to a total energy which is a variational upper bound to the correct energy.

Finally the coupled electron perturbation approach (CEPA) is a further modification in which some allowance is made for effects of multiple substitutions.[55,56]

A selection of some of the geometry studies done with wave functions, including correlation, is given in Table 9. The only molecule for which very high accuracy is achieved is H_2, for which a high-level wave function including the interelectronic distance has been explicitly used by Kolas and Wolniewicz.[57]

Table 9. Molecular Structures from Correlated Wave Functions

| Molecule | Geometry[a] | | Method |
	Theory	Experimental[b]	
H_2	0.7414	0.7416	Explicit[c] r_{12}
LiH	1.619	—	CI (elliptic basis)[d]
	1.608	1.595	CEPA[e]
BeH	1.345	1.343	CI[f]
BeH_2	1.334, 180	—	CI[g]
BH	1.247	1.236	CI[h]
BH_2	1.204, 128.6	1.18, 131	IEPA–PNO[i]
BH_3	1.192	—	IEPA–PNO[j]
	1.191	—	CEPA[e]
CH	1.118	1.120	CI[k]
$CH_2(^3B_1)$	1.081, 134.2	1.078, 136	IEPA–PNO[l]
$CH_2(^1A_1)$	1.116, 102.5	1.110, 101.4	IEPA–PNO[l]
CH_4	1.088	—	PNO–CI[m]
	1.091	1.085	CEPA[m]
NH	1.041	—	CI[n]
	1.040	1.038	OVC[o]
OH	0.974	—	CI[p]
	0.969–0.972	—	OVC[q]
	0.965	—	PNO–CI[r]
	0.981	0.971	CEPA[r]
OH_2	0.973, 103.3	0.957, 104.5	IEPA[s]
FH	0.920	—	CI[p]
	0.929	0.917	IEPA[s]

[a] Distances in angstroms, angles in degrees.
[b] For experimental sources, see Table 2.
[c] Ref. 57.
[d] R. E. Brown and H. Shull, *Int. J. Quantum Chem.* **2**, 663 (1968).
[e] R. Ahlrichs, *Theor. Chim. Acta* **35**, 59 (1974).
[f] P. S. Bagus, C. M. Moser, P. Goethels, and G. Verhaegen, *J. Chem. Phys.* **58**, 1186 (1973).
[g] R. P. Hosteny and S. A. Hagstrom, *J. Chem. Phys.* **58**, 4396 (1973).
[h] S. A. Houlden and I. Csizmadia, *Theor. Chim. Acta* **35**, 173 (1974).
[i] V. Staemmler and M. Jungen, *Chem. Phys. Lett.* **16**, 187 (1972).
[j] M. Gelus and W. Kutzelnigg, *Theor. Chim. Acta* **28**, 103 (1973).
[k] G. C. Liu, J. Hinze, and B. Liu, *J. Chem. Phys.* **59**, 1872 (1973).
[l] V. Staemmler, *Theor. Chim. Acta* **31**, 49 (1973).
[m] W. Meyer, *J. Chem. Phys.* **58**, 1017 (1973).
[n] S. V. O'Neil and H. F. Schaefer III, *J. Chem. Phys.* **55**, 394 (1971).
[o] G. Das, A. C. Wahl, and W. J. Stevens, *J. Chem. Phys.* **61**, 433 (1974).
[p] V. Bondybey, P. K. Pearson, and H. F. Schaefer III, *J. Chem. Phys.* **57**, 1123 (1972).
[q] W. J. Stevens, G. Das, A. C. Wahl, M. Krauss, and D. Neumann, *J. Chem. Phys.* **61**, 3686 (1974).
[r] Ref. 55.
[s] H. Lischka, *Theor. Chim. Acta* **31**, 39 (1973).

For all the other calculations errors are still in the region of a few thousandths of an angstrom. The general deficiency of the Hartree–Fock theory leading to bond lengths that are usually too short has been put right. However, it is still difficult to pick out clear trends since the calculations reported in Table 9 use a variety of different basis sets as well as a variety of different methods. There does appear to be some tendency for the IEPA method to predict bond lengths which are too long, particularly in the results reported for water and hydrogen fluoride. But no clear conclusions can be drawn until careful studies are made with the same basis and different wave function techniques, or until estimates are forthcoming about the limits of the various methods with infinitely flexible basis sets. Some very recent studies of Ahlrichs *et al.*[58] constitute a step in this direction.

4. Predictive Structures for Radicals and Cations

Having surveyed the level of success of the theory in reproducing structural data about which there is considerable experimental information, we now turn to studies in which structural predictions have been made ahead of experimental study. These have mostly been small ions and radicals. There have also been extensive studies of small intermolecular polymers, particularly those involving hydrogen bonding, but these will not be dealt with here.

Probably the simplest system for which a structural prediction is firmly made is the cation H_3^+. The ground state of this ion is found to have the three protons at the vertices of an equilateral triangle, the two electrons occupying a symmetrical two-electron three-center bonding molecular orbital. The fullest Hartree–Fock study[59] for H_3^+ gives a bond length of 0.868 Å. The most recent study using a wave function with correlation (and leading to an upper bound for the total energy) is due to Salmon and Poshusta,[60] who obtain 0.873 Å. A number of other CI studies give values in the range 0.868–0.878 Å as does an earlier study of Conroy[61] based on numerical techniques.

A number of ions in the set AH_n^+ (A = C, N, O, F) have been studied in detail. The entire series has been examined at the UHF/STO-3G, UHF/4-31G, and UHF/6-31G* levels.[23,26] The resulting angles (for those with C_{2v} and C_{3v} symmetries) are listed in Table 10, together with values obtained using fuller basis sets and/or correlated wave functions. The results show many features in common with those already noted for the neutral AH_n molecules. There is normally an increase in bond angle in going from STO-3G to 4-31G, but the value becomes smaller if polarization functions are added as in 6-31G*. There also appears to be a regular trend in which the bond angle of a cation is larger than that of the isoelectronic neutral system; this is presumably associated with extra repulsion between the less-screened hydrogen atoms in the cations. Several of the ions listed in Table 10 show notable geometrical structures. CH_3^+ is found to be uniformly planar, as expected. NH_2^+ is predicted

Table 10. Theoretical Bond Angles for Cations AH_n^+

Ion	Bond angle[a]			
	UHF/STO-3G	UHF/4-31G	UHF/6-31G*	Best
CH_2^+	136.0	141.8	138.8	—
CH_3^+	120	120	120	120[b]
NH_2^+	147.4	157.5	151.5	—
NH_3^+	120	120	120	—
OH_2^+	109.8	119.9	112.1	112.5[c]
OH_3^+	113.9	120	113.2	110.9[d]
FH_2^+	112.0	125.5	113.9	114.8[e]

[a] See Refs. 23 and 26 for STO-3G, 4-31G, and 6-31G* results.
[b] F. Driessler, R. Ahlrichs, V. Staemmler, and W. Kutzelnigg, *Theor. Chim. Acta* **30**, 315 (1973).
[c] Ref. 63.
[d] Ref. 65.
[e] Ref. 66.

to have a triplet ground state with a valence angle rather larger than that for methylene. NH_3^+ is predicted to have a planar structure in parallel with the methyl result, there is experimental evidence to support this.[62] The water cation OH_2^+ is predicted to have an angle opened up to about 112°. Here the UHF/6-31G* results are in close agreement with a PNO-CI study by Meyer.[53] A recent spectroscopic study by Lew and Heiber[63] has confirmed this opening, finding a valence angle of 110.5°. A large number of studies have been carried out for the hydronium ion OH_3^+ and it would appear that this ion is pyramidal, although a split-valence *sp* basis gives a planar structure. The inclusion of *d*-type polarization functions on oxygen is apparently necessary to give a pyramidal structure. Probably the most extensive study done to date is that of Lischka and Dyczmons.[64] Using a large basis, they obtain a valence angle of 113.5° at the Hartree–Fock level. They also made a study with the IEPA correlation scheme and found an increase in pyramidal character, the bond angle becoming 110.9°, and a corresponding small increase in the barrier.[58,64] Finally, a study of FH_2^+ also yields an angle greater than in the isoelectronic water molecule.[65]

A considerable amount of theoretical research has been carried out on the simple carbocations CH_5^+, $C_2H_3^+$, and $C_2H_5^+$. For protonated methane, extensive search with STO-3G basis led[31] to a most stable structure of the form IV with a three-center bond involving CH_3 and two hydrogen atoms. This is more stable than the trigonal bipyramid structure V. Similar results are obtained at

IV   V

the 4-31G level.[31] The calculated separation between the energies of these two optimized geometries is increased markedly (from 7 kcal/mole to 16 kcal/mole) if polarization functions are added to the basis.[66] A similar change is found if additional s-type basis functions are added at the centers of the bonds.[67] A more recent study including correlation still indicates IV to be most stable, although a C_{2v} structure is close.[68] The vinyl cation could exist either in a classical (C_{2v}) form VI or a "bridged" nonclassical (C_{2v}) form VII.

VI   VII

Early work[26,30] on these structures at the Hartree–Fock level with the STO-3G and 4-31G bases indicated that VI was substantially more stable than VII. However, addition of d functions sharply lowers the energy difference.[66] This presumably arises because the local symmetry about the carbon atoms is much lower in VII and the energy lowering due to d functions is correspondingly greater. A recent study[69] of $C_2H_3^+$ in which correlation corrections have been added indicates that the magnitude of the correlation energy is considerably greater in VII than in VI, leading to the final prediction that the nonclassical structure is the most stable. This appears to be the first system examined in which the inclusion of correlation makes a *major* difference in predicted equilibrium geometry. The results suggest that other nonclassical bonding structures may also have to be examined, carefully beyond the Hartree–Fock level. The ethyl cation $C_2H_5^+$ is another such example. At the Hartree–Fock level, the STO-3G and 4-31G bases predict[26,30] that the classical structure VIII is more stable than the bridged structure IX. Again the addition of d functions preferentially stabilized IX, making it more stable even at the Hartree–Fock level. Correlation corrections[69] widen the gap, again leading to a final prediction that the bridged form is most stable.

VIII   IX

The structures of some neutral radicals have also been studied by *ab initio* methods, including the small ones that are included in the AH_n series. For radicals, some indirect experimental evidence about structure is obtained from electron spin resonance, but direct measurements of bond lengths and angles by spectroscopic techniques are rare. Among the two heavy-atom radicals that have been studied are formyl, ethynyl, vinyl (ethenyl), ethyl, and methoxy. These show a number of interesting features.

Formyl is one of the few radicals that has been studied by microwave spectroscopy[70] and the resulting structure ($r_{CH} = 1.11$ Å, $r_{CO} = 1.17$ Å, $\theta_{HCO} = 127°$) can be directly compared with the UHF/STO-3G structure[31] ($r_{CH} = 1.10$ Å, $r_{CO} = 1.25$ Å, $\theta_{HCO} = 126°$). The rather poor agreement for the CO bond illustrates a deficiency in the UHF approach. As noted earlier in the paper, the UHF wave function with one excess α electron is not a pure doublet function but has some contamination from a neighboring quartet state. This is reflected in an expectation value for the spin-squared operator $\mathbf{S}^2$ which is greater than the ideal value of 0.75. Upon reoptimization of the geometry with an RHF method, the CO bond length fell to 1.19 Å.[31] (Actually the RHF technique used in Ref. 31 does not lead to the fully self-consistent wave function, but the results probably reflect true RHF structure quite accurately.) Rather similar effects are found with the ethynyl radical HCC. The UHF/STO-3G structure for this system has a CC bond length of 1.221 Å which is much greater than the corresponding theoretical value for acetylene (1.168 Å). Again this is a deficiency of the UHF method and is corrected if an RHF technique is applied.[71] The same applies to the vinyl radical which is also predicted to have an excessively long CC bond if a UHF technique is used.

A full geometrical study of the ethyl radical has been undertaken with the UHF/STO-3G method. In this case the contaminating effects of higher multiplicity are negligible. The most stable structure is found to be X, with a

X

significantly nonplanar radical center. The barrier to rotation of the methyl group, however, is quite small. Another noteworthy feature of the ethyl radical is that, unlike the ethyl cation, the energy of a bridged species is much higher, leading to a low probability of rearrangement. Finally, we may mention the methoxy radical H_3CO which has been investigated both by UHF[31] and RHF[72] techniques. An interesting feature of this system is the prediction of small distortions of a Jahn–Teller type[73] since the odd electron is assigned to an e-type molecular orbital if the molecular framework has C_{3v} symmetry.

5. Conclusions

The wide variety of theoretical studies of molecular geometries reviewed in this article allow some broad conclusions about the reliability and limitations of the various levels of molecular orbital theory.

Hartree–Fock theory with a minimal Slater-type basis set (such as STO-3G) is moderately successful in reproducing structural parameters for closed-shell ground states of neutral systems. Bond lengths are accurately given to about 0.03 Å and angles generally to within 4°. For all such systems discussed here, the HF/STO-3G theoretical model has given the correct molecular symmetry.

Expansion of the basis set to the double-zeta or split-valence level leads to some improvement in bond length predictions, but deficiencies in bond angles and dihedral angles. The HF/4-31G model clearly gives excessive relative stability to structures with high local symmetry.

If polarization functions are added to a split-valence basis, the predictions about angles are improved and the general features of a model such as HF/6-31G* appear to be close to those of the Hartree–Fock limit. The most notable deficiency at this level is a systematic underestimation of bond lengths, in most cases by about 0.01–0.02 Å.

Quantum mechanical methods allowing for correlation of the motion of electrons of opposite spin have not yet been sufficiently used in geometry studies for firm conclusions to be drawn. There are indications that the bond length deficiency of Hartree–Fock theory is corrected by systematic techniques such as those based on the pair approximation. But changes in geometry with correlation corrected appear to be small except possibly for nonclassical structures with abnormal valence.

ACKNOWLEDGMENT

The preparation of this article and some of the research described was supported in part by the National Science Foundation under Grant GP25617.

References

1. J. A. Pople, Molecular orbital methods in organic chemistry, *Acc. Chem. Res.* **3**, 217–223 (1970).
2. J. C. Slater, Atomic shielding constants, *Phys. Rev.* **36**, 57–64 (1930).
3. C. C. J. Roothaan, New developments in molecular orbital theory, *Rev. Mod. Phys.* **23**, 69–89 (1951).
4. J. A. Pople and R. K. Nesbet, Self-consistent orbitals for radicals, *J. Chem. Phys.* **22**, 571–574 (1959).
5. C. C. J. Roothaan, Self-consistent field theory for open shells of electronic systems, *Rev. Mod. Phys.* **32**, 179–185 (1960).
6. G. A. Segal, Alternative technique for the calculation of single determinant open-shell SCF functions which are eigenfunctions of S^2, *J. Chem. Phys.* **52**, 3530–3533 (1970).
7. W. J. Hunt, T. H. Dunning, and W. A. Goddard III, The orthogonality constrained basis set. Expansion method for treating of diagonal lagrange multipliers in calculations of electronic wave functions, *Chem. Phys. Lett.* **3**, 606–610 (1969).
8. J. S. Binkley, J. A. Pople, and P. A. Dobosh, The calculation of spin-restricted single determinant wavefunctions, *Mol. Phys.* **28**, 1423–1429 (1974).

9. D. A. Pierre, *Optimization Theory with Applications*, Wiley, New York (1969).
10. I. Shavitt and M. Karplus, Multicenter integrals in molecular quantum mechanics, *J. Chem. Phys.* **36**, 550–551 (1962).ʼ
11. J. M. Foster and S. F. Boys, A quantum variational calculation for HCHO, *Rev. Mod. Phys.* **32**, 203–304 (1960).
12. C. M. Reeves and R. Fletcher, Use of Gaussian functions in the calculation of wavefunctions for small molecules. III. The orbital basis and its effect on valence, *J. Chem. Phys.* **42**, 4073–4081 (1965).
13. K. O-ohata, H. Takota, and S. Huzinaga, Gaussian expansion of atomic orbitals, *J. Phys. Soc. Japan* **21**, 2306–2313 (1966).
14. W. J. Hehre, R. F. Stewart, and J. A. Pople, Self-consistent molecular orbital methods. I. Use of Gaussian expansions of Slater type orbitals, *J. Chem. Phys.* **51**, 2657–2664 (1969).
15. M. D. Newton, W. A. Lathan, W. J. Hehre, and J. A. Pople, Self-consistent molecular orbital theory, V. *Ab initio* calculations of equilibrium geometries and quadratic force constants, *J. Chem. Phys.* **52**, 4064–4072 (1970).
16. W. J. Hehre, R. Ditchfield, and J. A. Pople, Self-consistent molecular orbital methods. VIII. Molecular studies with least energy minimal atomic orbitals, *J. Chem. Phys.* **53**, 932–935 (1970).
17. R. M. Pitzer and D. P. Merrifield, Minimal basis wavefunctions for water, *J. Chem. Phys.* **52**, 4782–4287 (1970).
18. R. M. Stevens, Geometry optimization in the computation of barriers to internal rotation, *J. Chem. Phys.* **52**, 1397–1402 (1970).
19. W. J. Hehre, R. Ditchfield, R. F. Stewart, and J. A. Pople, Self-consistent molecular orbital methods. IV. Use of Gaussian expansion of Slater-type orbitals. Extension to second row molecules, *J. Chem. Phys.* **52**, 2769–2773 (1970).
20. L. C. Snyder and H. Basch, *Molecular Wave Functions and Properties*, Wiley, New York (1972).
21. R. Ditchfield, W. J. Hehre, and J. A. Pople, Self-consistent molecular orbital theory. IX. An extended Gaussian type basis for molecular orbital studies of organic molecules, *J. Chem. Phys.* **54**, 724–728 (1971).
22. A. Rauk, L. C. Allen, and E. Clementi, Electronic structure and inversion barrier of ammonia, *J. Chem. Phys.* **52**, 4133–4144 (1970).
23. P. C. Hariharan and J. A. Pople, Accuracy of AH_n equilibrium geometries by single determinant molecular orbital theory, *Mol. Phys.* **27**, 209–214 (1974).
24. G. Herzberg, *Spectra of Diatomic Molecules*, Van Nostrand, New York (1950).
25. G. Herzberg, *Electronic Spectra of Polyatomic Molecules*, Van Nostrand, New York (1965).
26. W. A. Lathan, W. J. Hehre, L. A. Curtiss, and J. A. Pople, Molecular orbital theory of the electronic structure of organic compounds. X. A systematic study of geometries and energies of AH_n molecules and cations, *J. Am. Chem. Soc.* **93**, 6377–6387 (1971).
27. P. E. Cade and W. M. Huo, Electronic structure of diatomic molecules. VI. A. Hartree–Fock wavefunctions and energy quantities for the ground states of the first-row hydrides, AH*, *J. Chem. Phys.* **47**, 614–648 (1967).
28. R. K. Nesbet, Approximate Hartree–Fock calculations for the hydrogen fluoride molecules, *J. Chem. Phys.* **36**, 1518–1533 (1962).
28. T. H. Dunning, R. M. Pitzer, and S. Aung, Near Hartree–Fock calculations on the ground state of the water molecule: energies, ionization potentials, geometry, force constants and one-electron properties, *J. Chem. Phys.* **57**, 5044–5051 (1972).
30. W. A. Lathan, W. J. Hehre, and J. A. Pople, Molecular orbital theory of the electronic structure of organic compounds. VI. Geometries and energies of small hydrocarbons, *J. Am. Chem. Soc.* **93**, 808–815 (1971).
31. W. A. Lathan, L. A. Curtiss, W. J. Hehre, J. B. Lisle, and J. A. Pople, Molecular orbital structure for small organic molecules and cations, *Prog. Phys. Org. Chem.* **11**, 175–261 (1974).
32. J. D. Dill, P. v. R. Schleyer, and J. A. Pople, Geometries and energies of small boron compounds, *J. Am. Chem. Soc.* **97**, 3402–3409 (1975).
33. A. Veillard, Relaxation during internal rotation ethane and hydrogen peroxide, *Theor. Chim. Acta* **18**, 21–33 (1970).

34. L. Radom, W. A. Lathan, W. J. Hehre, and J. A. Pople, Molecular orbital theory of the electronic structure of organic compounds. VIII. Geometries, energies and polarities of C_3 hydrocarbons, *J. Am. Chem. Soc.* **93**, 5339–5342 (1971).

35. W. J. Hehre and J. A. Pople, Geometries, energies and polarities of C_4 hydrocarbons, *J. Am. Chem. Soc.* **97**, 6941–6955 (1975).

36. A. Krantz, C. Y. Lin, and M. D. Newton, Cyclobutadiene. II. On the geometry of the matrix isolated species, *J. Am. Chem. Soc.* **95**, 2744–2746 (1973).

37. M. D. Newton, private communication.

38. L. Radom and J. A. Pople, Molecular orbital theory of the electronic structure of organic compounds. IV. Internal rotation in hydrocarbons using a minimal Slater type basis, *J. Am. Chem. Soc.* **92**, 4786–4795 (1970).

39. W. J. Hehre and L. Salem, Conformation of vinylic methyl groups, *Chem. Commun.* 754–757 (1973).

40. L. Radom, W. A. Lathan, W. J. Hehre, and J. A. Pople, Molecular orbital theory of electronic structure of organic compounds. XVII. Internal rotation in 1,2-disubstituted ethanes, *J. Am. Chem. Soc.* **95**, 693–698 (1974).

41. L. Radom, W. J. Hehre, and J. A. Pople, Molecular orbital theory of electronic structure of organic compounds. VII. A systematic study of energies, conformations and bond interactions, *J. Am. Chem. Soc,* **93**, 289–300 (1971).

42. L. Radom, W. J. Hehre, and J. A. Pople, Molecular orbital theory of the electronic structure of organic compounds. XII. Fourier component analysis of internal rotation potential functions in saturated molecules, *J. Am. Chem. Soc.* **94**, 2371–2381 (1972).

43. R. Seeger, unpublished.

44. D. Vitus and A. A. Bothner-By, private communication.

45. W. A. Lathan, L. Radom, P. C. Hariharan, W. J. Hehre, and J. A. Pople, Structures and stabilities of three-membered rings from *ab initio* molecular orbital theory, *Top. Current Chem.* **40**, 1–45 (1973).

46. J. S. Wright and L. Salem, Ring puckering and methylene rocking in cyclobutane, *Chem. Commun.* 1370–1371 (1969).

47. P. C. Hariharan, R. Ditchfield, and L. C. Snyder, private communication.

48. J. L. Nelson and A. A. Frost, A floating spherical Gaussian orbital model of molecular structure. X. C_3 and C_4 saturated hydrocarbons and cyclobutanes, *J. Am. Chem. Soc.* **94**, 3727–3731 (1972).

49. J. E. Kilpatrick, K. S. Pitzer, and R. Spitzer, The thermodynamics and molecular structure of cyclopentane, *J. Am. Chem. Soc.* **69**, 2483–2488 (1947).

50. D. Cremer and J. A. Pople, Pseudorotation in saturated five-membered ring compounds, *J. Am. Chem. Soc.* **97**, 1358–1367 (1975).

51. D. Cremer and J. A. Pople, unpublished.

52. W. Kutzelnigg, Molecular calculations involving electron correlation, in: *Selected Topics in Molecular Physics* (E. Clementi, ed.), p. 91, Verlag Chemie, Berlin (1972).

53. W. Meyer, Ionization energies of water from PNO-CI calculations, *Int. J. Quantum Chem.* **S5**, 341–348 (1971).

54. C. Edmiston and M. Krauss, Pseudonatural orbitals as a basis for the superposition of configurations. I. H_2^+, *J. Chem. Phys.* **45**, 1833–1839 (1966).

55. W. Meyer, PNY-CI and CEPA studies of electron correlation effects. II. Potential curves and dipole moment functions of the OH radical, *Theor. Chim. Acta* **18**, 21–33 (1970).

56. R. Ahlrichs, H. Lischka, V. Staemmler, and W. Kutzelnigg, PNO-CI (pair natural orbital configuration interaction) and CEPA (coupled electron pair approximation with pair natural orbitals). Calculation of molecular systems. I. Outline of the method for closed-shell states, *J. Chem. Phys.* **62**, 1225–1234 (1975).

57. W. Kolos and L. Wolniewicz, Improved theoretical ground state energy of the hydrogen molecule, *J. Chem. Phys.* **49**, 404–410 (1968).

58. R. Ahlrichs, F. Driessler, H. Lischka, V. Staemmler, and W. Kutzelnigg, PNO-CI (pair natural orbital configuration interaction) and CEPA-PNO (coupled electron pair approximation with pair natural orbitals) calculations of molecular systems. II. The molecules BeH_2, BH, BH_3, CH_4, CH_3^-, NH_3 (planar and pyramidal), H_2O, OH_3^+, HF and the Ne atom, *J. Chem. Phys.* **62**, 1235–1247 (1975).

59. M. E. Schwartz and L. J. Schaad, *Ab initio* studies of small molecules using 1*s* Gaussian basis functions. II. H_3^+, *J. Chem. Phys.* **47**, 5325–5334 (1967).
60. L. Salmon and R. D. Poshusta, Correlated Gaussian wavefunctions for H_3^+, *J. Chem. Phys.* **59**, 3497–3503 (1973).
61. H. Conroy, Molecular Schrödinger equation. X. Potential surfaces for ground and excited states of isosceles H_3^{++} and H_3^+, *J. Chem. Phys.* **51**, 3979–3993 (1969).
62. W. R. Harshbarger, Structure of the $^2A_1''$ state of NH_3^+, *J. Chem. Phys.* **56**, 177–181 (1972).
63. H. Lew and L. Heiber, Spectrum of H_2O^+, *J. Chem. Phys.* **58**, 1246–1247 (1973).
64. H. Lischka and V. Dyczmons, The molecular structure of H_3O^+ by the *ab initio* SCF method and with inclusion of correlation energy, *Chem. Phys. Lett.* **23**, 167–172 (1973).
65. G. H. F. Diercksen, W. von Niessen, and W. P. Kraemer, SCF LCGO MO studies on the fluoronium ion FH_2^+ and its hydrogen bonding interaction with hydrogen fluoride FH, *Theor. Chim. Acta* **31**, 205–214 (1973).
66. P. C. Hariharan, W. A. Lathan, and J. A. Pople, Molecular orbital theory of simple carbonium ions, *Chem. Phys. Lett.* **14**, 385–388 (1972).
67. V. Dyczmons, V. Staemmler, and W. Kutzelnigg, Near Hartree–Fock energy and equilibrium geometry of CH_5^+, *Chem. Phys. Lett.* **5**, 361–366 (1970).
68. V. Dyczmons and W. Kutzelnigg, *Ab initio* calculation of small hydrides including electron correlation. XII. The ions CH_5^+ and CH_5^-, *Theor. Chim. Acta* **33**, 239–247 (1974).
69. B. Zurawski, R. Ahlrichs, and W. Kutzelnigg, Have the ions $C_2H_3^+$ and $C_2H_5^+$ classical or non-classical structure, *Chem. Phys. Lett.* **21** 309–313 (1973).
70. J. A. Austin, D. H. Levy, C. A. Gottlieb, and H. E. Radford, Microwave spectrum of the HCO radical, *J. Chem. Phys.* **60**, 207–215 (1974).
71. J. S. Binkley, unpublished.
72. D. R. Yarkony, H. F. Schaefer III, and S. Rothenberg, Geometries of the methoxy radical (X^2E and A^2A_1 states) and the methoxide ion, *J. Am. Chem. Soc.* **96** 656–659 (1974).
73. H. A. Jahn and E. Teller, Stability of polyatomic molecules in degenerate electronic states. I. Orbital degeneracy, *Proc. Roy. Soc. London, Ser. A*, **161**, 220–235 (1937).

2

Barriers to Rotation and Inversion

Philip W. Payne
and
Leland C. Allen

1. Introduction

1.1. Relation to Other Chapters in Volumes 3 and 4

Approximately two-thirds of the chapters in these two companion volumes are devoted to methods of obtaining high-accuracy electronic wave functions for molecules and solids. The remaining third are concerned with particular chemical species or properties, and our chapter fits the latter category. Within this category the extensive literature on barriers offers two special opportunities of general interest to chemical theorists. First, it is possible to make rather definitive statements on the quality of wave functions required to yield quantitative predictions. Second, methods for analyzing *ab initio* wave functions to ascertain the physical origin of the barrier and provide a quantum mechanically well-defined, but simple picture of the mechanism have been more extensively developed for this topic than any other.

1.2. Other Reviews

The reviews most closely related to the present chapter are those by Pople,[1] Golebiewski and Parczewski,[2] Clark,[3] and Veillard.[4] Veillard gives

Philip W. Payne and Leland C. Allen • Department of Chemistry, Princeton University, Princeton, New Jersey

a compendium of computed barriers which is complete through 1972. His article also provides a good summary of the necessary quantum mechanical computational methods and he discusses methods of analysis and the accuracy of various basis sets. During the last three or four years a large number of new results have become available and it is now possible to carry the topics introduced by Veillard a great deal further.

1.3. Historical Notes

A delightful history of stereochemical ideas and conformational analysis, with special reference to internal rotation, has been written by Orville-Thomas.[5] A brief survey of early theoretical attempts to predict barrier heights and elucidate their origin has been given by Fink and Allen.[6] From the point of view of contemporary theoretical chemists the first pertinent chemical physics research on torsional barriers was two papers by Kenneth S. Pitzer[7] in 1936–1937. He introduced the threefold sinusoidal potential for ethane,

$$V = \tfrac{1}{2} V_0 (1 - \cos 3\theta)$$

and carried out a statistical mechanical calculation for the entropy. This, combined with heat of formation data, demonstrated the existence of an ethane barrier and he correctly established its value at approximately 3 kcal/mole. It seems appropriate, therefore, that his son Russel M. Pitzer should make the first complete and successful quantum mechanical barrier calculation for his Ph.D. thesis.[8] The calculation was performed for ethane and he obtained a barrier height close to 3 kcal/mole. His calculation was of special significance because it suggested that chemically important conformation problems involving only a few kilocalories might be accessible from molecular orbital theory. This exciting result was reported at a time of general discouragement in theoretical chemistry: the enormous number of multicenter six-dimensional electron–electron interaction integrals required for constructing *ab initio* wave functions seemed to present insurmountable problems, and binding energy calculations were proving to be in error by hundreds of kilocalories. The succeeding developments and ramifications of Russel Pitzer's 1962 calculation are the subject of this chapter.

2. Assessment of Computational Methods

There are four questions which characterize our assessment procedure. First, are barriers to rotation and inversion accessible within the Hartree–Fock approximation? Second, is the rigid rotor model adequate? Third, how sensitive are rotational barriers to the choice of a basis set? Fourth, are the answers to the first three questions different for second and third row atoms? These questions, of course, are governed by several important and well-established theorems for many-electron systems. It is also well known, however, that the degree to which such theorems can define their answers is quite limited, so it is the evidence from the large computational literature which provides our principal guide.

2.1. The Correlation Energy

One can deduce from purely theoretical considerations that the correlation energy should contribute little to rotation or inversion barriers. This conclusion is grounded in Brillouin's theorem and its generalizations.[9] Brillouin's theorem states that single excited determinants do not contribute to the correlation energy in first order. The extended Brillouin theorem says that inclusion of electron correlation cannot give first-order corrections to the expectation value of a one-electron operator. Specifically, following Goodisman and Klemperer,[10] consider a Hartree-Fock determinantal wave function $\langle\Phi|A|\chi\rangle$ must vanish. Therefore, for any one-electron operator A, $\langle A\rangle_\Psi =$ in first order, contains only double or higher excitations, the matrix element $\langle\Phi|A|\chi\rangle$ must vanish. Therefore, for any one-electron operator A, $\langle A\rangle_\Psi = \langle A\rangle_\Phi + \varepsilon^2\langle A\rangle_\chi$.

The relevance of the extended Brillouin theorem to rotational and inversion barriers has been demonstrated by Freed[11] and elaborated through a numerical procedure by Allen and Arents.[12] Starting with Stanton's proof[13] that Hartree–Fock wave functions satisfy the Hellmann–Feynman theorem,

$$\frac{\partial E(\lambda)}{\partial\lambda} = \frac{\partial}{\partial\lambda}\langle H(\lambda)\rangle = \left\langle\frac{\partial H}{\partial\lambda}\right\rangle$$

Freed showed that if λ is the coordinate associated with internal rotation or inversion, the barrier derivative $\partial E/\partial\lambda = \langle\partial H/\partial\lambda\rangle$ is the expectation value of a one-electron operator. The extended Brillouin theorem then permits one to conclude that barrier derivatives do not have a first-order dependence on electron correlation.

The expectation that correlation contributions to barriers should be small can be checked against those barrier calculations which have included electron correlation. These data are summarized in Table 1. Electron correlation contributes negligibly to the ethane rotational barrier,[14–16] but is about 15% of the rotational barrier in nitrogen tetroxide[17] and 1,3-butadiene.[18,19] The latter two molecules both have π systems, and these loosely bound π electrons

Table 1. Correlation Contributions to Rotation and Inversion Barriers[a]

Molecule[b]	E_{min}	ΔE, SCF	$\Delta(\Delta E)$, Correlated	Method[c]	Reference
N_2O_4		5.02	1.00	Gaussian lobe (8, 4, 1) → (4, 2/1) PNO–CI	17
NH_3 (inversion)	(−56.1953)	2.40	2.81	STO (4, 2/1, 2) all single, double CI	22
	(−56.2117)	9.84	−2.97	Basis (13, 7, 1/9, 1) → (6, 2, 1/3, 1) partial CI	21
	(−56.0076)	7.2	10.6	minimal STO basis, full CI	23
	(−56.2211)	5.9	0.34	(5, 4, 1/2, 1) STO basis, full CI	23
	(−56.2111)	4.7	0.4[d]	(9, 5/5) → (5, 3/3) plus d, p polarization PNO–CI and CEPA	20
			0.0[e]		
	(−56.2154)	5.2	+0.3[d]	(9, 5, 2, 1/5, 2) → (5, 3, 2, 1/3, 2) PNO–CI and CEPA	20
			−0.3[e]		
	(−56.2168)	6.2	−0.6	One-center expansion on N, analytic many-body perturbation theory	24
CH_3^- (inversion)	(−39.5129)	5.46	−2.7	(12, 6, 1/9, 1) → (6, 2, 1/3, 1) partial CI	21
	(−39.5200)	2.1	−1.2	(10, 6, 1/5, 1) → (6, 4, 1/3, 1) CEPA–PNO–CI	20
H_3O^+ (inversion)	(−76.3287)	0.8	0.7	(11, 7, 1/6, 1) → (7, 4, 1/4, 1) CEPA–PNO–CI	20
	(−76.3298)	1.30	0.75	(11, 7, 1/6, 1) → (5, 4, 1/3, 1) full CI	25
1,3-Butadiene (barrier relative to *trans* conformation)		7.6	−1.2	Gaussian lobe (10, 5/5) → (3, 2/1) partial CI	18
		7.1	−0.53	(7, 3/3) → (2, 1/1) perturbative correlation of all 2nd-order terms	19
Ethane		3.31	−0.01	Pitzer–Lipscomb (2, 1/1) STO perturbative correlation of all 2nd-order terms	14

Table 1 *(continued)*

Molecule[b]	E_{min}	ΔE, SCF	$\Delta(\Delta E)$, Correlated	Method[c]	Reference
Ethane *(cont'd)*		3.27	−0.13	Boys' localization of Pitzer–Lipscomb wave functions plus perturbative 2nd-order correlation	14
		3.21	+0.02	(10, 7, 2/6, 2) → (6, 4, 2/3, 2) correlation by density functional	15
		—	+0.14		16

[a] Values given in kilocalories per mole.
[b] Energy of SCF.
[c] Our convention for abbreviating the basis set is defined as follows: We first list the uncontracted primitives, on each atom, in descending order of atomic number. Primitive orbitals on a given atom are separated by commas and are listed in order of increasing angular momentum (s, p, d). Atoms are separated by slashes. For example, a basis consisting of 10 s-type, 6 p-type, and 2 d-type primitives on carbon, and 5 s-type and 1 p-type primitives on hydrogen would be abbreviated (10, 6, 2/5, 1). The process of contraction is indicated by an arrow, and we list the contracted basis in the same manner.
[d] Canonical molecular orbitals.
[e] Boys' localized orbitals.

are more sensitive to electron correlation. With inversion barriers the situation is somewhat different. The correlation part of the ammonia inversion barrier[20–24] is less than 0.6 kcal/mole, which is 10% of the total barrier. For the methyl anion[20,21] and hydronium ion,[20,25] correlation energy appears to be a significant part of the inversion barrier. Inversion barriers are more subject to electron correlation than rotation barriers because they involve large bond length and symmetry changes.

How trustworthy are the calculations reported in Table 1? We begin with a discussion of the ammonia inversion barrier, for which we have an experimental value of 5.8 kcal/mole.[26] The barriers computed by Ahlrichs *et al.*,[20] Dutta and Karplus,[24] and Stevens (extended basis)[23] are in good agreement with experiment and show only small correlation energy effects. The barrier calculations by Pipano *et al.*,[22] Kari and Csizmadia,[21] and Stevens (minimal basis)[23] are notably less accurate. The critical features differentiating these calculations are the choice of basis and molecular geometry. To evaluate the basis-set quality we compare the pyramidal conformer SCF energy with the near Hartree–Fock energy (−56.2219 hartrees) computed by Rauk *et al.*[27] This comparison immediately permits us to classify the Stevens minimal basis calculation ($E = -56.0076$) and the Pipano calculation ($E = -56.1953$) as unsatisfactory for obtaining an accurate correlation energy estimate. We attribute their poor SCF barriers and consequent exaggeration of correlation to the small basis set. An adequate description of ammonia inversion requires at least a double-zeta basis plus d functions on N. Despite a good basis set, however, Kari and Csizmadia have inexplicably obtained an incorrect barrier and we therefore must discount their unreasonably large correlation energy estimate (3 kcal/mole). Ahlrichs *et al.*[20] and Stevens both used the Rauk optimal geometry[27] with $\angle HNH = 107.2°$ and a contraction of R_{NH} from 1.89 a.u. (pyramidal) to 1.86 a.u. (planar). This discussion of the ammonia barrier brings out the need for extended basis sets and geometry optimization in studying inversion barriers.

We now turn to the methyl anion and hydronium ion inversion barriers. In all cases an excellent basis has been used and the geometries have been optimized. Again, the Kari–Csizmadia result[21] for the methyl anion appears incorrect on the grounds that a correlation energy of -2.7 kcal/mole in the inversion barriers is inconsistent with the much smaller correlation contributions (~ 1 kcal) reported for NH_3, CH_3^-, and H_3O^+ by other workers. It is expected that the correlation energy would contribute a greater percentage of the barrier in the methyl anion than in ammonia and in fact, electron correlation favors the planar structure by an amount equal to the total barrier. For the hydronium ion, correlation contributes about one-third of the total barrier. In both these molecules correlation is important only because the SCF barrier is already small. The correlation energies obtained by Ahlrichs *et al.*[20] should be accepted only qualitatively since these authors claim that the noise of their calculations is several tenths kcal/mole.

To summarize, the correlation energy changes associated with second-row inversion barriers and partial π rotational barriers are usually several tenths kcal/mole and may be as great as 1 kcal/mole. Whether or not this effect is important depends on the size of the SCF barrier. An estimation of the correlation energy contribution to the barrier is reliable only if one begins with a geometry-optimized extended-basis wave function.

2.2. Survey of Recent Barrier Calculations

The most powerful evidence for the adequacy of the Hartree–Fock approximation comes from comparing *ab initio* barrier heights with experimental data. These barriers are summarized in Table 2 for a representative sample of molecules. Our list of *ab initio* calculations is not exhaustive. We have generally tried to include all molecules on which parallel calculations have been done, so that one can assess the merits of different *ab initio* approaches. We have also included most barrier calculations involving third row atoms. But we have deliberately excluded from Table 2 any *isolated ab initio* calculations which cannot be compared with experimental data. The data prove that extended-basis geometry-optimized calculations can reproduce experimental barriers within a few percent accuracy. Minimal-basis rigid-rotor calculations can give qualitatively useful information about the barrier shape, but actual barrier heights may possess a significant error.

Table 2. Ab Initio Calculations of Rotation and Inversion Barriers

Molecule	Total energy, hartrees	Barrier, kcal/mole	Type	Contraction scheme	Comments	Reference
Ethane	—	2.93	—	—	Experimental, ir	61
	None	2.90	SCF-Xα	—	—	59
	−79.2587	3.21	GTO	(12, 7, 2/6, 2) → (6, 4, 2/3, 2)	Geometry optimized	15
	−79.2581	3.06	GTO	(10, 6, 2/5, 1) → (6, 4, 1/3, 1)	—	37
	−79.2390	3.07	GTO	(11, 7, 1/6, 1) → (5, 3, 1/3, 1)	Geometry optimized	39, 40
	−79.2377	3.65	GTO	(11, 7, 1/6, 1) → (5, 3, 1/3, 1)	Experimental geometry	39, 40
	−79.2346	3.71	GTO	(11, 7/6, 1) → (5, 3/3, 1)	Experimental geometry	39, 40
	−79.2151	3.41	GTO	(11, 7, 1/6) → (5, 3, 1/3)	Experimental geometry	39, 40
	−79.2098	3.42	GTO	(11, 7/6) → (5, 3/3)	Experimental geometry	39, 40
	−79.2059	2.87	STO	(9, 5/4) → (4, 3/2)	—	62
	—	3.06	STO	(9, 5/4) → (4, 3/2)	Calculate torsional distortion	37
	−79.2046	3.31	STO	—	—	63
	−79.1478	2.52	GLO	(10, 5/5) → (3, 1/1)	—	6
	−79.1475	2.58	GLO	(10, 6/4) → (3, 2/1)	—	64
	−79.1148	3.26	GTO	(8, 4/4) → (3, 2/2)	4-31G	65
	−79.0999	3.30	STO	(2, 1/1)	Full optimization	66
	−79.0997	3.12	STO	(2, 1/1)	Full optimization	37
	−79.0983	3.41	STO	(2, 1/1)	—	67
	−79.0980	3.5	STO	(2, 1/1)	Methane exponents	68
	−79.0906	3.01	GTO	(7, 3/3) → (5, 2/2)	Molecules-in-molecules	69
	−78.9912	3.31	STO	(2, 1/1)	—	14
	−78.9911	3.25	STO	(2, 1/1)	—	62
	−78.9093	2.6	STO	(2, 1/1)	Bond orbitals (not SCF)	70
	−78.8196	3.28	GTO	(5, 3/2) → (2, 1/1)	—	71
	−78.6795	2.82	GTO	(5, 3/3)	Bond orbitals (not SCF)	72
	−78.5704	3.45	GTO	(5, 2/2, 1)	—	73
	−78.5090	2.88	GTO	(5, 2/2)	—	73
	−78.4930	2.66	GTO	(5, 2/2)	Bond orbitals (not SCF)	72

continued overleaf

Table 2 (continued)

Molecule	Total energy, hartrees	Barrier, kcal/mole		Type	Contraction scheme	Comments	Reference
Ethane	−78.3276		2.96	GTO	$(10, 7/6) \rightarrow (4, 2/2)$	—	74
	−76.8762		3.8	FSGO	—	Triple—2 floating orbitals	75
	−76.5633		4.0	FSGO	—	Quadruple—3 floating orbitals; geometry optimized	75
	−76.5070		3.7	FSGO	—	Triple—2 floating orbitals; geometry optimized	75
	−75.8620		5.0	FSGO	—	Concentric double set of orbitals; geometry optimized	75
	−75.5134		6.3	FSGO	—	Minimal number of orbitals; geometry optimized	75
	−67.3473		5.17	FSGO	—	—	76
	−67.0276		3.09	FSGO	—	Concentric double Gaussian orbital set	41
	−67.0048		5.7	FSGO	—	—	41
	−67.005		5.7	FSGO	—	—	41
Hydrogen peroxide	—	*cis*	7.03	—	—	Experimental, ir	77
		trans	1.10				
	−150.8319	*cis*	10.9	STO	$(6, 3, 1/3, 1)$	—	78
		trans	0.72				
	−150.8219	*cis*	7.57	GTO	$(9, 5, 1/4, 1) \rightarrow (4, 3, 1/2, 1)$	Geometry optimized	48, 49
		trans	1.10				
	—	*cis*	12.1	GTO	$(9, 5, 1/4, 1) \rightarrow (4, 3, 1/2, 1)$	No geometry optimization	48, 49
		trans	0.27				
	−150.7993	*cis*	10.9	GTO	$(11, 7, 1/6, 1) \rightarrow (5, 3, 1/5, 1)$	Geometry optimized	46, 39
		trans	0.61				
	—	*cis*	14.7	GTO	$(11, 7, 1/6, 1) \rightarrow (5, 3, 1/5, 1)$	No geometry optimization	39
		trans	0.21				

Molecule	Energy	*cis*	*trans*	Type	Basis set	Notes	Ref.
Hydrogen peroxide	−150.7629	*cis* 19.1	*trans* 0.0	GTO	(11, 7/6) → (5, 3/3)	—	39
	−150.7910	*cis* 13.9	*trans* 0.24	GTO	(10, 5, 1/5) → (4, 2, 1/2)	—	47
	−150.7615	*cis* 10.3	*trans* 0.15	GTO	(9, 5/4) → (4, 3/2)	Geometry optimized	48, 49
	—	*cis* 17.9	*trans* 0.0	GTO	(9, 5/4) → (4, 3/2)	—	48, 49
	−150.7394	*cis* 17.0	*trans* 0.0	GTO	(10, 5/5) → (4, 2/2)	Optimized R_{OH}	47
	—	*cis* 18.4	*trans* 0.0	GTO	(10, 5/5) → (4, 2/2)	No geometry optimization	47
	−150.7231	*cis* 12.0	*trans* 0.0	GTO	(10, 5/5, 1) → (3, 1/1, 1)	—	47
	−150.7224	*cis* 13.2	*trans* 0.25	GTO	(10, 5/5, 1) → (3, 1/1, 1)	—	79
	−150.7049	*cis* 11.0	*trans* 0.0	GTO	(10, 5/5) → (3, 1/1)	Optimized ∠OOH	47
	—	*cis* 17.3	*trans* 0.0	GTO	(10, 5/5) → (3, 1/1)	No geometry optimization	47
	−150.7078	*cis* 13.7	*trans* 0.0	GLO	(10, 5/5) → (3, 1/1)	—	6
	−150.6067		*trans* 0.63	GTO	(7, 3, 1/4, 1) → (4, 2, 1/2, 1)	Geometry optimized	50
	−150.5752		*trans* 0.19	GTO	(7, 3/4, 1) → (4, 2/2, 1)	Geometry optimized	50
	−150.5539	*cis* 7.94	*trans* 0.65	GTO	(8, 4, 4) → (3, 2/2)	4-31G	80
	−150.5372		*trans* 0.0	GTO	(7, 3/4) → (4, 2/2)	Geometry optimized	50
	−150.2353	*cis* 9.5	*trans* 0.0	STO	(2, 1/1)	Exponent optimized	66
						trans geometry optimized	
	−150.2238	*cis* 11.72	*trans* 0.0	STO	(2, 1/1)	—	67

continued overleaf

Table 2 (continued)

Molecule	Total energy, hartrees	Barrier, kcal/mole		Type	Contraction scheme	Comments	Reference
Hydrogen peroxide	—	*cis*	13.79	STO	(2, 1/1)	Optimal bond orbitals	67
		trans	0.0				
	−150.2232	*cis*	13.0	STO	(2, 1/1)	Exponent optimized	81
		trans	0.0				
	−150.2231	*cis*	12.8	STO	(2, 1/1)	—	82
		trans	0.0				
	−150.1565	*cis*	11.8	STO	(2, 1/1)	—	83
		trans	2.2				
	−150.1467	*cis*	9.4	STO	(2, 1/1)	—	81
		trans	0.0				
	−150.0714	*cis*	8.8	GTO	(8, 4/4, 1) → (2, 1/1, 1)	—	84
		trans	0.54				
	−150.0026	*cis*	14.8	GTO	(8, 4/4) → (2, 1/1)	—	84
		trans	0.0				
	−149.2885	*trans*	0.19	GTO	(5, 2/2, 1)	—	73
	−149.2641	*cis*	15.9	GTO	(5, 2/2)	—	73
		trans	0.0				
	−127.6309	*cis*	11.3	FSGO	—	—	85
		trans	1.3				
	None	*cis*	58.0	SCF-Xα	—	Sphere radii scaled in proportion to covalent radii	60
		trans	0.0				
	None	*cis*	2.0	SCF-Xα	—	Equal sphere radii	60
		trans	23.0				
Hydrogen disulfide	—	≈6.8		—	—	Experimental, ir	86
	−796.1840	*cis*	9.33	GTO	(12, 9, 1/5, 1)	—	87
		trans	5.99				
	−793.9704	*cis*	7.7	GTO	(4, 3/4) → (4, 3/3)	—	84
		trans	2.2				

		cis		STO	(3, 2/1)	—	84
Hydrogen disulfide	−787.7848	*trans*	9.3 3.2				
Methanol	—		1.07	—	—	Experimental, microwave	88
	−115.0116		1.44	GTO	(13, 7/4) → (3, 2/2)	—	89
	−114.9957		1.06	GLO	(10, 5/5) → (3, 2/2)	—	6
	−114.9807		1.13	STO	(4, 2/2)	—	90
	−114.9355		1.35	GLO	(10, 5/5) → (3, 1/1)	—	6
	−114.9343		1.37	GLO	(10, 5/5) → (3, 1/1)	Repeat of Ref. 61, but with more stable SCF	91
	−114.9047		1.59	GTO	(5, 2/2)	—	73
	−114.8702		1.12	GTO	(8, 4/4) → (3, 2/2)	4-31G	65
	−114.5816		1.57	STO	(2, 1/1)	Bond orbitals	90
Methylamine	—		1.98	—	—	Experimental	92
	−95.1127		2.42	GLO	(10, 5/5) → (3, 1/1)	—	79
	−95.0680		2.13	GTO	(8, 4/4) → (3, 2/2)	4-31G, standard geometry	80
	−94.3235		2.02	GTO	(5, 2/2)	—	73
Borazane	−82.6308		3.06	GTO	(11, 7, 1/6, 1) → (5, 3, 1/3, 1)	—	40
	−82.6065		1.93	GTO	(10, 4, 1/4, 1) → (3, 2, 1/2, 1)	6-31G*, STO-3G, geometry optimized	43
	−82.4653		2.77	STO	(2, 1/1)	Exponent optimized geometry optimized	44
	−82.1323		2.47	GTO	(5, 3/3) → (3, 2/2)	—	93
	−81.5995		2.12	GTO	(6, 3/3) → (2, 1/1)	STO-3G, geometry optimized	43
Propane	—		$V_1 = 3.3$ $V_2 = 3.9$	—	—	Experimental, microwave	94
	—			—	—	—	—
	−118.0934		$V_1 = 3.48$ $V_2 = 4.32$	GTO	(8, 4/4) → (3, 2/2)	4-31G, experimental geometry	95
	−118.0921		$V_1 = 3.70$ $V_2 = 5.07$	GTO	(8, 4/4) → (3, 2/2)	4-31G, standard geometry	65

continued overleaf

Table 2 (continued)

Molecule	Total energy, hartrees	Barrier, kcal/mole		Type	Contraction scheme	Comments	Reference
Propane	-117.6776		$V_1 = 3.56$	GTO	$(5, 3/2) \rightarrow (2, 1/1)$	—	71
	-117.4379		$V_1 = 3.22$	GLO	$(5, 3/3)$	Bond orbitals	96
			$V_2 = 4.87$				
	-117.1721		$V_1 = 3.00$	GLO	$(5, 2/2)$	Bond orbitals	72
			$V_2 = 4.59$				
	-116.8857		$V_1 = 3.45$	GTO	$(6, 3/3) \rightarrow (2, 1/1)$	STO-3G, optimized $\angle$CCC	51
			$V_2 = 4.02$				
	-100.6475		$V_1 = 5.62$	FSGO	—	—	76
			$V_2 = 8.44$				
	-99.9969		$V_1 = 5.28$	FSGO	—	Geometry optimized	97
			$V_2 = 7.14$				
Hydrazine	—		3.14	—	—	Experimental	98
	-111.0743	*cis*	11.9	GLO	$(6, 4/3) \rightarrow (3, 1/1)$	—	99
		trans	3.7				
	-111.0301	*cis*	11.5	GTO	$(9, 3/3)$	—	100
		trans	4.7				
	-111.0021	*cis*	12.3	GTO	$(8, 4/4) \rightarrow (3, 2/2)$	4-31G	80
		trans	3.6				
	-110.8784	*cis*	12.0	GLO	$(7, 3/3) \rightarrow (3, 2/2)$	Basis includes floating functions, optimized $\angle$NNH	45, 101
		trans	1.6				
	—	*cis*	12.4	GLO	$(7, 3/3) \rightarrow (3, 2/2)$	Basis includes floating functions, No geometry optimization	45, 101
		trans	4.6				
	-110.8180	*cis*	13.1	GLO	$(7, 3/3) \rightarrow (3, 2/2)$	Optimized $\angle$NNH	45, 101
		trans	3.1				
	—	*cis*	13.2	GLO	$(7, 3/3) \rightarrow (3, 2/2)$	No geometry optimization	45, 101
		trans	3.6				
	-110.6853	*cis*	9.6	GTO	$(7, 3/3) \rightarrow (2, 1/1)$	—	102
		trans	3.7				

Molecule	Energy	Barrier	Orbital	Basis	Method	Ref.
Hydrazine	−110.1239	cis 11.1 trans 6.2	GTO	(5, 2/2)	—	73
Dinitrogen tetroxide	—	2.9	—	—	Experimental	103
	−407.7464	5.02	GTO	(8, 4/8, 4, 1) → (4, 2/4, 2, 1)	—	19
	−407.3650	11.6	GTO	(7, 3/7, 3)	Partial geometry optimization	104
	Not given	10.7	GTO	(5, 2/3, 2)	Partial geometry optimization	104
Propene	—	1.85	—	—	Experimental	105
	—	1.98	—	—	Experimental	106
	−117.0422	1.87	GTO	(9, 5/4) → (4, 3/2)	—	107
	−116.9266	1.25	GLO	(10, 5/5) → (3, 1/1)	—	108
	−116.4983	1.69	GTO	(5, 3/2) → (2, 1/1)	—	71
	−116.4884	1.35	GTO	(8, 4/4) → (2, 1/1)	STO-4G	51
	−116.3960	1.48	GTO	(5, 2/3)	—	109
	−116.2378	0.81	GLO	(5, 3/3)	—	72
	−115.9382	0.47	GLO	(5, 2/3)	—	72
	−115.6578	1.40	GTO	(6, 3/3) → (2, 1/1)	STO-3G	51
	−113.6651	3.58	GTO	(4, 3/2)	—	58
	−98.8507	4.21	FSGO	—	Full optimization	97
Acetaldehyde	—	1.16	—	—	Experimental, microwave	110
	−152.8550	1.09	GLO	(10, 5/5) → (4, 2/2)	—	111
	−152.6848	0.74	GTO	(8, 4/4) → (3, 2/2)	4-31G	65
	−152.2311	1.34	GTO	(5, 3/2) → (2, 1/1)	—	71
	−152.8708	1.19	GTO	(5, 2/2) → (3, 1/1)	—	112
	−150.9446	1.37	GTO	(6, 3/3) → (2, 1/1)	STO-3G	112
Acetone	—	0.78	—	—	Experimental, ir	113
	−191.6763	$V_1 = 0.75$ $V_2 = 2.22$	GTO	(8, 4/4) → (3, 2/2)	4-31G, standard geometry	65
	−190.6816	$V_1 = 0.99$	GTO	(5, 2/2) → (3, 1/1)	—	112
	−189.5316	$V_1 = 1.14$	GTO	(6, 3/3) → (2, 1/1)	STO-3G	112

continued overleaf

Table 2 (continued)

Molecule	Total energy, hartrees	Barrier, kcal/mole	Type	Contraction scheme	Comments	Reference
Glyoxal	—	13.7	—	—	$V = V_1 = 4V_2 + 9V_3$, experimental, ir	114
	—	$\Delta E_{t\to c}$ 3.2	—	—	Experimental	115
	-226.4703	$\Delta E_{t\to c}$ 6.4 barrier 7.9	GTO	(9, 5/4) → (4, 2/2)	—	116
	-226.3246	$\Delta E_{t\to c}$ 4.77 barrier 7.22	GTO	(9, 5/6, 4/4) → (3, 2/3, 2/2)	Geometry optimized	53
	-226.2477	$\Delta E_{t\to c}$ 2.99 barrier 3.23	GLO	(10, 5/5) → (3, 1/1)	—	117
	-226.2428	$\Delta E_{t\to c}$ 6.13 barrier 7.93	GTO	(8, 4/4) → (3, 2/2)	4-31G	65
Formamide	—	21.3	—	—	Experimental (aqueous solution)	118, 119
	-168.9629	21.73	GTO	(11, 7, 1/5, 3)	Geometry optimized	54
	-168.9173	19.49	GTO	(11, 7/5, 3)	Geometry optimized	54
	—	19.89	GTO	(11, 7/5, 3)	No geometry optimization	54
	-168.6776	24.7	GTO	(8, 4/4) → (3, 2/2)	4-31G	65
	-168.1837	20.3	GTO	(7, 3/3) → (2, 1/1)	—	120
1,3-Butadiene	—	$\Delta E_{t\to c}$ 2.5 barrier 7.16	—	—	Experimental, Raman torsion	121
	-154.8363	$\Delta E_{t\to g}$ 2.7 trans 6.0 cis 0.5	GTO	(9, 5/4) → (4, 2/2)	*Gauche* conformer at 40° is preferred to *cis*	52
	-154.8214	$\Delta E_{t\to g}$ 2.50 trans 7.6	GLO	(10, 5/5) → (3, 2/1)	—	18
	-154.6931	$\Delta E_{t\to g}$ 3.1 trans 6.6 cis 1.8	GTO	(7, 3/4) → (4, 2/2)	Geometry optimized, *gauche* conformer at 40° is preferred to *cis*	52

Molecule	Energy	ΔE	Method	Basis	Notes	Ref
1,3-Butadiene	—	$\Delta E_{t \to g}$ 3.4 trans 6.7 cis 1.8	GTO	(7, 3/4) → (4, 2/2)	No geometry optimization, *gauche* conformer is preferred to *cis*	52
	−154.4643	$\Delta E_{t \to c}$ 2.6 trans 7.14 cis 0.79	GTO	(7, 3/3) → (2, 1/1)	*Gauche* conformer at 35° is preferred to *cis*	19
	−153.0166	$\Delta E_{t \to g}$ 2.92 trans 6.61	GTO	(6, 3/3) → (2, 1/1)	STO-3G, geometry optimized	51
	—	$\Delta E_{t \to g}$ 2.05 trans 6.73	GTO	(6, 3/3) → (2, 1/1)	STO-3G, partial geometry optimization (∠CCC)	51
	—	$\Delta E_{t \to g}$ 5.66 trans 6.71	GTO	(6, 3/3) → (2, 1/1)	STO-3G	51
	−132.513	$\Delta E_{t \to c}$ 7.5 trans 40.8	FEGO	—	Floating elliptical Gaussian bond orbitals	122
Allene	−115.6977	82.1	GLO	(10, 5/5) → (3, 1/1)	∠HCH = 123.6°	122
	−115.3127	74.6	GTO	(2, 1/1)	—	124
	−114.8392	72.7	GTO	(5, 2/2)	∠HCH = 118.2°	125, 126
	−114.4194	91.9	GTO	(6, 3/3) → (2, 1/1)	STO-3G	51
	−98.9932	75.1	FSGO	—	Split core	127
Ammonia	—	5.8	—	—	Experimental, microwave	26
	−56.2219	5.1	GTO	(13, 8, 2/8, 2) → (8, 5, 2/4, 2)	Geometry optimized	27
	−56.2211	5.9	STO	(4, 2, 1/2, 1)	Geometry optimized	23, 128
	−56.2168	6.2	—	Analytic solution	Many-body perturbation theory	24
	−56.2154	5.2	GTO	(9, 5, 2, 1/5, 2) → (5, 3, 2, 1/3, 2)	—	20
	−56.2117	9.84	GTO	(13, 7, 1/9, 1) → (6, 2, 1/3, 1)	—	21
	−56.2111	4.7	GTO	(9, 5, 1/5, 1) → (5, 3, 1/3, 1)	—	20
	−56.1952	2.4	STO	(4, 2/2, 1)	—	22
	−56.0356	6.0	GLO	(7, 3/3) → (3, 2/2)	Basis includes floating functions at bond center	101
	None	−32	SCF-Xα	—	Sphere radii proportional to covalent radii	60
	None	−17	SCF-Xα	—	Intermediate radii	60

continued overleaf

Table 2 (continued)

Molecule	Total energy, hartrees	Barrier, kcal/mole	Type	Contraction scheme	Comments	Reference
Ammonia	None	5	SCF-Xα	—	Equal radii for spheres	60
	None	1	SCF-Xα	—	Interstitial spheres	60
Methyl anion	−39.5222	1.71	GTO	(10, 6, 2/6, 1) → (5, 3, 2/3, 1) + diffuse sp	Geometry optimized	129
	−39.5200	2.0	GTO	(10, 7, 1/5, 1) → (6, 4, 1/3, 1)	—	56
	−39.5129	5.46	GTO	(10, 6, 1/9, 1) → (6, 2, 1/3, 1)	—	21
	−39.5101	5.2	GTO	(10, 6, 2/5, 1)	—	130
	−39.5081	5.9	GTO	(10, 6, 1/8)	—	131
	−39.5056	5.18	GTO	(10, 6, 2/6, 1) → (5, 3, 2/3, 1)	No diffuse orbitals, geometry optimized	129
	−39.4922	2.7	GTO	(10, 6/8)	—	131
	−39.4799	1.22	GTO	40 primitives	—	132
	−39.4644	1.61	GTO	36 primitives	—	132
	−39.3885	4.38	GTO	28 primitives	—	132
	−39.0993	5.5	GTO	20 primitives	—	132
	−37.4031	4.17	GTO	12 primitives	—	132
Hydronium ion	—	Pyramidal	—	—	Experimental	133, 134
	−76.3418	0.96	GTO	(11, 7, 2/6, 2) → (7, 4, 2/4, 2)	Geometry optimized	135
	−76.3326	1.9	GTO	(9, 5, 2/4, 1) → (4, 3, 2/2, 1)	Geometry optimized	55
	−76.3295	1.30	GTO	(11, 7, 1/6, 1) → (5, 4, 1/3, 1)	Geometry optimized	25
	−76.3287	0.8	GTO	(9, 5, 2, 1/5, 2) → (5, 3, 2, 1/3, 2)	—	20
	−76.3244	1.0	GTO	(7, 4, 1/4, 1) → (5, 4, 1/3, 1)	—	136
	−76.3213	0.0	GTO	(7. 5/3, 1)	—	137
	−76.3162	0.0	GLO	(10, 5/5, 1) → (4, 2/2, 1)	—	138
	−76.3014	0.0	STO	(4, 2/2)	—	139
	−76.2995	0.0	GTO	(7, 4/4) → (5, 4/3)	—	136
	−76.2949	0.0	GTO	(10, 5/5) → (4, 2/2)	—	138
	−76.2222	0.3	GTO	(7, 3/4, 1) → (4, 3/2, 1)	Geometry optimized	140
	−76.2006	0.0	GTO	(8, 4/4) → (3, 2/2)	4-31G	141
	−75.8460	0.0	GTO	(5, 3/3)	—	137

Molecule	Energy	Barrier	Basis	Basis set	Description	Ref.
Phosphine	—	31.5	—	—	Experimental	142
	−342.4560	36.7	GTO	(12, 9, 2/5, 1) → (6, 4, 2/3, 1)	Geometry optimized	57
	−342.4517	39.2	GTO	(12, 9, 2/5, 1) → (6, 4, 2/3)	Geometry optimized	57
	−342.4512	35.8	GTO	(12, 9, 1/5, 1) → (6, 4, 1/3, 1)	Tight d orbitals, geometry optimized	57
	−342.4394	35.5	GTO	(12, 9, 1/5) → (6, 4, 1/3)	Tight d orbital, geometry optimized	55
	−342.4353	34.7	GTO	(12, 9, 1/5, 1) → (6, 4, 1/3, 1)	Diffuse d orbital, geometry optimized	55
	−342.4290	38.0	GTO	(12, 9, 1/5) → (6, 4, 1/3)	Diffuse d orbital, geometry optimized	55
	−342.4284	33.6	GTO	(12, 9/5, 1) → (6, 4/3, 1)	Geometry optimized	55
	−342.4042	30.4	GTO	(12, 9/5) → (6, 4/3)	Geometry optimized	55
	−341.3960	25.9	STO	—	One-center expansion	141
	−338.9185	37.6	GTO	(9, 7, 3/4) → (3, 3, 1/2)	—	142
	−338.8590	25.2	GTO	(9, 7/4) → (3, 3/2)	—	142
	None	30.9	—	Not given	—	143
	None	40.4	—	Not given	—	143
Methylphosphine	—	1.96	—	—	Experimental, microwave	144
	−381.213	1.83	GTO	(9, 5, 1/5, 2/3)	Uncontracted	145
	−381.157	1.71	GTO	(9, 5/5, 2/3)	Uncontracted	145
H_3PBH_3	—	2.47	—	—	Experimental, ir	146
	−364.8552	1.25	GTO	(9, 6, 3/6, 3/3) → (3, 2, 1/2, 1/1)	STO-3G+d orbitals, geometry optimized	147
	−364.8523	2.18	GTO	(9, 3/6, 3/3) → (3, 2, 1/2, 1/1)	STO-3G+d orbitals, geometry not optimized	147
	−364.7892	1.00	GTO	(9, 6/6, 3/3) → (3, 2/2, 1/1)	STO-3G, geometry optimized	147
	−364.7806	1.72	GTO	(9, 6/6, 3/3) → (3, 2/2, 1/1)	STO-3G, no geometry optimization	149
Methylsilane	—	1.67	—	—	Experimental	150
	−330.232	1.44	GTO	(12, 9, 1/10, 6, 1/5, 1) → (5, 3, 1/5, 3, 1/2, 1)	—	40
	−329.5769	1.44	STO	(3, 2, 1/2, 1/1)	Geometry optimized	151
	−329.5769	1.57	STO	(3, 2/2, 1/1)	Geometry optimized	151

The most important variables in an *ab initio* calculation are the choice of geometry and selection of a basis set. We first examine geometry effects and then discuss basis sets in the context of a molecule-by-molecule review of Table 2.

There are two important papers which deserve mention, even though they lie outside the mainstream of this review. Conformational energies for an exhaustive set of small organic molecules have been calculated at the 4-31G level by Radom, Hehre, and Pople.[28] *All* acyclic molecules containing only the atoms H, C, N, O, or F, and containing no more than three second row atoms, were examined. In general, heats of formation were reproduced within an accuracy of 1 kcal/mole. But this paper is not directed toward rotational barriers. Radom *et al.*[29] have made important new predictions by calculating rotational barriers for an extensive series of disubstituted ethanes using the 4-31G basis set but unfortunately, at present, there is no other theoretical or experimental work against which this work might be calibrated.

2.3. Geometry Optimization and Vibronic Coupling

Most *ab initio* calculations of rotational barriers have been done in the rigid-rotor approximation: bond lengths and bond angles are held constant as the dihedral angle is changed. But real molecules are not rigid rotors. The bond lengths and bond angles do depend on the internal rotation. If these changes are severe—more than 0.02 a.u. in bond length or 1° in bond angle—the rigid-rotor assumption can lead to significant errors in the computed barrier. We therefore consider two questions: What types of geometry distortions occur and what is their magnitude? And are these variations reliably predicted within the Hartree–Fock approximation?

We expect that the Hartree–Fock approximation will accurately represent geometry distortions since the energy gradient for distortion is the expectation value of a one-electron operator. The extended Brillouin theorem then guarantees that electron correlation will not contribute in first order to the shape of a potential surface.[11]

To study geometry distortions we examine spectroscopic evidence. If geometry distortions during internal rotation are large, one expects significant coupling of vibrational and torsional modes. This coupling is theoretically manifested as a nonzero barrier derivative (taken with respect to some vibrational coordinate). Kirtman[30,31] and Papousek[32] have derived equations which permit experimental analysis of vibration–torsion coupling for symmetric top molecules. (Mathematical complexity has prevented extension of the analysis to asymmetric tops.)

Experimental work has thus far been confined to ethane, methylsilane, and methanol. Susskind[33,34] has measured the torsional barrier of ethane in vibrationally excited states. He finds a rotational barrier of 3.33 kcal/mole in mode ν_{11}, which is an Eg deformation of methyl group angles. In the mode ν_{12}, which is an *Eg* rocking motion, the rotational barrier falls to 3.02 kcal/mole. Woods[35] has performed similar experiments on methanol. He found a 6% barrier increase in the first excited CO stretching mode, and a 13% barrier increase in the first excited CH_3 stretching mode.

Kirtman *et al.* have recently measured the distortional parameters of methylsilane[36] and ethane[37] using torsional satellites in the microwave spectra. The methylsilane geometry and distortion parameters were then calculated theoretically using a minimal exponent-optimized Slater orbital basis. The agreement with experiment is excellent:

	$\Delta R(Si-C)$	$\Delta R(C-H)$	$\Delta R(Si-H)$	$\Delta \angle SiCH$	$\Delta \angle SiH$
Experimental	0.011 Å	−0.0007 Å	0.0001 Å	0.35°	0.41°
Minimal-basis, Hartree–Fock	0.012 Å	−0.0005 Å	0.0002 Å	0.25°	0.37°

The rigid-rotor model gives a barrier of 1.57 kcal/mole. The distortion correction lowers the barrier by 54 kcal/mole, while harmonic averaging of zero point motion raises the barrier by 155 cal/mole. The predicted barrier is 1.67 kcal, in perfect agreement with the experimental value.[38]

To evaluate the vibration–torsion coupling for ethane, Kirtman expands the torsional potential V_3 in a multivariate Taylor series with respect to vibrational displacements S_i and averages over the vibrational wave function:

$$\langle V_3 \rangle = \sum_i (\partial V_3 / \partial S_i) \langle S_i \rangle + \tfrac{1}{2} \sum_{ij} (\partial^2 V_3 / \partial S_i \partial S_j) \langle S_i S_j \rangle + V_3^{(0)}$$

The single partials were evaluated in a double-zeta Slater basis $(9, 5/4) \rightarrow (4, 3/2)$, and double partial derivatives were computed in a minimal Slater basis. The calculated distortion terms are: CH stretch (−0.002 Å), CCH bending (−1.1°), CC stretch (0.0069 Å). If all derivatives are calculated in the minimal Slater basis, the CC stretch increases to 0.0094 Å in eclipsed ethane. In the $(9, 5/4) \rightarrow (4, 3/2)$ basis, a rigid-rotor barrier of 2.87 kcal/mole is calculated. Zero point averaging contributes 0.37 kcal/mole and distortion corrections contribute −0.16 kcal/mole, giving a net barrier of 3.06 kcal/mole. About half of the distortion energy is due to CC stretching, while the other half comes from contraction of the HCH angle. As with methylsilane, the calculated barrier derivatives and displacements are in excellent agreement with experiment.

To summarize, vibration–torsion coupling contributes a few tenths kcal/mole to the rotational barriers in ethane and methylsilane. The zero point correction is positive and somewhat larger than the negative distortion energy correction. For these molecules, even a minimal-basis molecular-orbital wave function does a good job estimating the barrier derivatives and distortional correction. The fact that zero point corrections often exceed distortional effects has several implications. First, the fair accuracy of rigid-rotor models comes in part from the partial cancellation of zero point corrections by distortional corrections. Second, optimization of geometry for all rotamers may not improve the agreement between theory and experiment unless one also allows for zero point averaging.

Why, then, is geometry optimization important? Hartree–Fock wave functions will accurately reproduce geometry distortions. If these distortions are small, zero point averaging will not matter. If distortions are large, one expects deviation from the experimental barrier unless zero point averages are included in the calculation. The vibrational coupling discussed above for rotation should be especially relevant to inversion barriers, where there may be strong contraction (0.03 a.u.) of transition state bond lengths.[20–25] Analysis of zero point corrections to inversion barriers would thus be strongly desirable.

We complete this discussion with a survey of geometry-optimized barrier calculations with the purpose of understanding the nature and scope of geometry distortions and learning how geometry variation affects the calculated barrier height.

The behavior of ethane, borazane, and methylsilane are representative of saturated molecules without lone pairs. In ethane, there is general agreement that geometry optimization lengthens the CC bond by about 0.01 Å, and opens the CCH angle by about 0.5°. The calculated distortions are not sensitive to the quality of the basis set in any regular way. Clementi and Popkie,[15] using an extended Gaussian basis, find $\Delta\theta = 0.3°$ and $\Delta R = 0.034$ a.u., whereas Kirtman *et al.*,[37] using a double-zeta STO basis, find $\Delta\theta = 1.1°$ and $\Delta R = 0.014$ a.u. Veillard[39,40] also employed an extended, polarized Gaussian basis and found $\Delta\theta = 0.3°$ and $\Delta R = 0.019$ a.u. These results are identical to the distortions Stevens obtained with a minimal Slater basis: $\Delta\theta = 0.3°$, $\Delta R = 0.02$ a.u. Although their barrier heights and absolute geometries are quite poor, FSGO wave functions appear to reproduce the ethane geometry distortions quite well. Blustin and Linnett[41] found $\Delta\theta = 0.5$ and $\Delta R = 0.012$ a.u., whereas Frost and Rouse[42] found $\Delta\theta = 0.5°$ and $\Delta R = 0.02$ a.u.

The patterns for borazane and methylsilane are similar. Dill *et al.*[43] have optimized the borazane geometry using a minimal basis and find $\Delta R = 0.042$ a.u. and $\Delta\theta = 0.2°$ for both HNB and BNH angles. Palke,[44] employing a minimal Slater basis, found comparable lengthening of the central bond, $\Delta R = 0.037$ a.u., but his angle changes $\Delta\theta(\text{HBN}) = 0.8°$ and $\Delta\theta(\text{HNB}) = 0.4°$ are somewhat larger. For methylsilane, Ewig *et al.*[36] find $\Delta R = 0.024$ a.u. and $\Delta\theta = 0.3°$ for both vicinal angles. As with ethane, the bond lengths to hydrogen do not change. These results show that a clear pattern exists for geometry distortions in the eclipsed conformers of saturated molecules without lone pairs. The central bond lengthens by about 0.02 a.u. and the vicinal angles open by 0.3°–0.5°. But there is a lack of reproducibility: Even the best basis sets guarantee only one-figure accuracy in distortional parameters.

Hydrazine and hydrogen peroxide provide examples of rotational barriers for saturated molecules with vicinal lone pairs. Jarvie and Rauk[45] have determined a detailed potential surface for rotation–inversion in hydrazine. Maintaining the HNH angle at its experimental value, they find the NN bond lengthens by only 0.008 a.u. on rotation into the *anti* conformer. Full geometry optimization gives a wrong sign for the perpendicular to *anti* barrier. This error can be traced to omission of *d* orbitals from their basis since strong torsion–inversion coupling was observed when all variable were optimized and a polarized basis is required for accurate description of amine inversion.

For hydrogen peroxide, Veillard[39,46] has found that the central OO bond lengthens by 0.050 a.u. in the *cis* rotamer, compared with *trans*. He calculated OOH angle openings of 5.2° and 4.3° in the *cis* conformer. Davidson and Allen[47] optimized the geometry without using polarization functions on O and obtained comparable results. They found $\Delta R = 0.050$ a.u. and $\Delta\theta$ (OOH) = 6°. Dunning and Winter,[48,49] working with an extended polarized basis, found an irregular variation of ±0.01 a.u. in R_{OO}, and predicted an expansion of 6°–7° in the OOH angle. In all cases, the *cis* barrier is overestimated by several kcal/mole if the OOH angle is not optimized. We note that the distortion for the *cis* barrier is not grossly basis-dependent and does not in particular require use of *d* orbitals. Ranck and Johansen[50] have carried out a detailed examination of the *trans* barrier in hydrogen peroxide and their results are summarized in Table 3. [Differences are between the stable conformer ($\phi = 120°$) and the *trans* conformer ($\phi = 180°$).] The central OO bond thus monotonically lengthens from the *trans* to the *cis* conformer, but the OOH angle is smallest at the equilibrium conformation. Minimal or extended and polarized basis sets yield very similar distortions, but intermediate basis sets produce markedly smaller changes.

1,3-Butadiene, glyoxal, and dinitrogen tetroxide (N_2O_4) provide examples of rotational barriers involving vicinal π systems. Radom and Pople,[51] using the minimal STO-3G basis, have optimized the CCC angle in 1,3-butadiene as a function of dihedral angle. They observed a monotonic opening from 124.3° in *trans* (180°) to 126.6° in *cis*. There is no special angle expansion near the barrier maximum at a 100° dihedral angle. Skancke and Boggs,[52] using a double-zeta basis, determined that the so-called *"cis"*

Table 3. Results of Ranck and Johansen Study of trans Barrier in Hydrogen Peroxide

Basis	$\Delta R_{OO}(120 \rightarrow 180)$, a.u.	$\Delta\theta$(OOH)	ΔR_{OH}, a.u.
$(5, 2/2) \rightarrow (2, 1/1)$	−0.020	1.63°	0.008
$(7, 3/4) \rightarrow (4, 2/2)$	−0.010	1.20°	0.003
$(7, 3/4, 1) \rightarrow (4, 2/2, 1)$	−0.012	0.98°	0.001
$(7, 3, 1/4, 1) \rightarrow (4, 2, 1/2, 1)$	−0.023	1.40°	0.001

conformer of butadiene is actually *gauche*, having a preferred dihedral angle of 40°. They find an angle opening of 3.2° between the *trans–gauche* barrier maximum and the *gauche* maximum. There is further opening of 1° upon rotation into the *cis* (0°) structure. The central CC bond lengthens by 0.026 a.u. as the *trans → gauche* barrier is traversed, and does not contract again in the *gauche* position. Sundberg and Cheung,[53] applying an extended basis to glyoxal, have found that the CC bond stretches monotonically by 0.02 a.u. during the *trans → cis* isomerization, and the CCO angle opens by 1.4°. In N_2O_4, due to its unusual π-only bonding, we find a dramatic contraction of the NN bond length as the barrier is traversed, this change is -0.22 a.u.

From these studies it is clearly important to optimize both vicinal angles and central bond lengths when one is studying vicinal π systems. But the relationship of these optimizations to the rotational barrier is not all clear since the bond lengths and bond angles change monotonically between *trans* and *cis* conformers, whereas the barrier is a maximum near a perpendicular orientation. For π systems, the factors influencing molecular geometry do not appear to be closely related to the barrier origin. The occurrence of the barrier maximum near a perpendicular conformation strongly suggests that a large part of the barrier comes from loss of partial π bonding and, therefore, geometry optimization may not give a significantly more accurate barrier. Basis-set variation can introduce more error than one might offset by optimizing geometry.

Christensen *et al.*[54] have optimized the bond lengths of formamide rotamers using an extended polarized basis. At the perpendicular conformation, which is a barrier maximum, R_{CN} lengthens by 0.100 a.u. and R_{OO} contracts by 0.016 a.u. Angle changes were not examined.

We have already noted the geometry changes associated with inversion barriers. The geometry-optimized inversion calculations all show a moderate contraction of bond lengths in the planar transition state. These contractions are estimated as 0.008 a.u. for the hydronium ion,[55] 0.03 a.u. for ammonia,[21,27] 0.037 a.u. for methyl anion,[56] and 0.068 a.u. for phosphine.[57] There are not enough data for evaluation of d orbital contributions to geometry changes.

In summary, most rotational barriers studied show changes of about 0.02 a.u. in bond lengths along the rotation axis, and changes of a few tenths of a degree in vicinal bond angles. Larger changes would be anticipated for sterically compressed systems such as doubly eclipsed propane and its congeners. All levels of molecular orbital calculation appear capable of reproducing geometry distortions to one significant figure. Geometry optimization will not necessarily improve the calculated barrier value since the barrier variation due to basis-set choice often exceeds the correction one would make by optimizing geometry and, if distortions are large, energy corrections due to zero point averaging will be comparable in magnitude to the distortional energy.

2.4. Discussion of Tabulated Barrier Calculations

The final issue to be explored in this section is the relationship between basis sets and calculated barrier heights. The nature of this relationship is best examined through detailed discussion of the barriers tabulated in Table 2. We first make several general observations:

(1)   One might expect the calculated rotational barrier to converge toward a particular value as basis set improvements lower the total energy. Such convergence will certainly occur if the total SCF energy lies very close to the Hartree–Fock limit—within a range comparable to the barrier itself. But it is generally impractical to carry *ab initio* calculations this close to the Hartree–Fock limit. For instance, even triple-zeta polarized basis sets commonly give total equilibrium energies which are several barrier heights above the true Hartree–Fock energy. And the variation principle itself is no help since basis-set improvements may affect two conformers quite differently. Turning to computational results immediately shows that there is, in fact, no correlation between total SCF energy and barrier heights. A lower total energy does not imply a more accurate rotational barrier. In Figure 1 we have plotted twenty-six ethane barrier heights versus total energies. The results appear random, both for this case and also for most other molecules in Table 2. The apparent randomness in total energy may be due to differing representations of inner shells. Even a slight change in inner-shell exponents can change the total energy by a few tenths of a hartree, while leaving the barrier almost invariant.

(2)   There is a tendency for all the calculations on a given barrier to jointly overestimate or jointly underestimate its height. If a barrier is governed by repulsions, there will be a driving force for charge redistribution in the

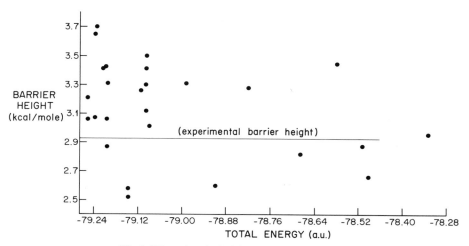

Fig. 1. Ethane barrier heights versus total energies.

high-energy conformer. If a barrier is due to special attractions, there will be a driving force for charge redistribution in the low-energy conformer. The indicated charge redistribution will always be imperfectly described, so that the energy of the low-energy conformer in the attractive case is overestimated. The energy of the high-energy conformer in the repulsive case is similarly overestimated. Inspection of Table 2 shows that ethane, acetaldehyde, and formamide calculations straddle both sides of the experimental barrier. The computed ethane barriers are about three-fourths on the high side, whereas acetaldehyde and formamide are most evenly distributed. If we ignore one questionable calculation,[58] all propylene calculations underestimate the barrier. The *trans* barrier in hydrogen peroxide is also underestimated. All other data in Table 2 exceed the experimental barriers.

(3)   The SCF-$X\alpha$[59,60] and FSGO (*f*loating *s*pherical *G*aussian *o*rbitals) methods are not trustworthy for quantitative work. The excellent ethane barrier (2.90 kcal) calculated via SCF-$X\alpha$ methods is an exception, not a norm. The hydrogen peroxide barrier is computed to be 58.0 kcal/mole (!), and the ammonia barrier has the wrong sign unless free parameters are suitably rigged.[60] FSGO and FEGO (*f*loating *e*lliptical *G*aussian *o*rbitals) calculations usually overestimate rotational barriers by several kcal/mole.

(4)   Computed barrier heights usually lie within a few tenths kcal/mole of each other, or else within 15% of the barrier height, whichever is greater.

Ethane. The experimental barrier height, measured by ir torsion analysis,[61] is 2.93 kcal/mole. As noted above, there is little correlation between the SCF energy and the calculated barrier height. Virtually all basis sets[62–74] give barriers lying within 0.5 kcal/mole of the experimental barrier. Polarization functions do not appear to be important. The FSGO calculations[41,42,75,76] are poor, yielding barriers 1–3 kcal/mole above the experimental value.

Hydrogen Peroxide. The far infrared spectrum[77] yields a *cis* barrier of 7.03 kcal/mole and a *trans* barrier of 1.10 kcal. All calculations overestimate the *cis* barrier and underestimate the *trans* barrier. The best results, obtained with a polarized basis and geometry optimization,[48,49] are 7.57 and 1.10 kcal/mole for the *cis* and *trans* barriers, respectively. Polarization functions and geometry optimization are required to obtain a *trans* barrier of more than a few tenths kcal/mole. Geometry optimization is required for the *cis* barrier (without geometry optimization the *cis* barrier usually lies in the range of 11–15 kcal/mole), but this barrier is not strongly influenced by polarization functions. The excellent results obtained with the 4-31G basis[80] (*cis* 7.94, *trans* 0.65 kcal/mole) and an FSGO basis[85] (*trans* 1.3 kcal/mole) are fortuitous in view of the much poorer results obtained with comparable basis sets.

Hydrogen Persulfide. A *cis* barrier of about 6.8 kcal/mole has been measured by ir torsion analysis.[86] A calculation which included *d* functions on sulfur[87] obtains a *trans* barrier, but agreement with the experimental *cis* value is mediocre.

Methanol. An experimental barrier height of 1.07 kcal/mole has been obtained from microwave splittings. Extended basis sets overestimate the barrier by 0.1 kcal/mole on the average, while minimal basis sets overestimate the barrier by 0.3 kcal/mole on the average. Polarized basis sets do not seem necessary.

Methylamine. An experimental barrier of 1.98 kcal/mole has been reported.[92] The split-out basis sets[73,80] give barriers within 0.15 kcal/mole of the experimental value, whereas the minimal contracted Gaussian lobe calculation exaggerates the barrier by almost 0.5 kcal/mole.

Borazane. There have been no experimental studies of internal rotation in borazane. From computed values[40,43,44,93] the barrier probably lies between 2 and 3 kcal/mole. As with ethane, there is little correlation between basis-set size and the computed barriers.

Propane. The propane molecule has two internal rotors. The first barrier V_1, refers to rotation of one methyl group when the other is held in a staggered arrangement. The second barrier V_2, refers to rotation of one methyl group when the other is held in an eclipsed orientation. Microwave analysis yields barriers of $V_1 = 3.3$ and $V_2 = 3.9$ kcal/mole.[94] The calculated values for V_1 range from 3.0 to 3.7 kcal/mole, with a concentration around 3.5 kcal/mole. There is no differentiation between results for minimal and extended basis sets. The second barrier V_2 is exaggerated by values up to 1.5 kcal/mole. We attribute this error to the strong overlap of α and δ hydrogens in the doubly eclipsed conformer, since opening the CCC angle by 4.7° lowers V_2 by 0.89 kcal/mole.[51] The FSGO results exceed the experimental barrier height by 2–3 kcal/mole because these bond orbitals are incapable of describing the rehybridization which occurs during internal rotation.

Hydrazine. The *anti* barrier in hydrazine has been measured as 3.14 kcal/mole,[98] but an experimental value for the *syn* barrier has not been reported. Except for the Pedersen–Morokuma work[73] (*anti* = 6.2 kcal) and the rather low barrier reported by Rauk and Jarvie[45] (*anti* = 1.6 kcal), most computed barriers lie within a few tenths kcal/mole of the experimental value. The high *syn* barriers (11–14 kcal/mole) remind one of hydrogen peroxide, where similar barriers were obtained before geometry optimization. Again, there seems to be no relationship between basis-set quality and the computed barrier heights. Polarization functions are not required for a qualitatively correct total energy curve, but they must be included and geometries must be optimized before the hydrazine calculations can be trusted for the *syn* barrier.

Nitrogen Tetroxide. An experimental value of 2.9 kcal/mole is reported for the rotational barrier in N_2O_4.[103] All calculations to date give unrealistically high values. Inclusion of polarization functions on N reduces the barrier by about 6 kcal/mole. In view of the strong contraction (0.2 a.u.) in the N–N bond, optimization of other parameters, such as the NNO angle, should give further improvement in the computed barriers. The large bond length change also implicates electron correlation as potentially important. Since the correlation energy should be greater for the most compact system, the correlation contribution should be negative and would bring the computed barriers into better agreement with experiment. The only estimate presently available for the correlation contribution is +1.0 kcal.[17]

Propylene. There are two experimental measurements of the propene barrier, both by Lide.[105,106] The most up-to-date value is 1.98 kcal/mole. This is in excellent agreement with the sole extended basis computation[107] (1.87 kcal/mole). All other calculations underestimate the barrier, except for FSGO (4.21 kcal/mole) and Zeeck's uncontracted (4, 3/2) calculation[58] (3.6 kcal/mole). When the number of basis functions is too small, the wave function and barrier height are unduly sensitive to the choice of exponents.

Acetaldehyde. An experimental barrier of 1.16 kcal/mole has been determined from microwave splitting.[110] The least accurate barrier (0.74)[72] was calculated for an idealized molecular geometry. All other calculations, using both minimal and extended basis sets, yield barriers within 0.2 kcal/mole of the experimental value.

Acetone. An experimental barrier of 0.78 kcal/mole for the first methyl group has been measured via ir torsion.[113] Calculated barriers are in reasonable agreement with experiment, the largest deviation being +0.36 kcal/mole. All barriers exceed the experimental value.

Glyoxal. There appears to be a clear difference between minimal basis sets and extended basis sets. The minimal basis set predicts a *cis–trans* energy separation of 2.99 kcal/mole, and a barrier (relative to *trans*) of 3.23 kcal/mole. Extended basis results yield values twice as large. The minimal basis results fit the experimental *cis–trans* difference[115] quite well, but all calculations underestimate the barrier.

Formamide. There is experimental uncertainty regarding the formamide rotational barrier because the best current measurement is in solution using dynamic NMR methods. In aqueous solution the barrier has been reported as 21.3 kcal/mole,[118,119] whereas it drops to 18 kcal/mole in acetone solution.[118] In view of the experimental uncertainties, most of the computed barriers are quite reasonable, ranging from 19.5 to 21.7 kcal/mole. The 4-31G barrier seems somewhat too high.[65] Inclusion of *d* functions raises the barrier by 2.35 kcal/mole.[54]

1,3-Butadiene. The experimental *cis–trans* energy difference is 2.5 kcal in favor of *trans*, and the barrier, relative to *trans*, is reported as 7.2 kcal/mole.[121] The Raman spectrum also indicates preference for a *gauche* conformer in the cisoid state, in agreement with the *ab initio* results which display a 40° dihedral angle in the cisoid structure. Most *ab initio* calculations yield a *gauche–trans* energy difference between 2.5 and 3.1 kcal/mole. Minimal basis (STO-3G) rigid-rotor calculations yield a 5.66 kcal *gauche–trans* separation.[51] Opening the CCC angle by about 4° lowers the STO-3G result to 2.05 kcal, and full geometry optimization gives a 2.92 kcal/mole *gauche–trans* difference. With the exception of a 40.8 kcal/mole barrier obtained using FEGO,[122] all the computed *trans* barriers are quite similar, ranging from 6.0 to 7.6 kcal/mole. There is no relationship between the computed barrier and basis-set quality. Polarization functions are probably unimportant. The barriers calculated for glyoxal lie only a few tenths kcal/mole above the calculated butadiene barriers. This is significant since 1,3-butadiene and glyoxal are isoelectronic.

Allene. There is no experimental value for the allene rotational barrier, but one may be encouraged by the consistency of three separate extended basis calculations: 72.9, 74.6, and 75.1 kcal/mole.[124–127] However, these numbers are inconsistent with the STO-3G result (92 kcal/mole)[51] and with Buenker's Gaussian lobe calculation (82.1 kcal/mole).[123]

Ammonia. The 5.8-kcal inversion barrier in ammonia was one of the first barriers to be measured by microwave spectroscopy.[18] It is important to include *d* functions in the nitrogen basis. Extended basis polarized wave functions generally reproduce the barrier within a few tenths kcal/mole. As noted earlier, the Kari–Csizmadia ammonia barrier of 9.84 kcal/mole is anomalous. Correlation corrections are also estimated to be no more than a few tenths kcal/mole. Provided one optimizes the planar geometry and includes *d* functions in the basis, amine inversion can be quantitatively ascertained by *ab initio* calculations.

Methyl Anion. No experimental barrier to inversion has been reported for the methyl anion, and the *ab initio* molecular orbital results are sharply divergent. Some calculated inversion barriers lie in the range 5–6 kcal/mole, others are in the range of 1–3 kcal/mole. Although geometry optimization is important, the energy change due to CH contraction is no more than several tenths of a kilocalorie. A definitive paper by Duke[129] shows that addition of diffuse orbitals to the basis set lowers the inversion barrier by almost 3.5 kcal/mole. Description of anion inversion requires a basis set with unusual radial flexibility. Correlation effects may contribute as much as −1 kcal/mole to the inversion barrier.

Hydronium Ion. The experimental inversion barrier has not been measured, but the ion is now known to be pyramidal.[133,134] As with the ammonia inversion barrier, *d* functions on oxygen are required. Optimization of the O–H bond length is not important. Hartree–Fock inversion barriers span the range 0.96–1.9 kcal/mole. Correlation energy contributions may be as great as −1 kcal/mole, so we predict an experimental barrier under 1 kcal/mole. There is no relationship between basis quality and the reported Hartree–Fock inversion barrier, as long as the basis is split-out and polarized.

Phosphine. The phosphine barrier to inversion has been experimentally assigned by NMR as 31.5 kcal/mole,[142] but this value may be in error by several kcal/mole. Basis sets which include *d* orbitals generally yield higher inversion barriers (35–39 kcal/mole) than basis sets without *d* orbitals (25–34 kcal/mole). This behavior follows the trend already set by ammonia and H_3O^+. In view of the sharp contraction in P–H bond lengths in the transition state, geometry optimization is important.

Methylphosphine. Microwave measurements yield an experimental barrier of 1.96 kcal/mole for internal rotation.[146] Calculations with and without *d* functions are both in good agreement (1.83 and 1.71, respectively) with the experimental result. The improvement due to *d* functions is only 0.1 kcal/mole, even for this third-row rotational barrier.

H_3PBH_3. An experimental barrier of 2.47 kcal/mole has been determined by ir torsion studies. STO-3G wave functions greatly underestimate the rotational barrier and the barrier is lowered still further by geometry optimization. The addition of *d* functions to this basis yields only a slight improvement. An extended basis appears necessary for this barrier.

Methysilane. The experimental rotational barrier has been reported as 1.67 kcal/mole.[150] A minimal basis yields a barrier of 1.57 kcal/mole which is lowered to 1.44 kcal/mole upon addition of a *d* function. A polarized triple-zeta level basis also yields a barrier of 1.44 kcal/mole. Palke *et al.*[36] point out that distortional and zero point corrections fully account for the 0.25 kcal differences between the calculated and experimental barriers. We conclude that the methysilane barrier is accessible without use of polarization functions or highly extended basis sets.

2.5. Extension to Large Molecules

Improvements in computational technology now permit *ab initio* calculations to be done for medium-sized molecules of biological interest.

Christoffersen has been one of the most active workers in this area and has been primarily concerned with barriers in simple peptides. Shipman and Christoffersen have reported barrier calculations on the glycine zwitterion,[152] polypeptides of glycine,[153] and the peptides *N*-methylacetamide and 2-formamidoacetamide.[154] All these calculations used the FSGO basis and from the small molecule FSGO work reviewed above, these barriers may be overestimated by several kcals/mole. Pullman *et al.*[155] have investigated the effect of hydration on internal rotation in *N*-methylacetamide, using an STO-3G basis set. As discussed previously, this basis leads to small molecule barrier errors in the order of 1 kcal/mole. Shipman *et al.*[156] report an *ab initio* analysis of the antibiotic Lincomycin using an FSGO basis.

Pople and Radom[157] have reported conformational maps for a number of double-rotor systems which simulate biologically important torsions. Calculations were made at the 4-31G level for C_2H_5OH, FCH_2OH, $CH_2(OH)_2$, NH_2CH_2OH, and NH_2CH_2CHO. At the minimal basis level conformational maps were calculated for the peptide and nucleoside fragments $OHCNHCH_2OH$, $OHCNHCH_2CHO$, and $NH_2CH_2CONH_2$.

Studies have also been carried out on psychoactive drugs. Acetylcholine computations have been published by Pullman and Port,[158,159] using the STO-3G basis, and by Christoffersen *et al.*[160,161] Comparison of these calculations shows that Christoffersen's FSGO method overestimates short-range nonbonded repulsions and consequently predicts the wrong conformer to be most stable. Port and Pullman[162] have done STO-3G calculations on serotonin and bufotenine, and report a conformational map for internal rotation of the alkylamine chain. Pullman and Berthod[163] have obtained a conformational energy map for γ-aminobutyric acid from an STO-3G wave function.

Internal rotation in nucleic acids has been studied using suitable model systems. Newton[164] has computed a conformational surface for rotation about the two O–P bonds in dimethylphosphate monoanion. He used a minimal basis of Slater-type orbitals. Clementi *et al.*[165,166] have investigated internal rotation in ribose phosphodiester ($C_{10}H_{19}O_8P$) and its monoanion using a minimal contracted Gaussian basis. Rotational barriers are similar for both the anion and the neutral species. *Ab initio* methods have also been extended to the 2-amino ethyl acetate cation.[167]

Almlöf[168] has reported a double-zeta level calculation on the rotational barrier in biphenyl. The minimum energy conformer is twisted 32° out of planarity. The calculated planar barrier to rotation is 1.2 kcal/mole, in good agreement with an experimental value[169] of 0.44–1.1 kcal/mole (solvent dependent). The computed perpendicular barrier is 4.5 kcal/mole. Pople *et al.*[170] have analyzed substituent effects on C–O barriers in para-substituted phenols using an STO-3G basis. They find that π-acceptor substituents raise the barrier, while π donors lower the barrier. Agreement with experimental data from far ir torsion studies is excellent. The barrier changes found by computation and experiment agree within a few hundredths kcal/mole.

Ab initio calculations of ring puckering have also begun. Despite the large amount of experimental data on ring inversion,[171] there have been few *ab initio* calculations, due to ambiguities in the definition of inversion coordinates and the size of the systems. Pople and Cremer[172] have recently proposed a general definition of ring coordinates which should simplify *ab initio* geometry searches. They have reported[173] a minimal basis (STO-3G) analysis of pseudorotation in cyclopentane, oxolane, 1,3-dioxolane, and cyclopentanone.

3. Methods for Analyzing Rotational Barrier Mechanisms

A principal role of theoretical chemistry is to find the origin of chemical "force laws." That is, to provide a relatively simple physical description, based on mathematically well-defined quantum mechanical quantities, of characteristic chemical entities such as rotation and inversion barriers, hydrogen bonds, or the shape of molecules.

Implicit in this definition of mechanistic understanding is the use and analysis of *ab initio* wave functions. And the first step, of course, must be demonstration that the level of *ab initio* solution employed is capable of reproducing experimentally observed magnitudes and trends. Section 2 of this chapter has been devoted to this aspect and it was found that an adequate representation of the empirical data could be obtained within the framework of the Hartree–Fock approximation. This is a fortunate circumstance because the Hartree–Fock method is the simplest mathematically and physically well-defined many-electron approximation and, therefore, it gives hope that a correspondingly simple physical picture of barriers may arise. Many researchers have been aware of this possibility and no less than fourteen distinctly different schemes have been investigated. A critical review of each is given below. Because a greater diversity of approaches has been employed in analyzing the rotational barrier mechanism than for any other chemical phenomenon and because acceptable accuracy can be achieved with the simplest possible *ab initio* wave function, the development of our understanding of barriers is an important case study in the use of quantum mechanics in chemistry.

3.1. Bond Orbitals and Localized Orbitals

3.1.1. Hartree Products

Our intuitive preference for interpreting chemistry in terms of bond interactions is most directly expressed by wave functions which are products of bond orbitals. Such wave functions permit a straightforward evaluation of electrostatic contributions to rotational barriers. In addition, we can evaluate the impact of the Pauli exclusion principle on rotational barrier problems by comparing energy expectation values calculated for symmetric and antisymmetric bond orbitals products. The Pauli principle is expected to manifest itself in two ways. First, antisymmetrization of the wave function introduces the exchange energy as a correction to the repulsion energy of a symmetric electron product. Second, antisymmetrization forces electrons of the same spin to occupy orthogonal spatial orbitals.

Bond orbital calculations for ethane, butadiene, and methanol indicate that the Pauli exclusion principle has an important influence on rotational barrier mechanisms. Results from these calculations are listed in Table 4. Sovers *et al.*[70] constructed energy-optimized bond orbitals for ethane in a minimal Slater basis. The mixing coefficient of $H(1s)$ with $C(sp^3)$ was independently optimized for each conformer. An antisymmetric wave function created from symmetrically orthogonalized bond orbitals yielded a rotational barrier of 2.6 kcal, whereas a Hartree product of the original bond orbitals gave a negative barrier of about −0.4 kcal. Stevens and Karplus,[174] using a similar wave function, examined the effects of geometry optimization. Their qualitative results are the same: the Hartree product of bond orbitals favored eclipsed ethane by 2.2 kcal, whereas the antisymmetric product of orthogonalized bond orbitals has a barrier of 3.1 kcal.

Kern, Pitzer, and Sovers[90] have treated methanol with their bond-orbital method and obtained results similar to those for ethane. Energy-optimized bond orbitals were constructed in a minimal Slater basis, bond orbitals for each conformer being independently optimized. A Hartree product wave function gave a barrier of −0.8 kcal, whereas the antisymmetric product of symmetrically orthogonalized bond orbitals produced a barrier equal to 0.95 kcal/mole (compared with an SCF barrier of 1.57 kcal/mole calculated with the same basis set).

Pincelli *et al.*[19] formed a Hartree product for butadiene by transfer of Magnasco–Perico localized orbitals[176,177] from ethylene. Although positive *trans → gauche* and *gauche → cis* barriers were obtained (0.4 and 0.2 kcal, respectively), their values are only a fraction of the barriers obtained from an antisymmetric orbital product or from an SCF calculation done in the ethylene basis.

We conclude from these studies that electrostatic interactions between bonds are not an important factor in the rotational barriers of these three molecules. The Pauli principle has been shown as necessary for an understanding of these barriers, but we cannot ascertain whether the barrier contributions come from electron exchange or from orbital orthogonality.

A calculation which does discriminate between exchange and orbital orthogonality has been presented by Christiansen and Palke.[175] Edmiston–Ruedenberg localized orbitals were constructed from SCF wave functions for a minimal Slater basis, and also for an extended Gaussian basis. These localized orbitals then define an energy-optimized Hartree product insofar as the Edmiston–Ruedenberg procedure minimizes the exchange energy, and hence minimizes the energy separation of symmetric and antisymmetric wave functions. With both basis sets, it was found that the Hartree product rotational barrier was within 0.4 kcal of the Hartree–Fock barrier. This result definitely rules out electron exchange as a source of the ethane rotational barrier. Christiansen and Palke then explicitly investigated the effects of orbital orthogonality via consideration of the partially antisymmetrized product function:

$$\psi = [\hat{A}(1s_1)(1s_2)(CC)][\hat{A}(CH_1)(CH_2)(CH_3)][\hat{A}(CH_4)(CH_5)(CH_6)]$$

The Edmiston–Ruedenberg (CH) orbitals were deorthogonalized by mixing them within their own linear space so as to optimize the energy of this wave

Table 4. Rotational Barriers Predicted by Hartree Product Wave Functions

Molecule	Hartree ψ barrier, kcal/mole	Antisymmetric ψ barrier, kcal/mole	Basis	Comments	Reference
Ethane	−0.4	2.6	Minimal STO	Bond orbitals from sp^3 on C plus H 1s, variationally determined mixing coefficient	70
	−0.5	2.6	Minimal STO	Bond orbitals from sp^3 on C plus H 1s, variationally determined mixing coefficient	174
	−2.2	3.1	Minimal STO	Same, plus geometry optimization	
	—	2.82	(5, 3/3) Gaussian lobe	Methane-optimized bond orbitals transferred	72
	—	2.66	(5, 2/2) Gaussian lobe	Methane-optimized bond orbitals transferred	72
	3.48	3.25	Minimal STO	Edminston–Ruedenberg localized orbitals	175
	3.28	2.87	Extended CGF	Edminston–Ruedenberg localized orbitals	175
	2.49	—	Minimal STO	Deorthogonalized Edmiston–Ruedenberg orbitals within each methyl rotor	175

Molecule			Basis	Method	Ref.
	−0.58	—	Minimal STO	Deorthogonalized Edmiston–Ruedenberg orbitals between rotors	175
Butadiene	*t-g* 0.4 kcal; *g-c* 0.2 kcal	—	$(7, 3/3) \to (2, 1/1)$ CGF	Magnasco–Perico localized orbitals transferred from ethylene; nonorthogonal	19
	—	*t-g* 2.1 kcal; *g-c* 2.7 kcal	$(7, 3/3) \to (2, 1/1)$ CGF	Symmetrically orthogonalized localized orbitals	19
Butadiene	—	*t-g* 7.14 kcal; *g-c* 0.79 kcal	$(7, 3/3) \to (2, 1/1)$ CGF	Normal SCF	19
Methanol	−0.18	—	$(2, 1/1)$ STO	Bond orbitals from sp^3 Pn C plus H 1s, variationally determined mixing coefficient for each conformer	90
	—	0.95	$(2, 1/1)$ STO	Symmetrically orthogonalized bond orbitals and reoptimized mixing coefficient	90
	—	1.57	$(2, 1/1)$ STO	SCF	90

function. The resultant loss of end-to-end orthogonality changed the ethane barrier to -0.85 kcal. A control calculation in which (CH) orbitals were only mixed within each rotor gave the same result as no mixing of Edmiston–Ruedenberg orbitals.

An important paper by Levy, Nee, and Parr[178] gives another approach to this problem which supports Christiansen and Palke's work. Bond orbitals for ethane were constructed from a linear combination of hydrogen $1s$ orbitals and tetrahedrally hybridized carbon orbitals (orbital exponents were optimal methane values). An orthogonalized Hartree product gave a rotational barrier of 3.04 kcal/mole—within 10% of the 2.77 kcal/mole barrier obtained by a determinantal wave function. This close correspondence between barrier heights for orthogonal Hartree products and the determinantal wave function is preserved over a wide range of hybridization parameters. A Hartree product of *nonorthogonal* bond orbitals on the other hand, yields a barrier of 1.0 kcal/mole. Their conclusion is that interorbital exchange does not contribute to the ethane barrier, but rather that bond orbital orthogonality is the predominant factor in generating the barrier.†

Bond orbital methods have been criticized on the grounds that total energies may lie as much as 0.2 a.u. above *or below* the Hartree–Fock energy calculated with the same atomic orbital basis—an error 10–100 times the barrier. On the other hand, bond orbital results cannot be fortuitous since all available calculations give reasonable values for the rotational barriers. The energy difference between a bond orbital calculation and an SCF calculation is principally caused by the superior sensitivity to electronic rearrangement in an SCF calculation. To the extent that bond orbital wave functions do reproduce barriers, we conclude that details of electronic polarization have minor influence on rotational barriers.

3.1.2. Interaction Energy between Bonds

The intuitive appeal of bond interactions has stimulated several attempts to partition the rotational barrier energy into interaction energies between bonds.

Pitzer[179] carried out an Edmiston–Ruedenberg analysis[180] on the 1963 Pitzer–Lipscomb[181] ethane wave function. From his list of localized orbital coefficients one sees that conformational dependence of localized orbital coefficients is most strongly manifested in the vicinal tail of CH bonds. The localized orbital tail density shifts toward the CC axis, but to the backsides (exterior) of the CH axes in eclipsed ethane. Opposite, but somewhat weaker, behavior is seen within the CH bond itself. Charge density shifts outward from C toward the inner side of the CH bond in eclipsed ethane. In balance, orbital tail effects dominate the net charge rearrangement in eclipsed ethane since there are three tails which overlap each CH bond. The CC localized orbital

†See note 1 of Notes Added in Proof, page 107.

shows a slight (<0.0005) displacement of electron density from the bond center.

Pitzer has tabulated energy components for the localized orbitals. As one rotates into the eclipsed conformer these changes are:

CC bond: $\Delta V_{ne} = 0.00042$ a.u. $\Delta T = -0.00059$ a.u.
$\Delta V_{ee} = 0.00017$ a.u.

CH bond: $\Delta V_{ne} = -0.00434$ a.u. $\Delta T = 0.00198$ a.u.
$\Delta V_{ee} = 0.00172$ a.u.

These components are subject to question since they predict $\Delta T = +0.01188$ a.u., whereas we know from the virial theorem that ΔT should be negative.

Bendazzoli, Bernardi, and Palmieri[182] have calculated Boys localized orbitals[192] for rotamers of propargyl alcohol and hydroxyacetonitrile using a contracted minimal set of Gaussian orbitals $(6, 3/3) \rightarrow (2, 1/1)$. The electronic energy of each rotamer is then partitioned as

$$E = \sum_R H_R + \sum_{R<S} (J_{RS} - K_{RS})$$

Unfortunately, most terms in this partition change by more than 0.01 hartree during internal rotation, and some components change by as much as 0.5 hartree. These localized orbital changes greatly outscale the rotational barriers, which are of magnitude 0.005 hartree.

Magnasco and Musso[183] have written wave functions as a determinantal product of self-consistent bond orbitals. Because these orbitals are not orthogonal (they have overlap $S = 1 + \Delta$), the one-electron density matrix is expressed as an overlap-independent term $p_0 = \Phi\Phi^\dagger$ and an overlap-dependent term $p_1 = \Phi\Delta S^{-1}\Phi^\dagger$. The Fock operator is likewise separable into an overlap-independent operator F_0 and the overlap-dependent operator G_1. Substitution of these partitions into the closed-shell Hartree–Fock energy expression leads to the energy separation

$$E = U + V_c + \mathrm{tr}(F_0 P_1 + G_1 P_1 - K_0 P_0)$$

where U is the internal energy of bond orbitals, V_c the Coulombic interaction between bonds, $F_0 P_1$ the unperturbed Fock operator acting on perturbed density matrix, $-K_0 P_0$ the exchange interaction between bonds, and $G_1 P_1$ the self-repulsion of perturbed electron density. The Magnasco–Musso method has been applied[67] to rotational barriers in ethane and hydrogen peroxide using a minimal Slater basis and rigid rotation (see Table 5). Results in both cases show that the rotational barrier is controlled by the $F_0 P_1$ term. That is, the rotational barrier comes from overlap-induced electron-density shifts. Further partitioning of the overlap matrix demonstrated that *vicinal overlap changes* are primarily responsible for the variation in $\mathrm{tr}(F_0 P_1)$. It is important to recall in

Table 5. *The Magnasco–Musso Method Applied to Rotational Barriers in Ethane and Hydrogen Peroxide*

Quantity	Ethane, kcal/mole	Hydrogen peroxide ($-cis$ barrier), kcal/mole
ΔV_c	−0.18	−22.88
$\Delta(F_0 P_1)$	6.11	39.06
$\Delta(G_1 P_1)$	−1.25	0.79
$\Delta(-K_0 P_0)$	−2.01	−3.17
Total ΔE	2.67	13.79
ΔE_{SCF}	3.41	11.72

this context that overlap changes involving vicinal tails are the dominant conformationally dependent feature of Edmiston–Ruedenberg localized orbitals.[180] (Conclusions analogous to those above have come from extended Hückel formulations of the Magnasco–Musso theory.[184,185]†)

3.1.3. Vicinal Interference Energies

England and Gordon have applied Ruedenberg's ideas of interference energy[186] to the analysis of rotational barriers in ethane,[187] hydrogen peroxide,[188] borazane,[189] methylamine,[190] methanol,[190] propene,[190] acetaldehyde,[190] formaldoxime,[191] and formic acid.[191] Although their work is based on INDO wave functions instead of *ab initio* wave functions, we discuss it here because their ideas could be easily developed as an *ab initio* theory. In fact, the interference energy, as used by England and Gordon, is closely related to the penetration energy defined by Magnasco and Musso.[183]

The analysis begins with Edmiston–Ruedenberg localized orbitals. If the nuclear changes are divided proportionally among the localized orbitals, the energy of the ith localized distribution can be written in the INDO approximation as

$$E_i = \sum_A U_i(A) + \sum_A \sum_{B \neq A} \beta_i(A, B) + \sum_j V_{ij}^{ne}(A, B) + \sum_j [V_{ij}^{ee}(A, B) + V_{ij}^{nn}(A, B)]$$

In this equation i and j are orbital indices, and A and B refer to atomic centers. $U(A)$ is the one-center energy on A, while $(V^{ne} + V^{ee} + V^{nn})$ defines the electrostatic energy R of a quasiclassical charge distribution. The interference energy $\beta(A, B)$ is simply the one-electron energy associated with an overlap distribution between atoms A and B. In the INDO approximation, the interference energy is proportional to the orbital bond order $P_i(A, B)$ via

$$\beta_i(A, B) = \sum_{\mu}^{A} \sum_{\nu}^{B} 2 C_{\mu i} C_{\nu i} \beta^0(A, B) S_{\mu\nu} = \beta^0(A, B) P_i(A, B)$$

†See note 2 of Notes Added in Proof, page 108.

The central claim made by England and Gordon is their assertion that INDO rotational barriers are successfully reproduced by certain terms of the *vicinal* interference energy. Their data are summarized in Table 6. It is true that the tabulated vicinal interference energies give an excellent qualitative fit to the INDO rotational barriers. But, with the exception of BH_3NH_3 and C_2H_6, the data are critically defective because vicinal interference energies involving the heavy atoms are omitted. Since the localized orbitals have comparable magnitude on both hydrogen and the heavy atoms, the indicated omissions are entirely arbitrary. The results to date on this potentially interesting approach are thus completely unconvincing. A start toward a better analysis has been made by application[193] of the Pitzer–Lipscomb *ab initio* wave function for ethane, where they compute a vicinal interference energy of 0.0072 a.u. within each (CH) localized orbital.

We now summarize the current standing of the vicinal interference idea. Because it is a one-electron energy, the vicinal interference energy is wholly contained within the term $tr(F_0P_1)$ which controlled the Musso–Magnasco

Table 6. *INDO Interference Energy Analysis of Barriers*[a]

Molecule	One center	Quasi-classical repulsion	Inter-ference energy	Net barrier	Reference
C_2H_6					
OG[b]	−0.49	−0.37	3.11	2.25	187
EG[b]	−1.44	+0.80	2.83	2.19	
H_2O_2, cis					
OG	—	—	6.45[c]	4.95	188
H_2O_2, trans					
OG	—	—	6.09[c]	4.15	188
BH_3NH_3					
OG	−1.11	1.19	1.89	1.97	189
EG	−3.19	3.23	1.77	1.81	
CH_3OH					
OG	—	—	1.07[c]	1.21	190
EG	—	—	0.64[c]	0.78	
CH_3NH_2					
OG	—	—	1.85[c]	1.91	190
EG	—	—	1.40[c]	1.56	
CH_3CHCH_2					
OG	—	—	1.98[c]	1.55	190
EG	—	—	1.01[c]	1.22	
CH_3, CHO					
OG	—	—	1.38[c]	0.77	190
EG	—	—	1.12[c]	0.60	

[a] Values given in kilocalories per mole.
[b] OG = optimized geometry; EG = experimental geometry.
[c] Does not include all interference energy terms.

barriers. We note that the vicinal interference energy is overlap dependent and this is consistent with tabulations of overlap populations which show that rotational barriers are usually accompanied by a decrease in vicinal bond order. Explicit studies of the electron density also reveal a distortion of electron density between eclipsed bonds. At present, however, vicinal interference is an interesting but untested approach. As a model for aiding chemical intuition it has the drawback that it is cast in abstract mathematical concepts that do not have a direct connection with spectroscopically accessible quantities.

3.2. N-Center Energy Partitions

Decomposition of the SCF energy into one-, two-, three-, and four-center terms has received considerable attention. Most analysis of this type has been done using INDO or CNDO wave functions because the zero differential overlap approximation eliminates all three- and four-center terms, leaving

$$\mathcal{E} = \sum_A E_A + \sum_{A<B} E_{AB}$$

where

$$E_{AB} = \sum_\mu \sum_\nu (2P_{\mu\nu}B_{\mu\nu} - \tfrac{1}{2}P^2_{\mu\nu}\gamma_{AB}) + Z_A Z_B R^{-1}_{AB}$$

$$- (P_{AA}Z_A + P_{BB}Z_B - P_{AA}P_{BB})\delta_{AB}$$

Again, we discuss the semiempirical approach here since it has a direct impact on *ab initio* theory. Specifically, the three- and four-center integrals of an *ab initio* theory can be unambiguously mapped onto one- and two-center energy terms if one adopts a suitable partition of bicentric charge distributions

$$\phi_A(1)\phi_B(1) \rightarrow \rho^A_{AB}(1) + \rho^B_{AB}(1)$$

This reduction of bicentric to monocentric terms can be done using either orbital or spatial criteria. Bader *et al.*[194] have constructed a unique spatial definition of atoms within molecules. Around each atom one can construct a surface such that the electron density is a local minimum with respect to displacements normal to that surface. Because such local minima appear to exist for all molecules studied to date and because the virial theorem is satisfied within the volume enclosed by each such surface, this surface offers a satisfactory and reasonable definition for an atom within a molecule. Local elements of bicentric charge distributions (and tails of monocentric distributions) are then assigned to various atoms in accord with the atomic region in which they lie. Unfortunately, Bader's prescription is difficult to implement computationally. One therefore partitions the bicentric distribution using an orbital criterion similar to that proposed by Mulliken for population analysis,

$$\rho_{AB}^A(1) \equiv W_A \phi_A(1)\phi_B(1)$$

$$\rho_{AB}^B(1) \equiv W_B \phi_A(1)\phi_B(1)$$

$$W_A + W_B = 1$$

Leibovicci and co-workers have applied the CNDO–INDO energy partition to the analysis of many rotational barriers.[195–204] In papers where ΔE_A and ΔE_{AB} are tabulated, one or two terms are often dominant. But many terms change by amounts comparable in magnitude to the barrier height. This high level of numerical noise precludes a definitive atom-by-atom analysis of rotational barrier mechanisms. This term-by-term instability occurs because charge-density rearrangements are exaggerated when the zero differential overlap approximation is invoked. The success of *ab initio* analogs will be critically dependent on the method one uses for partitioning the bicentric charge distributions.

Leibovicci *et al.* have also compared $\sum E_{AB}$ (bonded)$+\sum E_A$ and $\sum E_{AB}$ (nonbonded) with the potential energy curve for internal rotation. For molecules in which hyperconjugation gives partial double bond character to the rotating bond, ΔE (bonded) gives a good qualitative fit to the barrier with a maximal error of about 20%. These molecules include formic acid,[195] cyclopropyl acetaldehyde,[196] cyclopropyl acetone,[196] cyclopropyl carboxylic acid fluoride,[197] and formaldoxime.[198] Rotational barriers involving methyl rotors are well described by ΔE (nonbonded). The sum of nonbonded energies lies 10%–20% above the total energy for such molecules as propane,[199] dimethylamine,[199] dimethylether,[199] dimethylsilane,[199] dimethylsulfide,[199] dimethysulfone,[200] dimethysulfoxide,[201] dimethylphosphine,[199] amd $(CH_3)_2H_2P\cdot BH_3$.[203] In all these cases the sum of those nonbonded terms which involve hydrogen fits the rotational barrier within 20%. For molecules having internal rotation about a bond to a third row atom, nonbonded terms between hydrogen and the third row atom give a good qualitative fit to the rotational potential. We note that the bonded vs nonbonded analysis fails for $F_3P\cdot BH_3$,[204] since ΔE (nonbonded) exceeds the calculated barrier by 65%.

Koehler has applied a similar energy partition to the analysis of rotational barriers in formic acid[205] and hydrazine.[206] Using INDO, CNDO, and NDDO wave functions, he partitioned the energy as

$$E = \sum_A E_A + \sum_A \sum_{<B} E_{AB}$$

where

$$E_A = E_A^U + E_A^J + E_A^k$$

$$E_{AB} = E_{AB}^R + (E_{AB}^J + E_{AB}^N + E_{AB}^n) + E_{AB}^N$$

*Table 7. Terms Changing by Magnitudes
Comparable to the Rotational Barrier*

Type of term	Total ΔE, a.u.
C(1)	−0.07
H_3(1)	+0.02
C(1)–C(2)	+0.13
H_3(1)–H_3(2)	+0.01
H_3(1)–C(1)–H_3(2)	−0.04
H_3(1)–C(1)–C(2)–H_3(2)	+0.03

His data show this decomposition to be worthless for CNDO and INDO. At the NDDO level the formic acid barrier and the *trans* hydrazine barrier are qualitatively controlled by changes in the two-center exchange and resonance terms ΔE_{AB}^k and ΔE_{AB}^R. But the *cis* hydrazine barrier comes almost exclusively from the interatomic Coulombic component ΔE_{AB}^{J+V+N}.

Clementi[15] has computed an *ab initio* N-center expansion of the ethane barrier using an extended Gaussian basis $(12, 7, 2/6, 2) \rightarrow (6, 4, 2/3, 2)$. Musso and Magnasco[207] report a similar expansion for ethane using a minimal Slater basis. They report that two-center vicinal energies qualitatively explain the barrier, but Clementi's data, summarized in Table 7, show many types of terms. changing by magnitudes comparable to the rotational barrier, or even by substantially greater amounts.

3.3. Fourier Analysis

For barriers of low symmetry, Fourier analysis can help compare similar electronic features. This has proved most useful for the twofold component. Veillard[46] and Pople[51] were the first to apply this type of analysis to computed potentials for internal rotation. The expansion is

$$\Delta V(\phi) = \tfrac{1}{2}V_1 \cos \phi + \tfrac{1}{2}V_2 \cos 2\phi + \tfrac{1}{2}V_3 \cos 3\phi$$
$$+ V_1^1 \sin \phi + V_2^1 \sin \phi$$

with $\phi = 0$ for the *cis* conformation. V_1^1 and V_2^1 are required when there is a lack of symmetry around $\phi = 180°$ (e.g., substituted hydrazine). For coupled rotors (e.g., propane) the expansion is a function of two angles and explicitly includes coupling terms:

$$\tfrac{1}{2}V_3'(1 - \cos 3\phi_1)(1 - \cos 3\phi_2) + \tfrac{1}{4}V_3'' \sin 3\phi_1 \sin 3\phi_2$$

Some of the Fourier expansions carried out to date are collected in Table 8. In addition, an especially thorough list of Fourier components calculated from 4–31G wave functions has been tabulated by Random, Hehre, and Pople.[208]

Fourier decomposition of rotational barriers has several advantages. First, because the Fourier coefficients are simply a way of representing the total molecular energy changes, they have the same variational stability as the *ab initio* total energy. This is in contrast to fragment interaction energies and energy components, both of which may be unstable with respect to basis change or geometry optimization. The second advantage is the connection with experimental procedures. In an experimental measurement by far infrared or microwave spectroscopy it is customary to model the torsional potential by a truncated Fourier series. Another potential advantage is the possibility of transferring the components from one molecule to another. A first, apparently successful, test of this concept has been carried out by Radom[209] from a series of STO-3G calculations. Using rigid rotation Fourier components from ethane, propane, cyanoethane, *n*-butane, 1-cyanopropane, and 1,2-dicyanoethane, he estimated the rotational barrier of 2,3-dicyanobutane by superposition of the Fourier coefficients for each type of vicinal interaction. The estimated Fourier coefficients showed less than two percent deviation from the coefficients determined by a direct calculation.

To make progress in elucidating the physical origin of barriers it is clearly desirable that each Fourier component be identified with a well-defined electronic structure effect. For V_2 there is a highly useful interpretation which arises because of the creation or annihilation of partial π bonding when a $2p$ orbital is rotated through 90°. Hyperconjugation can often be associated with V_2. Pople[28,51,80] and co-workers have used V_2 to rationalize the conformation of such systems as

$$X - CH_2 - X'$$

where X, $X' = NH_2$, OH, and F. As indicated by the charge transfer arrows, there is a stabilizing effect produced by back donation from lone-pair orbitals at one end of the molecule into antibonding σ orbitals at the other. For example if $X = F$ and $X' = OH$ a maximum stabilization occurs when the FCO plane is perpendicular to the COH plane. In Pauling resonance terms this corresponds to the double-bond–no-bond structure $F^-C = O^+$. Another interesting example is $X = F$, $X' = NH_2$ and here the minimum energy occurs when the N lone pair is coplanar with the CF bond. V_1 is less easy to interpret uniquely and clearly, but can sometimes be interpreted as an across-the-molecule dipole–dipole interaction or as steric repulsion between bulky substituents, or as an intramolecular hydrogen bond. The interpretation of V_3, of course, is what almost all of the research on barrier origin has been directed toward (the review of which comprises a good part of this chapter).

3.4. Energy Components

It is well known that the binding of a diatomic molecule can be analyzed in terms of the balance between electrostatic attraction and increased kinetic

Table 8. Fourier Analysis of Rotational Barriers (kcal/mole)

Molecule	V_1	V_2	V_3	Basis set	Reference
Ethane	—	—	3.33	STO-3G; standard geometry	51
	—	—	2.67	STO-3G; CCC = 112°	209
	—	—	3.26	4-31G	65
Propane	—	—	3.45	STO-3G; geometry optimized	51
	—	—	2.91	STO-3G; CCC = 112°	209
Cyanoethane	—	—	2.90	STO-3G; CCC = 112°	209
N-Butane	4.84	2.65	3.91	STO-3G; CCC = 112°	209
	3.19	1.43	3.86	4-31G; standard geometry	210
1-Cyanopropane	1.66	1.23	3.66	STO-3G; CCC = 112°	209
1,2-Dicyanoethane	2.05	1.03	3.50	STO-3G; CCC = 112°	209
N-Propyl fluoride	0.48	0.60	4.66	4-31G; standard geometry	210
1,2-Difluorethane	4.68	2.72	4.08	4-31G; standard geometry	210
Butadiene	2.64	4.12	2.33	STO-3G; standard geometry	51
			($V_4 = -1.18$)		
	1.20	5.22	1.42	STO-3G; experimental geometry	51
			($V_4 = -0.54$)		
	0.73	5.70	1.20	STO-3G; geometry optimized	51
			($V_4 = -0.33$)		
	0.79	5.12	2.60	$(7, 3/3) \rightarrow (2, 1/1)$	19
			($V_4 = -0.86$)		
Allene	—	91.9	—	STO-3G; standard geometry	51

Formamide	—	23.89	—	4-31G; standard geometry	65
Formic acid	5.75	8.93	0.55	4-31G; standard geometry	65
Glyoxal	5.73	4.85	0.40	4-31G; standard geometry	65
	4.22	4.70	0.48	$(9,5/6,4/4) \rightarrow (3,2/3,2/2)$; Full geometry optimization	53
Hydrazine	15.52	3.80	2.26	$(6,3/3) \rightarrow (3,2/2)$ plus polarization $HNH = 100°$	45
Hydrazine	10.68	5.27	1.87	Same basis as above $HNH = 105°$	45
	6.43	8.46	1.28	Same basis as above $HNH = 110°$	45
	2.78	13.50	0.48	Same basis as above $HNH = 115°$	45
	0	20.08	0	Same basis as above $HNH = 120°$	45
Hydrogen peroxide	11.22	4.92	0.66	$(9,5/4) \rightarrow (4,3/2)$ plus polarization; experimental geometry	24
	8.70	4.86	0.50	$(9,5/4) \rightarrow (4,3/2)$ plus polarization; optimized R_{OO} at $\phi = 180°$	24
	7.00	4.06	0.26	Same basis as above; optimized R_{OO}; variable OOH	24
	6.96	4.16	0.30	Same basis as above; full optimization geometry	24

energy. With this motivation Allen[212] proposed that useful information about barrier mechanisms might be obtained from various combinations of the individual energy component changes across a barrier: ΔT, ΔV_{ne}, ΔV_{ee}, and ΔV_{nn}. In particular he sought combinations which would possess an intuitive physical utility and at the same time be as invariant as possible to changes in geometry and basis set. This led to the choice $\Delta V_{att} = \Delta V_{ne}$ and $\Delta V_{rep} = \Delta V_{ee} + \Delta T + \Delta V_{nn}$ because available data indicated that the combination $T + V_{nn}$ was best able to reduce geometry and basis-set sensitivity. Use of energy components was conceived as the energetic analog to Mulliken population analysis. (It may be noted that this latter scheme has retained its widespread usefulness in spite of its great basis-set sensitivity.) Because individual components generally change by amounts several times greater than the barrier and because no variation principle is applicable, it is clear, as in the case of population analysis, that a considerable amount of data on diverse systems must be accumulated before it is possible to assess the degree to which they can aid in the analysis of electronic structure calculations in general and barriers in particular. For barriers, individual components have been obtained now for twenty-two cases with multiple determinations for eleven, and these are given in Table 9. With these data we can ascertain where this method stands at present for barriers.

Before considering component combinations it is worthwhile to demonstrate and analyze the great sensitivity to geometry and basis present in the individual components; this is immediately evident in Table 9. A dramatic example is the 80 kcal/mole change in ΔV_{nn} that occurs with a CCH angle opening of only $0.5°$ when eclipsed ethane is geometry optimized.[39] Because Hartree–Fock wave functions satisfy the molecular viral theorem, we know that $T = -E$ at all extrema on the torsional potential surface. An accurate rotational barrier calculation should, therefore, indicate a negative kinetic energy change equal in magnitude to the barrier. But, in practice, calculated values for ΔT are usually positive, showing that ΔT is generally more sensitive than any other component. The error in ΔT is necessarily compensated by errors in ΔV_{ne}, ΔV_{ee}, and ΔV_{nn} since the sum of all four components is equal to the accurately calculated barrier. The known error of ΔT thus establishes a practical estimate on the uncertainty of *each* of the other components. Values are trustworthy only if $\Delta T \approx -\Delta E$ and the quantity $|\Delta E + \Delta T|$ provides a useful gauge to the error in all individual components.

The observed deviations in ΔT probably come from inadequate geometry optimization. If a molecule is not at its equilibrium geometry the molecular virial theorem takes the form

$$2T + V + \sum_i R_i \frac{dE}{dR_i} = 0$$

Table 9. Energy Components for Rotational Barriers (kcal/mole)

Molecule	Basis	ΔE	ΔT	ΔV_{nn}	ΔV_{ee}	$\Delta V_{ne} = \Delta V_{att}$	ΔV_{rep}	Reference
Ethane	$(10, 5/5) \rightarrow (3, 1/1)$, rigid rotation	2.56	8.15	4.69	9.36	−19.64	22.20	6
	Minimal STO, geometry optimized	3.3	6.2	−74.5	−54.8	126.4	−121.1	213
	Minimal STO, rigid rotation	3.28	12.61	4.70	16.67	−30.70	33.98	213
	$(10, 6/4) \rightarrow (3, 2/1)$, rigid rotation	2.57	8.43	4.69	9.53	−20.08	22.65	62
	$(5, 3/2) \rightarrow (2, 1/1)$, rigid rotation	3.28	—	4.78	12.83	—	—	69
	$(7, 3/3) \rightarrow (3, 1/1)$, rigid rotation	2.72	—	4.61	12.88	—	—	214
	FSGO, rigid rotation	6.0	11.0	4.8	2.1	−11.9	17.9	41
	Optimal FSGO, rigid rotation	5.9	16.2	4.8	9.1	−24.2	30.1	41
	Optimal single FSGO, geometry optimized	5.7	−6.2	−103.2	−100.9	+216.0	−210.3	41
	Double FSGO in CC-bond, complete optimization	3.1	−4.1	−167.2	−154.3	328.7	−325.6	41
Propylene	$(5, 2/2)$ Bond orbital, rigid rotation	0.47	−1.06	−12.43	−12.86	26.81	−26.34	72
	$(5, 3/3)$ Bond orbital, rigid rotation	0.81	−0.68	−12.43	−12.45	26.36	−25.55	72
	$(5, 2/3)$, rigid rotation	1.48	6.93	2.64	16.02	−24.11	+25.59	109
	$(10, 5/5) \rightarrow (3, 1/1)$, GLO, rigid rotation	1.25	6.16	−7.96	2.67	0.39	0.87	108

continued overleaf

Table 9 (continued)

Molecule	Basis	ΔE	ΔT	ΔV_{nn}	ΔV_{ee}	$\Delta V_{ne} = \Delta V_{att}$	ΔV_{rep}	Reference
Propylene	$(5, 3/2) \rightarrow (\hat{2}, 1/1)$, rigid rotation	1.69	—	−11.70	3.52	3.58	—	69
	$(9, 5/4) \rightarrow (4, 3/2)$, rigid rotation	1.87	5.08	−7.97	3.70	1.19	0.81	107
	FSGO, rigid rotation	4.21	−1.41	−87.8	−71.6	196.4	—	97
cis-1-Fluoropropylene	$(3, 1/1)$, rigid rotation	1.07	1.46	−45.54	−34.46	79.60	−78.53	108
	$(4, 3/2)$, rigid rotation	1.14	1.89	−45.56	−40.85	85.71	−84.52	107
trans-1-Fluoropropylene	$(3, 1/1)$, rigid rotation	1.34	7.18	−8.36	5.94	−3.41	4.75	107
	$(4, 3/2)$, rigid rotation	2.08	6.27	−8.35	10.04	−5.96	7.96	107
Ethyl fluoride	$(10, 6/4) \rightarrow (3, 2/1)$, rigid rotation	2.60	13.14	21.41	28.36	−60.31	62.91	62
	$(7, 3/3) \rightarrow (3, 1/1)$, rigid rotation	2.60	—	20.98	30.50	—	—	214
Methanol	$(10, 5/5) \rightarrow (3, 1/1)$, rigid rotation	1.36	6.21	1.12	8.15	−14.12	15.48	6
	$(10, 5/5) \rightarrow (3, 2/2)$, rigid rotation	1.03	2.26	1.10	3.13	−5.46	6.49	6
	$(13, 7/4) \rightarrow (3, 2/2)$, rigid rotation	1.44	—	1.71	2.49	—	—	89
Acetaldehyde	$(10, 5/5) \rightarrow (4, 2/2)$, rigid rotation	1.08	3.32	−5.00	−5.57	8.33	−7.25	111
	$(5, 3/2) \rightarrow (2, 1/1)$, rigid rotation	1.34	—	−7.21	−6.49	—	—	69
Hydrogen peroxide cis	$(10, 5/5) \rightarrow (3, 1/1)$ rigid rotation	13.66	−48.66	56.06	−94.51	100.77	−87.11	6

Hydrogen peroxide *cis*	Geometry optimized (11, 7/6) → (5, 3/3), plus *p* on H and *d* on O	10.6	−23.2	−232.5	−368.5	734.7	−724.1	39
	Experimental geometry barrier from 120° → 0° (11, 7/6) → (5, 3/3) plus *p* on H and *d*	20.7	20.7	76.5	−28.8	−47.7	58.3	39
Hydrazine (90° → 0°)	(10, 5/5) → (4, 2/1), plus *p* on H, barrier from 150° → 0°	13.24	−29.59	58.83	−91.83	80.82	67.58	79
	(10, 5/5) → (3, 2/2), rigid rotation	11.31	18.76	22.44	5.63	−35.52	46.83	99
	(7, 3/3) → (2, 1/1), rigid rotation	9.55	—	25.32	−5.09	—	—	102
Hydrazine (90° → 180°)	(10, 5/5) → (3, 2/2), rigid rotation	3.17	32.39	−15.71	+45.49	−58.98	62.15	99
	(7, 3/3) → (2, 1/1), rigid rotation	3.58	—	−16.50	55.39	—	—	102
Propane (*ss* → *es*)	(5, 3/2) → (2, 1/1), rigid model geometry	3.57	—	19.53	27.03	—	—	71
Propane (*es* → *es*)	FSGO, geometry optimized	5.29	−2.41	−75.02	−71.14	153.87	−148.58	97
	FSGO, geometry optimized	7.14	−6.78	−236.34	−268.98	549.24	−542.10	97
Methylamine	(10, 5/5) → (3, 1/1) GLO rigid rotation	2.41	9.14	2.55	11.33	−20.61	23.02	79
Fluoromethanol *cis* (60° → 0°)	(4, 2/2), rigid rotation; standard geometry	8.21	14.43	120.50	79.60	−206.32	214.53	99
Fluoromethanol *trans* (60° → 180°)	Rigid rotation; standard geometry	12.59	−14.98	−215.07	−212.96	455.60	−443.01	99
Hydroxymethyl radical *cis*	(10, 5/5) → (4, 2/2)	3.51	25.22	−15.00	12.42	−19.14	22.65	215

continued overleaf

Table 9 (continued)

Molecule	Basis	ΔE	ΔT	ΔV_{nn}	ΔV_{ee}	$\Delta V_{ne} = \Delta V_{att}$	ΔV_{rep}	Reference
Hydroxymethyl radical trans	(10, 5/5) → (4, 2/2)	1.82	20.91	−26.92	17.07	−9.22	11.04	215
Glyoxal cis	(10, 5/5) → (3, 1/1), rigid rotation standard geometry	3.23	38.36	−812.47	−831.22	1608.57	−1605.34	117
	(9, 5/6, 4/4) → (3, 2/3, 2/2), geometry optimized	2.1	4.8	−490.8	−504.5	992.6	−990.5	53
Glyoxal trans	(10, 5/5) → (3, 1/1), rigid rotation standard geometry	6.22	−11.95	455.11	483.2	−920.17	926.39	117
	(9, 5/6, 4/4) → (3, 2/3, 2/2), geometry optimized	6.9	−6.8	385.0	432.0	−803.3	810.2	53
HCOOH cis → 90°	(10, 5/5) → (3, 2/1), rigid rotation	13.00	1.69	−141.69	−105.60	258.60	−245.6	216
HONO trans → 90°	(10, 5/5) → (3, 2/1), rigid rotation	9.04	47.8	+115.5	80.9	−235.2	244.0	217
Allene	(7, 3/3) → (2, 1/1), rigid rotation	74.61	—	1.26	42.45	—	—	154
Phosphine inversion	(12, 9, 2/5, 1) → (6, 4, 2/3, 1), geometry optimized	36.7	−41.8	204.2	172.3	−298.0	344.7	57

which can be rearranged as

$$T = -E - \sum_i R_i \frac{dE}{dR_i}$$

Consider two common cases:

(1)   High-energy rotamers such as eclipsed ethane often experience repulsive forces which tend to lengthen the rotating bond and widen vicinal bond angles. If the eclipsed geometry is not optimized, $\sum_i R_i(dE/dR_i)$ will be *negative* and the kinetic energy will be overestimated for the high-energy rotamer. Similar reasoning applies if there are attractive interactions in the ground rotamer. Failure to optimize geometry results in a positive virial, and the ground conformer kinetic energy is underestimated. Either situation would explain why ΔT is too positive for most calculations made in the rigid-rotor approximation. Reports of geometry-optimized energy components for ethane,[41,213] propane,[97] and hydrogen peroxide[39] support the present claim that geometry optimization will improve ΔT.

(2)   Opposite behavior is expected for inversion barriers. Calculations on NH_3, CH_3^-, and PH_3 all indicate a slight contraction of bond lengths in the planar transition state. If one fails to optimize the geometry, there will be residual inward forces on outer atoms and $\sum_i R_i(dE/dR_i)$ will be positive. Use of rigid inversion geometries, therefore, underestimates the transition state kinetic energy; ΔT for inversion barriers is too negative if geometries are not optimized. This expectation has been corroborated by the PH_3 results of Lehn and Munsch,[57] who found that $\Delta T = -124.8$ kcal/mole without geometry optimization   [with   a $(12, 9, 2/5, 1) \rightarrow (6, 4, 2/3, 1)$ basis]   but $\Delta T = -41.8$ kcal/mole after geometry optimization.

Numerous attempts have been made to balance the energy components by scaling the wave function and nuclear coordinates to force satisfaction of the virial theorem. The concept of virial scaling was invented by Löwdin,[218] who observed that the virial theorem can be satisfied if all electronic and nuclear coordinates are scaled by a factor $\eta = -V/2T$, wherein V and T are the unscaled expectation values. The scaling algorithm is $[R] \rightarrow [R' = \eta^{-1}R]$. There are two objections to the use of scaled energy components. First, scaling destroys the self-consistency of the molecular orbitals. The scaled orbitals are not eigenfunctions of the scaled Fock matrix; so the scaled components are associated with an unphysical charge distribution. Second, scaling does not correspond to geometry optimization. Scaling causes a "breathing mode" expansion of all coordinates, leaving bond angles invariant, whereas bond angle variation is significant for geometry optimization of rotamers. The scaled system has a virial which *averages* to zero, but the geometry-optimized molecule has a virial which is zero term-by-term.

Use of combinations of components can reduce the geometry and basis-set sensitivity. One set of combinations[212] is $\Delta V_{\text{rep}} = \Delta V_{nn} + \Delta T + \Delta V_{ee}$ and $\Delta V_{\text{att}} = \Delta V_{ne}$. It would aid considerably the understanding of barrier mechanisms as well as suggesting possible models for predicting barriers if they could be classified as "repulsive dominant" or "attractive dominant," where the latter is defined as $|\Delta V_{\text{att}}| > |\Delta V_{\text{rep}}|$. Although perhaps not so severe, this combination set still has the problem that ΔV_{att} is a single component. However, attractive or repulsive dominance can be ascertained even without geometry optimization if ΔV_{ne} is sufficiently large, since ΔV_{ne} would override any changes in ΔV_{nn} that might be produced by geometry optimization. For barriers produced by interacting hydrogens, ΔV_{ne} needs to be approximately 100 kcal/mole greater than any other individual component; H_2O_2 is an example. In this case we can definitely say that the *cis* barrier is attractive dominant. This finding complements Pople's Fourier component analysis which indicated that V_2 dominates this barrier. The dominance of V_2 arises from the making and breaking of a partial π bond as the two oxygen p orbitals are rotated with respect to one another.

The relationship between energy component balance and steric effects has been emphasized in a recent article by Eilers and Liberles.[219] They point out that steric attraction can occur if changes in V_{ne} dominate changes in $(V_{nn} + V_{ee})$. Their argument is similar to the one we have given in the example of H_2O_2. Consider rotation from the less crowded to the more crowded conformation. Because the virial theorem is satisfied for both stable and metastable geometries $\Delta E = \frac{1}{2}\Delta V$. Eilers and Liberles reason that steric crowding will always lead to an increase in the magnitudes of V_{ne} and $(V_{nn} + V_{ee})$. If $-\Delta V_{ne} > \Delta(V_{nn} + V_{ee})$ steric attraction occurs and $\Delta E < 0$. If $\Delta(V_{nn} + V_{ee}) > -\Delta V_{ne}$ steric repulsion occurs and $\Delta E > 0$. Conversely, if $\Delta E < 0$ for rotation into the sterically crowded conformer, we may infer that $-\Delta V_{ne} > \Delta(V_{nn} + V_{ee})$. These arguments, of course, are correct only for the exact energy components.

Eilers and Liberles have calculated energy components for various conformers of *n*-butane and cylohexane, using a $(5, 3, 2) \rightarrow (2, 1/1)$ Gaussian basis. For *n*-butane, the *gauche* conformer lies 1.8 kcal/mole above the *trans* conformers, and the (*trans* $\rightarrow$ *gauche*) rotational barrier is 3.8 kcal/mole. Eilers and Liberles state that the *gauche* methyl groups in *n*-butane are sterically attractive, whereas the nonbonded interactions in cyclohexane remain sterically respulsive. These specific conclusions must be regarded as tentative at present because of their use of idealized geometries and a very small basis set. More generally, however, they have brought together spectroscopic and other computational evidence and used the concepts of attractive and repulsive component changes to present a coherent picture of conformational preference in *n*-butane, monosubstituted cyclohexanes, disubstituted ethanes, dithionite, and *trans* and *cis* allenes. This illustrates the opportunities for new chemical insight into conformational analysis potentially available from physically meaningful component combinations. Previous research by Wolfe[220] has already demonstrated this utility in study of the *gauche* effect—the tendency for molecules to adopt that structure which has a maximum number of *gauche* interactions between adjacent electron pairs and/or polar bonds.

Another possible combination set is $\Delta V_{elec} = \Delta V_{ne} + \Delta V_{ee}$ and $\Delta V_{tnn} = \Delta T + \Delta V_{nn}$. This combination retains the reduction in sensitivity inherent to $T + V_{nn}$ and matches this with an expected reduced sensitivity for ΔV_{elec} due to the opposite phase characteristics of V_{ne} and V_{ee}. ΔV_{tnn} may be viewed as a slightly modified nuclear framework potential (in a true Hartree–Fock solution $\Delta E = -\Delta T$ and ΔV_{nn} is much larger). ΔV_{elec} is the change in electronic potential.

This decomposition has the further advantage that the true sign of ΔV_{elec} can be obtained without an *ab initio* calculation if one can reliably estimate the barrier height and ΔV_{nn}. Use of the virial theorem leads to

$$\Delta E = \tfrac{1}{2}(\Delta V_{elec} + \Delta V_{nn})$$

A negative value of ΔV_{elec} indicates the presence of vicinal bonding.

In summary, (a) results in the literature to date—even those using extended bases—show a sensitivity and instability for individual components that usually renders them useless for analysis of barrier mechanisms; (b) in many cases a similar sensitivity is found or expected for combinations of components, but there are not enough cases yet to make a conclusive statement (in some, sufficiently large ΔV_{ne} will allow a definite statement about the attractive or repulsive dominance of a barrier); and (c) it is apparent that lack of geometry optimization has created a major difficulty in the utilization of energy components.

3.5. Hellmann–Feynman Theorems

The electrostatic theorem proposed by Hellmann[221] and Feynman[222] has greatly aided our understanding of chemical bonding since it establishes a correspondence between the quantum mechanical force $-\Delta_N E$ felt by a nucleus N and the classical force due to an electronic distribution $p(n) = \int d^3 r_2 \cdots d^3 r_N |\psi(r_1 \cdots t_n)|^2$:

$$-\nabla_N E = -\nabla_N(V_{nn}) + \eta_e \left[\int d\tau_1 \int d\tau_2 \, d\tau_3 \cdots d\tau_n \right.$$
$$\left. \times [\psi^*(1, 2, \ldots, n)\psi(1, 2, \ldots, n)][(r_1 - R_N)Z_N/r_{1N}^3] \right] \quad (1)$$

This correspondence is significant because it suggests that one can develop a classical description without sacrifice of quantum mechanical rigor. Moreover, the Hellmann–Feyman theorem prescribes how one might analyze molecular binding in terms of interactions between molecular subunits or in terms of individual molecular orbitals.

This conceptual simplicity has stimulated the application of Hellmann–Feynman formulas to the calculation and analysis of rotational barriers. Ruedenberg[223] attempted to rationalize the apparent proportionality between barrier heights and ΔV_{nn}. He assumed that the localized orbitals were conformation independent and "perfectly following." These assumptions have subsequently been shown to be invalid. Subsequent computations have also shown that there is, in fact, no simple relationship between barrier heights and ΔV_{nn}.

Other applications of Hellmann–Feynman formulas to *ab initio* wave functions have taken three general directions. First, the *differential* Hellmann–Feynman theorem, as written above, can be used to calculate the torque for nonequilibrium dihedral angles. Assumption of a shape function for the barrier then leads to an estimate of the barrier height. Second, the Hellmann–Feynman force can be integrated along a path connecting two conformations. This *integrated* Hellmann–Feynman theorem directly gives the barrier height. Third, Parr has proposed an *integral* Hellmann–Feynman theorem[224,225]:

$$(E_B - E_A) \int \psi_B^* \psi_A \, d\tau = \int \psi_B^* (H_B - H_A) \psi_A \, d\tau \tag{2}$$

It may not be initially obvious that Parr's formula is a theorem of the Hellmann–Feynman type. But Kim and Parr[225] have shown that use of a continuous parameter λ connecting structure A and B permits one to prove that the differential Hellmann–Feynman theorem is valid on any interval (λ_1, λ_2) for which the Parr formula holds. To prove that Eq. (2) is valid at $\lambda = \lambda_0 C(\lambda_1, \lambda_2)$ we expand Eq. (1) as a Taylor series about λ_0, take the limit as $\lambda \to \lambda_0$, and identify $\partial/\partial\lambda$ with ∇_N.

The three types of Hellmann–Feynman theorems noted were derived under the assumption that ψ is an exact wave function satisfying the Schrödinger equation. In practice, *ab initio* wave functions and energies $<E>$ are obtained via the variation theorem and do not strictly satisfy a local Schrödinger equation. We now consider the errors that can occur through use of exact Hartree–Fock wave functions, or through use of LCAO-MO approximate Hartree–Fock solutions. Löwdin[218] and Hurley[226] have both demonstrated that LCAO-MO wave functions generally do not satisfy the Hellmann–Feynman theorem. Löwdin has proven that expansion of the exact eigenstate in a finite set of basis functions necessarily introduces a derivative of the overlap integral into the Hellmann–Feynman formula:

$$\frac{\partial E}{\partial \alpha} = \left\langle \psi \left| \frac{\partial H}{\partial \alpha} \right| \psi \right\rangle - E \left\langle \psi \left| \frac{\partial S}{\partial \alpha} \right| \psi \right\rangle$$

Hurley has shown that the differential Hellmann–Feynman theorem is satisfied by a variationally determined wave function only if the parameter with respect to which E is differentiated is itself a valid variational parameter for ψ. An exact Hartree–Fock solution does not explicitly involve nuclear coordinates in the description of the electronic wave function, so Löwdin's and Hurley's theorems both predict satisfaction of the electrostatic theorem, Eq. (1). However, the basis functions of an LCAO wave function do depend on nuclear coordinates as parameters. Hurley's theorem applies and the differential Hellmann–Feynman theorem will not be satisfied unless basis function origin is made a variational parameter (in which case the basis orbitals no longer parametrically depend on nuclear coordinates). Löwdin's equation also proves that the differential Hellmann–Feynman theorem will break down for LCAO wave functions, because the overlap terms $\nabla_N S$ will not vanish. But Löwdin's equation does permit estimation of the Hellmann–Feynman error via $E\langle \psi | (\partial S/\partial \alpha) \psi \rangle$.

A similar analysis of the integral Hellmann–Feynmann theorem's validity has been developed by Epstein *et al.*[235] They have examined the conditions under which the energy difference $E_B - E_A$, predicted by the integral Hellmann–Feynman formula will reduce to the difference in expectation values $\langle \psi_B | H | \psi_B \rangle - \langle \psi_A | H | \psi_A \rangle$. They prove that optimization of both ψ_A and ψ_B within the same variational space is sufficient to guarantee agreement between the expectation value differences and the integral Hellmann–Feynman ΔE. In other words, *counterpoised* LCAO wave functions should satisfy the integral Hellmann–Feynman theorem. Otherwise, LCAO–Hartree–Fock wave functions need not satisfy the integral Hellmann–Feynman theorem. Musher[236] has criticized the usefulness of Hellmann–Feynman theorems on the grounds of a perturbation analysis using heuristic estimates of terms indicating potential errors of several hundred kilocalories per mole. In view of the rigorous theorems noted above, his estimates are not required and appear exaggerated.

When a counterpoised basis set is not used, error will be incurred by application of Hellmann–Feynman procedures to LCAO–MO–SCF wave functions. We now consider the accuracy of Hellmann–Feynman formulas in the context of computational experience. Inspection of the data in Table 10 proves that the integral Hellmann– Feynman theorem cannot reproduce the ΔE_{SCF} derived from a minimal basis wave function, but this formula has not yet been applied to extended basis wave functions. One would expect better

Table 10. Hellmann–Feynman Calculations of Barriers (kcal/mole)

Molecule	Barrier calculated	ΔE_{SCF}	Wave function	Type of calculation	Reference
NH_3 inversion	−13.8	10.5	Minimal STO[a] experimental geometry	Integral	227
	−16.0	11.6	Minimal STO[a] optimized geometry	Integral	227
	−4.22	−2.14	Double-zeta STO[a]	Integral	227
	12.5	0.0	Double-zeta STO[a] and bond polarization	Integral	227
CH_3OH	0.68	1.35	Gaussian lobe $(10, 5/5) \rightarrow (3, 1/1)$	Integral	228
CH_3NH_2	1.21	2.42	Gaussian lobe $(10, 5/5) \rightarrow (3, 1/1)$	Integral	228
CH_3CH_3	1.39	2.52	Gaussian lobe $(10, 5/5) \rightarrow (3, 1/1)$	Integral	228
	2.38	3.3	Minimal STO (Pitzer–Lipscomb, 1963)	Integral, rotate one CH_3 60°	229
	2.27	3.3	Minimal STO (Pitzer–Lipscomb, 1963)	Integral, rotate one CH_3 ±30°	229
	1.7	—	Nonvariational bond orbital, fit centroid	Differential, single point	230
	1.9	—	Nonvariational bond orbital, zero force in methane	Differential, single point	230
	0.9	—	Nonvariational bond orbital, fit populations	Differential, single point	230
	1.71	2.52	Gaussian lobe $(10, 5/5) \rightarrow (3, 1/1)$	Differential, single point	231
	2.7	2.6	Minimal STO, bond orbitals	Integrated	70
H_2O_2	*cis* 18.9	11.9	Minimal STO[a]	Integral	227
	trans −12.3	2.3			
	cis–trans diff. 24.5	16.5	Gaussian lobe $(5, 2/2) \rightarrow (3, 1/1)$	Integral	227, 228
	cis–trans diff. 32.6	13.1	Palke–Pitzer STO [*J. Chem. Phys.* **46**, 3948 (1967)]	Integral	232
	cis 11.8	—	Minimal STO bond orbitals optimized for water	Integral	234
	trans 1.0	—			

[a]Wave functions from Kaldor and Shavitt, *J. Chem. Phys.* **44**, 1823 (1966).

results than in the minimal basis case, on the grounds that extension of basis size will tend to mitigate dependence on nuclear coordinates. And, of course, a counterpoised minimal basis calculation will yield perfect agreement between the integral Hellmann–Feynman theorem and ΔE_{SCF}. It may be significant that the integral Hellman–Feynman formula appears to work best when the molecule has a high degree of azimuthal symmetry (e.g., ethane), because then the basis for one conformer closely resembles the basis for another conformer and a situation of *accidental counterpoise* is approached.

Whether or not rotational barriers can be reliably predicted by integration of the differential Hellmann–Feynman theorem is still an open question. The literature contains only two reports of such calculations. Sovers *et al.*[70] and Zülicke and Spangenberg[233] have calculated the Hellmann–Feynman forces using determinants built from symmetrically orthogonalized bond orbitals. In each case, the bond orbitals were constructed in a minimal Slater basis and mixing coefficients were variationally optimized. Sovers *et al.* chose an integration path involving dissociation of staggered ethane, rotation into eclipsed ethane at $R_{cc} = \infty$, and rejoining of the eclipsed fragments. They calculate a Hellmann–Feynman barrier height of 2.6 kcal/mole, in excellent agreement with their $\Delta\langle E \rangle = 2.7$ kcal/mole. Spangenberg and Zülicke have integrated the torsional force in hydrogen peroxide along a rigid rotation path between *trans* and *cis* rotamers. Their barrier heights were *cis* = 11.8 and *trans* = 1.0 kcal/mole, again in excellent agreement with the best *ab initio* calculations. More calculations are needed to test this approach.

Goodisman's discouraging results[230,231] on the ethane barrier are in sharp contrast to these. If one assumes a rotational barrier of form $\frac{1}{2}V_3 \cos 3\phi$, the Hellmann–Feynman torque obtained by differentiation is $T = -\frac{3}{2}V_3 \sin 3\phi$. The barrier height V_3 is then obtained from the torque $T(\phi)$, computed at some particular angle. Initially,[230] Goodisman used a nonvariational set of bond orbitals defined on the Pitzer–Lipscomb STO basis set. Bond polarities were chosen to fit SCF atomic populations or charge-density centroids or force constants, and all choices gave barriers between 0.9 and 1.9 kcal/mole. Although the Goodisman bond orbitals were not variationally optimized, that was not the source of error since the Sovers wave function (discussed above) was also defined on the Pitzer–Lipscomb STO basis. The Sovers paper[70] demonstrated that an arbitrary bond polarity can be assumed without destroying the agreement between $\Delta\langle E \rangle$ and the *integrated* Hellmann–Feynman barrier height, and that the particular bond polarities Goodisman assumed would give a barrier of 2.5 kcal from the integrated Hellmann–Feynman formula. Furthermore, Goodisman repeated his calculation using the Fink–Allen[6] LCAO-SCF wave function for ethane at $\phi = 30°$ and, despite the improved wave function, a low barrier was still obtained ($V_3 = 1.7$ kcal/mole). It seems likely that the error in this approach derives from the assumption of a perfectly sinusoidal rotational barrier. The Fink–Allen calculations showed that the ethane barrier is *steeper* than sinusoidal at $\phi = 30°$, and Goodisman's assumption of a sinusoidal barrier caused him to overestimate the contribution of electronic–nuclear torque to the barrier height. Since the electronic contribution *opposes* the barrier, the barriers are too low. It should be noted, therefore, that the poor barrier heights obtained by Goodisman's calculations do not constitute a proper test of the ability to accurately estimate Hellmann–Feynman forces from approximate wave functions.

In this regard it is worthwhile noting the recent promising work of Pulay.[234] He has computed intramolecular forces in excellent agreement with experiment using analytic differentiation of the Hartree–Fock energy expression. But forces calculated using the Hellmann–Feynman expectation value were often in error by an order of magnitude. More results using direct differentiation are eagerly awaited.

Apart from the question of computational accuracy, we now consider the usefulness of the integral Hellmann–Feynman theorem as an analytical tool for rotational barriers. Wyatt and Parr[229,237] have applied the integral Hellmann–Feynman theorem to the ethane rotational barrier and define the transition density $\rho(r) = \psi_B^*(r)\psi_A(r)$. They suggest that the rotational barrier can be understood in terms of the action of ΔH upon selected portions of the transition density. But there are three objections which can be raised against determination of such fragment interaction energies through analysis of local elements of the transition density.

First, in the integral Hellmann–Feynman approach ΔV_{nn} is calculated separately from other energy components and does not depend on the transition density. We know from our discussion of energy components, however, that ΔV_{nn} is quite sensitive to our choice of geometry, and full geometry optimization can change ΔV_{nn} by amounts much larger than the barrier. Therefore, we infer that a similar sensitivity is necessarily implicit within the transition density. The "local" electronic energy change which one might calculate from transition density fragments will be strongly dependent on small changes in molecular geometry.

Second, the definition of the transition density is quite arbitrary. Consider a molecule in two conformations A and B. The integral Hellmann–Feynman theorem will hold regardless of the spatial orientations of conformers A and B. But the transition density will be critically dependent on the orientation of molecule A *as a whole* with respect to molecule B *as a whole*. For an exact wave function, satisfaction of the integral Hellmann–Feynman theorem cannot depend on whether we rotate one end by 60°, as contrasted with rotating both ends 30° in opposite directions but the transition densities will differ. In a formal sense all transition densities are equally correct, but not all transition densities are equally interpretable. If approximate wave functions are used some particular choice of transition density may be more accurate than other choices. Because counterpoised basis sets satisfy the integral Hellmann–

Feynman theorem, we expect that transition densities will be most accurate when nuclei are moved the least.

The third question concerns the relationships of the integral Hellmann–Feynman theorem to local energy concepts. Any attempt to regionally divide the Hellmann–Feynman integrand simply regenerates the global energy. To understand this point more fully consider the complete expression for the integrand[229]:

$$\psi_A^*(T\psi_B) - \psi_B^*(T\psi_A^*) + (\Delta H)\psi_A^*\psi_B = (\Delta E)\psi_A^*\psi_B \tag{3}$$

If this integrand is integrated over a local region R, where R is not the whole space, the kinetic energy integrals do not cancel and we have

$$\Delta E_R = \left[\int_R \psi_A^*(T\psi_B) - \psi_B(T\psi_A^*) + (\Delta H)\psi_A^*\psi_B\right] \bigg/ \left[\int_R \psi_A^*\psi_B\right] = \Delta E \tag{4}$$

Whatever region is chosen, ΔE_R is just the global energy change. From SCF-LCAO determinantal wave functions defined on basis sets $\{u_i^A\}$ and $\{u_i^B\}$, the component of the transition energy associated with basis functions u_i^A and u_j^B may be reasonably defined by a bilinear expansion of the transition density

$$\psi_A^*(r)\psi_B(r) = \sum_{i,j} T_{ij} u_i^{A*}(r) u_j^B(r) \tag{5}$$

so that integration of Eq. (1) over all space yields

$$\Delta E_{ij} = \left[T_{ij}\int u_i^{A*}(r)(\Delta H)u_j^B(r)\,d^3r\right] \bigg/ \left[\int \psi^{A*}(r)\psi^B(r)\,d^3r\right] \tag{6}$$

and one concludes that partitioning of the integral Hellmann–Feynman barrier has meaning only in an orbital context.

3.6. Charge Distributions

3.6.1. *Mulliken Population Analysis*

In this widely used approach the charge distribution is characterized by the LCAO expansion coefficients in a single-determinant molecular orbital wave function,

$$\psi = (2N!)^{-\frac{1}{2}} \det\{\phi_1\phi_1 \cdots \phi_n\phi_n\}$$

where the kth molecular orbital is expanded in the basis χ

$$\phi_k = \sum_i C_{ik}\chi_i$$

and the overlap matrix S_{ij} is given as

$$S_{ij} = \int \chi_i^*(r)\chi_j(r)\,dr$$

With these definitions, the well-known total electronic charge is

$$Q = \sum_{i,j} \sum_{k}^{n} 2 C_{ik}^* C_{jk} S_{ij}$$

and the Mulliken population[238,239] in basis function χ_i is then the arbitrarily partitioned sum over molecular orbitals $Q_i = \sum_k 2|C_{ik}|^2$, with the corresponding overlap population between basis functions $Q_{ij} = \sum_k 2 C_{ik}^* C_{jk} S_{ij}$. Summing these latter two expressions over the basis functions on a given atom yields the "gross atomic changes" and the atomic overlap populations. When applied to rotational barriers, the primary concern is with the differences in Mulliken populations between two conformers. Decreases in gross atomic populations accompanied by increases in atomic overlap populations correspond to charge flow into the interatomic region, and are indicative of improving bonding or weakened repulsion. Conversely, increases in gross atomic populations at the expense of atomic overlap populations generally correspond to increased repulsions or weakened bonding.

A principal drawback of population analysis is its insensitivity to subtle polarizations of the electron density. This occurs because the spatial integration implicit in Mulliken population analysis necessarily averages all electron density shifts localized around a given atom to zero. A second defect of population analysis is its dependence on atomic coordinates, unless one uses a counterpoised basis set for the two conformers of interest. In particular, there is no way to distinguish population changes due to motion of the basis set from population changes due to genuine shifts of electron density. Third, population analysis is usually quite sensitive to the basis set. The difference in a given overlap population between a minimal basis calculation and an extended basis calculation is often an order of magnitude greater than the conformational dependence of that particular overlap population.

Population analysis has been applied to rotational barriers in three different ways. First, Kaufman[240] has suggested that molecular binding energy is qualitatively proportional to the total overlap population. This relationship would imply a negative proportionality between total overlap populations and rotational barrier heights. Second, changes in particular overlap populations or gross atomic populations have been correlated with hyperconjugation, or with special attractions or repulsions. Third, heuristic barrier theories based on the perturbation mixing of local orbitals have been proposed.[241–253] Population analysis affords a qualitative test of such orbital interaction theories. We consider the first two of these in this section.

The proposed proportionality between barrier heights and total overlap populations has now been tested in several cases. *Ab initio* calculations on HSSH,[254] HCOOH,[216] and HONO[217] yield good correlations between rotational barrier shape and total overlap populations. Wagner[102] has examined internal rotation in hydrazine using a minimal $(7, 3/3) \rightarrow (2, 1/1)$ basis. The total overlap population is indeed a maximum for the stable $(\phi = 90°)$ conformer in which nitrogen lone pairs are perpendicular. But the angular variation of the net overlap population is otherwise a poor indicator of the barrier shape. The atomic populations change monotonically from $|\phi = 0°|$ to $|\phi = 180°|$, showing no correlation with the energy minimum at $\phi = 90°$. We conclude from these preliminary studies that the total overlap population can be a qualitative indicator of conformational preference, but that there is no quantitative relationship between barrier heights and total overlap population changes. In particular, there is as yet no evidence that would warrant prediction of barrier magnitudes from total overlap population changes. The overlap population sum rule is likely to fail in the analysis of inversion barriers since bond length contraction in the planar transition state leads to an overlap population maximum.

Hyperconjugation manifests itself through overlap population changes amounting to a few hundredths of a charge as vicinal lone-pair electrons are donated into a bond. In both hydrogen persulfide[254] and hydrazine[102] the population changes are dominated by an increase in the heavy-atom overlap population when lone pairs are aligned with vicinal bonds. As lone pairs donate into the opposing bond there is a noticeable decrease in the NH and SH overlap populations.

A few studies have used population analysis to trace charge-density rearrangement. Lehn and Munsch[57] have reported atomic and overlap populations for phosphine inversion, using both $(12, 9/5)$ and $(12, 9, 2/5, 1)$ basis sets. The population transfer onto phosphorous during inversion appears to be quite basis sensitive. With the smaller basis, each hydrogen loses $0.076e^-$, whereas the loss is only $0.006e^-$ with the polarized basis. The large increase (0.297) in H–H overlap populations during inversion is doubtlessly due to the much shorter H–H distance in the planar state. In the planar intermediate there is a marked population transfer $(0.4e^-)$ from the $3s$ orbital of phosphorous into the p orbital along the C_3 axis. Because this p orbital is orthogonal to the hydrogen basis orbitals, the electron repulsions between P and H should be substantially reduced, thereby explaining the contraction of the P–H bond length in planar phosphine.

Allen and Basch[64] have compared the atomic populations of ethane and ethyl fluoride during internal rotation. As ethyl fluoride becomes eclipsed there is a population shift of -0.0168 off of the hydrogen *cis* to fluorine onto the C–H bonds *trans* to fluorine. There are no population changes on the fluoromethyl group. The charge flow away from fluorine-eclipsed H is clear evidence for a repulsive interaction between F and H.

The rotational barriers in acetaldehyde and propylene provide an interesting test case for population analysis. Liberles *et al.*,[71] using a minimal $(5, 3/2) \rightarrow (2, 1/1)$ basis, report the following group population shifts in acetaldehyde as the barrier is traversed: CH_3 (−0.0010), CH (0.0002), and O (0.0007). Davidson and Allen,[111] using a split-out (10, 5/5) → (4, 2/2) basis, find quite different group population shifts: CH_3 (−0.0068), CH_2 (0.0074), and O (−0.0006). Further, a breakdown of atomic population into σ and π contributions[111] shows a charge gain (0.0340) on the in-plane methyl C and H; the group population loss comes from a net population drop (−0.0408) on the out-of-plane methyl hydrogens. All methyl fragments apparently experience strong repulsions when *cis* to carbonyl, the stronger π repulsion being confor- mation determining. These results are in contradiction to the charge-density difference plots (see Section 3.6.2). In propene, Unland *et al.*,[109] using a (5, 2/3) uncontracted basis, find a decline of 0.0052 in the σ–H–CH_2 overlap population at the barrier top indicating an attractive H eclipsed-CH_2 interac- tion. But the net overlap population between out-of-plane methyl hydrogens and nonbonded carbons also falls by 0.0038, indicating stronger π–π repul- sions. Liberles *et al.*,[71] using a minimal (5, 3/2) → (2, 1/1) basis, have par- titioned the net atomic populations for propene into σ and π contributions. They find a π population transfer of 0.0042 from methyl H to methyl C in the H-eclipsed-H-conformer, which suggests a barrier interpretation based on π–π repulsions. On the other hand, net group populations show that shifts in σ density are of comparable magnitude to shifts in π density, suggesting that an argument based solely on π–π repulsions is unrealistic The population analysis results for acetaldehyde and propene demonstrates that population analysis cannot detect intricate polarization of charge, and may in fact engender erroneous interpretations of the rotational barrier.

Howell and Van Wazer[104] have reported a population analysis for the N_2O_4 barrier. These data show that an antisymmetric combination of nitrogen lone pairs donates charge into the N–N σ^* orbital, thereby stabilizing the planar N_2O_4 conformer. The mixing of lone-pair character into the N–N antibonding orbital also explains the unusually long N–N bond length found in planar N_2O_4 only.

3.6.2. *Charge-Density Difference Plots*

Charge-density difference diagrams for rotational barriers were first reported by Pitzer and Lipscomb[181] in 1963. They computed charge densities *ab initio* for staggered and eclipsed ethane using a minimal basis of Slater-type orbitals. Their data show that staggered ethane has a greater electron density than eclipsed ethane at the center of the carbon–carbon bond. The staggered excess is under 0.0005 e^-/a.u.^{3}. They find a comparable relaxation of electron density away from the interior side of eclipsed CH bonds toward the ends of the molecule.

The Pitzer–Lipscomb calculation has been criticized on the grounds that some of the integrals were done too crudely. Jorgensen and Allen,[272,273] therefore, recomputed the charge-density difference analysis using the more accurate wave functions of Fink and Allen.[6] Their results are in close agreement with those of Pitzer and Lipscomb, despite a 25% difference in the computed rotational barriers. In order to test the basis-set sensitivity of charge-density difference plots, Payne and Allen[257] have computed such plots from 4-31G wave functions for staggered and eclipsed ethane. Their data are in excellent agreement with earlier work despite replacement of the minimal basis by a split-valence basis. We conclude that charge-density difference plots for eclipsed minus staggered ethane are rather insensitive to the basis-set choice.

The papers of Allen *et al.*[255-257] examine the ethane charge distributions in somewhat more detail than the calculation of Pitzer and Lipscomb. In addition to relaxation across the CH bond, the charge density in eclipsed ethane is squeezed into the "holes" between geminal CH bonds. The charge-density difference results for ethane support the concept of Pauli repulsion. Specifically, the electron densities of eclipsed bonds act as if they cannot overlap, and in fact bend backward as the C–H bonds are brought into proximity. Payne and Allen[258] have recently calculated charge-density difference maps for eclipsed and staggered conformers of propane and ethylamine, using 4-31G level wave functions. The patterns are qualitatively similar to those for ethane: electron density is displaced from the center of the rotation axis and bonds appear to bend backward in the eclipsed conformer. We conclude that the rotational barrier mechanisms are parallel in these three molecules. All are controlled by Pauli repulsion.

Absar and Van Wazer[147] have determined electron-density difference maps for methylphosphine using a (9, 5, 1/5, 2, 3) uncontracted basis. They find a partial inversion of the phosphorous lone pair in the eclipsed conformation This result is similar to the partial inversion of the nitrogen lone pair in eclipsed ethylamine, found by Payne and Allen.[258]

Jorgensen and Allen[255] have calculated density difference maps for eclipsed (H eclipsing O) minus staggered (H eclipsing H) acetaldehyde. These data show a differential buildup of charge between hydrogen and oxygen in the stable eclipsed conformer. Such a buildup supports the belief that the acetaldehyde rotational barrier is due to attractive stabilization of the eclipsed conformer. The Jorgensen–Allen analysis is complemented by the work of Unland *et al.*[109] on propene internal rotation. Charge-density difference plots for H-eclipsed-H-conformer minus the H-eclipsed-C-conformer show that the H-eclipsing-C interaction is attractive, whereas the H-eclipsing-H interaction is repulsive.

Payne and Allen[258] have used charge-density difference analysis to elucidate the mechanism of substituent effects on rotational barriers. Difference maps between corresponding conformers of propane and ethane, or ethylamine and ethane, demonstrate a charge-density buildup in the carbon–carbon bond adjacent to the methyl or amine substituents. There is a proportionality between these increases in charge density and the increased barriers for internal rotation about this carbon–carbon bond. The Pauli repulsion of bonded distributions of charge which we have observed in density difference plots leads to a single conceptual explanation of many rotational barriers. Because functional groups are imperfectly shielded by their electron complement, electrons outside the group will move in a weak attractive potential. On electrostatic grounds, the electron distributions on functional groups are weakly polarized toward one another. Pauli repulsion of eclipsed bonds forces the charge density to redistribute against this electrostatic gradient, thereby creating an energy barrier. A charge-density buildup in the CC bond of propane or ethylamine is indicative of the strength of mutual polarization, so the Pauli repulsion model correctly predicts that stronger polarizations are associated with higher rotational barriers.

4. Semiempirical Models

In this section we discuss results from semiempirical calculations. Although the principal focus of this volume is *ab initio* techniques, a number of important ideas concerning the origin of barriers have come from very simple models. These include correlation diagrams and orbital mixing concepts obtained from Hoffmann's extended Hückel theory[259] and recent use of simple effective potentials by Benson,[260–262] Scheraga,[263–265] and Stillinger.[266,267]

4.1. Orbital Interaction Models

Using extended Hückel theory, Salem[247] has demonstrated that orbital symmetry rules qualitatively describe what happens when weakly interacting fragment orbitals mix according to the rules of perturbation theory. Fujimoto and Hoffmann,[250,252] Müller,[248] and Epiotis[244–246] have developed the perturbation theory of molecular conformation within this framework.

Perturbation theory based on the extended Hückel model is strictly valid, of course, only for a one-electron Hamiltonian but its concepts have been applied to the analysis of *ab initio* rotational barriers. Lowe[241–243] has explained the ethane rotational barrier in terms of the mixing of two doubly occupied π-type methyl orbitals. This mixing produces a bonding π orbital and an antibonding orbital with the antibonding combination destabilized more than the bonding combination is stabilized. Lowe therefore argues that eclipsed ethane has higher energy than staggered ethane because destabilizing π–π interactions are stronger in the eclipsed case. However, Epstein and Lipscomb[213] point out that Lowe's explanation of the ethane barrier requires that one account for the staggered–eclipsed energy difference solely in terms of

π-electron energies, while *ab initio* calculations show that all orbital energies change by magnitudes much greater than the barrier. Thus, there is no clear dominance of π orbitals.

Cremer *et al.*[253] have used an orbital interaction model to rationalize the ordering of rotational barriers in $(Me)_2X$ systems, and also to account for the differences between methyl rotational barriers in $MeXH$ and $(Me)_2X$. Each methyl group contains two electrons of π symmetry. If X is a 2-electron π donor, the doubly staggered $(Me)_2X$ conformer has six π-like electrons in a ring and is stabilized by π donation from X. This stabilization should show up as an increase in overlap population between the methyl hydrogens. Conversely, if X is a π-electron acceptor, the doubly staggered $(Me)_2X$ conformer has four π-like electrons in a ring and is antiaromatically destabilized. The π acceptor character of X will be manifest as a decrease in overlap population between the methyl hydrogens. Thus, when X is a π donor the barrier increment $\Delta E[(Me)_2X] - \Delta E[MeXH]$ should be large; when X is a π acceptor, the increment should be small. Cremer *et al.* calculated rotational barriers for $(CH_3)_2X$ and CH_3XH using the 4-31G basis and standard geometries $X = (CH_2, NH, O, C, C=CH_2, C=O, B-H, C^+-H)$. Their data show a good linear relationship between barrier increments and eclipsed–staggered differences in intermethyl π overlap population. (The intermethyl π overlap population is defined as the overlap population between two out-of-plane hydrogens on the same side of the molecule.) The linearity suggests that attractive $\pi-\pi$ interactions are responsible for the variation in barrier heights. However, recent charge-density difference plots constructed by Payne and Allen[257,258] for propane minus its constituent atoms show that charge is actually withdrawn from the intermethyl region into a tight zone around each methyl group. Further, the charge redistribution which occurs during rotation does not appear to involve attractive $\pi-\pi$ interactions. Instead, Cremer's results reflect the distance dependence of the overlap integral, so that further analysis is required to fully understand this problem.

Orbital interaction theories have also been invoked to explain *ab initio* rotational barriers in propylene,[249] aminoketones,[251] and fluoroethyl and chloroethyl cations.[268] For the 1-haloethyl cations, Hehre and Hiberty[268] observe that the methyl group and halogen lone pairs contribute 4 electrons of π symmetry, thereby creating an anti-aromatic system. The $\pi-\pi$ interaction theory predicts destabilization of the conformation in which out-of-plane methyl hydrogens are *cis* to halogen. However, their *ab initio* calculations, using STO-3G and 4-31G basis sets, show that there is no conformational preference.

It is instructive to analyze in detail the application of $\pi-\pi$ interaction theory to propene and acetaldehyde. We summarize the argument of Hehre and Salem.[249] As shown in Fig. 2 the interaction of two doubly occupied local π orbitals to form molecular π orbitals will be net destabilizing. Secondary

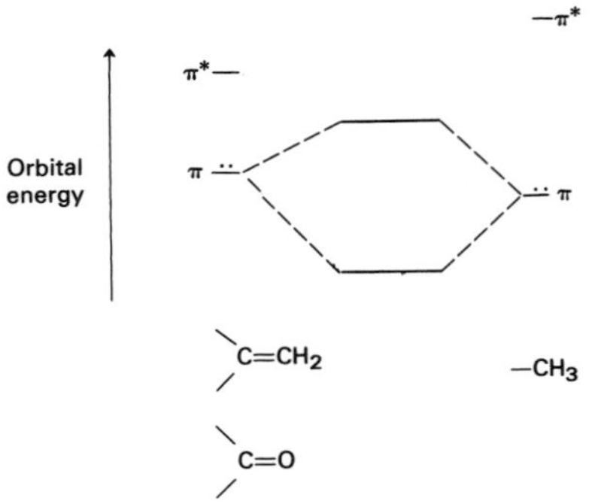

Fig. 2. Interaction of fragment π orbitals to form molecular π orbitals.

$\pi-\pi^*$ interactions will stabilize the occupied π levels. The conformational preference is then determined by the balance of these effects. Acetaldehyde and propylene are both unstable in the conformation which has the methyl π electrons *cis* to the double bond (a structure in which the occupied $\pi-\pi$ interactions are in fact strongest). This clever and reasonable analysis needs further study to make it a fully quantitative procedure. The barrier magnitude is a competition between energy level ordering and the different orbital amplitudes on O vs C in acetaldehyde. We use extended Hückel levels published by Jorgensen and Salem[269] with the π and π^* levels for C=O taken from formaldehyde, those for C=C taken from ethylene, and those for methyl taken from methane or pyramidal methyl radical. The π^* energies for C=O and C=C are nearly equal, but the C=O π level is intermediate between the C=C π energy and the methyl π energy. Thus for the four-electron term, the $\pi-\pi$ interactions will be stronger in acetaldehyde, leading to greater destabilization and hence to a higher barrier than in propene. On the other hand, there is a reduction in the destabilization of the two-electron term for the staggered conformation because of the low amplitude of O in π^*_{CO}. This latter effect implies a lower barrier for acetaldehyde than propene.†

4.2. Dominant Orbital Theories and Walsh–Mulliken Diagrams

Dominant orbital theories are distinguished by the claim that rotational barriers and inversion barriers can be explained by changes in a few orbital energies (usually not more than three). They are closely related to the π-interaction theories just discussed, since sufficiency of π-interaction theory would imply a dominant orbital concept involving π orbitals only. In connection with *ab initio* calculations, dominant orbital ideas have been investigated for the inversion barrier in phosphine,[57] and for rotational barriers in ethane, methylamine, and methanol,[6,79] allene and ethylene,[123] methylphosphine,[147] aminophosphorane,[270] allene,[125] and propene.[109] Fink and

†See note 3 of Notes Added in Proof, page 108.

Allen[79] showed that the Walsh–Mulliken orbital sum $\sum_i \varepsilon_i$ yields the wrong conformation and an upside-down curve of barrier height vs dihedral angle for hydrogen peroxide. The Walsh-Mulliken criteria also predicts a planar rather than pyramidal configuration for phosphine.

There are two reasons dominant orbital ideas cannot contribute to our quantitative understanding of rotational barriers. First, changes in the one-electron energies of nondominant orbitals often exceed the magnitude of the rotational barrier. Second, the electron repulsion energy which couples Hartree–Fock orbital energies to each other may change as much as the orbital energies.

4.3. Empirical Potentials

Up to this point, we have only considered quantum mechanical approaches to internal rotation and inversion barriers. But rotational barriers and inversion barriers can also be determined from empirical potential functions, with an accuracy which is quite competitive with *ab initio* methods. These empirical approaches are generally designated as molecular mechanics. A detailed discussion of molecular mechanics lies outside the scope of this review. We nonetheless present a brief survey of some important recent developments.

Lassettre and Dean[271] suggested in 1949 that empirical potential functions might be useful in conformational analysis. In particular, they suggested that barriers to internal rotation could be partitioned as

$$\Delta E = \Delta U_{tor} + \Delta U_{vw} + \Delta U_e$$

where ΔU_{tor} is some torsional component transferable from molecule to molecule, ΔU_{vw} is a nonbonded van der Waals potential, and ΔU_e is a nonbonded electrostatic energy. Scott and Scheraga[265] later developed a comprehensive set of empirical potentials especially suited to the conformational analysis of proteins and Lifson and Warshel invented the concept of a consistent force field.[272,273] Instead of using different potential functions to describe conformations, thermodynamic properties, and vibrational spectra, Lifson and Warshel developed a set of empirical potentials which *simultaneously* fit conformations, enthalpies of formations, and vibrational spectra.

Separation of the barrier into torsional, van der Waals, and electrostatic terms of van der Waals and electrostatic interactions. Transferability has been whatever is left over after comparison of the barrier with $(\Delta U_{vw} + \Delta U_e)$. But the central claim of molecular mechanics is that ΔU_{tor} is transferable between similar molecules, so that variations in barrier heights can be explained solely in terms of van der Waals and electrostatic interactions. Transferability has been computationally tested on a number of hydrocarbon barriers, and barriers accurate within a few tenths kcal/mole were found in all cases. Quantum mechanical models of barriers should take serious note of the assertion that there is an inherent torsional component associated with the rotating bond, a torsional component which does not depend on the particular substituents.

Recently, some researchers have used an empirical set of pairwise-additive interatomic potentials to define conformational energies. This approach necessarily describes the torsional barrier solely in terms of non-bonded interactions between vicinal substituents or bonds. A barrier theory based on interatomic potentials is superficially incompatible with the consistent force field method. The latter assigns a specific torsional component to the bond about which rotation occurs and neglects substituents. The former neglects the bond about which rotation occurs and explains the barrier solely in terms of substituent interactions.

An early attempt to define interatomic potentials was made by Scheraga,[263] who adopted a point-charge model based on partitioning of atomic charges (the PEM method). This method is comparable in accuracy to extended Hückel theory. Correct conformations are obtained, but energies are not quantitatively reliable. A variant on this procedure (the EPEN method) has been proposed by Shipman, Burgess, and Scheraga.[264] In the EPEN method, the conformational energy is decomposed as

$$E = U_e + U_{or} + U_d$$

where U_e is the electrostatic energy, U_{or} is the overlap repulsion energy, and U_d is a dispersion energy between heavy atoms. To calculate the electrostatic energy, Shipman *et al.* represent each electron pair by a point charge Q_i located along the bond- or lone-pair axis. The precise displacements are chosen by fitting dipole moments in H_2O, NH_3, CH_3OH and CH_3NH_2. The overlap repulsion term

$$U_{or} = \sum_i \sum_j Q_i Q_j A_{ij} \exp^{-B_{ij} R_{ij}}$$

is fit from crystal structures and from rotational barriers in ethane, propane, and *n*-butane. The dispersion term is fit to small molecule barrier heights and lattice energies. Computational tests of EPEN for barriers other than those used in parametrization are not yet available, so no practical assessment can be made regarding EPEN's merits.

Benson and Luria[260–262] have developed an electrostatic point-charge model which explains the enthalpies of a wide variety of hydrocarbons with spectacular success. They write the enthalpy of formation as

$$\Delta H_f = \sum_i^{\text{bond}} \Delta H_f^i + E_e + E_{NB}$$

where ΔH_f^i is a standard enthalpy of formation for the *i*th bond (defined by Benson and Luria and assumed transferable), E_e is an electrostatic interaction energy, and E_{NB} is a nonbonded interaction term. The electrostatic term is evaluated by placing point charges on the atoms, and the point charges are chosen so as to reproduce $\Delta(\Delta H)$ for certain isomerization reactions. For saturated hydrocarbons, each hydrogen atom bears a formal charge of $q = +0.0581$ a.u., and each carbon bears a charge which neutralizes the total charge on the attached hydrogen. Nonbonded interactions are evaluated from a potential proposed by Huggins[274]:

$$V_{HH} = (4.0 \times 10^5)e^{-5.4r} - 47r^{-6} - 98r^{-8} - 205r^{-10} \qquad \text{kcal, Å}$$

In general, Benson and Luria, using an idealized geometry, are able to fit the heats of formation of many hydrocarbons within 0.2 kcal/mole. Further, they give special attention to internal rotation in ethane and butane. The electrostatic contribution to the ethane barrier is 0.03 kcal/mole, whereas the Huggins nonbonded potential taken alone leads to a 3 kcal/mole barrier. The Huggins potential also successfully predicts the magnitude of the *trans–gauche* energy separation in *n*-butane. The electrostatic contribution is 0.15 kcal, but twisting of the end methyl groups by a few degrees leads to Huggins potential changes of about 1 kcal/mole.

Again, the quantum chemist is confronted with a special challenge. Benson's nonquantum mechanical results are too accurate to be fortuitous. His concept of a standard bond enthalpy corresponds to the transferability of Fock matrix fragments for a particular kind of bond. Electrostatic and nonbonded energy terms then add to the Fock matrix in such a way that their contributions to the energy converge in the first order of perturbation. Identification of Benson's model with a quantum formulation may help extend this appealing model to molecules containing heteratoms.

Interatomic potential functions have also been developed by Stillinger *et al.*[266,267] in conjunction with simulations of the liquid state. These potentials lead to an NH_3 inversion barrier of 4.1 kcal/mole, and shortening of the NH bond during inversion is correctly predicted. This description is in good agreement with both experimental and *ab initio* results. Two aspects should be noted however. First, it is known from *ab initio* data that charge migrates onto N during inversion. This migration should change the effective potential between N and H atoms. Second, Stillinger has noted[266] that a similar potential predicts a spurious *linear* structure for the water molecule.

To what extent are point charge models able to accurately represent the electrostatic contribution to conformational energy? This question has been addressed by Rein *et al.*[275] and by Tait and Hall.[276] Rein *et al.* partition the LCAO charge density onto atoms in the sense of Mulliken population analysis, and then use these atomic-charge densities to define atomic multipole moments. For H_2O_2 and CH_3OH, the monopole (point charge) approximation leads to an error of about 10% in the electrostatic interaction. On the other hand, the total electrostatic energy has converged within 1% of its actual value if the series

$$E_{el} = \sum_n \sum_{ij} A_{ij}^{(n)} r_{ij}^{-n}$$

is truncated at r^{-6}. Further conformational differences in electrostatic energy converge by $n = 4$. We conclude that it is inadequate to place a single point charge on each atom. But if one elects to make a multipole expansion about each atom, the electrostatic energy will rapidly converge.

Tait and Hall[276] have complemented Rein's work. They constructed FSGO bond orbital wave functions for LiH, CH_4, and H_2O and approximated the charge density by

$$p(r) = \sum_{ij} P_{ij}\phi_i(r)\phi_j(r) \approx \sum_{ij} P_{ij}S_{ij}\delta^3(R - r_{ij})$$

where r_{ij} is the maximum point of the basis product $\phi_i(r)\phi_j(r)$. The point-charge potential converges to the exact potential within $r = 3$ a.u., indicating that a multipole expansion would rapidly converge.

The discussion indicates that the electrostatic model of Benson and Luria may be less accurate when heteratoms are present in the molecule. For hydrocarbons, the charge distribution is rather uniform and is nearly isotopic around each atom. Uniformity means that the formal atomic charge will be small so that the total electrostatic energy will be only a few kcal/mole. Isotropy means that the atomic multipole moments, as defined by Rein, will be smaller than in H_2O_2 or CH_3OH (Rein's test cases), and we thus expect the monopole approximation to give an error much less than 10% of the total electrostatic energy. Therefore, it is reasonable that Benson and Luria are able to describe the electrostatic contributions to hydrocarbon enthalpies within an accuracy of 0.2 kcal/mole. But heteratoms will have greater formal charges, and atomic multipoles will become much more important in the analysis of electrostatic energy.

By way of summary, we note three ways in which *ab initio* theory can complement molecular mechanics. First, there is still much disagreement about the proper form for empirical interatomic potentials and Stillinger's work shows that conformational properties can be quite sensitive to the exact shape of the empirical potentials. *Ab initio* theory has important contributions to make in the standardization of nonbonded potential functions. Second, *ab initio* results can help compare and evaluate the claims of disparate models. Third, *ab initio* calculations can provide results for important cases that may not be easily accessible by experimental techniques.

References

1. J. A. Pople, Molecular orbital studies of conformation, *Tetrahedron* **30**, 1605–1615 (1974).
2. A. Golebiewski and A. Parczewski, Theoretical conformational analysis of organic molecules, *Chem. Rev.* **74**, 519–530 (1974).
3. D. T. Clark, Theoretical organic chemistry and ESCA, *Annu. Rep. Prog. Chem.*, **69(B)**, 40–83 (1973).

4. A. Veillard, *Ab initio* calculations of barrier heights, *in*: *Internal Rotation in Molecules* (W. J. Orville-Thomas, ed.), pp. 385–421, John Wiley and Sons, New York (1974).

5. W. Orville-Thomas, Internal rotation in molecules, *in*: *Internal Rotation in Molecules* (W. J. Orville-Thomas, ed.), pp. 1–18, John Wiley and Sons, New York (1974).

6. W. H. Fink and L. C. Allen, Origin of rotational barriers. I. Many-electron molecular orbital wavefunctions for ethane, methyl alcohol, and hydrogen peroxide, *J. Chem. Phys.* **46**, 2261–2275 (1967).

7. K. S. Pitzer, Thermodynamic functions for molecules having restricted internal rotations, *J. Chem. Phys.* **5**, 469–472 (1937); **5**, 473–479 (1937).

8. R. M. Pitzer, A Calculation of the Barrier to Internal Rotation in Ethane, Ph.D. dissertation, Harvard University, 1964, University Microfilms, Ann Arbor, Order No. 63-7842; *Diss. Abstr.* **25(2)**, 870 (1964).

9. S. Epstein, *The Variational Method in Quantum Chemistry*, Academic Press, New York (1974).

10. J. Goodisman and W. Klemperer, On error in Hartree–Fock calculations, *J. Chem. Phys.* **38**, 721–725 (1963).

11. K. F. Freed, Geometry and barriers to internal rotation in Hartree–Fock theory, *Chem. Phys. Lett.* **2**, 255–256 (1968).

12. L. C. Allen and J. Arents, Adequacy of the molecular orbital approximation for predicting rotation and inversion barriers, *J. Chem. Phys.* **57**, 1818–1821 (1972).

13. R. E. Stanton, Hellmann–Feynman theorem and correlation energies, *J. Chem. Phys.* **36**, 1298–1300. (1962).

14. B. Levy and M. C. Moireau, Correlation effect in the rotation barrier of ethane, *J. Chem. Phys.* **54**, 3316–3321 (1971).

15. E. Clementi and H. Popkie, Analysis of the formation of the acetylene, ethylene, and ethane molecules in the Hartree–Fock model, *J. Chem. Phys.* **57**, 4870–4883 (1972).

16. B. Zurawski and W. Kutzelnigg, Correlation energy of the internal rotation barrier in ethane, *Bull. Acad. Pol. Sci., Ser. Soc. Chim.* **22**, 363–366 (1974).

17. R. Ahlrichs and F. Keil, Structure and bonding in dinitrogen tetroxide (N_2O_4), *J. Am. Chem. Soc.* **96**, 7615–7620 (1974).

18. B. Dumbacker, *Ab initio* SCF and CI calculations on the barrier to internal rotation in 1,3-butadiene, *Theor. Chim. Acta* **23**, 346–359 (1972).

19. U. Pincelli, B. Cadioli, and B. Levy, On the internal rotation in 1,3-butadiene, *Chem. Phys. Lett.* **13**, 249–252 (1972).

20. R. Ahlrichs, F. Driessler, H. Lischka, V. Staemmler, and W. Kutzelnigg, PNO–CI (pair natural orbital configuration interaction) and CEPA–PNO (coupled electron pair approximation with pair natural orbitals) calculation of molecular systems. II. The molecules BeH_2, BH, BH_3, CH_4, CH_3^-, NH_3 (planar and pyramidal, H_2O, OH_3^+, HF, and the Ne atom), *J. Chem. Phys.* **62**, 1235–1247 (1975).

21. R. E. Kari and I. G. Csizmadia, Configuration interaction wavefunctions and computed inversion barriers for NH_3 and CH_3^-, *J. Chem. Phys.* **56**, 4337–4344 (1972).

22. A. Pipano, P. R. Gilman, C. F. Bender, and I. Shavitt, *Ab initio* calculations of the inversion barrier in ammonia, *Chem. Phys. Lett.* **4**, 583–584 (1970).

23. R. M. Stevens, CI calculations for the inversion barrier of ammonia, *J. Chem. Phys.* **61**, 2086–2090 (1974).

24. N. C. Dutta and M. Karplus, Correlation contribution to the ammonia inversion barrier, *Chem. Phys. Lett.* **31(3)**, 455–461 (1975).

25. G. H. F. Diercksen, W. P. Kraemer, and B. O. Roos, SCF-CI studies of correlation effects on hydrogen bonding and ion hydration. The systems: H_2O, $H^+ \cdot H_2O$, $Li^+ \cdot H_2O$, $F^- \cdot H_2O$, and $H_2O \cdot H_2O$, *Theor. Chim. Acta* **36**, 249–274 (1975).

26. J. D. Swalen and J. A. Ibers, A potential function for the inversion of ammonia, *J. Chem. Phys.* **36**, 1914–1918 (1962).

27. A. Rauk, L. C. Allen, and E. Clementi, Electronic structure and inversion barrier of ammonia, *J. Chem. Phys.* **52**, 4133–4144 (1970).

28. L. Radom, W. J. Hehre, and J. A. Pople, A systematic study of energies, conformations, and bond interactions, *J. Am. Chem. Soc.* **93**, 289–300 (1971).

29. L. Radom, W. A. Lathan, W. J. Hehre, and J. A. Pople, Internal rotation in 1,2-disubstituted ethanes, *J. Am. Chem. Soc.* **95**, 693–698 (1973).
30. B. Kirtman, Interactions between ordinary vibrations and hindered internal rotation. I. Rotational energies, *J. Chem. Phys.* **37**, 2516–2539 (1962).
31. B. Kirtman, Interactions between ordinary vibrations and hindered internal rotation. II. Theory of internal rotation fine structure in some perpendicular bonds of ethane-type molecules, *J. Chem. Phys.* **41(3)**, 775–788 (1964).
32. D. Papousek, High resolution infrared spectra of ethane-like molecules and the barrier to internal rotation, *J. Mol. Spectrosc.* **28**, 161–190 (1968).
33. J. Susskind, Theory of torsion–vibration–rotation interaction in ethane and analysis of the bond $\nu_{11} = \nu_4$, *J. Mol. Spectrosc.* **49**, 1–17 (1974).
34. J. Susskind, Torsion–vibration–rotation interaction in ethane: the bonds $\nu_{12} + \nu_4$, ν_8, and ν_6, *J. Mol. Spectrosc.* **49**, 331–342 (1974).
35. D. R. Woods, High-Resolution Infrared Spectra of Normal and Deuterated Methanol between $400\ cm^{-1}$ and $1300\ cm^{-1}$, Ph.D. dissertation, University of Michigan, 1970, University Microfilms, Ann Arbor, Order No. 70-21,819.
36. C. S. Ewig, W. E. Palke, B. Kirtman, Dependence of the CH_3SiH_3 barrier to internal rotation on vibrational coordinates: testing of models and effect of vibrations on the observed barrier height, *J. Chem. Phys.* **60**, 2749–2758 (1974).
37. B. Kirtman, W. E. Palke, and C. S. Ewig, private communication, 1975.
38. D. R. Herschbach, Calculation of energy levels for internal torsion and over-all rotation. III, *J. Chem. Phys.* **31**, 91–108 (1959).
39. A. Veillard, Relaxation during internal rotation in ethane and hydrogen peroxide, *Theor. Chim. Acta* **18**, 21–33 (1970).
40. A. Veillard, Distortional effects on the ethane internal rotation barrier and rotation barriers in borazane and methylsilane, *Chem. Phys. Lett.* **3**, 128–130 (1969).
41. P. H. Blustin and J. W. Linnett, Application of a simple molecular wavefunction. Part 2. The torsional barrier in ethane, *J. Chem. Soc., Faraday Trans. 2* **70**, 290–296 (1974).
42. A. A. Frost and Robert A. Rouse, A floating spherical Gaussian orbital model of molecular structure. Hydrocarbons, *J. Am. Chem. Soc.* **90**, 1965–1969 (1968).
43. J. D. Dill, P. v. R. Schleyer, and J. A. Pople, Geometries and energies of small boron compounds. Comparisons with carbocations, *J. Am. Chem. Soc.* **97**, 3402–3409 (1975).
44. W. E. Palke, Calculation of the internal rotation barrier and its derivatives in BH_3NH_3, *J. Chem. Phys.* **56**, 5308–5311 (1972).
45. J. O. Jarvie and A. Rauk, A theoretical study of the conformational changes in hydrazine, *Can. J. Chem.* **52**, 2785–2791 (1974).
46. A. Veillard, Distortional effects on the internal rotation in hydrogen peroxide, *Chem. Phys. Lett.* **4**, 51–52 (1969).
47. R. B. Davidson and L. C. Allen, Rotational barriers in hydrogen peroxide, *J. Chem. Phys.* **55**, 519–527 (1971).
48. T. H. Dunning, Jr. and N. W. Winter, private communication.
49. T. H. Dunning, Jr. and N. W. Winter, Hartree–Fock calculation of the barrier to internal rotation in hydrogen peroxide, *Chem. Phys. Lett.* **11**, 194–195 (1971).
50. J. P. Ranck and H. Johansen, Polarization functions and geometry optimization in *ab initio* calculations of the rotational barrier in hydrogen peroxide, *Theor. Chim. Acta* **24**, 334–345 (1972).
51. L. Radom and J. A. Pople, Internal rotation in hydrocarbons using a minimal Slater-type basis, *J. Am. Chem. Soc.* **92**, 4786–4795 (1970).
52. P. N. Skancke and J. E. Boggs, Molecular orbital studies of conformers and the barrier to internal rotation in 1,3-butadiene, *J. Mol. Struct.* **16**, 179–185 (1973).
53. K. R. Sundberg and L. M. Cheung, Potential energy curve in the *trans–cis* isomerization of glyoxal, *Chem. Phys. Lett.* **29**, 93–97 (1974).
54. D. H. Christensen, R. N. Kortzeborn, B. Bak, and J. J. Led, Results of *ab initio* calculations on formamide, *J. Chem. Phys.* **53**, 3912–3922 (1970).
55. P. A. Kollman and C. F. Bender, The structure of the H_3O^+ (hydronium) ion, *Chem. Phys. Lett.* **21**, 271–273 (1973).

56. F. Driessler, R. Ahlrichs, V. Staemmler, and W. Kutzelnigg, *Ab initio* calculations on small hydrides including correlation. XI. Equilibrium geometries and other properties of CH_3, CH_3^+, and CH_3^-, and inversion barrier of CH_3^-, *Theor. Chim. Acta* **30**, 315–326 (1973).

57. J. M. Lehn and B. Munsch, An *ab initio* SCF–LCAO–MO study of the phosphorous pyramidal inversion process in phosphine, *Mol. Phys.* **23**, 91–107 (1972).

58. E. Zeeck, *Ab initio* calculation of the barrier to free rotation of the methyl group in propylene (in German), *Theor. Chim. Acta* **16**, 155–162 (1970).

59. U. Wahlgren and K. H. Johnson, Determination of the internal rotation barrier in ethane by the SCF-Xα scattered wave method, *J. Chem. Phys.* **56**, 3715–3716 (1972).

60. U. Wahlgren, Calculations of potential barriers using the SCF-Xα method, *Chem. Phys. Lett.* **20**, 246–249 (1973).

61. S. Weiss and G. Leroi, Direct observations of the infrared torsional spectrum of C_2H_6, CH_3CD_3, and C_2D_6, *J. Chem. Phys.* **48**, 962–967 (1968).

62. P. A. Christiansen and W. E. Palke, Ethane internal rotation barrier, *Chem. Phys. Lett.* **31**, 462–466 (1975).

63. W. E. Palke, Calculations of the barrier to internal rotation in ethyl fluoride: a comparison with ethane, *Chem. Phys. Lett.* **15**, 244–247 (1972).

64. L. C. Allen and H. Basch, Theory of the rotational barriers in ethyl fluoride and ethane, *J. Am. Chem. Soc.* **93**, 6373–6377 (1971).

65. L. Radom, W. A. Lathan, W. J. Hehre, and J. A. Pople, Internal rotation in some organic molecules containing methyl, amino, hydroxyl, and formyl groups, *Aust. J. Chem.* **25**, 1601–1612 (1972).

66. R. M. Stevens, Geometry optimization in the computation of barriers to internal rotation, *J. Chem. Phys.* **52**, 1397–1402 (1970).

67. G. F. Musso and V. Magnasco, Localized orbitals and short-range molecular interactions. III. Rotational barriers in C_2H_6 and H_2O_2, *J. Chem. Phys.* **60**, 3754–3759 (1974).

68. R. M. Pitzer, Calculation of the barrier to internal rotation in ethane with improved exponential wavefunctions, *J. Chem. Phys.* **47**, 965–967 (1967).

69. W. von Niessen, A theory of molecules in molecules. II. The theory and its application to the molecules Be–Be, Li_2–Li_2, and to the internal rotation in C_2H_6, *Theor. Chim. Acta* **31**, 111–135 (1973).

70. O. J. Sovers, C. W. Kern, R. M. Pitzer, and M. Karplus, Bond-function analysis of rotational barriers: ethane, *J. Chem. Phys.* **49**, 2592–2599 (1968).

71. A. Liberles, B. O'Leary, J. E. Eilers, and D. R. Whitman, Methyl rotation barriers and hyperconjugation, *J. Am. Chem. Soc.* **94**, 6894–6898 (1972).

72. J. R. Hoyland, *Ab initio* bond-orbital calculations. I. Application to methane, ethane, propane, and propylene, *J. Am. Chem. Soc.* **90**, 2227–2232 (1968).

73. L. Pedersen and K. Morokuma, *Ab initio* calculations of the barriers to internal rotation of CH_3CH_3, CH_3NH_2, CH_3OH, N_2H_4, H_2O_2, and NH_2OH, *J. Chem. Phys.* **46**, 3941–3947 (1967).

74. E. Clementi, H. Kistenmacher, and H. Popkie, On the SCF–LCAO–MO and the SCH-Xα-SW approximations: Computation of the barrier to internal rotation for ethane, *J. Chem. Phys.* **58**, 4699–4700 (1973).

75. J. L. Nelson and A. A. Frost, Local orbitals for bonding in ethane, *Theor. Chim. Acta* **29**, 75–83 (1973).

76. R. E. Christoffersen, D. W. Gensen, and G. M. Maggiora, *Ab initio* calculations on large molecules using molecular fragments. Hydrocarbon characterization, *J. Chem. Phys.* **54**, 239–252 (1971).

77. R. H. Hunt, R. A. Leacock, C. W. Peters, and K. T. Hecht, Internal rotation in hydrogen peroxide: the far infrared spectrum and the determination of the hindering potential, *J. Chem. Phys.* **42**, 1931–1946 (1965).

78. C. Guidotti, U. Lamanna, M. Maestro, and R. Moccia, Barriers to the internal rotation and observables of the ground state for hydrogen peroxide, *Theor. Chim. Acta* **27**, 55–62 (1972).

79. W. H. Fink and L. C. Allen, Origin of rotational barriers. I. Methylamine and improved wavefunctions for hydrogen peroxide, *J. Chem. Phys.* **46**, 2276–2284 (1967).

80. L. Radom, W. H. Hehre, and J. A. Pople, Fourier component analysis of internal rotation potential functions in saturated molecules, *J. Am. Chem. Soc.* **94**, 2371–2381 (1972).

81. W. E. Palke and R. M. Pitzer, On the internal rotation potential in H_2O_2, *J. Chem. Phys.* **46**, 3948–3950 (1967).

82. P. F. Franchini and C. Vergani, SCF calculation with minimal and extended bases sets for H_2O, NH_3, CH_4, and H_2O_2, *Theor. Chim. Acta* **13**, 46–55 (1969).

83. U. Kaldor and I. Shavitt, LCAO–SCF computations for hydrogen peroxide, *J. Chem. Phys.* **44**, 1823–1829 (1966).

84. I. H. Hillier, V. R. Saunders, and J. F. Wyatt, Theoretical study of the electronic structure and barriers to rotation in H_2O_2 and H_2S_2, *Trans. Faraday Soc.* **66**, 2665–2670 (1970).

85. B. V. Cheney and R. E. Christoffersen, *Ab initio* calculations on large molecules using molecular fragments. Oxygen-containing molecules, *J. Chem. Phys.* **56**, 3503–3518 (1972).

86. G. Winnewisser, M. Winnewisser, and W. Gordy, *Bull. Am. Phys. Soc.* **2**, 312 (1966).

87. A. Veillard and J. Demuynck, Barrier to internal rotation in hydrogen persulfide, *Chem. Phys. Lett.* **4**, 476–478 (1970).

88. E. V. Ivash and D. M. Dennison, The methyl alcohol molecule and its microwave spectrum, *J. Chem. Phys.* **21**, 1804–1816 (1953).

89. L. M. Tel, S. Wolfe, and I. G. Csizmadia, Near-molecular-Hartree–Fock wavefunctions for CH_3O^-, CH_3OH, and $CH_3OH_2^+$, *J. Chem. Phys.* **59**, 4047–4060 (1973).

90. C. W. Kern, R. M. Pitzer, and O. J. Sovers, Bond-function analysis of rotational barriers: methanol, *J. Chem. Phys.* **60**, 3583–3587 (1974).

91. S. Rothenberg, Localized orbitals for polyatomic molecules. I. The transferability of the C–H bond in saturated molecules, *J. Chem. Phys.* **51**, 3389–3396 (1969).

92. D. R. Lide, Jr., Structure of the methylamine molecule. I. Microwave spectrum of CD_3ND_2, *J. Chem. Phys.* **27**, 343–352 (1941).

93. D. R. Armstrong and P. G. Perkins, Calculation of the electronic structures and the gas-phase heats of formation of $BH_3 \cdot NH_3$ and BH_3CO, *J. Chem. Soc. A* **1969**, 1044–1048 (1969).

94. J. R. Hoyland, Internal rotation in propane. Reanalysis of the microwave spectrum and quantum-mechanical calculations, *J. Chem. Phys.* **49**, 1908–1912 (1968).

95. P. W. Payne and L. C. Allen, Charge density difference analysis. Comparison of rotational barriers in ethane and propane, submitted to *J. Am. Chem. Soc.* (1977).

96. J. R. Hoyland, Barriers to internal rotation in propane, *Chem. Phys. Lett.* **1**, 247–248 (1967).

97. P. H. Blustin and J. W. Linnett, Applications of a simple molecular wavefunction. I. Floating spherical Gaussian orbital calculations for propylene and propane, *J. Chem. Soc., Faraday Trans. 2* **70**, 274–289 (1974).

98. T. Kasuya and T. Kojima, Internal motions of hydrazine, *J. Phys. Soc. Japan* **18**, 364–368 (1963).

99. W. H. Fink, D. C. Pan, and L. C. Allen, Internal rotation barriers for hydrazine and hydroxylamine from *ab initio* LCAO–MO–SCF wavefunctions, *J. Chem. Phys.* **47**, 895–905 (1967).

100. A. Veillard, Quantum mechanical calculations on barriers to internal rotation. I. Self-consistent field wavefunctions and theoretical potential energy curves for the hydrazine molecule in the Gaussian approximation, *Theor. Chim. Acta* **5**, 413–421 (1966).

101. J. O. Jarvie, A. Rauk, and C. Edmiston, The effect of bond function polarization on the LCAO–MO–SCF calculation of bond angles and energy barriers, *Can. J. Chem.* **52**, 2778–2784 (1974).

102. E. L. Wagner, *Ab initio* versus CNDO barrier calculations. I. N_2H_4 and N_2F_4, *Theor. Chim. Acta* **23**, 115–126 (1971).

103. R. G. Snyder and I. C. Hisatsune, Infrared spectrum of dinitrogen tetroxide, *J. Mol. Spectrosc.* **1**, 139–150 (1957).

104. J. M. Howell and J. R. Van Wazer, Electronic structure of dinitrogen tetroxide and diboron tetrafluoride and an analysis of their conformational stabilities, *J. Am. Chem. Soc.* **96**, 7902–7910 (1974).

105. D. R. Lide, Jr., and D. E. Mann, Microwave spectra of molecules exhibiting internal rotation. I. Propylene, *J. Chem. Phys.* **27**, 868–876 (1957).

106. D. R. Lide, Jr., and D. Christiansen, Microwave structure of propylene, *J. Chem. Phys.* **35**, 1374–1378 (1962).
107. A. D. English and W. E. Palke, Calculation of barriers to internal rotation in propene and monofluoropropenes, *J. Am. Chem. Soc.* **95**, 8536–8538 (1973).
108. E. Scarzafaza and L. C. Allen, Rotational barriers in propene and its fluoro derivatives, *J. Am. Chem. Soc.* **93**, 311–314 (1971).
109. M. L. Unland, J. R. Van Wazer, and J. H. Letcher, *Ab initio* calculation of the barrier to internal rotation in propylene using a Gaussian basis self-consistent field wavefunction, *J. Am. Chem. Soc.* **91**, 1045–1052 (1969).
110. R. W. Kilb, C. C. Lin, and E. B. Wilson, Jr., Calculation of energy levels for internal torsion and over-all rotation. II. CH_3CHO type molecules; acetaldehyde spectra, *J. Chem. Phys.* **26**, 1695–1704 (1957).
111. R. B. Davidson and L. C. Allen, Attractive nature of the rotational barrier in acetaldehyde, *J. Chem. Phys.* **54**, 2828–2830 (1971).
112. N. L. Allinger and Sister M. J. Hickey, Acetone, *ab initio* calculations, *Tetrahedron* **28**, 2157–2161 (1972).
113. W. G. Fately and F. A. Miller, Torsional frequencies in the far infrared. II. Molecules with two or three methyl rotors, *Spectrochim. Acta* **18**, 977–993 (1962).
114. W. G. Fately, R. K. Harris, F. A. Miller, and R. E. Witkowski, Torsional frequencies in the far infrared. IV. Torsions around the C–C single bond in conjugated molecules, *Spectrochim. Acta* **21**, 231–244 (1965).
115. G. N. Currie and D. A. Ramsay, The 4875 Å band system of *cis* glyoxal, *Can. J. Phys.* **49**, 317–322 (1971).
116. U. Pincelli, B. Cadioli, and D. J. David, A theoretical study of the electronic structure and conformation of glyoxal, *J. Mol. Struct.* **9**, 173–176 (1971).
117. T. K. Ha, *Ab initio* calculation of *cis–trans* isomerization in glyoxal, *J. Mol. Struct.* **12**, 171–178 (1972).
118. B. Sunners, L. H. Piette, and W. G. Schneider, Proton magnetic resonance measurements of formamide, *Can. J. Chem.* **38**, 681–688 (1960).
119. H. Kamei, Nuclear magnetic double-resonance study of the hindered internal rotation in formamide, *Bull. Chem. Soc. Japan* **41**, 2269–2273 (1968).
120. M. Penicaudet and A. Pullman, An *ab initio* quantum-mechanical investigation on the rotational isomerism in amides and esters, *Int. J. Pept. Protein Res.* **5**, 99–107 (1973).
121. L. A. Carreira, Determination of the torsional potential function of 1,3-butadiene, *J. Chem. Phys.* **62**, 3851–3854 (1975).
122. P. Th. van Duijnen and D. B. Cook, *Ab initio* calculations with ellipsoidal Gaussian basis sets, *Mol. Phys.* **21**, 475–483 (1971).
123. R. J. Buenker, Theoretical study of the rotational barriers of allene, ethylene, and related systems, *J. Chem. Phys.* **48**, 1368–1379 (1968).
124. J. M. André, M. C. André, and G. Leroy, Barrier to internal rotation in allene, *Chem. Phys. Lett.* **3**, 695–698 (1969).
125. L. J. Schaad, The internal rotation barrier in allene, *Tetrahedron* **26**, 4115–4118 (1970).
126. L. J. Schaad, L. A. Burnelle, and K. P. Dressler, The excited states of allene, *Theor. Chim. Acta* **15**, 91–99 (1969).
127. L. J. Weimann and R. E. Christoffersen, *Ab initio* calculations on large molecules using molecular fragments. Cumulenes and related molecules, *J. Am. Chem. Soc.* **95**, 2074–2083 (1973).
128. R. M. Stevens, Accurate SCF calculation for ammonia and its inversion motion, *J. Chem. Phys.* **55**, 1725–1729 (1971).
129. A. J. Duke, A Hartree–Fock study of the methyl anion and its inversion potential surface: use of an augmented basis set for this species, *Chem. Phys. Lett.* **21**, 275–282 (1973).
130. P. Millie and G. Berthier, SCF wavefunction in Gaussians for methyl and vinyl radicals, *Int. J. Quantum Chem.* **2**, 67–73 (1968).
131. R. E. Kari and I. G. Csizmadia, Potential-energy surfaces of CH_3^+ and CH_3^{-11}, *J. Chem. Phys.* **50**, 1443–1448 (1969).

132. R. E. Kari and I. G. Csizmadia, Near molecular Hartree–Fock wavefunction for CH_3^{-11}, *J. Chem. Phys.* **46**, 4585–4590.

133. R. Grahn, A theoretical study of the H_3O^+ ion, *Arkiv. Fys.* **19**, 1417 (1961).

134. M. Fournier, G. Mascherpa, D. Rousselet, and J. Potier, Assignment of the vibrational frequencies of the oxonium ion, *C. R. Acad. Sci., Ser. C* **269**, 279–282 (1969).

135. J. Lischka, *Ab initio* calculations on small hydrides including electron correlation. IX. Equilibrium geometries and harmonic force constants of HF, OH^-, H_2F^+, and H_2O and proton affinities of F^-, OH^-, HF, and H_2O, *Theor. Chim. Acta* **31**, 39–48 (1973).

136. M. Allevena and E. Le Clech, A conformational study of the H_3O^+ ion by an MO–SCF *ab initio* calculation, *J. Mol. Struct.* **22**, 265–272 (1974).

137. J. W. Moskowitz and M. C. Harrison, Gaussian wavefunctions for the 10-electron systems. III. OH^-, H_2O, H_3O^+, *J. Chem. Phys.* **43**, 3550–3555 (1965).

138. P. A. Kollman and L. C. Allen, A theory of the strong hydrogen bond, *ab initio* calculations on HF_2^- and $H_5O_2^+$, *J. Am. Chem. Soc.* **92**, 6101–6107 (1970).

139. G. Alagona, R. Cimiraglia, and U. Lamanna, Theoretical investigations of the solvation process. III. STO double-Z SCF calculations on the hydrated $H_5O_2^+$, *Theor. Chim. Acta* **29**, 93–96 (1973).

140. J. Almlöf and U. Wahlgren, *Ab initio* studies of the conformation of the oxonium ion in solids, *Theor. Chim. Acta* **28**, 161–168 (1973).

141. M. D. Newton and S. Ehrenson, *Ab initio* studies on the structures and energetics of inner and outer-shell hydrates of the proton and the hydroxide ion, *J. Am. Chem. Soc.* **93**, 4971–4990 (1971).

142. R. E. Weston, Vibrational energy level splitting and optical isomerism in pyramidal molecules of the type XY_3, *J. Am. Chem. Soc.* **76**, 2645–2648 (1954).

143. R. Moccia, One-center basis set SCF MO's. II. NH_3, NH_4^+, PH_3, and PH_4^+, *J. Chem. Phys.* **40**, 2176–2192 (1964).

144. L. J. Aarons, M. F. Guest, M. B. Hall, and I. H. Hillier, Theoretical study of the geometry of PH_3, PF_3, and their ground ionic states, *J. Chem. Soc., Faraday Trans. 2* **69**, 643–647 (1973).

145. A. Rauk, L. C. Allen, and K. Mislow, Pyramidal inversion, *Angew. Chem.* **82**, 453–468 (1970).

146. T. Kojima, E. L. Breig, and C. C. Lin, Microwave spectrum and internal barrier of methylphosphine, *J. Chem. Phys.* **35**, 2139–2144 (1961).

147. I. Absar and J. R. Van Wazer, Rotational barrier and electronic structure of monomethyl-phosphine from *ab initio* LCAO–MO–SCF calculations, *J. Chem. Phys.* **56**, 1284–1289 (1972).

148. J. R. Durig, Y. S. Li, L. A. Carreira, and J. D. Odom, Microwave spectrum, structure, dipole moment, and barrier to internal rotation of phosphine–borane, *J. Am. Chem. Soc.* **95**, 2491–2496 (1973).

149. J. R. Sabin, On the barrier to internal rotation in phosphineborane, *Chem. Phys. Lett.* **20**, 212–214 (1973).

150. D. R. Herschbach, Calculation of energy levels for internal torsion and over-all rotation. III, *J. Chem. Phys.* **31**, 91–108 (1959).

151. C. S. Ewig, W. E. Palke, and B. Kirtman, Dependence of the CH_3SiH_3 barrier to internal rotation on vibrational coordinates: testing of models and effect of vibrations on the observed barrier height, *J. Chem. Phys.* **60**, 2749–2758 (1974).

152. L. L. Shipman and R. E. Christoffersen, *Ab initio* calculations on large molecules using molecular fragments. Characterization of the zwitterion of glycine, *Theor. Chim. Acta* **31**, 75–82 (1973).

153. L. L. Shipman and R. E. Christoffersen, *Ab initio* calculations on large molecules using molecular fragments. Polypeptides of glycine, *J. Am. Chem. Soc.* **95**, 4733–4744 (1973).

154. L. L. Shipman and R. E. Christoffersen, *Ab initio* calculations on large molecules using molecular fragments. Model peptide studies, *J. Am. Chem. Soc.* **95**, 1408–1416 (1973).

155. A. Pullman, G. Alagona, and J. Tomasi, Quantum mechanical studies of environmental effects on biomolecules. IV. Hydration of *N*-methylacetamide, *Theor. Chim. Acta* **33**, 87–90 (1974).

156. L. L. Shipman, R. E. Christoffersen, and B. V. Cheney, *Ab initio* calculations on large molecules using molecular fragments. Lincomycin model studies, *J. Med. Chem.* **17**, 583–589 (1974).

157. J. A. Pople and L. Radom, Internal rotation potentials in biological molecules, *in*: *The Jerusalem Symposium on Quantum Chemistry and Biochemistry, Vol. 5, Conformation of Biological Molecules and Polymers* (E. D. Bergmann and B. Pullman, eds.), Academic Press, New York (1973).

158. A. Pullman and G. N. J. Port, An *ab initio* SCF molecular orbital study of acetylcholine, *Theor. Chim. Acta* **32**, 77–79 (1973).

159. G. N. J. Port and A. Pullman, Acetylcholine, gauche or trans? A standard *ab initio* SCF investigation, *J. Am. Chem. Soc*, **95**, 4059–4060 (1973).

160. D. W. Genson and R. E. Christoffersen, *Ab initio* calculation on large molecules using molecular fragments, electronic and geometric characterization of acetylcholine, *J. Am. Chem. Soc.* **95**, 362–368 (1973).

161. R. E. Christoffersen, D. Spangler, G. G. Hall, and G. M. Maggiora, *Ab initio* calculations on large molecules using molecular fragments. Evaluation and extension of initial procedures, *J. Am. Chem. Soc.* **95**, 8526–8536 (1973).

162. G. N. J. Port and B. Pullman, An *ab initio* SCF molecular orbital study on the conformation of serotonin and bufotenine, *Theor. Chim. Acta* **33**, 275–278 (1974).

163. B. Pullman and H. Berthod, Molecular orbital studies on the conformation of GABA Cγ-aminobutyric acid. The isolated molecule and the solvent effect, *Theor. Chim. Acta* **36**, 317–328 (1975).

164. M. D. Newton, A model conformational study of nucleic acid phosphate ester bonds. The torsional potential of dimethyl phosphate monoanion, *J. Am. Chem. Soc.* **95**, 256–258 (1973).

165. E. Clementi and H. Popkie, Study of the electronic structure of molecules. Barriers to internal rotation in polynucleotide chains, *Chem. Phys. Lett.* **20**, 1–4 (1973).

166. G. C. Liu and E. Clementi, Additional *ab initio* computations for the barrier to internal rotation in polynucleotide chains, *J. Chem. Phys.* **60**, 3005–3010 (1974).

167. J. Koller, S. Kaiser, and A. Azman, *Ab initio* calculation on 2-amino-ethylacetate ion, *Z. Naturforsch.* **28A**, 1745 (1973).

168. J. Almlöf, *Ab initio* calculations on the equilibrium geometry and rotation barriers in biphenyl, *Chem. Phys.* **6**, 135–139 (1974).

169. R. J. Kurland and W. B. Wise, The proton magnetic resonance spectra and rotational barriers of 4,4'-disubstituted biphenyls, *J. Am. Chem. Soc.* **86**, 1877–1879 (1964).

170. L. Radom, W. J. Hehre, J. A. Pople, G. L. Carlson, and W. G. Fately, Torsional barriers in para-substituted phenols from *ab initio* molecular orbital theory and far infrared spectroscopy, *J. Chem. Soc. D* **1972**, 308–309 (1972).

171. V. M. Guttins, W. Wyn-Jones, and R. F. M. White, Ring inversion in some six-membered heterocyclic compounds, *in*: *Internal Rotation* (W. Orville-Thomas, ed.), John Wiley and Sons, New York (1974).

172. D. Cremer and J. A. Pople, A general definition of ring puckering coordinates, *J. Am. Chem. Soc.* **97**, 1354–1358 (1975).

173. D. Cremer and J. A. Pople, Pseudorotation in saturated five-membered ring compounds, *J. Am. Chem. Soc.* **97**, 1358–1367 (1975).

174. R. M. Stevens and M. Karplus, A test of the closed-shell overlap-repulsion model for the ethane barrier, *J. Am. Chem. Soc.* **94**, 5140–5141 (1972).

175. P. A. Christiansen and W. E. Palke, Ethane internal rotation barrier, *Chem. Phys. Lett.* **31**, 462–466 (1975).

176. V. Magnasco and A. Perico, Uniform localization of atomic and molecular orbitals. I, *J. Chem. Phys.* **47**, 971–981 (1967).

177. V. Magnasco and A. Perico, Uniform localization of atomic and molecular orbitals. II, *J. Chem. Phys.* **48**, 800–808 (1968).

178. M. Levy, T. S. Nee, and R. G. Parr, Method for direct determination of localized orbitals, *J. Chem. Phys.* **63**, 316–318 (1975).

179. R. M. Pitzer, Localized molecular orbitals for ethane, *J. Chem. Phys.* **41**, 2216–2217 (1964).
180. C. Edmiston and K. Ruedenberg, Localized atomic and molecular orbitals, *Rev. Mod. Phys.* **35**, 457–465 (1963).
181. R. M. Pitzer and W. N. Lipscomb, Calculation of the barrier to internal rotation in ethane, *J. Chem. Phys.* **39**, 1995–2004 (1963).
182. G. L. Bendazzoli, F. Bernardi, and P. Palmieri, Group function analysis of the barriers to internal rotation on propargyl alcohol, and hydroxyacetonitrile, *J. Chem. Soc. Faraday Trans. 2* **69**, 579–584 (1973).
183. V. Magnasco and G. F. Musso, Localized orbitals and short-range molecular interactions. I. Theory, *J. Chem. Phys.* **60**, 3744–3748 (1974).
184. V. Magnasco and G. F. Musso, On factors contributing to rotational barriers, *Chem. Phys. Lett.* **9**, 433–436 (1971).
185. V. Magnasco and G. F. Musso, Simple model of short-range interactions. III. Ethane, propane, and butane, *J. Chem. Phys.* **54**, 2925–2935 (1971).
186. K. Ruedenberg, The physical nature of the chemical bond, *Rev. Mod. Phys.* **34**, 326–376 (1962).
187. W. England and M. S. Gordon, Localized charge distributions. I. General theory, energy partitioning, and the internal rotation barrier in ethane, *J. Am. Chem. Soc.* **93**, 4649–4657 (1971).
188. W. England and M. S. Gordon, Localized charge distributions. II. An interpretation of the barriers to internal rotation in H_2O_2, *J. Am. Chem. Soc.* **94**, 4818–4823 (1972).
189. W. England and M. S. Gordon, Localized charge distributions. The internal rotation barrier in borazane, *Chem. Phys. Lett.* **15**, 59–64 (1972).
190. M. S. Gordon and W. England, Localized charge distributions. V. Internal rotation barriers in methylamine, methyl alcohol, propene, and acetaldehyde, *J. Am. Chem. Soc.* **95**, 1753–1760 (1973).
191. M. S. Gordon, Localized charge distributions. VI. Internal rotation in formaldoxime and formic acid, *J. Mol. Struct.* **23**, 399–410 (1974).
192. S. F. Boys, Construction of molecular orbitals to be approximately invariant for changes from one molecule to another, *Rev. Mod. Phys.* **32**, 296–299 (1960).
193. W. England and M. S. Gordon, On energy localization of approximate molecular orbitals, *J. Am. Chem. Soc.* **91**, 6846–6866 (1969).
194. R. F. W. Bader, Molecular fragments of chemical bonds? *Acc. Chem. Res.* **8**, 34–40 (1975).
195. C. Leibovicci, Electronic structure and the origin of energy differences between rotational isomers, *J. Mol. Struct.* **10**, 333–342 (1971).
196. M. Pelissier, A. Serafini, J. Devanneaux, J. F. Labarre, and J. F. Tocanne, Theoretical conformational analysis of cyclopropylcarboxaldehyde, cyclopropyl methyl ketone, and *cis* and *trans* 2-methyl cyclopropyl methyl ketones, *Tetrahedron* **27**, 3271–3284 (1971).
197. M. Pelissier, C. Leibovicci, and J. F. Labarre, Theoretical conformation analysis of the acid fluoride of cyclopropanecarboxylic acid, *Tetrahedron Lett.* **1971**, 3759–3762 (1971).
198. C. Leibovicci, Electronic structure and the origin of energy differences between rotational isomers. II. Formaldoxime, *J. Mol. Struct.* **15**, 249–255 (1973).
199. B. Robinet, C. Leibovicci, and J. F. Labarre, On the electronic origins of barriers to methyl rotation, CNDO/2 calculations on $(CH_3)_2XHn$ (X = C, Si, N, P, O, S) molecules, *Chem. Phys. Lett.* **15**, 90–95 (1972).
200. G. Robinet, F. Crasnier, J. F. Labarre, and C. Leibovicci, Theoretical conformational analysis of dimethylsulfone, *Theor. Chim. Acta* **25**, 259–267 (1972).
201. G. Robinet, C. Leibovicci, and J. F. Labarre, Theoretical conformational analysis of dimethylsulfoxide, *Theor. Chim. Acta* **26**, 257–265 (1972).
202. C. Leibovicci, Electronic structure and origin of energy differences between rotational isomers. III. Methyltrifluorosilane, *J. Mol. Struct.* **18**, 303–307 (1973).
203. F. Crasnier, J. F. Labarre, and C. Leibovicci, Theoretical conformational analysis of Lewis adducts. II. CNDO/2 calculations versus microwave data for methylphosphine borane $(CH_3)H_2P\cdot BH_3$, *J. Mol. Struct.* **14**, 405–412 (1972).

204. J. F. Labarre and C. Leibovicci, Electronic structure of Lewis acid–base complexes. I. Electronic structure and molecular conformation of the molecules $F_3P\cdot BH_3$ and $F_2HP\cdot BH_3$, *Int. J. Quantum Chem.* **6**, 625–637 (1972).
205. H. J. Koehler and F. Birnstock, Conformational analysis by energy partitioning in the CNDO, INDO and NDDO formalisms, *Z. Chem.* **12**, 196–198 (1972).
206. H. J. Koehler, Conformational analysis by energy partitioning in the CNDO, INDO, and NDDO formalisms. The rotational barrier of hydrazine and the interaction of adjacent lone pair orbitals, *Z. Chem.* **13**, 157–159 (1973).
207. G. F. Musso and V. Magnasco, Nonadditivity of interbond interactions and the rotation barrier in ethane. A preliminary investigation, *Chem. Phys. Lett.* **23**, 79–82 (1973).
208. L. Radom, W. J. Hehre, and J. A. Pople, Fourier-component analysis of internal rotation potential functions in saturated molecules, *J. Am. Chem. Soc.* **94**, 2371–2381 (1972).
209. L. Radom and P. J. Stiles, An additivity scheme for conformational energies in substituted ethanes, *J. Chem. Soc. D* **1974**, 190–192 (1974).
210. W. A. Latham, L. Radom, W. J. Hehre, and J. A. Pople, Molecular orbital theory of the electronic structure of organic compounds. XVIII. Conformations and stabilities of trisubstituted methanes, *J. Am. Chem. Soc.* **95**, 699–703 (1973).
211. T. Dunning and N. W. Winter, private communication, 1975.
212. L. C. Allen, Energy component analysis of rotational barriers, *Chem. Phys. Lett.* **2**, 597–601 (1968).
213. I. R. Epstein and W. N. Lipscomb, Comments on the barrier to internal rotation in ethane, *J. Am. Chem. Soc.* **92**, 6094–6095 (1970).
214. D. T. Clark and D. M. J. Lilley, A non-empirical LCAO–MO–SCF investigation of cross sections through the potential energy surface for the $[C_2H_4Cl]^+$ systems; comparison with the $[C_2H_5^+]$ and $[C_2H_4F]^+$ systems, *Tetrahedron* **29**, 845–856 (1973).
215. T. K. Ha, Theoretical study of the internal rotation and inversion in hydroxymethyl radical, *Chem. Phys. Lett.* **30**, 379–382 (1975).
216. M. E. Schwartz, E. F. Hayes, and S. Rothenberg, Theoretical study of the barriers to internal rotation in formic acid, *J. Chem. Phys.* **52**, 2011–2014 (1970).
217. M. E. Schwartz, E. F. Hayes, and S. Rothenberg, Theoretical study of the barriers to internal rotation in nitrous acid, *Theor. Chim. Acta* **19**, 98–101 (1970).
218. P.-O. Löwdin, Scaling problem, virial theorem, and connected relations in quantum mechanics, *J. Mol. Spectrosc.* **3**, 46–66 (1959).
219. J. E. Eilers and A. Liberles, A quantum mechanical approach to conformational analysis, *J. Am. Chem. Soc.* **97**, 4183–4188 (1975).
220. S. Wolfe, The gauche effect, some stereochemical consequences of adjacent electron pairs and polar bonds, *Acc. Chem. Res.* **5**, 102–111 (1972).
221. H. Hellmann, *Einführing in Die Quantenchemie*, Franz Denticke and Co., Leipzig (1937).
222. R. P. Feynman, Forces in molecules, *Phys. Rev.* **56**, 340–343 (1939).
223. K. Ruedenberg, Hindered rotation, Hellman–Feynman theorem and localized molecular orbitals, *J. Chem. Phys.* **41**, 588–589 (1964).
224. R. G. Parr, Theorem governing changes in molecular conformation, *J. Chem. Phys.* **40**, 3726 (1964).
225. H. Kim and R. G. Parr, Integral Hellmann–Feynman theorem, *J. Chem. Phys.* **41**, 2892–2897 (1964).
226. A. C. Hurley, The molecular orbital interpretation of bond length changes following excitation and ionization of diatomic molecules, *in: Molecular Orbitals in Chemistry, Physics, and Biology* (P.-O. Löwdin, ed.), pp. 161–190, Academic Press, New York (1964).
227. M. P. Melrose and R. G. Parr, Some integral Hellmann–Feynman calculations on hydrogen peroxide and ammonia, *Theor. Chim. Acta* **8**, 150–156 (1967).
228. W. H. Fink and L. C. Allen, Numerical test of the integral Hellmann–Feynman theorem, *J. Chem. Phys.* **46**, 3270–3271 (1967).
229. R. E. Wyatt and R. G. Parr, Theory of the origin of the internal rotation barrier in the ethane molecule. II, *J. Chem. Phys.* **44**, 1529–1545 (1966).

230. J. Goodisman, Barrier to internal rotation in ethane using the Hellmann–Feynman theorem, *J. Chem. Phys.* **45**, 4689–4696 (1966).
231. J. Goodisman, Postscript to barrier to internal rotation in ethane by Hellmann–Feynman theorem, *J. Chem. Phys.* **47**, 334–335 (1967).
232. S. M. Rothstein and S. M. Blinder, The internal Hellmann–Feynman theorem applied to hydrogen peroxide, *Theor. Chim. Acta* **8**, 427–430 (1967).
233. L. Zülicke and H. J. Spangenberg, On calculating the internal rotation potential in hydrogen peroxide, *Theor. Chim. Acta* **8**, 139–147 (1966).
234. P. Pulay, *Ab initio* calculation of force constants and equilibrium geometries in polyatomic molecules. II. Force constants of water, *Mol. Phys.* **18**, 473–480 (1970).
235. S. T. Epstein, A. C. Hurley, R. E. Wyatt, and R. G. Parr, Integrated and integral Hellmann–Feynman formulas, *J. Chem. Phys.* **47**, 1275–1286 (1967).
236. J. I. Musher, On Parr's theorem, *J. Chem. Phys.* **43**, 2145–2146 (1965).
237. R. E. Wyatt and R. G. Parr, Theory of the internal-rotation barrier in the ethane molecule. I, *J. Chem. Phys.* **43S**, 217–227 (1965).
238. R. Mulliken, Electronic population analysis on LCAO–MO molecular wavefunctions. I, *J. Chem. Phys.* **23**, 1833–1840 (1955).
239. R. Mulliken, Electronic population analysis on LCAO–MO molecular wavefunctions. II. Overlap populations, bond orders, and covalent bond energies, *J. Chem. Phys.* **23**, 1841–1846 (1955).
240. J. J. Kaufman, Mulliken population analysis in CNDO and INDO LCAO–MO–SCF methods, *Int. J. Quantum Chem., Symp.* **4**, 205–208 (1971).
241. J. P. Lowe, A simple molecular orbital explanation for the barrier to internal rotation in ethane and other molecules, *J. Am. Chem. Soc.* **92**, 3799–3800 (1970).
242. J. P. Lowe, The barrier to internal rotation in ethane, *Science* **179**, 527–532 (1973).
243. J. P. Lowe, The Woodward–Hoffmann approach, the extended Hückel method, and the barrier to rigid internal rotation in ethane, *J. Am. Chem. Soc.* **96**, 3759–3764 (1974).
244. N. D. Epiotis, Attractive nonbonded interactions in organic molecules, *J. Am. Chem. Soc.* **95**, 3087–3096 (1973).
245. N. D. Epiotis, D. Bjorkquist, L. Bjorkquist, and S. Sarkanen, Attractive nonbonded interactions in 1-substituted propenes. Consequences for geometric and conformational isomerism, *J. Am. Chem. Soc.* **95**, 7558–7562 (1973).
246. N. D. Epiotis, S. Sarkanen, D. Bjorkquist, L. Bjorkquist, and R. Yates, Open shell interactions, nonbonded attraction, and aromaticity. Implications for regiochemistry, *J. Am. Chem. Soc.* **96**, 4075–4084 (1974).
247. L. Salem, Intermolecular orbital theory of the interaction between conjugated systems. I. General theory, *J. Am. Chem. Soc.* **90**, 543–552 (1968).
248. K. Müller, Slow inversion at pyramidal nitrogen: configuration and conformation of *N,N*-dialkoxy-alkylamine in terms of a semi-empirical MO model, *Helv. Chim. Acta* **53**, 1112–1127 (1970).
249. W. J. Hehre and L. Salem, Conformation of vinylic methyl groups, *J. Chem. Soc. D* **1973**, 754 (1973).
250. R. Hoffmann, C. C. Levin, and R. A. Moss, On steric attraction, *J. Am. Chem. Soc.* **95**, 629–631 (1973).
251. C. C. Levin, R. Hoffmann, W. J. Hehre, and J. Hudec, Orbital interaction in amino ketones, *J. Chem. Soc., Perkin Trans. 2* **1973**, 210 (1973).
252. H. Fuijimoto and R. Hoffmann, Perturbation of molecules by static fields, orbital overlap, and charge transfer, *J. Phys. Chem.* **78**, 1874–1880 (1974).
253. D. Cremer, J. S. Binkley, J. A. Pople, and W. J. Hehre, Molecular orbital theory of the electronic structure of organic compounds. XXI. Rotational potentials for geminal methyl groups, *J. Am. Chem. Soc.* **96**, 6900–6903 (1974).
254. M. Schwartz, Theoretical study of the barrier to internal rotation in hydrogen persulfide, HSSH, *J. Chem. Phys.* **51**, 4182–4186 (1969).
255. W. L. Jorgensen and L. C. Allen, Charge distribution characteristics of attractive dominant barriers, *Chem. Phys. Lett.* **7**, 483–485 (1970).

256. W. L. Jorgensen and L. C. Allen, Charge density analysis of rotational barriers, *J. Am. Chem. Soc.* **93**, 567–574 (1971).
257. P. W. Payne and L. C. Allen, Charge density difference analysis. Comparison of internal rotation in ethane and methylamine, *J. Am. Chem. Soc.*, to be published (1977).
258. P. W. Payne and L. C. Allen, Charge density difference analysis. Internal rotation in ethylamine, *J. Am. Chem. Soc.*, to be published (1977).
259. R. Hoffmann, An extended Hückel theory. Hydrocarbons, *J. Chem. Phys.* **39**, 1397–1412 (1963).
260. S. W. Benson and M. Luria, Electrostatics and the chemical bond. I. Saturated hydrocarbons, *J. Am. Chem. Soc.* **97**, 704–709 (1975).
261. S. W. Benson and M. Luria, Electrostatics and the chemical bond. II. Unsaturated hydrocarbons, *J. Am. Chem. Soc.* **97**, 3337–3342 (1975).
262. M. Luria and S. W. Benson, Electrostatics and the chemical bond. III. Free radicals, *J. Am. Chem. Soc.* **97**, 3342–3346 (1975).
263. H. A. Scheraga, Calculations of conformations of polypeptides, *Adv. Phys. Org. Chem.* **6**, 103–185 (1968).
264. L. L. Shipman, A. W. Burgess, and H. A. Scheraga, A new approach to empirical intermolecular and conformational potential energy functions. I. Description of model and derivation of parameters, *Proc. Natl. Acad. Sci. USA* **72**, 543–547 (1975).
265. R. A. Scott and H. A. Scheraga, Conformational analysis of macromolecules. III. Helical structures of polyglycine and poly-L-alanine, *J. Chem. Phys.* **45**, 2091–2101 (1966).
266. A. Rahman, F. H. Stillinger, and H. L. Lemberg, Study of a central force model for liquid water by molecular dynamics, *J. Chem. Phys.* **63**, 5223–5230 (1975).
267. F. H. Stillinger, Construction and use of central force fields for the theory of polyatomic fluids, to be published.
268. W. J. Hehre and P. C. Hiberty, Theoretical approaches to rearrangements in carbocations. I. The haloethyl system, *J. Am. Chem. Soc.* **96**, 2665–2678 (1974).
269. W. L. Jorgensen and L. Salem, *The Organic Chemist's Book of Orbitals*, Academic Press, New York (1973).
270. J. M. Howell, *Ab initio* calculations of the rotational barrier in PH_4NH_2, *Chem. Phys. Lett.* **25**, 51–54 (1974).
271. E. Lassettre and L. Dean, An electrostatic theory of the potential barriers hindering rotation around single bonds, *J. Chem. Phys.* **17**, 317–352 (1949).
272. S. Lifson and A. Warshel, Consistent force field for calculations of conformations, vibrational spectra, and enthalpies of cyclohexane and *n*-alkane molecules, *J. Chem. Phys.* **49**, 5116–5129 (1968).
273. A. Warshel and S. Lifson, Crystal structures, sublimation energies, molecular and lattice vibrations, molecular conformations, and enthalpies of alkanes, *J. Chem. Phys.* **53**, 582–594 (1970).
274. M. L. Huggins, Interaction between nonbonded atoms, *in*: *Structural Chemistry and Molecular Biology* (A. Rich and N. Davidson, eds.), pp. 761–768, W. H. Freeman, San Francisco (1968).
275. R. Rein, T. J. Swissler, V. Renugopalakrishnan, and G. R. Pack, Some refinements in the electrostatic theory of rotational potential functions, *in*: *The Jerusalem Symposium on Quantum Chemistry and Biochemistry*, Vol. 5 (E. D. Bergmann and B. Pullman, eds.), Academic Press, New York (1973).
276. A. D. Tait and G. G. Hall, Point charge models for LiH, CH_4, and H_2O, *Theor. Chim. Acta* **31**, 311–324 (1973).

Note Added in Proof

1. The contrasting roles of orbital orthogonality and electron exchange are further clarified in a recent paper by Levy.[277] Levy has found that application of Edmiston–Ruedenberg exchange

localization *without* orbital orthogonality constraints generates localized orbitals that are nearly orthogonal. This result helps rationalize the apparent dependence of some barrier models on electron exchange energy. Exchange energy minimization has little intrinsic importance for rotational barrier mechanisms; but exchange energy minimization tends to orthogonalize orbitals, and orthogonality is important for the barrier mechanism.

2. Brunck and Weinhold[278] attribute rotational barriers in ethane, methylamine, and methanol to vicinal mixing between bonds and antibonds. Their key step is expansion of the INDO Hamiltonian matrix in a basis set of local bonding and antibonding orbitals. Since the rotational barriers disappear if the pseudomolecular orbitals are constructed as a linear combination of local bonding orbitals, they claim that vicinal mixing between bonds and antibonds is at the heart of barriers.

This model has a strong intuitive appeal. In order to establish its credibility, further work is needed on the following problems: First, the definitions of local bonding and antibonding orbitals are arbitrary. It is not clear that the model would hold up under small adjustments in the bond orbitals. Second, the balance between INDO matrix elements is often quite different from that in an *ab initio* theory. Third, a barrier model should not be sensitive to geometry optimization if total energy is insensitive. Because INDO barrier heights are poor when geometries are optimized, any barrier model derived from INDO wave functions is tentatively best.

3. Another interesting paper that is formulated within the framework of a one-electron orbital theory and addresses a long-standing problem is that of Salem, Hoffmann, and Otto on barriers in substituted ethanes.[279]

277. M. Levy, Unconstrained exchange localization and distant orbital tails, *J. Chem. Phys.* **65**, 2473–2475 (1976).
278. T. K. Brunck and F. Weinhold, Quantum-mechanical origin of barriers to internal rotation about single bonds, *J. Am. Chem. Soc.* (1977), in press.
279. L. Salem, R. Hoffmann, and P. Otto, The energy of substituted ethanes: Asymmetry orbitals, *Proc. Nat. Acad. Sci. (U.S.A.)*, **70**, 531–532 (1970).

3

Hydrogen Bonding and Donor–Acceptor Interactions

Peter A. Kollman

1. Introduction

Ab initio calculations have played an important role in the development of the theory of the hydrogen bond. First, the calculations have been able to predict properties of hydrogen-bonded complexes prior to experimental observation. Second, the *ab initio* calculations have been the standard against which semiempirical MO calculations and model theories could be evaluated. Finally, these calculations have provided an important framework for understanding the chemical properties of the hydrogen bond, and its relation to other donor–acceptor interactions and to covalent bonds.

This chapter will focus on the above roles of *ab initio* calculations in H-bond theory. No attempt will be made to exhaustively review the literature in this area since this has been recently done by Schaad[1] (through 1973). Other reviews of H-bond theory (post-1970) are those by Kollman and Allen,[2] Lin,[3] Murthy and Rao,[4] Ratner and Sabin,[5] and Murrell.[6] Earlier reviews include those of Bratoz,[7] Coulson,[8] Pimentel and McClellan,[9] and Sokolov.[10] Huggins, who was the first to propose the existence of hydrogen bonds, has written an interesting historical account of the early theory of the hydrogen bond.[11] There is a great variation in the focus of the different review articles: Schaad and Kollman and Allen concentrating on the *ab initio* calculations, Murthy and Rao on semiempirical MO approaches, and Lin,

Peter A. Kollman • Department of Pharmaceutical Chemistry, School of Pharmacy, University of California, San Francisco, California

Murrell, Bratoz, and Coulson on perturbation theory approaches to different components of the hydrogen-bond energy.

2. Theoretical Methods

2.1. *Ab Initio* Methods for Studying H-Bond Potential Surfaces

A majority of the *ab initio* calculations on hydrogen bonded systems have been carried out using the Roothaan single-determinant self-consistent field method[12] on a basis set consisting of atomic-centered Gaussian or Slater orbitals. By finding the minimum energy in the potential surface for the hydrogen bonded complex A—H···B one is able to predict its structure, and by comparing the total energy for the hydrogen-bonded complex relative to the isolated molecules A—H and B one can predict the energy of formation of the complex. The nature of the potential surface of the complex allows one to predict infrared and Raman spectra. The wave functions for the complex can be used to compute other molecular properties and to compare these with the molecular properties of the monomers.

Applications of the Karplus–Kolker variation-perturbation method[13] to dimer wave functions have allowed comparison of the proton magnetic resonance of the H-bonded complex with the monomers. Electron hole potential[14] or multideterminantal (configuration interaction[15]) wave functions allow one to study excited states of hydrogen bonded complexes and compare these with the excited state properties of the monomers. In addition, the configuration interaction calculations are capable (in principle) of giving a more precise estimate of the H-bond energy. Thus, by using the rather well-established techniques of *ab initio* quantum chemistry, one can gain insight into the structure, energy, and spectral properties of hydrogen bonded complexes.

The limitations of the theoretical methods used to study hydrogen bonding have been clearly pointed out by Schaad,[1] but a brief review is in order here. Single-determinant molecular orbital theory does a very good job in determining ground-state molecular structure and this is also the case for hydrogen bonded complexes. The slopes of potential surfaces (e.g., A–H stretching) are usually somewhat overestimated; such defects preclude quantitative ir predictions for H-bonded complexes but the qualitative features are generally satisfactory.

There is an important difference between studying the energy of covalent bonds and donor–acceptor (e.g., hydrogen) bonds with molecular orbital theory. As is well known,[16] single-determinant theory makes gross errors in reproducing covalent bond dissociation because it forces the wave function to have significant amounts of ionic character, even at infinite separation between the fragments. [H_2 dissociates to $\frac{1}{2}(H + H) + \frac{1}{2}(H^+ + H^-)$ and F_2 is predicted not to be stable relative to 2F atoms.] In donor–acceptor complex dissociation [e.g., $H_3^+ \rightarrow H_2 + H^+$, $(H_2O)_2 \rightarrow 2H_2O$, $BH_3NH_3 \rightarrow BH_3 + NH_3$], electron pairs are conserved going from the complex to the separated fragments and the dissociation process is correctly treated. However, there are other factors which limit single-determinant theory from predicting quantitatively correct dissociation energies for donor–acceptor complexes and we now consider these.

Perturbation theory approaches to the study of weak intermolecular complexes have concluded that the main terms in the hydrogen bond energy are: (1) electrostatic energy, which is the interaction between the fixed nuclei and electron distribution of the individual molecules; (2) exchange repulsion energy, the intermolecular Pauli principle repulsion between electrons of like spin; (3) polarization energy, the attractive interaction between the polarizable charge cloud of one molecule and the permanent multipoles of the other molecule; (4) charge transfer energy, a quantum mechanical attraction, due to the fact that the wave function for the complex contains a contribution from a term in which charge is transferred from one molecule to the other; and (5) dispersion energy, the second-order induced dipole-induced dipole attraction which is present even in rare-gas–rare-gas interactions.

Margenau and Kestner[17] have proved that a single-determinant wave function is incapable of yielding a dispersion attraction between two molecules. In addition, a wave function which does not quantitatively reproduce the multipole moments for the isolated molecules will not correctly predict the electrostatic energy (1). Deviations from the correct multipole moments and static polarizabilities will show up as errors in reproducing the polarization energy (3), and errors in ionization potentials will certainly contribute to errors in the calculation of the charge transfer energy (4). A single-determinant wave function, using a very large basis set approaching the Hartree–Fock limit,* is capable of reproducing molecular dipole moments, static polarizabilities, and ionization potentials to within 10%.[21] The error in the H-bond energy may be (fortuitously) less than this because near Hartree–Fock wave functions tend to overestimate dipole moments [and electrostatic energies (1)], but underestimate polarizabilities [polarization and charge transfer energies, (3) and (4)]. Most *ab initio* hydrogen bond studies have used limited basis sets which are far from the Hartree–Fock limit and which lead to much larger errors in the individual energy components. We will examine these problems further.

At this point one can summarize as follows:

(1) Donor–acceptor complex studies with single-determinant wave functions do not suffer from the same fundamental correlation problem that

---

*See Ref. 18 for a description of the methodology of SCF theory, and Ref. 19 for a survey of applications.

covalent bond formation studies do. The evidence at hand,[20]* not totally conclusive, indicates that the *intramolecular* correlation energy depends mainly on the number of electron pairs and thus there should be little change in this quantity upon donor–acceptor complex formation.

(2) *Intermolecular* correlation effects, which lead to the dispersion energy cannot be treated at a single-determinant level and one probably needs very extensive configuration interaction to calculate these accurately *ab initio* even for small H-bonded complexes.†

(3) The ability of a single-determinant wave function to accurately calculate the four remaining contributions to the H-bond energy depends upon the accuracy with which this wave function represents the molecular properties (multipole moments, polarizabilities, and ionization potentials) of the isolated molecules.

2.2. Methods for Evaluating the H-Bond Energy Components

Prior to the advent of *ab initio* calculations, the definitive studies in calculating the H-bond energy were those of Coulson and Danielsson[22] and Tsubomura[23] in the 1950s and Murrell and van Duijneveldt[24] in the late 1960s and early 1970s. Semiempirical calculations to determine the main contributions to the H-bond energy were carried out. Coulson[22] and Tsubomura[23] carried out such calculations for an O—H···O linkage and van Duijneveldt[24] for a general A—H···B linkage. Coulson and Tsubomura separately calculated electrostatic (−6 kcal/mole), polarization + charge transfer (−8 kcal/mole), dispersion (−3 kcal/mole), and exchange repulsion (+8.4 kcal/mole) at the minimum energy O···O distance in ice. Coulson's conclusion[25] was that the hydrogen bond should not be thought of as due to a single energy component with all the components playing significant roles at an O···O separation of 2.76 Å. He also pointed out that longer H bonds [R(O···O) greater than 2.9 Å] were "essentially electrostatic." van Duijneveldt[24] has carried out some very interesting model calculations using a perturbation theory formalism and a three-center four-electron model A—H···B of the hydrogen bond. The model parameters used were the percent

*For example, Schaad[1] points out that the protonation of H_2 ($H_2 + H^+ \rightarrow H_3^+$), where $-\Delta E = 97$ kcal/mole, has a correlation contribution of 1–3 kcal/mole. Also, it appears from the SCF calculations of Kollman and Bender,[20] J. Kaufman (personal communication), and B. Roos (personal communication) that the correlation contribution to the proton affinity of water is less than 5 kcal/mole out of a total $-\Delta E$ of 166 kcal/mole. However, it is not clear whether this correlation error will have a similarly small fractional contribution to weak H-bonding interactions, which are themselves only ~5 kcal/mole.
†See, for example, Schaefer's[21] description of the computational effort involved in calculating the disperson attraction between two He atoms.

s character in the A—H bond and the B lone pair and the "ionicity" of the A—H bond. The energy of the H bond was calculated to be the sum of the "first-order" energies [electrostatic (Coulomb) and exchange repulsion] and the "second-order" energies (induction, second-order exchange, charge transfer, and dispersion). Unlike Coulson's calculations, where the various energy contributions were each evaluated in a different manner, each of the terms in van Duijneveldt's treatment was well defined from the perturbation theory expansion. van Duijneveldt also found nontrivial contributions to the hydrogen bond energy for each of the different energy components.

Morokuma,[26] Dreyfus and Pullman,[27] and Kollman and Allen,[28] carrying out *ab initio* SCF calculations, decomposed the total H-bond energy into components (Morokuma's decomposition scheme being the most complete of the three). His decomposition scheme is as follows: Ψ_A^0 and Ψ_B^0 are the wave functions for the isolated molecules, E_A^0 and E_B^0 the isolated molecule energies, $(\Psi_A^0\Psi_B^0)$ is the Hartree product of the isolated molecule wave functions (antisymmetric with respect to the intramolecular electron exchange, but not intermolecular exchange), $\mathscr{A}(\Psi_A^0\Psi_B^0)$ is the completely antisymmetrized (Hartree–Fock) product of the isolated molecule wave functions, and $\hat{H}$ is the total molecular Hamiltonian; $\Psi_A\Psi_B$ is the SCF optimized Hartree product wave function for the complex and $\mathscr{A}(\Psi_A\Psi_B)$ is the SCF optimized Hartree–Fock product for the complex. The energy components are:

$$E_{\text{electrostatic}} = \langle \Psi_A^0\Psi_B^0 | \hat{H} | \Psi_A^0\Psi_B^0 \rangle - E_A^0 - E_B^0$$

$$E_{\text{exchange}} = \langle \mathscr{A}\Psi_A^0\Psi_B^0 | \hat{H} | \mathscr{A}\Psi_A^0\Psi_B^0 \rangle - \langle \Psi_A^0\Psi_B^0 | \hat{H} | \Psi_A^0\Psi_B^0 \rangle$$

$$E_{\text{polarization}} = \langle \Psi_A\Psi_B | \hat{H} | \Psi_A\Psi_B \rangle - \langle \Psi_A^0\Psi_B^0 | \hat{H} | \Psi_A^0\Psi_B^0 \rangle$$

$$E_{\text{charge transfer}} = \langle \mathscr{A}\Psi_A\Psi_B | \hat{H} | \mathscr{A}\Psi_A\Psi_B \rangle - E_{\text{electrostatic}}$$

$$- E_{\text{exchange}} - E_{\text{polarization}} - E_A^0 - E_B^0$$

$$E_{\text{total}} = \langle \mathscr{A}\Psi_A\Psi_B | \hat{H} | \mathscr{A}\Psi_A\Psi_B \rangle - E_A^0 - E_B^0$$

Although the correspondence between the Morokuma[26] and van Duijneveldt[24] energy terms is not exact, Morokuma's "charge transfer" energy appears to include both the "charge transfer" and "second-order exchange" of van Duijneveldt. The conceptual basis for Dreyfus and Pullman's[27] and Kollman and Allen's[28] decomposition was the same as Morokuma's but they only broke down the energy into three (electrostatic, exchange, and polarization + charge transfer)[27] and two terms (electrostatic + exchange, and polarization + charge transfer),[28] respectively. As stated above, these SCF calculations are incapable of determining the attractive dispersion energy between molecules, but the decomposition of the total energy into components allows one to get a clearer understanding of the nature of the hydrogen bond (as we shall see in Section 4).

3. Observable Properties of Hydrogen-Bonded and Other Donor–Acceptor Complexes

3.1. Structure and Binding Energy

Typical hydrogen bonds between neutral molecules in the gas phase are weak interactions ($-\Delta E = 3$–10 kcal/mole), and thus these intermolecular interactions should have a relatively small effect on intramolecular geometry parameters. In *ab initio* calculations on weak H-bonded complexes where the fixed constraint of intramolecular geometry upon intermolecular interaction is relaxed,[29] one finds a relatively small increase of interaction energy ($<10\%$) and an even smaller decrease ($<5\%$) of intermolecular separation, compared to the results where intramolecular constraints are retained. Thus, most of the potential surfaces discussed in Sections 1–4 have assumed fixed monomer geometry (either experimental or energy optimized) upon H-bond formation.

In Section 4 we examine strong H bonds, which usually involve an ion and a neutral compound. Here the interaction energies are quite large ($-\Delta E = 20$–50 kcal/mole) and the structural parameters of the complex are often quite different from those of the molecules from which it is formed.

In Section 5 we consider a number of intermolecular interactions not involving "hydrogen bonds" in order to put the H-bond potential surfaces into better perspective.

3.1.1. The Water Dimer $(H_2O)_2$

Owing to the importance of hydrogen bonds in determining the liquid and solid structure of water, the O—H$\cdots$O hydrogen bond has received the most attention from theoreticians, both in the earlier electrostatic and semiempirical quantum mechanical theories and in later *ab initio* calculations. Unfortunately the condensed phases of water that are of most interest to experimental physical chemists are not easily amenable to *ab initio* calculations. On the other hand, the water dimer $(H_2O)_2$, which has been studied extensively by various theoretical methods, was not studied experimentally until 1957 by Pimentel and van Thiel.[30] They examined the concentration dependence of O—H infrared stretching and bending region in N_2 matrices at 20 K and assigned certain bands to the dimer species. They considered three different structures for the dimer: linear, cyclic, and bifurcated (Fig. 1); and because they observed only two O–H stretching bands, they concluded that the dimer must be centrosymmetric and hence have a cyclic structure.

Some ten years later, when the advent of the 7094 generation computers and the use of Gaussian functions[31] to represent atomic orbitals made potential surface calculations on polyatomic nonlinear molecules feasible, *ab initio* calculations on the water dimer began to appear in the literature. The first

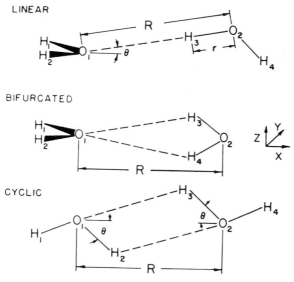

Fig. 1. The three water dimer structures considered in the original water dimer study by van Thiel *et al.*[30] and by theoretical studies.[32,33] A further variation in the cyclic structure involves out of plane movement of hydrogens 1 and 4 and the minimum energy cyclic structure found in Ref. 33 has a center of symmetry with H_1 and H_4 out of plane by 40°.

of these by Morokuma and Pedersen,[32] using a very small Gaussian basis [(5, 3/3) without contraction],* found the linear dimer with $R = 2.68$ Å and $\theta = 17°$ to be the lowest energy structure, $(-\Delta E = 12.6$ kcal/mole) with the bifurcated next in energy, and the cyclic structure as the least stable. The next calculation to appear was that of Kollman and Allen using a $(10, 5/5) \rightarrow (3, 1/1)$ Gaussian basis.* They found a similar order of stabilities with linear more stable than bifurcated or cyclic and a $-\Delta E$ (for the linear structure of 5.27 kcal/mole,[33] $R = 3.00$ Å, and $\theta = 25°$). Further calculations followed, but the basic conclusions about the structure of the dimer have remained: The minimum energy structure of the water dimer is linear, with the O—O distance R greater than that of ice I and liquid water (the most accurate calculations find $R = 3.00$ Å),[39] and the angle θ between 25 and 57 degrees (the external hydrogens on the two oxygens are *trans* with respect to each other). The optimum O—H⋯O bond is linear within ±5°. A summary of these calculations is presented in Table 1; they are ordered according to the total energies.

*The notation is as follows: (5, 3/3) means $5s$ and $3p$ Gaussians on oxygen and $3s$ Gaussians on hydrogen. $(10, 5/5) \rightarrow (3, 1/1)$ means $10s$ and $5p$ Gaussians on oxygen and $5s$ functions on hydrogen contracted to $3s$ and $1p$ functions on oxygen and $1s$ function on hydrogen. For further information on basis sets, see Chapter 1, Volume 3 of this series, written by T. H. Dunning.

Table 1. Water Dimer Calculations

Basis set	E_t, a.u.	μ, D[a]	R, Å[b]	θ, degrees[b]	$-\Delta E$, kcal/mole[c]	Other structures
STO-3G[d,e]	-149.94	1.71	2.74	57	5.88	No other local minima in 6-dimensional surface
STO-4G[f]	-151.010	1.82	2.73	58	6.09	
(5, 3/3)[g]	-151.118	—	2.68	17	12.6	$-\Delta E$(bifurcated) = 9.2 kcal/mole; $-\Delta E$(cyclic) = 7.8 kcal/mole
STO[h]	-151.420	—	2.78	54	6.55	—
431G[i]	-151.828	2.61	2.85	37	8.1	
(10, 5/5) → (3, 1/1)[j]	-151.957	2.48	3.00	25	5.27	$-\Delta E$(bifurcated) = 4.45 $-\Delta E$(cyclic) = 4.0
(10, 5, 1/4, 1) → (5, 3, 1/2, 1)[k]	-152.091	2.19	3.00	40	4.73	$-\Delta E$(bifurcated) = 3.6
(11, 7, 1/6, 1) → (5, 4, 1/3, 1)[l]	-152.112	2.20	3.00	38	4.83	$-\Delta E$(cyclic) = 3.66
(11, 7, 1/6, 1) → (4, 3, 1/3, 1)[m]	-152.118	—	3.00	30	4.60	—
(11, 7, 2/6, 2) → (4, 3, 2/2, 2)[m]	-152.12561	—	3.00	30	3.90	—
(13, 8, 2, 1/6, 2, 1) → (8, 5, 2, 1/4, 2, 1)[m]	-152.13781	—	3.00	30	3.88	

[a] Calculated dipole moment for H_2O monomer.
[b] See Fig. 1.
[c] Calculated dimerization energy.
[d] STO-3G means a three Gaussian fit to a Slater orbital (exponents from Slater's rules or energy optimized).
[e] Ref. 34. [f] Ref. 35. [g] Ref. 32. [h] Ref. 36. [i] Refs. 103 and 104. [j] Ref. 33. [k] Ref. 37. [l] Ref. 38. [m] Ref. 39.

It was gratifying for the theoreticians to have the matrix ir work of Tursi and Nixon[40] appear in 1970, since these workers were able to resolve the four peaks in the O—H stretching region of the water dimer and conclude, on this basis, that the minimum energy dimer structure in inert matrices was linear. However, this pleasure wore off quickly when ir measurements of water in CCl_4 by Magnusson[41] were interpreted in terms of a cyclic structure for the dimer. Kollman and Buckingham[42] and Atkins and Symons[43] suggested alternative explanations for the CCl_4 spectra, but the question of the dimer structure in CCl_4 was not settled.

To the rescue came supersonic nozzle beam techniques,[44] which allowed nonequilibrium amounts of water dimer and polymers to be generated in the gas phase. This technique is a theoretician's dream because it allows direct comparison with calculations without the perturbations (however subtle) of inert matrices or liquid CCl_4. Dyke and Muenter were able to show that the dimer structure was linear, with an O—O separation of 2.98 ± 0.04 Å.[45] It appears, from the *ab initio*[35,39] and semiempirical[46] potential functions, that there are no other local minima in the six-dimensional water dimer surface. These nozzle beam experiments were very important because they provided important support for the assumption that *ab initio* single-determinant calculations should be capable of good accuracy in structure prediction, for complexes as well as small molecules.

In contrast to the water dimer *structures*, where all the *ab initio* calculations predict qualitatively similar results, the calculated hydrogen bond *energy* varies over a relatively large range. However, excluding the first calculation,[32] all $-\Delta E$ for the dimerization reaction fall in the range 3.9–8.6 kcal/mole, which are not wildly wrong values considering that the H-bond energy in ice is 6.7 kcal/mole.*

Unfortunately, the experimental values for the water dimer H-bond energy are as uncertain as the theoretical calculations; the only estimates are 5.0 kcal/mole from second virial coefficient measurements,[48] which are rather insensitive to the depth of the intermolecular potential, and ir studies of H_2O vapor ($-\Delta H_{300} = 5.2 \pm 1.5$ kcal/mole).[49] Thus it is likely that the best theoretical values from the *ab initio* calculations are as "good" as the experimental values; however, to directly compare a dimerization energy at 0 K (measured from the depth of the potential well) with a dimerization enthalpy at 300 K requires all the vibrational and rotational constants of the monomer and dimer. We postpone such a comparison until Section 2.2.

In view of the considerable dependence of the calculated water dimer H-bond energy on basis set, it is worth digressing from structural questions to try to understand this dependence in terms of the concepts introduced in Section 1.2.

---

*This is the lattice energy of ice at 0 K. The enthalpy/H bond is 5.7 kcal/mole at 0 K and 5.5 kcal/mole at 298 K, see Eisenberg and Kauzmann.[47]

There are three levels of basis sets[50] which have been extensively applied to hydrogen bonding: minimum Slater bases, double zeta bases, and "double zeta plus polarization" bases. The highest in total energy is the Slater (usually expanded in linear combinations of Gaussian functions) basis. This basis set reproduces the dipole moment of the water monomer quite well (1.71 D calculated[34] vs 1.85 experimental) but generally (for bonds involving atoms of the first full row of the periodic table and hydrogen), somewhat underestimates electric moments and charge separation. Since it is very limited it greatly underestimates molecular polarizabilities but compensates for this by exaggerating "charge transfer effects": Because this basis set is so limited, the energy of a single water molecule is lowered by 6 kcal/mole by the presence of the basis functions of the second water *without* the nuclei of the second water.[51] The reasonable H-bond energy (6 kcal/mole) calculated for the H_2O dimer with the Slater basis is due to a cancellation of these substantial errors.

The "double zeta" basis sets overestimates the H-bond energy mainly because they exaggerate electrostatic attractions. The calculated dipole moments are 2.5 D and typical $\Delta E \sim -8$ kcal/mole.[33] It is not clear how well these calculations reproduce the charge transfer energy, but they probably underestimate the polarization energy. The fact that the calculation by Kollman and Allen[33] found a much lower H-bond energy ($-\Delta E = 5.3$ kcal/mole) was due to the severe contraction of their basis set. This led to an even smaller polarization energy to compensate for the exaggerated electrostatic energy.

Addition of polarization functions to the basis set lowers the monomer dipole moment to ~ 2.2 D and increases the polarizability, although the polarizability is still significantly smaller than the experimental value for water monomer.* Polarization functions decrease the H-bond energy in $(H_2O)_2$ to ~ 4.7 kcal/mole.[37,38]

Further expansions of the basis set by Popkie *et al.*[39] have led to a continuing decrease in the dimerization energy. The best calculation, including orbitals of f symmetry on oxygen and orbitals of d symmetry on hydrogen† led these workers to conclude that the Hartree–Fock dimerization energy of water is 3.9 ± 0.25 kcal/mole and Popkie *et al.* estimate that the correlation energy contribution to the dimerization energy is 0.3–0.5 kcal/mole.[39] Thus, at this

---

*Our calculations (unpublished) on the static polarizability of H_2O with "double zeta" and double zeta plus a single d on O and a single p on H basis sets all find a significantly smaller α than experiment.

†It is not clear that functions of this symmetry are required to lower the dimerization energy of water to 3.9 kcal/mole. We see that the last two entries in Table 1 have nearly the same dimerization energy and the former has no f functions on O or d functions on H. The importance of 2 flexible d functions on the heteroatom in reproducing the one-electron properties of water is clear from the studies of Neumann and Moskowitz.[52] Rauk *et al.*[53] have shown the importance of 2 d functions on nitrogen to an adequate representation of the NH_3 inversion barrier.

point the best *ab initio* calculation of the dimerization of water predicts a dimerization energy of 4.3 ± 0.35 kcal/mole. It is somewhat difficult to estimate the error in this value, since the dipole (μ) and quadrupole (θ) moments and static polarizabilities ($\alpha, \beta, \ldots$) of the water monomer are not reported. As was emphasized at the end of Section 1.1, there may be errors in the dimerization energy due to *intramolecular* correlation (errors in the Hartree–Fock representation of μ, θ, and α) which will affect the electrostatic, polarization, charge transfer, and exchange repulsion energy terms and due to *intermolecular* correlation (dispersion attraction). The Popkie *et al.*[39] estimate appears to be based on intramolecular correlation effects; Hankins *et al.*[37] have estimated 0.8 kcal/mole for the intermolecular (dispersion) energy by using the known dispersion attraction between Ne atoms and comparing the polarizabilities and ionization potentials of Ne and H_2O. If we assume that the two correlation effects are additive, then the best theoretical estimate for the dimerization energy of $(H_2O)_2$ is 5.1 kcal/mole.

3.1.2. Water Polymers

The two most important *ab initio* studies of water polymers $(H_2O)_n$, $n > 2$ have been those by Del Bene and Pople[35] and by Hankins, Moskowitz, and Stillinger.[37] Del Bene and Pople examined the potential surface for water trimer, tetramer, pentamer, and hexamer with an STO-4G basis and concluded that the minimum energy for these polymers was a cyclic (nonpolar) structure. This conclusion was subsequently verified by the nozzle beam work of Dyke and Muenter.[45] Del Bene and Pople's predicted structure for the water trimer has been questioned by Lentz and Scheraga,[54] who carried out "double zeta plus polarization" calculations on linear and cyclic water trimers and found the cyclic trimer to be less stable than a sequential linear trimer. However, the geometry search carried out by Lentz and Scheraga was not extensive enough to conclude that this basis set does predict the linear trimer to be more stable than the cyclic structures. At this point, one can conclude that theory and experiment are in agreement that the minimum energy structure for the water trimer is cyclic with three hydrogen bonds.

Why is the water dimer predicted to be linear and the higher water polymers predicted to be cyclic? In the dimer, one linear O—H···O bond is more stabilizing than the two very nonlinear O—H···O interactions in either the bifurcated or cyclic structures. In the tetramer or $n$mer ($n \geq 4$) a cyclic structure allows the formation of n nearly linear O—H···O hydrogen bonds, while a sequential linear structure allows only $n - 1$. In the trimer the three hydrogen bonds in the cyclic structure are forced by geometrical constraints to be significantly nonlinear but these bonds appear to be still stronger than the two linear O—H···O bonds in the sequential trimer, providing an *ab initio* estimate of the stability of *ménage à trois* over *deux*.

Hankins *et al.*[37] examined the energy of the three water trimer structures found in the structure of ice I and termed these structures as follows

double donor   double acceptor   sequential

based on the hydrogen bonding pattern of the central water molecule. These authors pointed out the qualitative difference between the nonadditivities of the H-bond energies in the three different trimers. The nonadditivity in a trimer is the difference between the interaction energy calculated by adding up the three calculated $H_2O \cdots H_2O$ two-body interaction energies and comparing this to the interaction energy calculated for the trimer relative to three isolated monomers. Hankins *et al.* found an *attractive* three-body energy of -1.36 kcal/mole for the sequential trimer, 0.87 kcal/mole (repulsive) for the "double donor," and 0.35 kcal/mole for the "double acceptor." Del Bene and Pople[55] compared the nonadditivities using three different basis sets (STO-4G, LEAO, and 431G) and found some significant differences in the actual energies, but the greater stability of sequential H-bonded structures than expected on the basis of individual H-bond energies was supported by their work.

The cyclic polymers of water found to be most stable by Del Bene and Pople[35] are cyclic structures with sequential $O \cdots H—O \cdots H—O \cdots$ hydrogen bonds. Thus, these structures are stabilized by energy nonadditivity in a manner similar to linear structures with sequential H bonds. In fact, Del Bene and Pople find the optimized cyclic hexamer of H_2O to have an $O—O$ separation of 2.44 Å and a stabilization energy of 72 kcal/mole relative to $6H_2O$ [recall that they found $R(O—O) = 2.73$ Å and $-\Delta E = 6.1$ kcal/mole for $(H_2O)_2$]. This nonadditivity of 35 kcal/mole in the hexamer is unrealistically large and the $O—O$ separation is too small, but it should be clear from the above that cyclic structures of water polymers are quite stable because of (a) additional stabilization of the H bonds due to nonadditivity effects and (b) the formation of one more H bond than linear structures of the same $n$mer.

3.1.3. Hydrogen Fluoride Dimers and Polymers

Early experiments on hydrogen fluoride vapor were interpreted in terms of an equilibrium between monomers, dimers, tetramers, and hexamers. The hexamer was structurally characterized in 1969 by electron diffraction as a cyclic structure with $R(F—F) = 2.53$ Å and $\theta(FFF) = 104°$.[56]

Ab initio calculations on the hydrogen fluoride dimer were published in 1970 by Kollman and Allen (KA)[57] and Kraemer and Diercksen (KD).[58] Both found a linear structure more stable than a cyclic one, with $\theta(HFF) = 160°$

(KA) and 140° (KD), $R(F–F) = 2.88$ Å (KA) and 2.85 Å (KD), and a linear hydrogen bond $\theta(F—H\cdots F) = 180°$. Subsequent nozzle beam experiments by Dyke and Klemperer[44] confirmed the predicted structure; these workers found a dipolar dimer with $R(F–F) = 2.80$ Å.

Del Bene and Pople[59] carried out STO-4G calculations on HF polymers and found, as in the case of H_2O polymers, that cyclic structures were the most stable for $n \geq 3$. Kollman and Allen carried out a very limited number of calculations on the cyclic HF tetramer and hexamer.[60] Both of these studies found a stabilizing nonadditivity in the H-bond energy as had been the case in the cyclic water polymers. Dyke, Howard, and Klemperer confirmed the Del Bene and Pople structural predictions for HF polymers, finding that trimers and higher polymers were nonpolar.

A subsequent paper on the HF dimer by Lischka[61] was very important because it examined the effect of electron correlation effects on the dimerization energy and minimum-energy F—F distance with a large enough basis set to adequately reproduce the polarizability and dipole moment of the H—F monomer. None of the three correlation methods employed (independent electron-pair approximation, coupled electron-pair approximation, and pair natural-orbital configuration interaction) predicted any significant effect on the H-bond energy or F—F distance relative to the SCF values (calculated to be 3.46 kcal/mole and 2.91 Å). Since Lischka constrained the dimer to have $\theta(H\cdots F—H) = 0°$, the calculated R is expected to be too large and the ΔE too small (by about 0.8 kcal/mole).*

A more complete search of the dimer potential surface, as had been done by Del Bene and Pople at the STO-4G level, was carried out by Yarkony *et al.*[62] using a double zeta plus polarization basis. They found a dimer with a near-linear hydrogen bond, $\theta(FH\cdots F) = 5°$, $R(F–F) = 2.8$ Å, $\theta(H\cdots F—H) = 128°$ and $-\Delta E = 4.6$ kcal/mole, quite similar to that of Kraemer and Diercksen.

3.1.4. Other Nonionic H-Bonded Complexes

A large number of other H-bonded structures have been studied by *ab initio* methods but at this point we will discuss only a few for which gas phase (or matrix ir) experimental information is available.

Carboxylic acids and amides form well-characterized cyclic dimers in the gas phase, and the calculations on cyclic formic acid dimer[63] and formamide dimer[64] find reasonable agreement with experimental structures. The calculated dimerization energies are, as expected, exaggerated in the formamide dimer calculation (which was studied using a small basis set) and in quite good

---

*This estimate is based on a comparison of the energies of the $\theta = 0$ and optimum geometries in Ref. 62.

agreement with the experimental ΔH for formic acid, where a large basis set ("double zeta plus polarization") was used.

The characterization of the complex formed between NH_3 and HCl was the first *ab initio* study of a neutral H bond,[65] and the subsequent experimental[66] verification of gas-phase NH_4Cl was a feather in the cap for E. Clementi and *ab initio* quantum chemistry. A subsequent matrix ir study by Pimentel[67] verified the qualitative features of the structure, i.e., the proton in the hydrogen bond was roughly halfway between the position it had in monomeric HCl and the position it would have in crystalline $NH_4^+ \cdots Cl^-$. Kollman *et al.*[68] compared the structure of $NH_3 \cdots HCl$ and $H_2O \cdots HCl$ as calculated *ab initio* with a 431G basis set and concluded that gas phase $H_2O \cdots H—Cl$, unlike $N_3N \cdots HCl$ would have intramolecular structural parameters similar to isolated H_2O and HCl; this was consistent with Pimentel's matrix ir study of $H_2O \cdots HCl$.[69]

3.1.5. Strong Hydrogen Bonds

The *energy* of ion–neutral hydrogen bonds have, for the most part, been much easier to characterize than neutral–neutral bonds, but the *structure* of these bonds comes exclusively from solid-state data. A direct comparison between the calculations on the gas phase species and the environmentally perturbed solid-state structure must be done cautiously.

The most ambitious set of *ab initio* calculations in this area has been that by Newton and Ehrenson,[70] who studied $H_{2n+1}O_n^+$ ions ($n = 1$–5) and $H_{2n-1}O_n^-$ anions ($n = 1$–5) with a 431G basis set. The agreement with the gas-phase enthalpies found by Kebarle[71] was generally good, and the minimum energy structures were consistent with available x-ray structures containing clusters of hydrated protons or hydroxide anions. A series of other *ab initio*[72] calculations on $H_5O_2^+$, the simplest H-bonded cation, all found a dimerization energy of 32–36 kcal/mole, a minimum energy O—O distance of 2.4 Å, and a proton potential with either a single minimum or very shallow double minimum.

Despite the fact that the properties of negative ions are poorly represented by single-determinant wave functions, the *ab initio* SCF calculated energies of gas-phase hydrogen bonding between F^- and H_2O[73] and Cl^- and H_2O are in near quantitative agreement with experimental values[74] ($\Delta E = -23$ and -13 kcal/mole for $F^- \cdots HOH$ and $Cl^- \cdots HOH$, respectively).

The smallest system with a strong hydrogen bond $(FHF)^-$ has been the subject of the first *ab initio* calculations (in the early 1960s) on a hydrogen bond. Bessis and Bratoz[75] and Erdahl[76] carried out valence bond calculations on this ion, and Clementi and McLean[77] did SCF calculations. None of these workers evaluated the hydrogen bond energy, but later calculations by McLean and Yoshimine,[78] Kollman and Allen,[79] Noble and Kortzeborn,[80] and Almlöf[81] all used SCF calculations to examine the potential surface for FHF^-, and all found the minimum energy to be a linear $D_{\infty h}$[F–H–F]$^-$ structure with F—F distance as 2.25–2.28 Å, in excellent agreement with observed crystal structures of the bifluoride ion.

The calculated energy of the hydrogen bond FHF$^-$ → HF + F$^-$ varied from 52 kcal/mole (Kollman and Allen) to 28 kcal/mole (McLean and Yoshimine) with the energy decreasing monotonically with decreasing total energy. Unfortunately, there is no experimental gas-phase energy for the H bond in the bifluoride ion, but the number of theoretical and experimental studies comparing HOH and HF as proton donors indicate that the reaction enthalpy of F$^-$ + HOH → F$^-$···HOH, $|\Delta H| = 23$ kcal/mole, is a lower bound for the absolute value of the H-bond strength in the bifluoride ion.

Where comparisons can be made, *ab initio* SCF calculations do predict the structures of strong H-bonded systems in very good agreement with experiment. The calculated H-bond energies vary (depending on basis set size) much more for anionic than cationic H bonds, but basis sets capable of giving near Hartree–Fock energies reproduce the experimental ΔE quite well.

3.1.6. Other Noncovalent Intermolecular Interactions

To put calculations on hydrogen bonding in proper perspective, it is necessary to compare them with other noncovalent interactions. The weak dispersion attraction between two He atoms cannot be found at the SCF level, but extensive CI calculations by Bertoncini et al.[82] and Schaefer et al.[83] have been able to reproduce the experimental well depth of the He–He interaction quite well. A stronger interaction exists between He···HF,[84] Ne···HF,[85] and Ne···HOH[86]; these interactions might be called "hydrogen bonds," although there is no classical electrostatic interaction between them. These SCF calculations on rare-gas···HA interactions might contain large relative errors since the interaction energies are very small. Dispersion or intramolecular correlation effects have not been included. Experimental nozzle beam studies by Klemperer on Ar···HCl and Ar···HF[87] indicate that θ(Ar–HCl) is 140°, so one is pressed to call this interaction a B···H—A hydrogen bond. If hydrogen bonding (or polarization) forces were dominant, one would expect that θ(Ar–HCl) would be 180°; the observed θ indicates that dispersion attraction (which would favor $\theta = 0°$) may be playing an important role in determining the observed geometry.*

Alkali halides form well-characterized cyclic dimers in the gas phase and the structure of (LiF)$_2$ was studied theoretically by Baskin et al.[88] using a "double zeta plus polarization" basis. The agreement with the experimental dimerization energy was good, and the cyclic structure was of lower energy

---

*However, Novick et al.[87] point out that the minimum energy structure may well still be the $\theta = 180°$ one and that they are only observing an average θ. Losonszy et al.[85] found a minimum energy at $\theta = 180°$ for Ne···HF in their theoretical SCF calculations (without dispersion terms).

than the linear dimer [in contrast to $(HF)_2$]. Calculations on the dimerization of LiH (not known experimentally) indicated that the correlation contribution to the dimerization energy was negligible,[89] similar to Lischka's result for $(HF)_2$.

If we infer, from comparing the experimental and calculated ΔE for LiF dimerization and the calculated very small correlation energy contribution to LiH dimerization, that closed-shell reaction energies should be well represented in the SCF approximation, we are then puzzled by the calculations of Lipscomb *et al.*[90] and Ahlrichs[91] indicating that the correlation contribution to the dimerization of B_2H_6 is quite large [$-\Delta E(SCF) = 19–20$ kcal/mole; $-\Delta E$ (exp) = 36 kcal/mole]. Despite the fact that both are closed-shell–closed-shell interactions there may be a fundamental reason why the correlation energy is different for four-electron three-center F—Li—F and F—H···F bonds and two-electron three-center B—H—B bonds. It may be that the former interactions are dominated by electrostatics and the latter by specific orbital interactions.

Cation-hydration interactions have been studied by Clementi,[92] Diercksen and Kraemer,[93] Kollman and Kuntz,[94] and Schuster and Preuss[95] by *ab initio* SCF methods. The double zeta plus polarization and more extended bases predict interaction energies in good agreement with the experimental gas-phase interaction energies.

3.2. Spectroscopic Properties

3.2.1. Vibrational Spectroscopy

The single most dramatic experimental evidence of hydrogen bond formation of a compound A—H occurs in the A—H stretching region of the infrared spectra. In the presence of increasing base B, there appears a broad intense red-shifted A—H stretch as the narrower peak due to the "free" A—H absorption disappears. Hydrogen bonding also causes blue shift in intramolecular bending modes (involving the A—H bond) and gives rise to new intermolecular peaks at low frequencies.

A number of *ab initio* calculations[2] have calculated the minimum-energy A—H distance, the A—H force constant K_s, the A—H "intensity enhancement" $|\partial\mu/\partial r|^2$, the intramolecular bending force constant K_b, and the A···B stretching force constant K_h in the hydrogen-bonded dimer. By comparing the distance, frequencies, and intensities of the intermolecular modes with those calculated for the isolated molecule, one sees that the *ab initio* calculations do qualitatively reproduce the increased A—H bond length, the red shift of the A—H stretch, the blue shift of the bend, the increased intensity of the stretch, and the decreased intensity of the bend. The calculated A···B stretching frequencies ($\sim$200 cm^{-1}) are in reasonable agreement with those observed in the far ir.[2]

However, as clearly pointed out by Schaad,[1] the above calculations cannot be rigorously compared with experiments because the normal modes of the H-bonded complex have not been used, only "bond stretching" and "bond bending" motions. Thus, the calculations by Curtiss and Pople on the potential surface of HCN···HF[96] and $(H_2O)_2$[97] were of great interest. These authors used the quadratic force constants from the *ab initio* potential surface and then the *FG* matrix method, outlined by Wilson, Decius, and Cross,[98] to determine the normal modes and vibrational frequencies of the complex. The 431G calculation gave very good agreement with the experimental intermolecular modes found by Thomas[99] for HCN···HF, but the red shift of the H—F stretch was significantly underestimated (149 cm^{-1} calculated, 251 cm^{-1} experimental). This underestimate of the red shift has been found in all the earlier *ab initio* calculations of the A—H force constant dependence on H bonding. It is possibly partially due to the well-known tendency of the Hartree–Fock wave functions to overestimate force constants. If the A—H force constant is too large, a calculation of the perturbation by :B will yield too small a red shift. However, *ab initio* studies by Lischka[61] on $(HF)_2$, including correlation (his independent-pair and correlation-pair methods find H—F monomer stretching force constants close to the experimental), find less than a 0.01-Å increase in the H—F bond length and very small shifts (0–2% change in K_s) for the H—F stretch relative to the monomer. The fact that the normal mode analysis was not done for $(HF)_2$ and that Lischka did not use the optimum dimer geometry to compute the red shift of the H—F stretch preclude a quantitative estimate of the predicted red shift and why it appears to be underestimated in the calculations. It should be noted that infrared experiments on HF vapor by Huong and Couzi[100] lead one to expect a 5% decrease in K_s in the HF dimer.

The Curtiss and Pople study on $(H_2O)_2$ at the 431G level gave qualitative agreement with the intramolecular modes assigned by Tursi and Nixon but also underestimated the red shift in the O—H stretching frequency. Since this basis set overestimates the H-bond energy, these authors also evaluated a few of the diagonal force constants with an extended basis set,* and found significantly reduced intermolecular force constants. Thus, they predict that their intermolecular frequencies from the *FG* matrix analysis (using the 431G potential surface) will be somewhat too large (these have not been studied experimentally).

These vibrational frequencies for $(H_2O)_2$ are very useful in another way because they allow a more precise comparison between the calculated and experimental dimerization energies for water. As has been clearly pointed out by Schaad,[1] the *ab initio* ΔE_0 is related to the ΔH_{298} experimentally observed

---

*Their basis was of "double zeta" quality plus a set of *d* functions on oxygen (631G*); the 431G basis is "double zeta" for the valence shell.

Table 2. Enthalpies of Dimerization for Simple Hydrides

Energy/Enthalpy	$(H_2O)_2$	$(D_2O)_2$	$(HF)_2$	$(NH_3)_2$	$(CH_3OH)_2$
ΔE_0	−5.1	−5.1	−5.4[a]	−3.3[b]	−4.7[c]
$\Delta E_{trans.}$	−0.888	−0.888	−0.888	−0.888	−0.888
$\Delta E_{rot.}$	−0.888	−0.888	−0.296	−0.888	−0.888
$\Delta E_{vib.}$ (intra)	−0.149[d]	−0.113[d]	−0.149[e]	−0.149[e]	−0.149[e]
$\Delta E_{vib.}$ (inter)	4.180	3.910	2.368[f]	3.552[f]	3.552[f]
$\Delta(VB)$	−0.592	−0.592	−0.592	−0.592	−0.592
ΔH^0_{298} (calc.)	−3.4	−3.7	−5.0	−2.3	−3.7
ΔH^0_{298} (exp.)	−5.2 ± 1.5[g]	(−5.2 ± 1.5)[g]	−7.0 ± 1.5[h]	−4.1 ± 0.4[i]	−4.5 ± 0.4[i]

[a] Using the calculations by Lischka[61] and Yarkony et al.,[62] we estimate $\Delta E = -4.2$ kcal/mole and add the same correlation contribution as for $(H_2O)_2$.
[b] Calculations by Kollman and Allen[33,102] find dimerization energies for $(H_2O)_2$ and $(NH_3)_2$ of 5.3 and 2.7 kcal/mole; calculations by Kollman et al.[103] using a different basis set find 8.1 and 4.3 for the two dimerization energies. Thus we have taken the Hartree–Fock dimerization energy of water (3.9 kcal/mole), divided by 1.9 (the average ratio in the above two sets of calculations) and added 1.2 kcal/mole correlation energy to arrive at the tabulated ΔE_0.
[c] Del Bene[34] and Topp and Allen[104] find methanol to form weaker dimers than water by ratios of 0.95 and 0.90, respectively. Thus, we have used the average ratio (0.925) and added 1.2 kcal/mole correlation correction.
[d] From Tursi and Nixon's[40] experimental data.
[e] Assumed to be the same as for $(H_2O)_2$.
[f] Assumed a classical contribution, 0.592 kcal/mole, for each intermolecular vibrational mode.
[g] Ref. 49. [h] Ref. 105. [i] Ref. 106. [j] Ref. 107.

in the following way:

$$\Delta H_{298} = \Delta E_0 + \Delta(PV) + \Delta E_{rot.} + \Delta E_{trans.} + \Delta E_{vib.}$$

where $\Delta E_{rot.}$, $\Delta E_{vib.}$, and $\Delta E_{trans.}$ are the differences between the dimer rotational, translational, and vibrational energies, respectively, at 298 and the corresponding internal energies for two monomers. If we use the intermolecular vibrational frequencies from Curtiss and Pople[91] the intramolecular frequencies of Tursi and Nixon,[40]* and our previous estimate for $\Delta E_0(-5.1$ kcal/mole) we find that $-\Delta H_{298}$ is predicted to be 3.5 kcal/mole, smaller than the currently accepted experimental value (5.2 ± 1.5 kcal/mole). An overestimate of the two intermolecular vibrational frequencies by Curtiss and Pople might increase the calculated $|\Delta H|$ by ~0.2 kcal/mole, but our estimate for the correlation contribution to the dimerization energy (1.2 kcal/mole) is probably an upper bound for this value. The theoretical calculations on $(H_2O)_2$ suggest that the experimental $-\Delta H_{298}$ is nearer the lower error limit (3.7 kcal/mole) than the reported value (5.2 kcal/mole)[49] and closer to the estimate by Bollander et al. (3.6 kcal/mole) which was based on applications of Cluster theory.[101]

By estimating vibrational, rotational, and translational corrections to some other ab initio calculated dimerization energies, we can compare them to water dimerization and available experimental data [this is done in Table 2 for

---

*If we used the intramolecular frequencies from Curtiss and Pople[91] we would find that the contribution of ΔH is +0.163 kcal/mole rather than −0.149 kcal/mole found using the experimental frequencies (Table 2).

$(NH_3)_2$, $(HF)_2$, and $(CH_3OH)_2$]. To calculate these dimerization enthalpies we first attempted to estimate what the *ab initio* dimerization energy would be if one had near Hartree–Fock wave functions for the dimers. We then used the same correlation correction to the H-bond energies as in $(H_2O)_2$, assumed that all the intermolecular modes were classically excited (kT), and that the correction to the energy for the intramolecular modes was the same as for $(H_2O)_2$. The largest error in these estimates probably comes in the estimation of the Hartree–Fock dimerization energy since the available *ab initio* calculations on these dimers [$(NH_3)_2$ and $(CH_3OH)_2$] are relatively crude. The calculated enthalpies for $(H_2O)_2$, $(HF)_2$, and $(CH_3OH)_2$ are in reasonable agreement with experiment, considering the large uncertainties in the experimental values for the first two dimers. Only the ΔE for ammonia dimerization is very different from the experimental value, but more accurate calculations on its dimerization are needed before one should suggest reexamination of the experimental enthalpy.

A number of *ab initio* potential surfaces for the bifluoride ion have been published, and the theoretical attempts to reproduce the experimental spectra have followed two lines: Noble and Kortzeborn, and Kollman and Allen attempted to fit their potentials (and those of McLean and Yoshimine, and Erdahl) to the "experimental" force field proposed by Ibers. The agreement with these force constants was rather poor, even for the very accurate wave function of McLean and Yoshimine.

Almlöf[81] was more clever. He examined enough points on the two-dimensional linear HF_2^- surface $F\overset{r_1}{-}H\overset{r_2}{-}F$ to fit the surface to a high-order polynomial in r_1 and r_2. He then calculated the vibrational frequencies directly and found excellent agreement with experiment with hydro and deutero bifluoride ions. His results indicate that the potential surface was far more anharmonic than was implied by the experimental force fields and suggests that the use of "experimental" force fields to fit force constants in any *ab initio* calculation is a less fruitful approach than direct calculation of the vibrational frequencies. In fact, this approach would be very interesting to try for $(H_2O)_2$, $(HF)_2$, or $HCN\cdots HF$ where, as mentioned above, the calculations based on harmonic force fields underestimate the red shift in the A—H frequency. That inclusion of anharmonicity in the A—H frequency in the monomer and dimer A—H stretches will increase the red shift from the harmonic value might be expected in view of Sandorfy's[108] conclusion that weak H bonds increase the anharmonicity of the A—H stretch (strong H bonds, like HF_2^- have an *increased* anharmonicity).

3.2.2. Electronic Spectroscopy

Another well-characterized experimental property of the hydrogen bonding of ketones is the blue shift of the $n \rightarrow \pi^*$ excited state and a red or blue shift of the $\pi \to \pi^*$ slate. Both Del Bene[109] and Iwata and Morokuma[110] have shown that a minimal basis set plus limited configuration interaction or electron hole potential SCF is capable of reproducing the blue $n \to \pi^*$ shift in good agreement with experiment. In fact, Del Bene showed that the $n \to \pi^*$ shift observed upon dissolving acetone in water is accurately reproduced by the calculation when one water is H bonded to the carbonyl[111]; Iwata and Morokuma's[112] calculations on $H_2CO \cdots HOH$ and $H_2CO \cdots 2HOH$ indicated that two waters bonded to the two carbonyl lone pairs should give roughly twice the blue shift of one water. Iwata and Morokuma[112] have also examined the source of the often observed red shift of the $\pi \to \pi^*$ transition on H bonding using the model system acrolein–H_2O and have found that the *red* shift depends on the presence of conjugation to the carbonyl group (water H bonding causes a blue shift in the $\pi \to \pi^*$ transition of formaldehyde). Both results could be rationalized by comparing the negative charge of the oxygen in the ground and excited states. If the oxygen is more negative in the excited states, H bonding is stronger in this state and one calculates and observes a red shift for the acrolein $\pi \to \pi^*$ transition upon H bonding. If the oxygen is less negative in the excited state, we expect H bonding to cause a blue shift. Blue shifts are predicted for the $\pi \to \pi^*$ transition of formaldehyde and the $n \to \pi^*$ transitions of formaldehyde and acrolein.

Although electronic spectra are not as sensitive to H bonding as vibrational spectra, extension of the theoretical calculations to other transitions and other compounds (e.g., amides)[113] is clearly desirable. Because it is often not so easy to relate observed transitions to specific excitations (e.g., note the complexity of the uv spectrum of ethylene), calculations on the effect of H bonding on the spectra coupled with experimental studies of the effects of various H-bonding solvents on the spectra may play an important role in a better understanding of the transitions.

3.2.3. Magnetic Resonance Properties of H-Bonded Systems

The characteristic downfield shift of *n*uclear *m*agnetic *r*esonances (NMR) of protons involved in hydrogen bonding has played an important role in experimental studies of H bonds. A consideration of the electron-density shifts in the linear water dimer $H_2O \cdots HOH$ would indicate that the two protons on the proton acceptor and the H-bonding proton of the proton donor should be deshielded and the other proton on the proton donor water shielded relative to isolated H_2O. This assumes that the diamagnetic contribution to the chemical shift dominates the paramagnetic, but this may not be a bad approximation for protons. The only *ab initio* calculation[114] on NMR chemical shifts of an H-bonded system uses Diercksen's water dimer wave function[38] and the Karplus–Kolker perturbation method[13] to calculate the proton chemical shifts in the water dimer, finding the following chemical shifts

$$+4.0 \; H \quad\quad H—O$$

relative to monomeric H_2O. The downfield shift of the hydrogen bonded proton is reasonable (but small); however, the results for the other protons are puzzling. Unfortunately, it is unlikely that one could ever see isolated proton resonance for a water dimer so no direct experimental test is possible. The large gauge dependence of the above calculations indicates that calculations employing gauge-invariant atomic orbitals[115] would be very interesting, both for proton and for N^{15}, O^{17}, and F^{19} nuclei of molecules involved in H bonding.

Electric field gradients at deuterons have been examined theoretically and experimentally for H-bonded systems.*

3.3. Summary

Although we have discussed in detail only some of the hydrogen bonded complexes studied by *ab initio* methods, we have concentrated on those for which there are adequate experimental data. A comparison of H-bond structures calculated using a "double zeta plus polarization" basis for $(HF)_2$ and $(H_2O)_2$ and those determined experimentally, indicates that *ab initio* structure predictions for weak H-bond structure can be quantitatively reliable. There have been a large number of dimers studied with cruder basis sets, and these give qualitatively correct structure predictions (underestimating the A···B distance by a consistent amount). In the area of gas-phase structure of H-bonded dimers, it is probably fair to say that theory is leading experiment; many structural predictions have been made which await experimental testing. In addition, there are interesting questions posed in experimental studies which can be examined theoretically. For example, a recent study of ether···HCl complexes[116] suggested that the structure of R_2O···2 HCl was

$$R_2O \quad \begin{matrix} HCl \\ \\ HCl \end{matrix}$$

rather than R_2O···HCl···HCl, a result in qualitative agreement with potential

---

*See Ref. 2 for further analyses of these studies.

surfaces of $H_2O\cdots HCl$.[103,104] Direct calculations on the above $2:1$ complex are clearly feasible.

Hydrogen-bond energy predictions are an area where the theory is somewhat less satisfactory because the predicted ΔE are far more basis-set dependent than the structure. As the basis set is improved beyond the double zeta level, the calculated dimerization energy follows a smooth downward trend, but only for $(H_2O)_2$ and $(HF)_2$ has the Hartree–Fock limit been approached. The estimated Hartree–Fock dimerization energy for $(H_2O)_2$ is 3.9 kcal/mole; for $(HF)_2$ 4.2 kcal/mole. Only for $(H_2O)_2$ has a vibrational analysis been carried out, allowing a direct calculations of ΔH_{300} (dimerization) from the theoretical methods. This number appears to be sufficiently accurate to warrant reexamination of the experimental value ($\Delta H_{298} = -5.2 \pm 1.5$ kcal/mole), since it is likely that the lower bound of that value is closer to the true enthalpy than the mean value. Similarly, the HF dimer ought to be examined, because the experimental ΔH_{298} is certainly too high. Methanol and ammonia dimerization ought to be examined theoretically with double zeta plus polarization bases, in order to compare with the more reliable experimental values for these two dimers. Thus, in the area of H-bond dimerization energies, theory and experiment are in a position to work in a very complementary fashion. An example of this is the recent reexamination of the dimerization energy of HCN,[117] caused by a theoretical calculation of the value.[118] In Table 3, we attempt to give a summary of *ab initio* calculations on those gas-phase complexes where there exist experimental data. Although not exhaustive, this table should summarize the "state of the art." The list is not complete, but contains examples of the different *types* of H-bonded (and other donor–acceptor) complexes for which experimental and theoretical studies are available. The reader should be aware that not all the experimental–theoretical comparisons involve the same energy term (e.g., ΔH°_{298} vs ΔE_0).

In the areas of vibrational spectroscopy and electronic spectroscopy it is clear that studies of H-bonded complexes may play the same important role in the unraveling of spectra that theoretical studies on isolated molecules have. Normal mode analysis of such species as $H_5O_2^+$, $(NH_3)_2$, and $(CH_3OH)_2$ would be of considerable interest. The effect of H bonding on excited states of amides, amines, sulfoxides, sulfides, and quinones will be useful subjects of theoretical studies, whether by *ab initio* or appropriately parameterized CNDO–CI methods.[126]

In the area of NMR, *ab initio* calculations have a long way to go to have an important impact on experimental studies, both because of the difference in the nature of the experiment (NMR compared to uv or ir) and because of the difficulty in *a priori* predictions of small chemical shifts. However, the GIAO method of Ditchfield[114] as applied to ^{13}C chemical shifts in organic molecules might prove to be useful. It certainly would be interesting to have some reliable chemical shift calculations on simple H-bonded systems.

Table 3. *Comparison of Calculated and Experimental Energies and A···B Distances for Gas-Phase Complexes*

Complex	Calculated		Experimental	
	$-\Delta E_0(\Delta H_{298})$, kcal/mole	R, Å	$-\Delta E_0(\Delta H_{298})$, kcal/mole	R, Å
He···He[a]	0.024	3.0	0.021 ± 0.003	2.95 ± 0.05
HSH···SH$_2$[b]	1.45	4.4	1.7 ± 0.3	—
H$_2$NH···NH$_3$[c]	2.3	3.3	4.1 ± 0.4	—
HOH···OH$_2$[d]	3.4	3.0	5.2 ± 1.5	2.98
CH$_3$OH···OHCH$_3$[e]	3.7[f]	3.0	4.5 ± 0.4	—
NCH···NCH[g]	3.7	3.2	3.8 ± 0.2	—
HF···HF[h]	5.0	2.8	7.0 ± 1.5	2.80
ClH···O(CH$_3$)$_2$[i]	8.6	3.1	7.1 ± 0.8	—
HOH···Cl^{-j}	11.4	3.3	13.1	—
HCONH$_2$···OCHNH$_2$[k]	14–19	—	14.0	2.9
HCOOH···OCHOH[l]	16.2	2.7	14.0	2.7
HOH···F^{-m}	22.2	2.5	23.3	—
H$_2$OH$^+$···OH$_2$[n]	32.2	2.4	36.0	—
Li$^+$···OH$_2$[o]	34.1	1.9	34.0	—
LiF···LiF[p]	62.8	2.22, 2.65	58.9	—

[a]See Ref. 21 for an extensive discussion of the theoretical and experimental well depth for He···He.
[b]Theory, Ref. 119; experimental, Ref. 120.
[c]Theory, Table 2 this paper; experimental, Ref. 106.
[d]Theory, Table 2 this paper; experimental, Ref. 49.
[e]Theory, Table 2 this paper; experimental, Ref. 107.
[f]Note that CH$_3$OH is "predicted" to have a higher $-\Delta H$ of dimerization than H$_2$O, despite the *ab initio* results quoted earlier. This is because (see Table 2) we have used the Curtiss and Pople predictions for $\Delta E_{vib.}$ (inter) (4.18 kcal/mole) of (H$_2$O)$_2$ and, since no value was available, the classical value for $\Delta E_{vib.}$ (inter) (3.55 kcal/mole) for (CH$_3$OH)$_2$.
[g]Theory, Ref. 118; experimental, Ref. 117.
[h]Theory, Table 2 this paper; experimental, Ref. 105.
[i]Theory, Ref. 104; experimental, Ref. 121.
[j]Theory, Ref. 73; experimental, Ref. 74.
[k]Theory, Ref. 64; experimental, see Ref. 122, this is not really a gas-phase value but should be close since the H bonds are stronger than the intermolecular crystal effects.
[l]Theory, Ref. 63; experimental, Ref. 123.
[m]Theory, Ref. 73; experimental, Ref. 74.
[n]Theory, Ref. 72; experimental, Ref. 71.
[o]Theory, Ref. 73; experimental, Ref. 124.
[p]Theory, Ref. 88; experimental, Ref. 125, the distances are, respectively, the Li–Li and F–F distances in the D_{2h} minimum-energy geometry.

4. Generalizations about the Hydrogen Bond

4.1. H-Bond Structure

The *ab initio* calculations have been able to give structural information on hydrogen-bonded systems matched in detail only by solid-state neutron diffraction. Since the latter includes environmental effects, it is clear that the potential surfaces from *ab initio* calculations give the best information on the structure of the isolated hydrogen bond.

4.1.1. The A–B Distance

Considering the general H bond X—A—H···B—Y, Kollman and Allen[102] and later Kollman *et al.*[103] pointed out that for fixed X—A—H, the distance between fragments $R(\text{A—B})$ is nearly constant for a series of related B—Y molecules, although within a series of differently hybridized B molecules[103] (e.g., HCN, H_2CHN, and H_3N) there is a small but systematic dependence on the *p* character of the lone pair donated (the shorter A—B bonds having a greater *p* character in the B lone pair). The A—B distance is much more strongly a function of the nature of the A—H bond than the nature of B (see Table 4) because exchange repulsion is a very steep function of distance and the A—H bond pair is much less polarizable than the lone pair. The less tightly bound the B lone pair, the larger the exchange repulsion, but the greater the charge transfer and polarization energies. Allen[127] has shown that the product of the "lone-pair" length and ionization potential are approximately constant for the hydrides of a given row of the periodic table. Although the above generalizations must be considered tentative because they are based on wave functions of only double zeta quality, they receive some support from the more exact calculations and experiments on $(\text{HF})_2$ and $(H_2O)_2$, where

Table 4. Geometrical Parameters for First and Second Row Hydrides[a]

Electron donor	Parameter	Proton donor					
		HF	H_2O	NH_3	HCl	H_2S	PH_3
HF	R	2.69	2.94	3.22	3.37	3.68	4.05
	θ	42	60	65	39	34	76
H_2O	R	2.64	2.85	3.24	3.17	3.59	4.18
	θ	6	37	53	13	22	52
NH_3	R	2.67	2.93	3.28	3.13	3.52	4.32
	θ^b	—	—	—	—	—	—
HCl	R	3.42	3.70	4.10	4.05	4.40	4.80
	θ	71	82	75	70	75	87
H_2S	R	3.36	3.66	4.00	4.09	4.39	4.70
	θ	68	78	88	71	73	87
PH_3	R	3.40	3.74	4.06	4.01	4.40	4.80
	θ^b	—	—	—	—	—	—

[a] R calculated to ± 0.05 Å, θ to $\pm 5°$; distances in Å; angles in degrees.
[b] Assumed to be 0.

the stronger proton donor and weaker acceptor H—F have a shorter A—B distance (2.8 Å) than the weaker donor and stronger acceptor $H_2O[R(O-O) = 3.0$ Å$)]$. Recall that both interactions are calculated to be approximately the same strength. A further insight into the R dependence could come from an analysis of the localized orbitals and energy components of a number of hydrides of varying proton donor and acceptor strengths.

4.1.2. H-Bond Directionality

The question of the optimum position of approach of the A—H bond toward the proton acceptor B—Y has been examined in some detail. For complexes with ammonia as proton acceptor, the A—H bond points along the C_3 axis; recall that the directionality in the water dimer ($\theta = 40°$) and HF dimer ($\theta = 52°$) appear to indicate an intrinsic "lone-pair" directionality of the hydrogen bond. Further support for this comes from studies of carbonyl donors, where the A—H bond approaches at a near 60° angle to the C=O bond. Pimentel and McClellan[9] noted a number of years ago that H bonds seemed to point in the directions of "lone pairs," but more recent examinations of crystal structures of C=O[128] and R_2O[128,129] donors have argued against ascribing any lone-pair directionality for these proton acceptors in the solid state. On the other hand, Del Bene[130] has pointed out that the equilibrium geometries calculated for gas-phase dimers are close to those predicted for considering an A—H bond pointing toward a hybridized lone pair.

What is the cause of this directionality? Is it incipient covalent bond formation? Kollman[131] has proposed that the key to understanding the different H-bond directionalities lies in the differences in the orbital energies. For example, in HF the highest occupied orbital is of π symmetry with the σ lone-pair orbital 0.10 a.u. higher in energy. An A—H bond gains more charge-transfer energy by approaching the π orbital than the σ orbital and this causes the H—F$\cdots$H angle to deviate from the 180° one would expect from the optimum dipole–dipole interaction energy. Kollman *et al.*[103] later showed that electrostatic potential considerations could predict $\theta(H-X\cdots H)$ to be $\neq 180°$. van Duijneveldt[132] has pointed out that a consideration of the dipole [which would predict $\theta(H-F\cdots H)$ to be 180°] and quadrupole (which would predict a T-shaped dimer) of HF can qualitatively rationalize the structure of the HF dimer. One might also use quadrupole and higher moments to examine $\theta(Y-B\cdots H)$ in other electron donors; it appears that electrostatic considerations can explain some of the lone-pair directionality observed.[103]

The most definitive study on the lack of well-defined lone-pair directionality in water dimer was the study by Hankins *et al.*[137] By moving the external hydrogen of the proton donor perpendicular to the original symmetry plane of the dimer, they studied the dependence of the energy and found virtually no difference in the energy for θ from -40 to $+40$ degrees. In the dimer with the external hydrogen *cis* ($\theta = -40°$) to the proton-acceptor hydrogens, the energy is higher than the $\theta = 0°$ geometry, even though in this configuration the O—H is pointing toward a lone pair (see Fig. 1). Thus, external atom effects play an important role in H-bond structure and appear to be as important as lone-pair effects in determining structure.

In summary, the directionality found in calculations of gas-phase H bonds is close to that predicted from hybridization considerations,* although the nature of the force causing this directionality appears to as much electrostatic as "covalent bonding."†

4.1.3. Linearity of the H Bond

In the absence of intramolecular constraints (such as those found in the H_2O cyclic trimer[35] and 1,3 propanediol[133]), one might expect the A—H···B angle to be exactly 180°. A number of potential-surface studies on simple dimers show this not to be the case. While the deviations are very small (less than 5°) and the energy differences between $\theta = 180°$ and the optimum angle less than 0.1 kcal/mole, the effect is real. This effect can also be rationalized by electrostatic considerations: for example, the HF dimer is predicted to look as follows[62]:

$$\underset{\underset{\text{H}}{5°}}{\overset{\overset{\text{H}}{52°}}{\text{F}\cdots\text{F}}}$$

If the H···F—H angle is $\neq 180°$ there is less H···H repulsion for a slightly nonlinear H bond, and this causes the H-bonded hydrogen to tilt away from the other hydrogen. No such tilt is found in the bifluoride ion, where $\theta(\text{FHF}) = 180°$. A consideration of repulsion of the external atoms on the basic A—H···B fragment of other dimers can be used to rationalize other calculated deviations from $\theta(\text{A—H···B}) = 180°$.

4.1.4. Effect of Multiple H Bonds

The calculations on sequential polymeric H-bonded structures [e.g., $(H_2O)_6$][35] compared with $(H_2O)_2$, indicate a decreased O···O length [2.73 Å for $(H_2O)_2$ and 2.44 Å for $(H_2O)_6$] of the same order of magnitude ex-

---

*For example, in $(HF)_2$ and $(H_2O)_2$ the "lone-pair" angle found in localized orbital studies or Nyholm–Gillespie considerations would predict θ (H—F···H) ≥ 71° and θ (H—O···H) ≥ 57° and the best *ab initio* surfaces find θ (H—F···H) = 52° and θ (H—O···H) = 40°.

†Although "charge transfer" might be considered the same as covalent bonding, it has been argued by Kollman[131] that the charge-transfer term favors θ (H—F···H) = 90° and so it alone cannot explain the $(HF)_2$ directionality.

perimentally observed when going from the dimer $[R(O\cdots O)=2.98\text{ Å}^{(45)}]$ to ice I $[R(O\cdots O)=2.76\text{ Å}].^{(47)}$

4.1.5. Proton Potential Functions

Motion of the proton between A and B in an A—H···B bond has been the subject of a number of studies. In strong symmetric H bonds such as HF_2^- and $H_5O_2^+,^{(79)}$ there is a single proton well for F—F (or O—O) distances less than 2.4 Å, with a double well appearing as the A–B distance exceeds 2.4 Å. In neutral H bonds, such as $H_3N\cdots HF,^{(102)}$ only one well is found (with the hint of an inflection point) until one gets to N···F distances much greater than the minimum energy. Studies of coupled-proton motion in formic acid dimer and guanine–cytosine$^{(63)}$ find that the nature of the potential function (single or double well) is very basis-set dependent. Such proton motion is also likely to be very solvent dependent so that gas phase calculations may not adequately represent proton motion in condensed phases. Newton and Ehrenson$^{(70)}$ point this out clearly in their studies of the hydrated proton. Rapid proton shuffling occurs between water molecules in the center of the complex but not on the edge. Meyer *et al.*$^{(134)}$ have studied the correlation effect on the proton motion in $H_5O_2^+$; at $R(O—O)=2.74$ Å, where there is a double well potential, correlation effects lower the barrier by about 50%.

In virtually all neutral dimers,$^{(104)}$ the increase in the A—H length upon H-bond formation A—H···B is quite small (less than 0.05 Å), the one exception so far studied theoretically being $H_3N\cdots HCl,^{(65,104)}$ where the proton position in the H bond is halfway between completely transferred to NH_3 and completely fixed on HCl. Only a single minimum is found in this proton potential at the minimum-energy N—Cl distance. Morokuma *et al.*$^{(112)}$ have found significant differences between the proton-potential function in ground and various excited states of the hydrogen maleate ion [which has a strong intramolecular H bond $R(O—O)=2.4$ Å].

4.2. Contributions to the H-Bond Energy

In Section 1.2 we described the methods employed in decomposing the hydrogen-bond energy into its different components: electrostatic, exchange-repulsion, polarization, charge transfer, and dispersion. A comparison of the different results obtained in perturbation and SCF calculations on hydrogen-bonded systems has been reviewed elsewhere,$^{(2)}$ but a summary of the important results follows: the Coulson estimates,$^{(25)}$ van Duijneveldt's perturbation calculations,$^{(24)}$ and Morokuma's SCF calculations$^{(26)}$ agree that all five components contribute significantly to the hydrogen-bond energy. The different SCF calculations separately determining the energy components indicate that the magnitude of the components is very basis-set dependent.[110,112] This is true both because the different basis sets represent the monomer properties (dipole and quadrupole moment and polarizability) with different accuracy and because the components vary rapidly with A···B distance and the different basis sets predict different minimum energy A···B distances. Of all the separate components, the electrostatic energy is the only one which parallels the total H-bond energy when one compares a number of different H-bonded complexes. This was noted by van Duijneveldt in his perturbation calculations on Ne, H_2O, H_3N, and HCN as proton acceptors and HF, H_2O, and H_3N as proton donors.[24] Morokuma, in his SCF calculations on carbonyl–water H bonds, πH bonds, and acrolein excited-state H bonds has pointed out[109] that the relative strength of these H bonds is reflected in the electrostatic energies.

It should be emphasized that this electrostatic energy is the total electrostatic interaction, not just a dipole–dipole interaction energy; by itself this component of the energy would predict an ever-increasing attraction as the fragments approach each other. The fact that the electrostatic energy often comes reasonably close to predicting the total H-bond energy implies a significant cancelation between the Pauli exchange repulsion and the polarization, charge transfer, and dispersion energies. In this context, interesting approaches by Bowers and Pitzer,[135] Fink,[136] and von Niessen[137,138] should be mentioned. Bowers and Pitzer used a bond-orbital approach to determine energy components for the water dimerization and found results similar to those derived from SCF calculations. Fink has studied Li···He interactions using an energy component approach. This method allows "freezing" of one part of the wave function while allowing the rest to change; the method might prove a useful complement to Morokuma's approach for other intermolecular interactions. von Niessen has applied his method of molecules in molecules to $(HF)_2$ and $H_3N···HOH$, transferring the localized orbitals from the individual molecules and watching them change as the two molecules approach each other. He finds that $(HF)_2$[139] is dominated by electrostatic effects, with little energy gained by allowing the orbitals to relax; in $H_3N···HOH$,[140] changes in the nitrogen lone pair and O—H bonding orbital play a significant role in the attraction energy.

Kollman et al. have made use of the fact that the electrostatic energy often approximates the total H-bond energy and the results found in the studies of protonation sites by Bonaccorsi et al.[139] They compare the electrostatic potential surrounding a wide variety of proton donors and acceptors. By examining the electrostatic potential at a reference distance (2 Å) from the proton along the A—H line and 2.12 Å (for atoms of the first row) along the B lone-pair direction, they were able to order the relative H-bond forming ability of first and second row hydrides, C—H proton donors, "π" proton acceptors, differently hybridized proton acceptors (H_3N, H_2CHN and HCN), ionic proton acceptors (F^-), and ionic proton donors (H_3O^+). The difference in the electrostatic potential surrounding water dimer and water monomer allowed one to rationalize the difference in the nonadditivities for the different water trimers (sequential, double donor and double acceptor), and the qualitative difference between $\theta(H \cdots B - Y)$ for the first and second row donors was also explicable in terms of the electrostatic potential.

The electrostatic potential at a fixed point from the proton acceptor B contains indirect information about the electrostatic polarization and charge-transfer energies. The more negative the potential, the greater the electrostatic interaction with proton donor H—A. The greater the potential, the greater it is able to polarize the H—A molecule, and the greater the potential the easier it is for partial donation of electrons to H—A to take place.

4.3. Charge Redistribution and Charge Transfer

The charge redistribution in H-bonded complexes compared with that found in isolated molecules has been studied extensively, both using Mulliken populations[2] and charge-density difference plots.[27,38,57,79] The basic findings using the Mulliken atomic populations are supported by the charge-density difference maps and are as follows: in a X—A—H$\cdots$B—Y complex compared to the isolated XAH and BY molecules, X, A, and B gain electrons and Y and H lose electrons. There is also a small amount of charge transfer ($\sim$0.05 electrons) from BY to XAH in a moderately strong ($\Delta E = 5$–10 kcal/mole) hydrogen bond. These charge redistributions are consistent with one's expectations that the polarization effects will increase the existing polarities of the molecules as follows,

$$X - A - H \qquad B - Y$$

and H and Y will lose electrons to the other atoms. An example of a charge-density difference map is presented in Fig. 2. As pointed out in Section 4.1, the minimum energy R in an A—H$\cdots$B dimer is mainly a function of A—H; however, the charge redistribution (polarization and charge transfer) are much more strongly a function of B. Kollman and Allen[102] and later Allen[127] showed that at fixed $R(A-B)$ the charge redistribution effects were almost the same as for H_2N—H$\cdots$B, HOH$\cdots$B, and FH$\cdots$B. When B was changed from HF to H_2O to H_3N [all at the same $R(A-B)$], there was a systematic and large increase in charge redistribution.

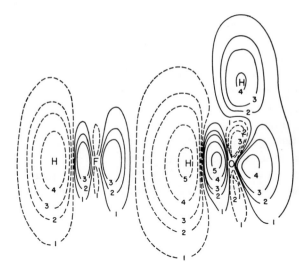

Fig. 2. Electron-density difference map of HFHOH in molecular plane. An increase of electron density upon dimer formation is represented by (——) and an equal decrease by (– – –). The contours are:

1. $0.001e^-/a_0^3$;
2. $0.003e^-/a_0^3$;
3. $0.005e^-/a_0^3$;
4. $0.01e^-/a_0^3$;
5. $0.02e^-/a_0^3$.

See Ref. 57 for details of calculation.

Dreyfus and Pullman[27] have compared the charge changes due to exchange repulsion with those due to polarization and charge transfer and have concluded that a large part of the decrease in electron density near the H-bonding proton is due to exchange repulsion. This shows an important difference between covalent and hydrogen bonds, since covalent bond formation is usually accompanied by charge buildup between the nuclei involved in the bond. Yamabe and Morokuma[140] have separated the charge-distribution changes in H-bonded species involving C=O groups and H_2O into exchange polarization and charge-transfer effects and find, in contrast to the magnitude of the polarization and charge transfer energies, the actual charge redistribution due to polarization is significantly larger than that due to charge transfer. This gives some support to the conclusion (based only on Mulliken populations) by Kollman and Allen[102] that the amount of "charge shift" (polarization) was usually significantly larger than the amount of charge transfer in weakly bound dimers.

The charge densities found in the bifluoride ion[79] are consistent with the "deshielding" of the proton in this species relative to the proton in HF, despite the negative charge on the molecule. The observed downfield PMR shift of FHF^- relative to HF could thus be rationalized by simple deshielding arguments.

A number of models of H bonding have focused on charge-transfer effects as the key explanation of the H bond,[141] but evidence from the *ab initio* calculations to date shows that a rationalization of qualitative structural features and a comparison of relative energies of H bonds can also be done within an electrostatic model.[104] Charge transfer plays a role in spectroscopic properties such as the A—H intensity enhancement upon H-bond formation, but polarization effects appear to be at least as important in contributing to the intensity enhancement.*

"Charge transfer" has been used here to refer to electron transfer from the proton acceptor to a donor (or, in valence bond language a contribution of a dative structure $A^{(-)}: H—B^{(+)}$ to the overall *ground-state* wave function). Spectroscopically, "charge transfer" refers to an electronic excitation where the excited state has an electron transferred from the electron donor molecule to the acceptor. Now a typical Mulliken donor–acceptor complex, such as pyridine–I_2, has an obvious low-lying transition which involves charge transfer from pyridine to I_2. Considering the water dimer, which has a $-\Delta H$ comparable to the pyridine–I_2 complex. Kollman and Allen have shown that upon H-bond formation,[142] the electron donor orbitals are lowered and those of the proton donor raised. Thus, as one can see from Fig. 3a one might expect the lowest energy charge-transfer transition to be from the *proton donor* lone pair to the *proton acceptor* antibonding orbital, the opposite direction to which "charge transfer" takes place in the ground state. In H-bonded complexes like $H_3N\cdots HF$ (Fig. 3b) the change in orbital energies as the two molecules come together is too small to change the highest occupied orbital from mainly N lone pair, but again the lowest energy "charge-transfer" transition based on the orbital energies would be from electron acceptor to electron donor. The above simple analysis based on orbital energies needs to be examined with CI calculations, but if valid its implications are interesting. First, charge-transfer transitions (from electron donor to acceptor) have hardly ever been observed in H-bonded systems because these transitions are *blue* shifted relative to intramolecular transitions. Second, in homomolecular dimers, like $(H_2O)_2$, the lowest energy transition should be from the lone pair of the proton donor water to the O–H antibonding orbital of the electron donor. Even though this transition is red shifted relative to the monomer bands, it would be very weak because of the very small overlap between the donor and acceptor orbitals.

4.4. The Inductive Effect on H Bonds and Proton Affinities

Del Bene has studied a series of H bonds $X—A—H\cdots B—Y$, where X and Y have been varied through the series of H, CH_3, NH_2, OH, and F.[130] As one goes through the above series from X = H to X = F, $X—A—H$ progressively becomes a better proton donor. As one goes from Y = H to Y = F, B becomes a poorer proton acceptor. Both of these trends were rationalized on the basis of electrostatic effects. The more electronegative X became, the more positive the proton became and $X—A—H$ became a better proton donor; the more

---

*See Refs. 2 and 7 for a more extensive description of "charge-transfer" models in H bonding; see also a recent analysis of H-bond energies with a charge transfer model by Ratajczak.[138]

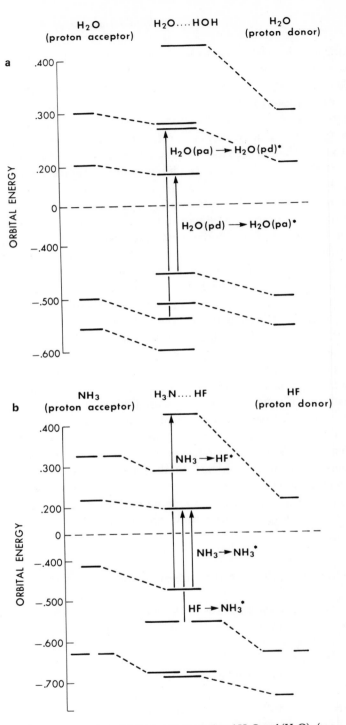

Fig. 3. (a) Molecular orbital energies for isolated H_2O and $(H_2O)_2$ (see Ref. 103). (b) Molecular orbital energies for NH_3, HF, and $H_3N \cdots HF$ (see Ref. 103).

electronegative Y, the less the electrons on B and the smaller the H$\cdots$B attraction.

It is interesting to compare the inductive effect on proton affinities with that found in hydrogen bonding. Johansson *et al.*[143] have studied a number of protonation reactions by *ab initio* methods and found that in protonation, $H^+ + B—Y \rightarrow (H—B—Y)^+$; $B—Y$ with $Y = CH_3$ and NH_2 had a higher affinity than $B—H$; with $Y = OH$ and F, $B—Y$ has a lower proton affinity than $B—H$. Thus in proton affinities the ability of the substituent to disperse the positive charge outweighs simple electrostatics. Comparatively $B—Y$ with $Y = CH_3$ and NH_2 ($B—CH_3$ and $B—NH_2$) are better acceptors toward a proton than $B—H$, but poorer acceptors in a hydrogen bond $A—H\cdots B$. A further interesting example of this is the different CH_3 inductive effect in CH_3F as compared with that in CH_3OH or CH_3NH_2.[144] In all three cases, each methyl compound is calculated (and found) to have a higher proton affinity than its parent hydride. However, in hydrogen bonding to H_2NH, CH_3F is a *better* proton acceptor than HF, whereas CH_3OH and CH_3NH_2 are poorer than H_2O and H_3N, respectively. This can be nicely rationalized by comparing the electrostatic potentials at 2.12 Å from the heteroatom. The potentials for CH_3NH_2 and CH_3OH are less negative potentials than the potentials in NH_3 and H_2O. Thus, the nature of the CH_3 inductive effect depends on the base as well as the particular process (protonation or H bonding) involved. Comparing CH_3F and HF, both electrostatic and inductive effects make CH_3F a better proton acceptor toward either H^+ or H—X. Comparing CH_3NH_2 with NH_3 and CH_3OH with H_2O, one finds that electrostatics causes the methyl compounds to form *weaker* H bonds to a "weak acid" like ammonia, but for strong acids like H^+, the inductive effect "wins out" and the methyl compounds are *stronger* proton acceptor than the hydrides. This trend is also beautifully illustrated in the different CH_3 effect on the experimental proton affinities (in kcal/mole): $PA(HF) = 114$, $PA(CH_3F) = 151$; $PA(H_2O) = 164$; $PA(CH_3OH) = 181$; $PA(NH_3) = 207$, $PA(CH_3NH_2) = 216$.[145] The electrostatic and inductive (polarization + charge transfer) effects are working in the same direction for the fluoro compounds so $\Delta PA = 37$ kcal/mole. For the nitrogen and oxygen electron donors, the electrostatic and inductive effects oppose each other so ΔPA is smaller.

4.5. What Makes a Hydrogen Bond Unique?

This question has been considered by Murrell and Coulson in terms of the energy components, and these authors concluded that it is the smallness of the exchange repulsion that distinguishes H bonds from other weak noncovalent interactions. A recent study of Lathan and Morokuma[146] allows us to compare hydrogen bonds with the weak donor–acceptor interaction etherdicyanocarbonyl

In this complex, as in the H-bonded case, the electrostatic energy is the key in determining both the orientation of approach and the net interaction energy. At the minimum energy distance the exchange repulsion is small (of the same order of magnitude as observed in H-bonding calculation), although this exchange repulsion appears to have a steeper distance dependence than found in H-bonded complexes. One might still argue that less exchange repulsion is the key to H bonds because observed A—H···B distances are significantly smaller than van der Waals radii, whereas $D:A$ distances in donor–acceptor complexes are close to the sum of D and A radii. However, these explanations are not particularly satisfying when it appears, from the evidence above, that electrostatics is the key term in determining much of H-bond *and* donor–acceptor complex structure. A comparison of energy components for "double zeta plus polarization" calculations on $(LiF)_2$ and $(HF)_2$ in both linear [which is the lowest energy structure for $(HF)_2$] and cyclic [which is the optimum for $(LiF)_2$] might give further insight into this question.*

 Kollman *et al.*[103] have proposed that the sign of the electrostatic potential surrounding A—H and B be used as index of whether one should call an interaction a "hydrogen bond." If the electrostatic potential 2 Å from H along the A—H line is not positive (extrapolation of the proton donors examined in this study F—H, O—H, N—H, C—H, Cl—H, S—H and P—H indicates it is not for H_2B—H or H_3Si—H) or if the electrostatic potential 2.12 Å from B is not negative, then the interaction should not be considered a hydrogen bond. Rare gas electron donors are on the borderline of being called a hydrogen bond according to the above definition. As noted previously, structural studies on Ar···H—Cl yield a $\theta(ArClH)$ angle of 45°, indicating that "electrostatic and polarization forces" may not be dominant here and that the observed structure is a compromise between H-bonding forces (favoring $\theta = 0$) and dispersion (favoring $\theta = 180°$). Thus rare gas···H—A interactions appear to be on the borderline of being a hydrogen bond.

 What about the electron deficient hydrogen bonds in the boron hydrides discussed above? It appears that they are an interesting case of closed-shell–closed-shell attractions *not* dominated by electrostatic forces, which have played a major role in all of the interactions discussed to this point. In fact, it may be conceptually useful to consider three classes of donor–acceptor interactions: (1) electrostatic dominated, such as H bonds and many other dative

---

*We (Peter Kollman and D. M. Hayes) are currently examining this question.

complexes; (2) dispersion dominated (rare gas interactions); and (3) orbital interaction dominated, such as $2BH_3 \rightarrow B_2H_6$. Pimentel's three-center MO model, which attempted to put boron hydride bonding and ordinary hydrogen bonding in the same framework, focused on the correct physical effect for the boron hydrides, but the wrong one for hydrogen bonds. The very weak H bond in F···HF* in contrast to FHF⁻, is a clear demonstration that H bonds, unlike boron hydrides, contain a very important contribution from electrostatic attraction.

4.6. The Impact of the *Ab Initio* Calculations on Semiempirical and Model Calculations

Semiempirical MO studies of H bonding have been extensively reviewed elsewhere, but *ab initio* theory has played an important role in these calculations by providing a good reference point in the (often) absence of experimental data. van Duijneveldt compared his calculations with *ab initio* studies in order to better assess areas of greater reliability of the model. Finally, Anderson[146] has made significant use of *ab initio* H-bond potentials in his development of empirical potential functions for the hydrogen bond.

Allen[127] has recently developed a simple model for H bonding based on *ab initio* calculations on the second and third row hydrides in an attempt to focus on the features of the proton donor and proton acceptor molecules which are most important in determining hydrogen bond strengths. He has chosen the bond dipole, $\hat{\mu}(A-H)$ as the crucial feature in the proton donor and ΔI, the difference between the ionization potential for the proton acceptor and a reference (noble gas) atom as the crucial feature of the proton acceptor.

5. Summary

It should be clear from the above that *ab initio* calculations have played an important role in studies of hydrogen bonding. First, these studies have had a significant impact on experimental work on hydrogen bonding; and second, they have provided us with a better understanding of the nature of the hydrogen bond.

The major impact on experiment has, of course, been on gas-phase studies and it is a significant challenge for *ab initio* methods to contribute meaningfully to liquid-state and solid-state intermolecular interactions. For liquid structure studies, the role of the *ab initio* calculations will most likely be as a provider of

---

*Very accurate calculations by Bender *et al.*[147] indicate that, unlike F⁻, neutral F is a very weak proton acceptor and the minimum energy for the FHF complex is a very little perturbed HF molecule, hydrogen bonding to F atom, $R(F-F) = 3.2$ Å. See also Noble and Kortzeborn.[80]

potential functions for Monte Carlo[149] or molecular dynamics studies.[150] However, incorporation of the surrounding dielectric into the Hamiltonians as carried out by Newton[151] in his studies of the hydrated electron and Hylton, Christofferson, and Hall[152] in their studies of He might prove generally useful. In the solid state, interesting approaches by Santry (using semiempirical MO theory) and Almlöf (using *ab initio* methods for a few of the molecules and electrostatic terms for the rest of the lattice[154]) show promise.

Further insight into the nature of the H bond and its relation to other noncovalent interactions is likely to come from further studies (employing extended basis sets) analyzing electrostatic potentials and energy components. In this context, it is worth mentioning the recent studies of Lucchese and Schaefer.[155] These authors examined amine–halogen (F_2, Cl_2, and FCl) interactions and found that the relative interaction energies of the halogen were stronger with ammonia than with trimethylamine, just as was found in the amine–HF and amine–HNH_2 interactions discussed in Section 4.4. This, plus the calculated orientation of interaction of amine with FCl leads one to conclude that the amine–X_2 and amine–HX interactions are dominated by electrostatic effects. However, as pointed out by Lucchese and Schaefer, Nagakura's[156] experiments on $NH_3 \cdots I_2$ and trimethyl amine–I_2 found a significantly more negative ΔH for the TMA$\cdots I_2$ interaction (-12 vs -4 kcal/mole). It will be an interesting challenge to *ab initio* theory to characterize the nature of these different interactions more precisely.*

Finally, the localized orbital approach that Gordon *et al.*[159] have taken in studying H bonds with semiempirical theory could be carried out at the *ab initio* level in order to get further insight into the nature of the charge redistribution accompanying H-bond formation.†

References

1. L. J. Schaad, Theory of the hydrogen bond, *in: Hydrogen Bonding* (M. D. Joesten and L. J. Schaad, eds.), Marcel Dekker, Inc., New York (1974), Chap. 2.
2. P. A. Kollman and L. C. Allen, The theory of the hydrogen bond, *Chem. Rev.* **72**, 283 (1972).
3. S. H. Lin, *in: Physical Chemistry—An Advanced Treatise, Vol. 5* (H. Eyring, D. Henderson, and W. Jost, eds.), Chap. 8, p. 439, Academic Press, New York (1970).
4. A. S. N. Murthy and C. N. R. Rao, Recent theoretical studies of the hydrogen bond, *J. Mol. Struct.* **6**, 253 (1970).
5. M. A. Ratner and J. R. Sabin, The wave mechanical treatment of hydrogen bonded systems *in: Wave Mechanics—The First Fifty Years* (S. S. Chissick, W. C. Price, and T. Ravendale, eds.), Butterworths, London (1973).
6. J. N. Murrell, The hydrogen bond, *Chem. Britain* **5**, 107 (1969).
7. S. Bratoz, Electronic theories of hydrogen bonding, *Adv. Quantum Chem.* **3**, 209 (1967).
8. C. A. Coulson, The hydrogen bond—review of the present position, *Research (London)* **10**, 149 (1957).

*However, it appears from the calculations by Hanna[157] and Lefevre *et al.*[158] that the benzene–I_2 interaction is not significantly stabilized by charge transfer in its ground state.

†See Note Added in Proof, page 151.

9. G. C. Pimentel and A. L. McClellan, *The Hydrogen Bond*, W. H. Freeman, San Francisco (1960).
10. N. D. Sokolov, De la théorie de la liaison hydrogène, *Ann. Chim. Phys.* **10**, 497 (1965).
11. M. L. Huggins, 50 years of hydrogen bond theory, *Angew. Chem. Int. Ed. Eng.* **10**, 147 (1971).
12. C. J. Roothaan, New developments in molecular orbital theory, *Rev. Mod. Phys.* **23**, 69 (1951).
13. H. J. Kolker and M. Karplus, Theory of nuclear magnetic shielding in diatomic molecules, *J. Chem. Phys.* **41**, 1259 (1964).
14. S. Iwata and K. Morokuma, Extended Hartree–Fock theory for excited states, *Chem. Phys. Lett.* **16**, 192 (1973).
15. J. Del Bene, R. Ditchfield, and J. A. Pople, Self-consistent molecular orbital methods. X. Molecular orbital studies of excited states with minimal and extended basis sets, *J. Chem. Phys.* **55**, 2236 (1971); I. Shavitt, A general approach to configuration interaction, *in*: *Modern Theoretical Chemistry, Vol. 3: Methods of Electronic Structure Theory* (H. Schaefer, ed.), Chap. 6, Plenum Publishing, New York (1977).
16. A. C. Wahl and G. Das, Multiconfiguration self-consistent field method, *in*: *Modern Theoretical Chemistry, Vol. 3: Methods of Electronic Structure Theory* (H. Schaefer, ed.), Chap. 3, Plenum Publishing, New York (1977).
17. H. Margenau and N. R. Kestner, *The Theory of Intermolecular Forces*, Pergamon Press, Oxford (1969).
18. F. Pilar, *Elementary Quantum Chemistry*, McGraw-Hill, New York (1968).
19. H. F. Schaefer, *The Electronic Structure of Atoms and Molecules*, Addison-Wesley, Reading, Massachusetts (1972).
20. P. A. Kollman and C. F. Bender, The structure of the H_3O^+ (hydronium) ion, *Chem. Phys. Lett.* **21**, 271 (1973).
21. H. F. Schaefer, D. R. McLaughlin, F. E. Harris, and B. J. Alder, *Phys. Rev. Lett.* **25**, 988 (1970).
22. C. A. Coulson and U. Danielsson, Ionic and covalent contributions to the hydrogen bond I and II, *Ark. Fys.* **8**, 239–245 (1954).
23. H. Tsubomura, The nature of the hydrogen bond. I. The delocalization energy in the hydrogen bond as calculated by the atomic orbital method, *Bull. Chem. Soc. Japan.* **27**, 445 (1954).
24. F. B. van Duijneveldt and J. N. Murrell, Some calculations on the hydrogen bond, *J. Chem. Phys.* **46**, 1759 (1967); J. G. C. M. van Duijneveldt-van de Rijdt and F. B. van Duijneveldt, Perturbation calculations on the hydrogen bonds between some first row atoms, *J. Am. Chem. Soc.* **93**, 5644 (1971) and references therein.
25. C. A. Coulson, The hydrogen bond, *in*: *Hydrogen Bonding* (D. Hadzi, ed.), p. 339, Pergamon Press, London (1959).
26. K. Morokuma, Molecular orbital studies of hydrogen bonds. III. C=O···H—O hydrogen bond in $H_2CO···H_2O$ and $H_2CO···2H_2O$, *J. Chem. Phys.* **55**, 1236 (1971).
27. M. Dreyfus and A. Pullman, A non-empirical study of the hydrogen bond between peptide units, *Theor. Chim. Acta* **19**, 20 (1970).
28. P. A. Kollman and L. C. Allen, An SCF partitioning method for the hydrogen bond, *Theor. Chim. Acta* **18**, 399 (1970).
29. P. A. Kollman and L. C. Allen, Hydrogen bonded dimers and polymers involving hydrogen fluoride, water and ammonia, *J. Am. Chem. Soc.* **92**, 753 (1970); *Chem. Rev.* **72**, 283 (1972).
30. M. van Thiel, E. D. Becker, and G. Pimentel, Infrared studies of hydrogen bonding of water by the matrix isolation technique, *J. Chem. Phys.* **27**, 486 (1957).
31. T. H. Dunning and P. J. Hay, Basis sets for molecular calculations, *in*: *Modern Theoretical Chemistry, Vol. 3: Methods of Electronic Structure Theory* (H. Schaefer, ed.), Chap. 1, Plenum Publishing, New York (1977).
32. K. Morokuma and L. Pedersen, Molecular Orbital Studies of hydrogen bonds. An *ab initio* calculation for dimeric H_2O, *J. Chem. Phys.* **48**, 3275 (1968).
33. P. A. Kollman and L. C. Allen, Theory of the hydrogen bond: electronic structure and properties of the water dimer, *J. Chem. Phys.* **51**, 3286 (1969).

34. J. Del Bene, Theoretical Study of open chain dimers and trimers containing CH_3OH and H_2O, *J. Chem. Phys.* **55**, 4633 (1971).

35. J. Del Bene and J. A. Pople, Theory of molecular interactions. I. Molecular orbital studies of water polymer with a minimal Slater-type basis, *J. Chem. Phys.* **52**, 4858 (1970).

36. K. Morokuma and J. R. Winick, Molecular orbital studies of hydrogen bonds: dimeric H_2O with the Slater minimal basis set, *J. Chem. Phys.* **52**, 1301 (1970).

37. D. Hankins, J. W. Moskowitz, and F. H. Stillinger, Water molecule interactions, *J. Chem. Phys.* **53**, 4544 (1970).

38. G. H. F. Diercksen, SCF–MO–LCGO studies on hydrogen bonding. The water dimer, *Theor. Chim. Acta* **335** (1971).

39. H. Popkie, H. Kistenmacher, and E. Clementi, Study of the structure of molecular complexes. IV. The Hartree–Fock potential for the water dimer and its application to the liquid state, *J. Chem. Phys.* **59**, 1325 (1973).

40. A. Tursi and E. Nixon, Matrix isolation study of water dimer in solid nitrogen, *J. Chem. Phys.* **52**, 154 (1970).

41. L. B. Magnusson, Infrared absorbance by water dimer in carbon tetrachloride solution, *J. Phys. Chem.* **74**, 4221 (1970).

42. P. A. Kollman and A. D. Buckingham, The structure of the water dimer, *Mol. Phys.* **21**, 567 (1971); L. B. Magnusson, The structure of the water dimer, *Mol. Phys.* **21**, 571 (1967).

43. P. W. Atkins and M. C. R. Symons, Infrared spectrum of water dimer in carbon tetrachloride solution, *Mol. Phys.* **23**, 831 (1972).

44. T. R. Dyke, B. J. Howard, and W. Klemperer, Radiofrequency and microwave spectrum of the hydrogen fluoride dimer: a nonrigid molecule, *J. Chem. Phys.* **56**, 2442 (1972).

45. T. R. Dyke and J. S. Muenter, Microwave spectrum and structure of hydrogen bonded water dimer, *J. Chem. Phys.* **60**, 2929 (1974).

46. L. Shipman and H. A. Scheraga, Structure, energetics and Dynamics of the water dimer, *J. Phys. Chem.* **78**, 2055 (1974).

47. D. Eisenberg and W. Kauzmann, *The Structure and Properties of Water*, Oxford University Press, New York (1969).

48. J. D. Lambert, Association in polar vapours and binary vapour mixtures, *Discuss. Faraday Soc.* **15**, 226 (1953); J. S. Rowlinson, The lattice energy of ice and the second virial coefficient of water vapour, *Trans. Faraday Soc.* **47**, 120 (1951).

49. H. A. Gebbie, W. J. Burroughs, J. Chamberlain, J. E. Harries, and R. J. Jones, Dimers of the water molecule in the earth's atmosphere, *Nature* **221**, 143 (1969).

50. T. H. Dunning and P. J. Hay, Basis sets for molecular calculations, *in*: *Modern Theoretical Chemistry, Vol. 3: Methods of Electronic Structure Theory* (H. Schaefer, ed.), Chap. 1, Plenum Publishing, New York (1977).

51. A. Johansson, P. Kollman, and S. Rothenberg, An application of the functional Boys–Bernardi counterpoise method to molecular potential surfaces, *Theor. Chim. Acta* **29**, 167 (1973).

52. D. Neumann and J. W. Moskowitz, One electron properties of near-Hartree–Fock wave functions, I. Water, *J. Chem. Phys.* **49**, 2056 (1968).

53. A. Rauk, L. C. Allen, and E. Clementi, Electronic structure and inversion barrier of ammonia, *J. Chem. Phys.* **52**, 4133 (1970).

54. B. Lentz and H. A. Scheraga, Water molecule interactions. Stability of cyclic polymers, *J. Chem. Phys.* **58**, 5296 (1973).

55. J. Del Bene and J. A. Pople, Theory of molecular interactions. III. A comparison of studies of H_2O polymers using different molecular orbital basis sets, *J. Chem. Phys.* **58**, 3605 (1973).

56. J. Janzen and L. S. Bartell, Electron diffraction study of polymeric gaseous hydrogen fluoride, *J. Chem. Phys.* **50**, 3611 (1969).

57. P. A. Kollman and L. C. Allen, Theory of the hydrogen bond: *ab initio* calculations on hydrogen fluoride dimer and the mixed water–hydrogen fluoride dimer, *J. Chem. Phys.* **52**, 5085 (1970).

58. G. H. F. Diercksen and W. P. Kraemers, SCF–MO–LCGO studies on hydrogen bonding: the hydrogen fluoride dimer, *Chem. Phys. Lett.* **6**, 419 (1970).

59. J. Del Bene and J. A. Pople, Theory of molecular interactions. II. Molecular orbital studies of HF polymer using a minimal Slater-type basis, *J. Chem. Phys.* **55**, 2296 (1971).
60. L. C. Allen and P. A. Kollman, Cyclic systems containing divalent hydrogen symmetrically placed between sp^2 hybridized electron rich atoms. A new form of chemical bond?, *J. Am. Chem. Soc.* **92**, 4108 (1970).
61. H. Lischka, *Ab initio* calculations on inter-molecular forces. III. Effect of electron correlation on the hydrogen bond in the hydrogen fluoride dimer, *J. Am. Chem. Soc.* **96**, 4761 (1974).
62. D. R. Yarkony, S. V. O'Neil, H. F. Schaefer III, C. P. Baskin, and C. F. Bender, Interaction potential between two rigid HF molecules, *J. Chem. Phys.* **60**, 855 (1974).
63. E. Clementi, J. Mehl, and W. von Niessen, Study of the electronic structure of molecules. XII. Hydrogen bridges in the guanine–cytosine pair and in the dimeric form of formic acid, *J. Chem. Phys.* **54**, 508 (1971).
64. M. Dreyfus, B. Maigret, and A. Pullman, A non-empirical study of hydrogen bonding in the dimer of formamide, *Theor. Chim. Acta* **17**, 109 (1970).
65. E. Clementi and J. N. Gayles, Study of the electronic structure of molecules. VII. Inner and outer complex in the NH_4Cl formation from NH_3 and HCl, *J. Chem. Phys.* **47**, 3837 (1967) and references therein.
66. P. Goldfiner and G. Verhaegen, Stability of the gaseous ammonium chloride molecule, *J. Chem. Phys.* **50**, 1467 (1969).
67. B. S. Ault and G. C. Pimentel, Infrared spectra of the ammonia–hydrochloric acid complex in solid nitrogen, *J. Phys. Chem.* **77**, 1649 (1973).
68. P. Kollman, A. Johansson, and S. Rothenberg, A comparison of HCl and HF as proton donors, *Chem. Phys. Lett.* **24**, 199 (1974).
69. B. S. Ault and G. C. Pimentel, Infrared spectrum of the water–hydrochloric acid complex in solid nitrogen, *J. Phys. Chem.* **77**, 57 (1973).
70. M. D. Newton and S. Ehrenson, *Ab initio* studies on the structures and energetics of inner and outer shell hydrates of the proton and the hydroxide ion, *J. Am. Chem. Soc.* **93**, 4971 (1971).
71. P. Kebarle, S. K. Searle, A. Zolla, J. Scarborough, and M. Arshadi, The solvation of the hydrogen ion by water molecules in the gas phase. Heats and entropies of solutions of individual reactions $H^+(H_2O)_{n-1} + H_2O \rightarrow H^+(H_2O)_n$, *J. Am. Chem. Soc.* **89**, 6393 (1967).
72. W. Kraemer and G. Diercksen, SCF–LCGO–MO calculations on hydrogen bonding. The hydrogen fluoride dimer, *Chem. Phys. Lett.* **5**, 463 (1970); P. A. Kollman and L. C. Allen, A theory of the strong hydrogen bond. *Ab initio* calculations on HF_2^- and $H_5O_2^+$, *J. Am. Chem. Soc.* **92**, 6101 (1970).
73. H. Kistenmacher, H. Popkie, and E. Clementi, Study of the structure of molecular complexes. V. Heat of formation for the Li^+, Na^+, K^+, F^- and Cl^- complexes with a single water molecule, *J. Chem. Phys.* **59**, 5842 (1973).
74. M. Arshadi, R. Yamdagni, and P. Kebarle, Hydration of the halide negative ions in the gas phase. II. Comparison of hydration energies for the alkali positive and halide negative ions, *J. Phys. Chem.* **74**, 1475 (1970).
75. S. Bratoz and G. Bessis, Study of the structure of the FHF ion by the method of configuration interaction, *C. R. Acad. Sci., Ser. C* **249**, 1881 (1959).
76. R. M. Erdahl, Valence bond theory and the electronic structure of molecules, Ph.D. thesis, Princeton University, 1965.
77. E. Clementi and A. D. McLean, SCF–LCAO–MO wave functions for the bifluoride ion, *J. Chem. Phys.* **36**, 745 (1962).
78. A. D. McLean and M. Yoshimine, Tables of linear molecules, *IBM J. Res. Dev.* **11**, 169 (1967).
79. P. A. Kollman and L. C. Allen, Theory of the strong hydrogen bond. *Ab initio* calculation on HF_2^- and $H_5O_2^+$, *J. Am. Chem. Soc.* **92**, 6101 (1970).
80. P. Noble and R. Kortzeborn, LCAO–MO–SCF studies of HF_2^- and the related unstable systems HF_2 and HeF_2, *J. Chem. Phys.* **52**, 5375 (1970).
81. J. Almlöf, Hydrogen bond studies 71. *Ab initio* calculations of the vibrational structure and equilibrium geometry in HF_2^- and DF_2^-, *Chem. Phys. Lett.* **17**, 49 (1972).

82. P. Bertoncini and A. C. Wahl, *Ab initio* calculation of the helium–helium potential at intermediate and large separations, *Phys. Rev. Lett.* **25**, 991 (1970).

83. H. F. Schaefer, D. R. McLaughlin, F. E. Harris, and B. J. Alder, Calculation of the attractive He pair potential, *Phys. Rev. Lett.* **25**, 988 (1970).

84. H. Lischka, *Ab initio* calculations on intermolecular forces. The systems He$\cdots$HF and He$\cdots$H$_2$O, *Chem. Phys. Lett.* **20**, 448 (1973).

85. M. Losonszy, J. W. Moskowitz, and F. H. Stillinger, Hydrogen bonding between neon and hydrogen fluoride, *J. Chem. Phys.* **61**, 2438 (1974).

86. M. Losonszy, J. W. Moskowitz, and F. H. Stillinger, Hydrogen bonding between neon and water, *J. Chem. Phys.* **59**, 3264 (1973).

87. S. E. Novick, P. Davies, S. J. Harris, and W. Klemperer, Determination of the structure of ArHCl, *J. Chem. Phys.* **59**, 2273 (1973); S. J. Harris, S. E. Novick, and W. Klemperer, Determination of the structure of ArHF, *J. Chem. Phys.* **60**, 3208 (1974).

88. C. P. Baskin, C. F. Bender, and P. A. Kollman, Dimers of lithium fluoride and sodium hydride, *J. Am. Chem. Soc.* **95**, 5868 (1973).

89. P. A. Kollman, S. Rothenberg, and C. F. Bender, A theoretical prediction of the existence and properties of the lithium hydride dimer, *J. Am. Chem. Soc.* **94**, 8016 (1972).

90. D. S. Marynick, J. H. Hall, and W. N. Lipscomb, Energy of formation of B$_2$H$_6$ from 2BH$_3$ near the Hartree–Fock limit, *JCP* **61**, 5460 (1974).

91. R. Ahlrichs, Correlation contribution to the dimerization of BH$_3$ and LiH, *Theor. Chim. Acta* **35**, 59 (1974).

92. E. Clementi and H. Popkie, Study of the structure of molecular complexes. I. Energy surface of a water molecule in the field of a lithium positive ion, *J. Chem. Phys.* **57**, 1077 (1972).

93. G. H. F. Diercksen and W. P. Kraemer, SCF–MO–LCGO studies on the hydration of ions: the systems H$^+$ H$_2$O, Li$^+$ H$_2$O and Na$^+$ H$_2$O, *Theor. Chim. Acta* **23**, 387 (1972); SCF–MO–LCGO studies on the hydration of ions: The system Li$^+$ 2H$_2$O, *Theor. Chim. Acta* **23**, 393 (1972).

94. P. A. Kollman and I. D. Kuntz, The hydration number of Li$^+$, *J. Am. Chem. Soc.* **96**, 4766 (1974).

95. P. Schuster and H. W. Preuss, *Ab initio* calculations on the hydration of monatomic cations (LCAO-MO studies of molecular structure VII), *Chem. Phys. Lett.* **11**, 35 (1971).

96. L. A. Curtiss and J. A. Pople, Molecular orbital calculation of some vibrational properties of the complex between HCN and HF, *J. Mol. Spectrosc.* **48**, 413 (1973).

97. L. A. Curtiss and J. A. Pople, *Ab initio* calculation of the vibrational force field of the water dimer, *J. Chem. Phys.* (in press).

98. E. B. Wilson, J. C. Decius, and P. C. Cross, *Molecular Vibrations*, McGraw-Hill, New York (1955).

99. R. K. Thomas, Hydrogen bonding in the gas phase: the infrared spectrum of complexes of hydrogen fluoride with hydrogen cyanide and methyl cyanide, *Proc. R. Soc. London, Ser. A* **325**, 133 (1971).

100. P. V. Huong and M. Couzi, The ir spectrum of gaseous hydrogen fluoride, *J. Chem. Phys.* **66**, 1309 (1969).

101. R. W. Bolander, J. L. Cassner, and J. T. Zung, Semi-empirical determination of the hydrogen bond energy for water to the dimer, clusters in the vapor phase. I. General theory and application to the dimer, *J. Chem. Phys.* **50**, 4402 (1969).

102. P. A. Kollman and L. C. Allen, The nature of the hydrogen bond. Dimers involving the electronegative atom of the first row, *J. Am. Chem. Soc.* **93**, 4991 (1971).

103. P. A. Kollman, J. McKelvey, A. Johansson, and S. Rothenberg, Theoretical Studies of hydrogen bonded dimers: complexes involving HF, H$_2$O, NH$_3$, HCl, H$_2$S, PH$_3$, HCN, HNC, HCP, CH$_2$NH, H$_2$CS, H$_2$CO, CH$_4$, CF$_3$H, C$_2$H$_2$, C$_2$H$_4$, C$_6$H$_6$, F$^-$ and H$_3$O$^+$, *J. Am. Chem. Soc.* **97**, 955 (1975).

104. W. Topp and L. C. Allen, Structure and properties of hydrogen bonds between the electronegative atoms of the second and third rows, *J. Am. Chem. Soc.* **96**, 5291 (1974).

105. E. U. Franck and F. Meyer, HF III, Specific heat and association in the gas phase at low pressure, *J. Electrochem.* **63**, 577 (1959).

106. J. E. Lowder, Spectroscopic studies of hydrogen bonding in ammonia, *J. Quant. Spectrosc. Radiat. Transfer* **10**, 1085 (1970).

107. A. D. H. Clague, G. Govil, and H. J. Bernstein, Medium effects in nuclear magnetic resonance. VII. Vapor phase studies of hydrogen bonding in methanol and methanol–trimethyl amine mixtures, *Can. J. Chem.* **47**, 625 (1969).

108. A. Foldes and C. Sandorfy, Anharmonicity and hydrogen bonding. III. Examples of strong bonds. General discussion, *J. Mol. Spectrosc.* **20**, 262 (1966).

109. J. Del Bene, Molecular orbital theory of the hydrogen bond. IV. The effect of hydrogen bonding on the $n \rightarrow \pi^*$ transition in dimers $ROH \cdots OCH_2$, *J. Am. Chem. Soc.* **95**, 6517 (1973).

110. S. Iwata and K. Morokuma, Molecular orbital studies of hydrogen bonds. V. Analysis of the hydrogen-bond energy between lower excited states of H_2CO and H_2O, *J. Am. Chem. Soc.* **95**, 7563 (1973).

111. J. Del Bene, On the blue shift of the $n \rightarrow \pi^*$ band of acetone in water, *J. Am. Chem. Soc.* **96**, 5643 (1974).

112. S. Iwata and K. Morokuma, Molecular orbital studies of hydrogen bonds. VI. Origin of red shift of $\pi \rightarrow \pi^*$ transitions: *trans*-acrolein–water complex, *J. Am. Chem. Soc.* **97**, 966 (1975).

113. K. Morokuma, S. Iwata, and W. Lathan, Molecular interactions in ground and excited states, *in*: *The World of Quantum Chemistry* (R. Daudel and B. Pullman, eds.), D. Reidel Publishing Co., Dordrecht-Holland (1974).

114. M. Jaszunski and A. J. Sadlej, Proton magnetic shielding in the water molecule, *Theor. Chim. Acta* **27**, 135 (1972).

115. A. P. Zens, P. D. Ellis, and R. Ditchfield, The carbon-13 nuclear magnetic resonance chemical shifts of the fluoroallenes. A comparison between theory and experiment, *J. Am. Chem. Soc.* **96**, 1309 (1974).

116. S. D. Christian and B. M. Keenan, Complexes of hydrogen chloride with ethers in carbon tetrachloride and heptane. Effects of induction of the basicity of ethers, *J. Phys. Chem.* **78**, 432 (1974).

117. H. D. Mettee, Vapor-phase dissociation energy of $(HCN)_2$, *J. Phys. Chem.* **77**, 1762 (1973).

118. A. Johansson, P. A. Kollman, and S. Rothenberg, The electronic structure of the hydrogen cyanide dimer and trimer, *Theor. Chim. Acta* **26**, 97 (1972).

119. J. R. Sabin, Hydrogen bonds involving sulfur. I. The hydrogen sulfide dimer, *J. Am. Chem. Soc.* **93**, 3613 (1971).

120. J. E. Lowder, L. A. Kennedy, K. G. P. Sulzman, and S. S. Penner, Spectroscopic studies of hydrogen bonding in hydrogen sulfide, *J. Quant. Spectrosc. Radiat. Transfer* **10**, 17 (1970).

121. G. Govil, A. D. H. Clague, and H. J. Bernstein, Medium effects in NMR. VI. Vapor phase studies of hydrogen bonding between dimethyl ether and hydrogen chloride, *J. Chem. Phys.* **49**, 2821 (1968).

122. H. Poland and H. A. Scheraga, Energy parameters in polypeptides. I. Charge distributions and the hydrogen bond, *Biochem.* **6**, 3791 (1967).

123. A. D. H. Clague and H. J. Bernstein, Heat of dimerization of some carboxylic acids in the vapor phase determined by a spectroscopic method, *Spectrochim. Acta* **25A**, 593 (1969).

124. I. Dzidic and P. Kebarle, Hydration of alkali ions in the gas phase. Enthalpies and entropies of Reactions $M^+(H_2O)_{n-1} + H_2O \rightarrow M^+(H_2O)_n$, *Chemistry* **74**, 1466 (1970).

125. M. Eisenstadt, P. Rothberg, and P. Kusch, Molecular composition of alkali fluoride vapors, *J. Chem. Phys.* **29**, 797 (1958).

126. W. Hug and I. Tinoco, Electronic spectrum of nucleic acid bases. I. Interpretation of the in-plane spectra with the aid of all-valence electron MO-CI (configuration interaction) calculations, *J. Am. Chem. Soc.* **95**, 2803 (1973).

127. L. C. Allen, A simple model of hydrogen bonding, *J. Am. Chem. Soc.* **97**, 6921 (1976).

128. J. Donohue, Selected topics in hydrogen bonding, *in*: *Structural Chemistry and Molecular Biology* (A. Rich and N. Davidson, eds.), W. H. Freeman, San Francisco (1968).

129. J. Kroon, J. A. Kanters, J. G. C. M. van Duijneveldt-van de Rijdt, F. B. van Duijneveldt, and J. A. Vliegenthart, $O-H \cdots O$ hydrogen bonds in molecular crystals. A statistical and quantum chemical analysis, *J. Mol. Struct.* **24**, 109 (1975).

130. J. Del Bene, Molecular orbital theory of the hydrogen bond. VII. Series of dimers having ammonia as the proton acceptor, *J. Am. Chem. Soc.* **95**, 5460 (1973).

131. P. A. Kollman, A theory of hydrogen bond directionality, *J. Am. Chem. Soc.* **94**, 1837 (1972).

132. F. van Duijneveldt, personal communication to P. A. Kollman.

133. A. Johansson, P. Kollman, and S. Rothberg, An *ab initio* molecular orbital study of intramolecular H-bonding: 1,3-propanediol, *Chem. Phys. Lett.* **18**, 276 (1973).

134. W. Meyer, W. Jakubetz, and P. Schuster, Correlation effects on energy curves for proton motion. The cation $[H_5O_2]^+$, *Chem. Phys. Lett.* **21**, 97 (1973).

135. M. J. T. Bowers and R. M. Pitzer, Bond orbital analysis of the hydrogen bond in the linear water dimer, *J. Chem. Phys.* **59**, 163 (1973).

136. W. H. Fink, Approach to partially predetermined electronic structure. The Li–He interaction potential, *J. Chem. Phys.* **57**, 1822 (1972).

137. W. von Niessen, A Theory of molecules in molecules III. Application to the interaction between 2FH molecules, *Theor. Chim Acta* **31**, 297 (1973).

138. W. von Niessen, A theory of molecules in molecules. IV. Application to the hydrogen bonding interaction in $NH_3 \cdot H_2O$, *Theor. Chim. Acta* **32**, 13 (1974).

139. R. Bonaccorsi, A. Pullman, E. Scrocco, and J. Tomasi, N vs. O proton affinities of the amide group: *ab initio* electrostatic molecular potentials, *Chem. Phys. Lett.* **12**, 622 (1972).

140. S. Yamabe and K. Morokuma, *J. Am. Chem. Soc.* **97**, 4458 (1975).

141. H. Ratajczak and W. Orville-Thomas, Charge transfer theory and vibrational properties of the hydrogen bond, *J. Mol. Struct.* **19**, 237 (1972).

142. P. A. Kollman, J. F. Liebman, and L. C. Allen, The lithium bond, *J. Am. Chem. Soc.* **92**, 1142 (1970).

143. A. Johansson, P. Kollman, and J. Liebman, Substituent effects on proton affinities, *J. Am. Chem. Soc.* **96**, 3750 (1974).

144. P. Kollman, unpublished.

145. J. L. Beauchamp, Ion cyclotron resonance, *Ann. Rev. Phys. Chem.* **22**, 527 (1971); M. S. Foster and J. L. Beachamp, unpublished.

146. W. Lathan and K. Morokuma, Molecular orbital studies of electron donor–acceptor complexes. I. Carbonyl cyanide–ROR and tetracyanoethylene–ROR complexes, *J. Am. Chem. Soc.* **97**, 3615 (1975).

147. C. F. Bender, C. W. Bauschlicher, and H. F. Schaefer, Saddle point geometry and barrier height for $H + F_2 \rightarrow HF + F$, *J. Chem. Phys.* **60**, 3707 (1974).

148. G. Anderson, Semi-empirical study of hydrogen bonding in the diaquohydrogen ion $H_5O_2^+$, *J. Phys. Chem.* **77**, 2560 (1973).

149. H. Kistenmacher, H. Popkie, E. Clementi, and R. O. Watts, Study of the structure of molecular complexes. VII. Effect of correlation energy corrections to the Hartree–Fock water–water potential on Monte Carlo simulations of liquid water, *J. Chem. Phys.* **60**, 4455 (1974).

150. F. Stillinger and A. Rahman, Improved simulation of liquid water by molecular dynamics, *J. Chem. Phys.* **60**, 1545 (1974).

151. M. D. Newton, *Ab initio* Hartree–Fock calculations with inclusion of a polarized dielectric; formalism and application to the ground state hydrated electron, *J. Chem. Phys.* **58**, 5833 (1973).

152. J. Hylton, R. E. Christoffersen, and G. G. Hall, A model for the *ab initio* calculation of some solvent effects, *Chem. Phys. Lett.* **24**, 501 (1974).

153. J. Bacon and D. P. Santry, Molecular orbital theory for infinite systems: hydrogen bonded molecular crystals, *J. Chem. Phys.* **56**, 2011 (1972).

154. J. Almlöf, Ave Kvick, and J. O. Thomas, Hydrogen bond studies. 77. Electron density and distribution in α glycine: X–N difference Fourier synthesis vs. *ab initio* calculations, *J. Chem. Phys.* **59**, 3901 (1973).

155. R. R. Lucchese and H. F. Schaefer III, Charge transfer complexes. NH_3–F_2, NH_3Cl_2, NH_3–ClF, $N(CH_3)_3$–F_2, $N(CH_3)_3$–Cl_2 and $N(CH_3)_3$–ClF, *J. Am. Chem. Soc.* (submitted).

156. S. Nagakura, Molecular complexes and their spectra: the molecular complex between iodine and triethylamine, *J. Am. Chem. Soc.* **80**, 520 (1958).

157. M. Hanna, Bonding in donor acceptor complexes. I. Electrostatic contribution to the ground state properties of benzene–halogen complexes, *J. Am. Chem. Soc.* **90**, 285 (1968).
158. R. Lefevre, D. V. Radford, and P. S. Stiles, The degree of charge transfer in the ground state of molecular π complexes, *J. Chem. Soc. London, Ser. B*, 1297 (1968).
159. M. S. Gordon, D. E. Tallman, C. Monroe, M. Steinback, and J. Ambrust, Localized orbital studies of hydrogen bonding. II. Dimers containing H_2O, NH_3, HF, H_2CO and HCN, *J. Am. Chem. Soc.* **97**, 1326 (1975).

Note Added in Proof

Since the submission of this article (April, 1975) there have been a large number of additional interesting applications of electronic structure theory to the study of noncovalent interactions. Umeyama and Morokuma[160] have completed two manuscripts which represent very important advances. Kitaura and Morokuma[161] have further developed the Morokuma energy decomposition method to calcùlate separately the charge transfer and higher-order effects. Kollman and Kuntz,[162] Noell and Morokuma,[163] McCreery, Christoffersen, and Hall,[164] and Newton[165] have carried out interesting applications of various methods of studying solvation of molecules.

There have been a number of large-scale configuration interaction studies by Diercksen, Kraemer, and Roos[166] and Matsuoka et al.,[167] that have confirmed the reasonableness of our estimate of ~1.2 kcal/mole for the contribution of correlation energy to the dimerization energy of $(H_2O)_2$. In addition, in a systematic study of the dimerization energy of the first row hydrides with a 631G basis set, Dill et al.[168] find a dimerization energy for NH_3 (2.9 kcal/mole) very close to the estimate we used for ΔE_0 in our thermodynamic analysis in Table 2 (p. 126).

Ditchfield[169] has carried out very interesting studies employing gauge invariant atomic orbitals (as suggested above) in studying the NMR properties of the water dimer. He finds that deshielding effects on the H-bonded proton have important contributions from diamagnetic terms (as would be expected from charge density difference plots as in Fig. 2) and induced currents in the proton acceptor. The external proton of the proton donor in $(H_2O)_2$ is predicted to be shielded by H-bonding, which charge density plots would lead one to expect.

Interesting studies on intramolecular H-bonding include those by Dietrich et al.[170] and Newton and Jeffrey.[171]

Proton affinities have been a subject of interest for Del Bene and Vaccaro,[172] Umeyama and Morokuma,[173] and Kollman and Rothenberg.[174]

Trenary, Schaefer, and Kollman[175] have studied a series of radical-hydride noncovalent interactions, and Kollman[176] has attempted to organize all the types of noncovalent complexes discussed above in a single framework.

160. H. Umeyama and K. Morokuma, Molecular orbital studies of electron donor–acceptor complexes. IV. Energy decomposition analysis for halogen complexes: H_3N—F_2, H_3N—Cl_2, H_3N—ClF, CH_3H_2N—ClF, H_2CO—F_2 and F_2—F_2, *J. Am. Chem. Soc.* **99**, 330 (1977); The origin of hydrogen bonding, *J. Am. Chem. Soc.* **99**, 1316 (1977).
161. K. Kitaura and K. Morokuma, *Int. J. Quant. Chem.* **10**, 325 (1976).
162. P. Kollman and I. D. Kuntz, The hydration of NH_4F, *J. Am. Chem. Soc.* **98**, 6820 (1976).
163. J. O. Noell and K. Morokuma, A fractional charge model in the MO theory and its application to molecules in solutions and solids, *J. Phys. Chem.* **80**, 2675 (1976).
164. J. McCreery, R. E. Christoffersen, and G. G. Hall, *J. Am. Chem. Soc.* **98**, 7191, 7198 (1976).
165. M. Newton, *J. Phys. Chem.* **79**, 2795 (1975).
166. G. Diercksen, W. Kraemer, and B. Roos, *Theor. Chim. Acta* **36**, 249 (1975).
167. O. Matsuoka, E. Clementi, and M. Yoshimine, *J. Chem. Phys.* **64**, 1361 (1976).
168. J. Dill, L. C. Allen, W. C. Topp, and J. A. Pople, *J. Am. Chem. Soc.* **97**, 7220 (1975).
169. R. Ditchfield, *J. Chem. Phys.* **65**, 3123 (1976).
170. S. Dietrich, S. Rothenberg, E. C. Jorgensen, and P. Kollman, A theoretical study of intramolecular H-bonding in phenols and thiophenols, *J. Am. Chem. Soc.* **98**, 8310 (1976).

171. M. Newton and G. Jeffrey, The stereochemistry of the α-hydroxy carboxylic acids and related systems, *J. Am. Chem. Soc.* (in press).

172. J. Del Bene and A. Vaccaro, A molecular orbital study of protonation. Substituted carbonyl compounds, *J. Am. Chem. Soc.* **98**, 7526 (1976).

173. H. Umeyama and K. Morokuma, *J. Am. Chem. Soc.* **98**, 4400 (1976).

174. P. Kollman and S. Rothenberg, A theoretical study of basicity: Proton affinities, Li$^+$ affinities and H-bond affinities of simple molecules, *J. Am. Chem. Soc.* **99**, 1333 (1977).

175. M. Trenary, H. F. Schaefer, and P. Kollman, A novel type of complex, *J. Am. Chem. Soc.* (in press).

176. P. Kollman, A general analysis of noncovalent interactions, *J. Am. Chem. Soc.* (in press).

4

Direct Use of the Gradient for Investigating Molecular Energy Surfaces

Péter Pulay

1. Gradient Method Versus Pointwise Calculations

A great deal of chemical and spectroscopical processes involve the relative motion of atomic nuclei. For most low-energy processes the Born–Oppenheimer fixed-nuclei approximation is sufficient: the nuclear motion takes place on an effective potential surface which is the sum of the electronic energy and the nuclear repulsion as a function of the nuclear coordinates. One of the main fields of quantum chemical activity is the study of these surfaces. Complete characterization of a multidimensional potential surface is a very complex task. Often, however, the nuclear motion takes place in the vicinity of a reference configuration, and the surface can be adequately characterized by a power series expansion, i.e., by its derivatives with respect to the nuclear coordinates. Traditionally, these derivatives have been evaluated from a pointwise calculation of the energy, followed by a fitting procedure. This method has some serious drawbacks both in efficiency and in numerical accuracy. Indeed, Hartree[1] observes that "the differentiation of a function specified only by a table of values ... is a notoriously unsatisfactory process, particularly if higher derivatives than the first are required" (see Gerratt and Mills[2] for examples).

Another way to determine the characteristics of potential surfaces is by direct analytical calculation of energy derivatives from the wave function.

Péter Pulay • Eötvös L. University, Department of General and Inorganic Chemistry, Budapest

Although such methods require significant programming effort and, in principle, yield the same information as pointwise calculations, they have considerable advantages in practical calculations.

For several years this author has stressed the particular usefulness of the direct calculation of the *first* derivative (followed by numerical differentiation if higher derivatives are needed). This allows the efficient and numerically accurate calculation of molecular geometries, force constants, reaction paths, etc. Furthermore, its calculation is particularly simple for a large class of wave functions: general SCF wave functions. It is the purpose of this chapter to discuss the theoretical basis and the computational implications of such calculations and to present some recent applications. Moreover, the practicability of directly calculating higher derivatives is discussed.

The terms gradient and forces will be used interchangeably for the set of the first derivatives of the total energy with respect to the nuclear coordinates, as they differ only in sign. If not otherwise specified, the nuclear coordinates are chosen as simple Cartesian coordinates of the nuclei. The discussion is based on the energy surface of a general polyatomic system. Methods essentially applicable only to diatomic systems, such as virial theorem methods,[3] are not considered.

Direct calculation of the gradient offers two obvious advantages:

(1)   The analytic calculation of the gradient greatly enhances the information obtainable about the energy surface from a single SCF calculation because there are $3N-6$ (N is the number of nuclei) independent forces versus a single energy value (note that *all* first derivatives can be determined at the same time). Calculation of the gradient is thus equivalent to $3N-6$ energy calculations but usually takes much less computer time.* Moreover, there is a savings in human effort if the gradient is available, e.g., automatic determination of molecular geometries and force constants becomes much simpler.

(2)   High numerical accuracy is attainable. Note that the numerical first differentiation of the energy is particularly dangerous because small differences of large energy values must be taken. Subsequent differentiations are less susceptible to numerical errors because the changes in forces are of the same magnitude as the forces themselves. The net saving in the number of differentiations also improves the numerical accuracy. Examples for the need of high numerical accuracy are the small but important changes taking place during internal rotation[4-7] or the study of coupling force constants.[8] It is easily seen, for example, that equilibrium geometries can be more sharply determined by the condition of vanishing forces than by that of minimum energy.

---

*It is clear that this advantage is more prominent for large molecules because the computational effort for gradient calculation does not increase in comparison with the SCF procedure as the number of nuclei increases.

2. Calculation of the Energy Gradient from SCF Wave Functions

2.1. Structure of the Wave Function

Let Ψ be a normalized linear combination of orthogonal configurations Φ_k,

$$\Psi = \sum_k A_k \Phi_k \tag{1}$$

with

$$\langle \Phi_k | \Phi_l \rangle = \delta_{kl}; \qquad \sum_k A_k^2 - 1 = 0 \tag{2}$$

The configurations Φ_k are fixed linear combinations of Slater determinants constructed from n orthogonal orbitals φ_i which are linear combinations of m basis functions χ_r

$$\varphi_i = \sum_r^m C_{ri} \chi_r \tag{3}$$

From the orthonormality of the orbitals

$$\mathbf{C}^+\mathbf{S}\mathbf{C} = \mathbf{I}_n \tag{4}$$

where $S_{pq} = \langle \chi_p | \chi_q \rangle$ and $\mathbf{I}_n$ is the n by n unit matrix. Let us restrict ourselves for brevity of the formulas to real configurations, orbitals, and basis functions; this is not a substantial restriction. The basis functions are defined by their general functional form which is regarded as fixed, and by a set of nonlinear parameters p_t, the most important of which are the positions of the orbital centers x_r and the orbital exponents ζ_r. It is assumed that the basis functions depend explicitly on the nuclear coordinates X_a through the parameters p_t. This dependence is discussed in detail in Section 2.3.

Let us denote the expectation value of the energy for the wave function (1) with arbitrary values of the parameters A_k, C_{ri}, and p_t and nuclear coordinates X_a by

$$\langle \Psi | H | \Psi \rangle = \mathscr{E}(X_a, A_k, C_{ri}, p_t) \tag{5}$$

For Ψ to be an SCF wave function the energy should be stationary with respect to the linear parameters A_k and C_{ri} under the orthonormality conditions (2) and (4). We subtract these constraints, multiplied by Lagrangian multipliers from $\mathscr{E}$, and set the derivatives of the resulting functional to zero:

$$\frac{\partial}{\partial A_k}\left\{ \mathscr{E} - \eta\left(\sum_l A_l^2 - 1\right) \right\} = 0 \tag{6}$$

$$\frac{\partial}{\partial C_{ri}}\left\{ \mathscr{E} - \mathrm{Tr}\left(\boldsymbol{\varepsilon}(\mathbf{C}^+\mathbf{S}\mathbf{C} - \mathbf{I}_n) \right) \right\} = 0 \tag{7}$$

Since the procedure for obtaining optimum (or stationary) values for A_k and C_{ri} are the subject of Volume 3, Chapters 3 and 4, we only note that both η and the n by n matrix ε are determined by the SCF process. This is all we need for the subsequent derivation of the gradient. Note that η becomes the optimum energy for the final solution. The Hermitian (in our case symmetric) matrix ε can be made diagonal by a suitable unitary transformation if all orbitals have the same occupation number; it then becomes the matrix of orbital energies multiplied by the occupation number. Specifically, for a single-determinant wave function with doubly occupied orbitals ε is the double of orbital energies. Generally,

$$\varepsilon_{ij} = \tfrac{1}{2}\langle \varphi_i | F_i + F_j | \varphi_j \rangle \tag{8}$$

if the Fock matrices F_i and F_j are normalized to the shell occupation number, as advocated by Hinze.[9]

From (6) and (7) we obtain, for the derivative of $\mathscr{E}$ with respect to the linear parameters,

$$\frac{\partial \mathscr{E}}{\partial A_k} = 2\eta A_k \tag{9}$$

and

$$\frac{\partial \mathscr{E}}{\partial C_{ri}} = 2(\mathbf{SC}\varepsilon)_{ri} \tag{10}$$

2.2. First Derivative of the SCF Energy

Let us denote the SCF energy with stationary (for the ground-state optimum) linear parameters $C_{ri}(X_a)$ and $A_k(X_a)$ by

$$E(X_a) = \mathscr{E}(X_a, A_k(X_a), C_{ri}(X_a), p_t(X_a))$$
$$= \langle \Psi(A_k(X_a), C_{ri}(X_a), p_t(X_a)) | H(X_a) | \Psi(A_k(X_a), C_{ri}(X_a), p_t(X_a)) \rangle \tag{11}$$

In the following we shall denote the differentiation with respect to the nuclear coordinate X_a by the superscript a. The derivative of E with respect to X_a is the sum of four terms, corresponding to the dependence of E on X_a directly, through A_k, C_{ri}, and p_t. We shall deal with these terms [(A)–(D)] separately in the following:

(A) $$-f_a^{\mathrm{HF}} = \langle \Psi | H^a | \Psi \rangle \tag{12}$$

This is the negative *Hellmann–Feynman force*[10,11] (see Section 2.4). Since it is only the one-electron part of the Hamiltonian which depends on the nuclear coordinates, H^a is a one-electron operator, and the Hellmann–Feynman force is simply the expectation value of the field operator times the nuclear charge:

$$\frac{\partial}{\partial X_a}\left[\sum_j \frac{Z_a}{r_{ja}} - \sum_b \frac{Z_a Z_b}{R_{ab}}\right] = Z_a \sum_j \frac{x_j - X_a}{r_{ja}^3} - Z_a \sum_b \frac{Z_b(X_b - X_a)}{R_{ab}^3}$$

(B)
$$\sum_k \frac{\partial \mathscr{E}}{\partial A_k} A_k^a = 2\eta \sum_k A_k A_k^a = \eta \frac{\partial}{\partial X_a} \sum_k A_k^2 = 0 \tag{13}$$

where (9) and (2) have been used to prove that this term vanishes.

(C)
$$\sum_{ri} \frac{\partial \mathscr{E}}{\partial C_{ri}} C_{ri}^a = 2 \sum_{ri} (SC\varepsilon)_{ri} C_{ri}^a = 2\,\mathrm{Tr}(SC\varepsilon C^{a+}) \tag{14}$$

according to (10). Equation (14) can be cast in a simpler form. To this end we differentiate the orthonormality condition

$$C^{a+}SC + C^+ S C^a = -C^+ S^a C \tag{15}$$

Multiplying (15) by the matrix ε and taking the trace we obtain

$$2\,\mathrm{Tr}(\varepsilon C^{a+}SC) = 2\,\mathrm{Tr}(SC\varepsilon C^{a+}) = -\mathrm{Tr}(\varepsilon C^+ S^a C) = -\mathrm{Tr}(C\varepsilon C^+ S^a) = -\mathrm{Tr}(RS^a) \tag{16}$$

where the identities $\mathrm{Tr}(AB) = \mathrm{Tr}(BA)$ and $\mathrm{Tr}(A^+) = \mathrm{Tr}(A)$ have been used and the matrix $R = C\varepsilon C^+$ has been introduced. Equation (16) shows that we do not need the derivatives of the orbital coefficients C_{ri} for the evaluation of (14); the easily calculable matrices R and S^a are sufficient. *This is the key step in deriving the analytical form of the gradient.* Because of its connection with the change in the density matrix as the nuclear coordinates change, the negative of (14) is called the *density force.*[8]

Note that the density force arises because the overlap matrix depends, through the nonlinear parameters p_t, on the nuclear coordinates,

$$S^a = \sum_t \frac{\partial S}{\partial p_t} p_t^a \tag{17}$$

Therefore, each nonlinear parameter p_t dependent on X_a has a contribution to the density force: $\mathrm{Tr}[R(\partial S/\partial p_t)]p_t^a$.

(D)
$$\sum_t \frac{\partial \mathscr{E}}{\partial p_t} p_t^a \tag{18}$$

Leaving the discussion of $p_t^a = \partial p_t/\partial X_a$ to Section 2.3 we note that, as the SCF energy expression is linear in the basis integrals, $\partial \mathscr{E}/\partial p_t$ presents a problem in differentiating these integrals with respect to the basis-set parameters. Therefore, we call the negative of (18) the *integral force*. The integral force requires most of the numerical work in calculating the gradient; we shall return to it in Section 2.5.

The sum of the three nonvanishing contributions, the Hellmann–Feynman force, the density force, and the integral force, will be referred to as the expression for the *exact force* because this sum *should be equal* (within the limits of numerical accuracy) *to the negative derivative obtained from pointwise SCF calculations.*

The general form of the SCF energy expression is[9]

$$E = \sum_{ij} \gamma_{ij} \sum_{rs} C_{ri} C_{sj} \langle r|h|s \rangle$$

$$+ \tfrac{1}{2} \sum_{ijkl} \Gamma_{ij,kl} \sum_{rstu} C_{ri} C_{sj} C_{tk} C_{ul} J_{rstu} + \Omega \tag{19}$$

where the values of the orbital density matrices γ_{ij} and $\Gamma_{ij,kl}$ depend on the forms of the configurations and on the CI coefficients A_k. In (19) h is the one-electron Hamiltonian

$$\langle r|h|s \rangle = \langle \chi_r |h| \chi_s \rangle$$

$$J_{rstu} = \langle \chi_r(1)\chi_s(1)|1/r_{12}|\chi_t(2)\chi_u(2) \rangle$$

and Ω is the nuclear repulsion energy. In the case of a closed-shell Hartree–Fock determinant the only nonvanishing elements are $\gamma_{ii} = 2$, $\Gamma_{ii,jj} = 4$, and $\Gamma_{ij,ji} = -2$ and the energy formula (19) simplifies to

$$E = \sum_{rs} D_{rs} \langle r|h|s \rangle + \tfrac{1}{2} \sum_{rstu} D_{rs} D_{tu} (J_{rstu} - \tfrac{1}{2} J_{rstu}) + \Omega \tag{20}$$

where $\mathbf{D}(= 2\mathbf{CC}^+)$ is the density matrix. Using Eqs. (12)–(18), the derivative of (20) becomes

$$E = \sum_{rs} D_{rs} \langle r|h^a|s \rangle + \Omega^a - \mathrm{Tr}(\mathbf{RS}^a)$$

$$+ 2 \sum_{rs} D_{rs} \langle r^a|h|s \rangle + 2 \sum_{rstu} D_{rs} D_{tu} (J_{r^a stu} - \tfrac{1}{2} J_{r^a tsu}) \tag{21}$$

Here, the integrals containing the derivatives of the basis functions are, e.g., defined as

$$J_{r^a stu} = \langle \chi_r^a \chi_s|1/r_{12}|\chi_t \chi_u \rangle \quad \text{and} \quad \langle r^a|h|s \rangle = \sum_t \left\langle \frac{\partial \chi_r}{\partial p_t}|h|\chi_s \right\rangle p_t^a \tag{22}$$

Usually, a basis function depends only on a small number of nonlinear parameters, and the sum (22) contains only a few terms.

Equation (21) was first derived by Bratož[12] in 1958 as a first step toward his analytical second derivative formula. This was programmed and successfully used by Bratož and Allavena[13–15] for one-center basis sets. As seen from formulas (11)–(18) the gradient reduces to the Hellmann–Feynman force in this case because the basis set does not depend on the nuclear coordinates. Moccia[16] has given an analogous formula for the derivative of the SCF energy with respect to an orbital exponent; in this case the first two terms of (21) (the Hellmann–Feynman force) are, of course, absent. This expression was also obtained by Gerratt and Mills,[2] in a slightly different form by Pulay,[8] and recently by Bloemer and Bruner.[17] The above generalization to multiconfiguration SCF wave functions is due to Meyer,[18] following an earlier generalization to open-shell wave functions by him and Pulay.[19] However, only the open-shell case has, thus far, been programmed.[19]

Gradients for exponent optimization were first used by Moccia,[16] and then for geometry optimization and force constant calculation by Pulay[8,20] (see also Klauss[21] and Vladimiroff[22]).

The complexity of (21) as compared to the usual first-order energy expression in the perturbed Hartree–Fock theory[23] is due to the use of basis functions which explicitly depend on the nuclear coordinates. This introduces a perturbation in the two-electron part of the Hamiltonian.

2.3. Definition of the Basis Set in a Distorted Molecule

There are three reasonable ways to define the dependence of the nonlinear parameters on the nuclear coordinates.[24] The form of the integral force critically depends upon this definition.

(i) Define an arbitrary but physically reasonable and simple functional dependence of p_t on the nuclear coordinates and evaluate the integral force (18) with this definition of p_t^a.

(ii) Hold the parameter constant; then $p_t^a = 0$, and the contribution of p_t to both the integral and density force vanishes.

(iii) Optimize the parameter p_t. Since the overlap matrix in the orthonormality condition (4) generally depends on p_t, it is the functional $\mathscr{E} - \mathrm{Tr}(\boldsymbol{\varepsilon}(\mathbf{C}^+\mathbf{SC} - \mathbf{I}_n))$ which is to be optimized using the method of Lagrangian multipliers.* This yields,

$$\frac{\partial \mathscr{E}}{\partial p_t} - \mathrm{Tr}\left(\boldsymbol{\varepsilon}\mathbf{C}^+\frac{\partial \mathbf{S}}{\partial p_t}\mathbf{C}\right) = \frac{\partial \mathscr{E}}{\partial p_t} - \mathrm{Tr}\left(\mathbf{R}\frac{\partial \mathbf{S}}{\partial p_t}\right) = 0 \tag{23}$$

If all nonlinear parameters obey either condition (ii) or (iii), the total integral force is

$$\sum_t \frac{\partial \mathscr{E}}{\partial p_t}p_t^a = \sum_t \mathrm{Tr}\left(\mathbf{R}\frac{\partial \mathbf{S}}{\partial p_t}\right)\frac{\partial p_t}{\partial X_a} = \mathrm{Tr}(\mathbf{RS}^a) \tag{24}$$

and thus just compensates for the density force (16), the exact force is then

---

*The author is very grateful to W. Meyer (Mainz) for pointing out the necessity of including the second term of this expression, $\mathrm{Tr}(\boldsymbol{\varepsilon}(\mathbf{C}^+\mathbf{SC} - \mathbf{I}_n))$. Hurley[25] omits this term; his derivation of the Hellmann–Feynman theorem for floating functions is, therefore, valid only if there are no orthonormality constraints on the orbitals, a condition which rarely occurs in practical calculations. Pople *et al.*[55] derive the Hellmann–Feynman theorem for electric and magnetic perturbations. In these cases (in contrast to nuclear coordinate changes) it is physically reasonable to assume the nonlinear parameters of the basis set independent of the perturbation. According to (ii) this implies the validity of the Hellmann–Feynman theorem.

equal to the Hellmann–Feynman force. Basis sets which contain only nonlinear parameters obeying (ii) or (iii) are called floating functions* by Hurley,[25] and wave functions built from floating functions are termed stable with respect to the nuclear coordinates by Hall.[26] As the calculation of the gradient is particularly simple for stable wave functions, it is important to compare floating basis sets to the more common basis sets defined by (i). Unfortunately, it turns out that both methods (ii) and (iii) increase the computational effort out of proportion to the gain obtained by the simplification in the gradient calculation. Optimization of basis function positions and exponents has been used with success by Frost[27] for subminimal basis sets. However, application of this method to extended basis sets does not seem practical. The other approach, using fixed basis functions, requires an unproportional extension of the basis set to yield physically reasonable potential surfaces. In order that the wave function be able to adapt itself to infinitesimal changes in the nuclear coordinates it is necessary to include the polarization functions $\partial \chi_r / \partial x_r$, $\partial \chi_r / \partial y_r$, and $\partial \chi_r / \partial z_r$, for every basis function χ_r which is significantly populated; this increases the computational time by about two orders of magnitude.

It is indeed definition (i) which has been intuitively adopted for calculating potential surfaces in most cases. The simplest physically reasonable dependence of the nonlinear parameters on the nuclear coordinates is the following[24]: (a) the exponents of the basis functions are constant and (b) the positions of the basis functions are linear combinations of the nuclear position vectors (as with the usual special case, this includes functions moving rigidly with a particular nucleus).

With fixed orbital coefficients the above conditions describe properly the bulk of the change which takes place during the distortion of the nuclear framework: the atomic part of the electron density follows the nuclei. In order to describe the remaining finer changes by the orbital coefficients C_{ri}, the basis set must be sufficiently flexible. Restriction (a) requires the presence of members with different exponents for every important type of basis function. Restriction (b) can be compensated for by adding polarization functions to the basis set, particularly in the valence region where deviations from atomic behavior are largest (note that functions centered on other nuclei also provide a certain polarization effect).

It has been argued[28–30] that reoptimization of orbital exponents at every nuclear position is important (note that Ref. 29 incorrectly equates the gradient method with methods using the Hellmann–Feynman forces). Recent results do not support this view, although it may hold for limited one-center expan-

---

*This is actually a slight generalization of Hurley's definition which requires *all* nonlinear parameters to be optimized. In this form the definition is not useful, as the number of nonlinear parameters of any given function can be increased without limit by considering wider classes of functions. In the following the term floating function will be used for basis functions with optimized positions and exponents.

sions[28] where the wave function has no other means of adapting itself to the changing nuclear positions. For extended multicenter basis sets the futility of exponent optimization, as compared to the extension of the basis set, has been clearly demonstrated by Cade and Huo.[31] Exponent optimization for extended molecular basis sets is extremely time consuming. This usually forces one to only partially carry out the optimization, resulting in an uneven potential surface from which higher derivatives cannot be evaluated. The results of Cade *et al.*[32] on the nitrogen molecule illustrate this danger. In their high-quality Hartree–Fock calculation these authors obtain a slightly uneven potential curve due to incomplete exponent optimization. The calculated anharmonicity parameter $\omega_e x_e$ varies wildly with the choice of points on the curve, the maximum and minimum values being 2.85 and 39.75 cm^{-1} vs the experimental result of 14.2. A fixed-exponent $(11s, 6p, 2d, 1f)$ Gaussian basis set yields a Hartree–Fock value[33] of $\omega_e x_e = 13.2$ cm^{-1} which is quite stable against modifications in the basis set. The agreement is similar to that obtained by Cade and Huo[31] for hydrides.

2.4. Hellmann–Feynman Forces and Their Limitations

Owing to the simplicity of their evaluation, there have been numerous suggestions[2,34–41] in favor of using the Hellmann–Feynman forces as an alternative to customary energy evaluation; some authors propose differentiating them further to obtain force constants or integrating them to obtain energy differences. The Hellmann–Feynman theorem,[10,11] i.e., the equality of the exact force and the Hellmann–Feynman force

$$\frac{\partial}{\partial X_a} \langle \Psi | H | \Psi \rangle = \left\langle \Psi \left| \frac{\partial H}{\partial X_a} \right| \Psi \right\rangle \tag{25}$$

with $\langle \Psi | \Psi \rangle = 1$, holds for the exact wave function and for certain types of fully optimized wave functions, e.g., for the Hartree–Fock limit. This latter fact was implied by Hurley[25] and pointed out by Hall,[26] Cohen and Dalgarno,[42] and Stanton,[43] and later extended to more general SCF wave functions by Tuan[44] and Coulson.[45] Note that the discussion in Section 2.3 furnishes us with a proof that all SCF wave functions in their limit, obey the Hellmann–Feynman theorem. To show this we choose a complete basis set with parameters that are independent of the nuclear coordinates. It is possible, although not practical, to calculate an SCF-limit wave function with such a basis set and according to Section 2.3 it will obey the Hellmann–Feynman theorem.

For approximate wave functions the Hellmann–Feynman theorem holds only if the wave function is stable. However, in Section 2.3 it was shown that, in general, it is not practical to work with stable wave functions. The question naturally arises whether Hellmann–Feynman forces calculated from nonstable wave functions are of any value. Unfortunately the answer is no. This was pointed out by Salem and Wilson,[46] among others. As they show, for a wave function which is correct to order ε, the expectation value of $\partial H/\partial X$ (which does not generally commute with the Hamiltonian) is correct only to order ε. The approximate energy, however, is correct to order ε^2 and this is also valid under rather plausible assumptions for the exact forces[2] rendering Hellmann–Feynman forces uncompetitive even for highly accurate wave functions. Numerical tests[47-51,20] show that Hellmann–Feynman forces are generally very unreliable, even for very good wave functions. Kern and Karplus,[48] for example, calculated the Hellmann–Feynman forces in hydrogen fluoride with the wave function of Clementi (16 Slater functions) and obtained a very large force 0.78 a.u. = 6.42 mdyn on the fluorine nucleus at the experimental H–F distance.

An example given by Salem and Wilson[46] helps our understanding of why Hellmann–Feynman forces are so sensitive. Consider a neutral atom in a uniform electric field F. If the electronic charge distribution is taken as spherical in a first approximation, then the electrons do not exert a force on the nucleus and the total Hellmann–Feynman force is ZF instead of the correct value of zero. Note that the exact force behaves correctly even with this approximate spherical wave function. If the polarization of the electron cloud in the field is properly taken into account, the electrons will exert a Hellmann–Feynman force on the nucleus that cancels the ZF term. It is clear from this example that a necessary condition for obtaining usable Hellmann–Feynman forces is the correct description of the small, and for most purposes unimportant, polarization of the atomic cores.

Another difficulty with the Hellmann–Feynman forces is that their sum and torque do not vanish identically if they are calculated from nonstable wave functions.* For example, in a heteronuclear diatomic molecule AB the Hellmann–Feynman force on atom A is not necessarily equal and opposite to that on B. This leads to ambiguities in the calculation of the forces in internal coordinates which are the relevant physical quantities. The transformation proposed by Lazarev and Kovalev[52] is quite arbitrary and depends on the nuclear masses, i.e., it gives different results for isotopic species. The reasonable agreement with the experimental force constants of ammonia,[52] obtained using a simple wave function and the Hellmann–Feynman theorem, must be regarded as fortuitous.

2.5. Computational Aspects

First, we rewrite (21) in a form more suited to numerical calculation. To this end we rearrange the fourfold sum so that only one integral appears in

---

*The sum and torque of the exact forces vanish if the basis functions guarantee the invariance of the molecular energy against rigid translations and rotations of the nuclear framework.

each term. To take advantage of the fact that once, say, $J_r{}^a{}_{stu}$ has been calculated, the calculation of $J_{rs}{}^a{}_{tu}$, $J_{rst}{}^a{}_u$, etc., is simple, we restrict the indices in the usual way and obtain

$$E^a = \sum_{r>s} D_{rs}[\langle r|h^a|s\rangle + \langle r^a|h|s\rangle + \langle r|h|s^a\rangle](2-\delta_{rs})$$

$$+ \Omega^a - \mathrm{Tr}(\mathbf{RS}^a) + \frac{1}{2}\sum_{\substack{r>s,t>u \\ rs>tu}} (2D_{rs}D_{tu} - \tfrac{1}{2}D_{rt}D_{su} - \tfrac{1}{2}D_{ru}D_{st})$$

$$\times (J_r{}^a{}_{stu} + J_{rs}{}^a{}_{tu} + J_{rst}{}^a{}_u + J_{rstu}{}^a)(2-\delta_{rs})(2-\delta_{tu})(2-\delta_{rs,tu}) \qquad (26)$$

The basis functions are defined according to Section 2.3, i.e., their exponents are fixed and their position vectors are linear combinations of the nuclear coordinates

$$x_r = \sum_a \mu_{ra} X_a \qquad (27)$$

Here μ_{ra} is the *coefficient of following* for the basis set positional coordinate x_r and the nuclear coordinate X_a. As the basis functions are usually placed either on the nuclei or on the bonding line between the nuclei, the sum (27) contains at most two nonvanishing terms. [Note that if general nuclear coordinates like bond lengths and angles are used, Eq. (27) may contain many more nonvanishing terms. This shows that it is advantageous to first calculate the gradients in Cartesian coordinates.] The derivatives p_i^a in (22) are just our μ_{ra}. At the cost of a little storage requirement we can save the multiplications by μ_{ra} in (22), introducing the forces f_r acting on the position vectors of the basis functions

$$-f_r = \frac{\partial \mathscr{E}}{\partial x_r} = -\mathrm{Tr}\left(\mathbf{R}\frac{\partial \mathbf{S}}{\partial x_r}\right) + 2\sum_s D_{rs}\langle r'|h|s\rangle$$

$$+ 2\sum_{stu}(D_{rs}D_{tu} - \tfrac{1}{2}D_{rt}D_{su})J_{r'stu} \qquad (28)$$

where $r' = \partial\chi_r/\partial x_r$. The force acting in the direction of the nuclear coordinate X_a is then

$$f_a = -E^a = f_a^{\mathrm{HF}} + \sum_r f_r \mu_{ra} \qquad (29)$$

with f_a^{HF} denoting the Hellmann–Feynman force.

Two points are worth noting in the calculation of the forces according to (26) or (28) and (29).

(a)   The derivatives of the basis integrals can be summed up immediately after being calculated, thus no external storage is required.

(b)   It is possible to calculate the forces for all nuclear coordinates* by going through the summation in (26) *only once*, multiplying the integrals containing $\partial\chi_r/\partial x_r$ by the appropriate coefficients of following μ_{ra} and adding the resulting term to f_a. The possibility of calculating *all forces at the same time* is one of the most advantageous features of the gradient method: the amount of information obtained from gradient calculations is proportional to the number of nuclei N, while the computational requirement (in units of computer time necessary to obtain the SCF wave function) remains approximately constant.

As to the calculation of the integrals containing the derivatives of the basis functions $\partial\chi_r/\partial x_r$, they are easily reduced to integrals over functions with quantum number $l' = l \pm 1$; for both Gaussian and Slater functions the radial part (apart from a possible change in the main quantum number) remains essentially unchanged. For Slater functions, $1p$, and $2d$ types which are normally not used, may appear.

Although the calculation of the forces is simple in principle, it requires a considerable amount of computer time since all derivatives of the two-electron integrals must be evaluated. From a single integral J_{rstu} 12 integrals arise by differentiation with respect to the x, y, and z coordinates of the four functions. This does not require 12 times as much computer time as the calculation of the integrals for the Roothaan process because these integrals are quite similar. Still, with the MOLPRO[53] computer program the ratio $T_{forces}/T_{integral}$ is about 3. (The ratio of the gradient calculation time to the total SCF time depends on several factors, in particular on the contraction of the basis set; with MOLPRO this ratio varies between 1 for small uncontracted basis sets and 2.5 for large heavily contracted basis sets.) There have been recent trends to reduce the integral computation time in comparison to the SCF iteration time.[54] Doubtless this will enhance the usefulness of force calculations.

Table 1 shows that the characteristics of potential curves calculated via the exact forces for a near Hartree–Fock calculation on the hydrogen molecule agree with those from a pointwise calculation. The data in this table will also be useful in discussing the procedure used to obtain force constants.

As previously mentioned, the sum and torque of the exact forces should vanish if the basis set guarantees that the molecular energy is invariant against translations and rotations of the nuclear framework as a whole. Translational invariance is easily attained by fixing the basis functions to the nuclei. For rotational invariance one must make sure that the basis set cannot carry torques; this usually requires that the atomic basis functions exhibit rotational invariance with respect to their centers. The latter is equivalent to the require-

---

*From the condition of vanishing sum and torque it is possible to determine the forces for six (five for linear molecules) nuclear coordinates. The resulting savings is, however, usually small and does not compensate for the loss of a valuable check. Only when most basis functions are situated on a single nucleus can a significant savings be attained; for a molecule like ethane the savings is less than 10%.

Table 1. Geometry and Force Constants of the Hydrogen Molecule Near the
Hartree–Fock Limit (7s4p2d/2b Gaussian Basis Set)[a]

R	1.3	1.35	1.4	1.45	1.5
E	−1.1318488	−1.1332234	−1.1334654	−1.1327485	−1.1312196
dE/dR	0.0401356	0.0155367	−0.0052808	−0.0229076	−0.0378342

Property	Points used	Fitting	From force	From energy
R_e	All	Quartic	1.386507	1.3865
$(dE/dR)_{1.4}$	All	Quartic	−0.0052808	−0.005283
$(d^2E/dR^2)_{1.4}$	All	Quartic	0.382641	0.38267
$(d^3E/dR^3)_{1.4}$	All	Quartic	−1.27294	−1.2824
$(d^4E/dR^4)_{1.4}$	All	Quartic	4.3248	4.288
$(d^2E/dR^2)_{1.4}$	1.35, 1.45	Linear	0.38444	—
$(d^3E/dR^3)_{1.4}$	1.35, 1.4, 1.45	Quadratic	−1.2763	—

[a] In atomic units.

ment that for a given exponent and n, l quantum numbers, all functions with the possible m quantum numbers should be included. This requirement can be reduced to a subgroup of the rotational group only if the molecule preserves a certain symmetry for all geometries investigated. For example, in a planar π system different exponents may be used for p_σ and p_π functions as long as the system remains strictly planar. Note that in a Gaussian lobe basis set, rotational invariance is only approximately fulfilled, even if the basis set contains all functions to a given l quantum number and exponent. This may cause a very small net torque.[56]

Recently application of the gradient method to semiempirical calculations of the CNDO type has become popular.[17,57–61] It is appropriate to note that gradient calculation offers even more advantages in semiempirical than in *ab initio* work. Integral evaluation, which must be redone for gradient calculation, is only a minor time consumer in CNDO, resulting in gradients at virtually no extra cost.[57,59]

2.6. Transformation of Cartesian Forces and Force Constants to Internal Coordinates

Programming considerations dictate the calculation of the forces first in Cartesian coordinates. However, the relevant quantities are the forces and force constants in internal coordinates, typically in internal valence coordinates, i.e., bond lengths and angles, out-of-plane angles are dihedral angles, or a linear combination of them. This section deals with this transformation.[8]

Let $\mathbf{X} = (X_1, \ldots, X_{3N})^+$ denote the set of Cartesian displacement coordinates from a reference configuration, and $\mathbf{q} = (q_1, \ldots, q_M)^+$ the chosen set of internal displacement coordinates. The forces are defined in internal coordinates as $\varphi_i = (\partial E/\partial q_i)_0$ with all q_j, $j \neq i$, fixed. It is clear from this definition that *the internal forces are unambiguously defined only if the internal coordinates form a complete but nonredundant set*, requiring $M = 3N - 6$ ($3N - 5$ for linear molecules). In general, it makes sense to speak of an internal force φ_i only if all coordinates q_i are defined and independent.

Let us write the energy as a quadratic function of $\mathbf{X}$ and $\mathbf{q}$:

$$E = E_0 - \sum_a f_a X_a + \tfrac{1}{2} \sum_{a,b} K_{ab} X_a X_b = E_0 - \mathbf{f}^+ \mathbf{X} + \tfrac{1}{2} \mathbf{X}^+ \mathbf{K} \mathbf{X} \tag{30}$$

and

$$E = E_0 - \sum_i \varphi_i q_i + \tfrac{1}{2} \sum_{i,j} F_{ij} q_i q_j = E_0 - \boldsymbol{\varphi}^+ \mathbf{q} + \tfrac{1}{2} \mathbf{q}^+ \mathbf{F} \mathbf{q} \tag{31}$$

where $\mathbf{f}$ and $\boldsymbol{\varphi}$ are the column vectors of the forces, and $\mathbf{K}$ and $\mathbf{F}$ are the harmonic force constant matrices in Cartesian and internal coordinates, respectively. Deviating from the usual procedure, let us retain the quadratic terms in the expression of the internal coordinates by the Cartesian ones:

$$q_i = \mathbf{B}_i \mathbf{X} + \tfrac{1}{2} \mathbf{X}^+ \mathbf{C}^i \mathbf{X} \tag{32}$$

where $\mathbf{B}_i$ is the ith row of the matrix which relates $\mathbf{q}$ and $\mathbf{X}$ to first order,[62] $\mathbf{q} = \mathbf{B}\mathbf{X}$; $\mathbf{B}$ can be easily determined.[63] Substituting (32) into (31) and comparing the coefficients of like terms we obtain

$$\mathbf{f} = \mathbf{B}^+ \boldsymbol{\varphi} \tag{33}$$

and

$$\mathbf{K} = \mathbf{B}^+ \mathbf{F} \mathbf{B} - \sum_i \varphi_i \mathbf{C}^i \tag{34}$$

We cannot directly express $\boldsymbol{\varphi}$ and $\mathbf{F}$ from (33) and (34) because $\mathbf{B}^+$, being a $3N$ by M matrix, has no inverse. However, a set of $M \times 3N$ matrices can be defined, all symbolically denoted by $\mathbf{B}^{\pm 1}$, having the property

$$\mathbf{B}^{\pm 1} \mathbf{B}^+ = \mathbf{I}_M \tag{35}$$

Indeed, if $\mathbf{m}$ is any matrix for which $(\mathbf{B}\mathbf{m}\mathbf{B}^+)$ is not singular* then

$$\mathbf{B}^{\pm 1} = (\mathbf{B}\mathbf{m}\mathbf{B}^+)^{-1} \mathbf{B}\mathbf{m} \tag{36}$$

satisfies (35). Multiplying (33) and (34) by $\mathbf{B}^{\pm 1}$ we get

$$\boldsymbol{\varphi} = \mathbf{B}^{\pm 1} \mathbf{f} \tag{37}$$

and

$$\mathbf{F} = \mathbf{B}^{\pm 1} \mathbf{K} \mathbf{B}^{-1} + \sum_i \varphi_i \mathbf{B}^{\pm 1} \mathbf{C}^i \mathbf{B}^{-1} \tag{38}$$

where $\mathbf{B}^{-1}$ is the transpose of (36).

---

*If $\mathbf{B}\mathbf{m}\mathbf{B}^+$ is singular for every m then the transformation to internal coordinates is singular and the chosen set of internal coordinates is not appropriate. This is the case, e.g., if we try to describe the out of plane distortion of the BF_3 molecule by the sum of the three FBF angles.

By explicitly including the six translational and rotational coordinates in the set $\mathbf{q}$, it can be shown that any of the matrices (36) yields the same internal forces if the sum and torque of the Cartesian forces vanish. They also yield the same internal force constants if $\mathbf{K}$ obeys the translational and rotational invariance condition. The simplest choice of $\mathbf{m}$ is obviously the unit matrix $\mathbf{I}_{3N}$. Note that $\boldsymbol{B}$ *must be calculated at the actual nuclear configuration*. This condition corresponds to the use of curvilinear internal coordinates instead of rectilinear ones. In general, curvilinear coordinates are preferable for two reasons: (a) the anharmonic force field is much more diagonal in curvilinear valence coordinates; and (b) serious difficulties are encountered in the calculation of force constants in rectilinear coordinates at nonequilibrium reference geometries (see Section 3.2.1).

The second term on the right-hand side of (38) is usually neglected because it is in the equilibrium geometry that the second derivatives of the energy are of main interest. However, as shown in Section 3, in quantum chemical calculations it may be preferable to determine the force constants at the experimental geometry, where the *theoretical* φ_i are usually not zero. In these cases the full equation (38) must be used. For example, consider a diatomic molecule AB which is initially oriented along the z axis, with forces $f_A = -f_B$ acting on the nuclei. The Cartesian force constant for the coordinate X_A is clearly nonvanishing: $K_{XX} = -f_A/(Z_A - Z_B)$. Introducing the new coordinates R_{AB} and β, the angle of the molecule with the z axis in the xz plane, the first term in (38) yields a spurious nonzero rotational force constant $F_{\beta\beta} = R_{AB}^2 K_{XX}$. The second term, however, exactly compensates for this.* Note that Cartesian and internal force constants may lead to different harmonic vibrational frequencies if they are not calculated at the theoretical equilibrium geometry. This represents an inherent ambiguity in the definition of harmonic vibrational frequencies at a nonequilibrium geometry. The fact that a diatomic molecule may have nonzero rotational frequency in Cartesian coordinates, but not in valence coordinates, is a strong argument in favor of the latter.

3. Applications

3.1. Molecular Geometries and Reaction Paths

The gradient greatly facilitates the determination of molecular geometry. If a reasonable approximation F_0 to the *theoretical* force constant matrix is

---

*Recently Thomsen and Swanstrøm[64] have attributed the difficulties in the transformation of Cartesian force constants (the apparent violation of rotational invariance) to cubic and quartic terms in the potential energy. This is not correct because these terms cannot contribute to the quadratic terms in internal coordinates, provided the transformation is smooth and nonsingular. Their difficulties are due to the neglect of the second term in (38).

known and the potential surface is approximately quadratic, the force relaxation method[8] can be used

$$\mathbf{q}_{i+1} = \mathbf{q}_i + \Delta\mathbf{q}_i = \mathbf{q}_i + \mathbf{F}_0^{-1}\boldsymbol{\varphi}_i \qquad (39)$$

where $\mathbf{q}_i$ and $\boldsymbol{\varphi}_i$ are the vectors of internal forces and forces in the ith step. (It is essential to use internal coordinates here because only these allow one to estimate $\mathbf{F}_0$.) Equation (39) would, of course, deliver the final equilibrium geometry in the first step if the potential surface were strictly quadratic and if $\mathbf{F}_0$ was exactly the force constant matrix appropriate to the theoretical surface. In practice only a few steps are required. Note that all coordinates are simultaneously optimized in this procedure and the final geometry does not depend on $\mathbf{F}_0$, which only controls the rate of convergence.

If great precision is not required, the zeroth (extrapolation) step may suffice. For example, in methylamine[65] a C, N, $7s\,3p$/H, $3s$/$1b$ Gaussian basis set yields 0.0836 mdyn Å/rad force in the direction of the CH_3A' rocking coordinate q_9 if the axis of the CH_3 group coincides with the C–N bond. Assuming that F_0 is diagonal and $F_{99} = 1.0$ mdyn Å/rad^2 (the experimental value 0.82 is increased by 20% to allow for the consistent overestimation of diagonal deformation force constants by similar basis sets), we obtain $\Delta q_9 = 4.8°$ and the tilting of the methyl group *toward* the nitrogen lone pair is $(6^{1/2}/3)\Delta q_9 = 3.9°$, a value which agrees well with the microwave result $(3.5°)$.[66,67]

Transformation of the geometry to Cartesian coordinates, which is necessary for the next iteration step, may be carried out to first order by the matrix $\mathbf{B}^{-1}$, the transpose of (36)

$$\Delta\mathbf{X} = \mathbf{B}^{-1}\Delta\mathbf{q} \qquad (40)$$

The force relaxation method is linearly convergent. Quadratic convergence can be achieved by utilizing the information (which is contained in the forces of the $i+1$th step) about the error of the force constant matrix in the ith step. Meyer[68] suggests the use of

$$\mathbf{F}_{i+1} = \mathbf{F}_i - (\boldsymbol{\varphi}_{i+1}\Delta\mathbf{q}_i^+ + \Delta\mathbf{q}_i\boldsymbol{\varphi}_{i+1}^+ - c\,\Delta\mathbf{q}_i\Delta\mathbf{q}_i^+)/b \qquad (41)$$

where $b = \Delta q_i^+ \Delta q_i$ and $c = \boldsymbol{\varphi}_{i+1}^+\Delta q/b$. The idea behind this formula is that $\mathbf{F}_{i+1}$ satisfies $\Delta\boldsymbol{\varphi}_i = \boldsymbol{\varphi}_{i+1} - \boldsymbol{\varphi}_i = \mathbf{F}_{i+1}\Delta\mathbf{q}_i$; a necessary condition for the true force constant matrix. This is also the basis of the variable metric optimization methods,[69,70] where the correction is performed directly on the inverse force constant matrix. The formula given by Murtagh and Sargent[70] is particularly simple:

$$\mathbf{F}_{i+1}^{-1} = \mathbf{F}_i^{-1} - \mathbf{z}_i\mathbf{z}_i^+/c_i \qquad (42)$$

with $\mathbf{z}_i = \Delta\mathbf{q}_i + \mathbf{F}_i^{-1}\Delta\boldsymbol{\varphi}_i$ and $c_i = \Delta\boldsymbol{\varphi}_i^+\mathbf{z}_i$. Both methods[69,70] employ a scale factor α so that the correction $\Delta\mathbf{q}$ is α times that in (39). The factor α is determined to yield minimum energy along the direction $\mathbf{F}_i^{-1}\varphi_i$. Given the costliness of the energy evaluation and the availability of good guesses to the force constant matrix, the use of scale factors does not seem effective in *ab initio* calculations. No comparison of the force constant updating methods (41) and (42) is available. McIver and Komornicki[57] discuss the relative virtues of optimization schemes working directly with Cartesian forces. These methods require no input (the matrices $\mathbf{B}$ and $\mathbf{F}_0$) but they converge more slowly and are thus not well suited to *ab initio* work.

On the basis of formulas (39)–(41) a fully automatic geometry search program has been implemented in the MOLPRO system.[53] In addition to the geometry it yields an improved set of force constants. An example for convergence is shown in Table 2 for the lowest $^2A_{1g}$ and $^2B_{3g}$ states of ethylene positive ion,[71] using a slightly contracted C, 7s, 3p/H, 3s/1b Gaussian basis set. The geometry obtained should be accurate enough for interpretation of the vibrational structure in photoelectron spectra; in particular, the large change in R_{CC} upon ionization from the $1b_{3g}$ orbital contradicts the modified assignment of Baker *et al.*[72] Other examples of geometry determination by gradient optimization include H_2O^+,[73] CH_4^+,[74] H_2CO,[75] HCO,[76] H_2CNH,[77] and the interesting molecule N_2O_2.[78]

For all its virtues it should be realized that gradient optimization is no panacea. In rigid molecules having a single well-defined energy minimum it converges quickly. For very anharmonic potential surfaces or in the case of multiple minima it may be necessary to first carry out a pointwise mapping of the potential surface to roughly localize the minima. Another difficulty arises in systems with very low force constants in certain directions, i.e., in cases when the lowest eigenvalue of the force constant matrix is near zero. In such systems small forces on the nuclei do not ensure that we are close to the equilibrium geometry. The best way to deal with these cases is probably to determine the force constants for the suspected "weak" coordinates, although large corrections to the geometry remain unreliable because of anharmonicity. Finally, it

Table 2. Geometry Determination for the Lowest $^2A_{1g}$ and $^2B_{3g}$ States of Ethylene Positive Ion (Restricted Hartree–Fock Wave Function, C, 7s, 3p/H, 3s/CH, CC, 1b Gaussian Basis Set, Contracted to C, 5s, 3p)[a]

Step number	$^2A_{1g}$			$^2B_{3g}$		
	r_{CC}, Å	r_{CH}, Å	β_{HCH}, degrees	r_{CC}, Å	r_{CH}, Å	β_{HCH}, degrees
0	1.337	1.085	117.3	1.337	1.085	117.3
1	1.392	1.100	134.0	1.288	1.140	103.0
2	1.407	1.088	138.7	1.229	1.135	103.2
3	1.426	1.083	139.8	1.250	1.134	103.3

[a] The forces in the last step indicate that the geometry has converged to the decimals given.

must be kept in mind that gradient optimization is symmetry conserving: if the starting geometry has incorrect high symmetry this will persist.

It is often required that some coordinates be held fixed during geometry optimization, notably for the calculation of adiabatic reaction paths. In this case the only mathematical requirement for the coordinate to be fixed is that its value should change monotonically along the reaction path. It is, however, numerically advantageous to have the fixed coordinate approximately parallel to the reaction path.* As an example for adiabatic reaction path determination we mention the calculation of the rotational barrier in nitrous acid.[79] Furthermore, it may be advantageous to determine the molecular geometry by fixing some parameters at their experimental values. For example, in water the bond angle at the Hartree–Fock limit is 106.4° (Ermler and Kern[80] give 106.1°; Dunning *et al.*[81] 106.6°; a numerically more accurate gradient optimization by Meyer[82] yields 106.4°) while the latest experimental value is 105.02°.[83] The discrepancy of 1.4° diminishes to 0.8° if the bond lengths are constrained to the experimental values, as a simple calculation using the quadratic force constants near the potential minimum shows.

When working with internal coordinates, constrained geometry optimization means that the corresponding rows and columns of $\mathbf{F}^{-1}$ are simply set to zero. In gradient optimization working directly in Cartesian coordinates, define the reduced Cartesian forces

$$\mathbf{f}_r = [\mathbf{I}_{3N} - \mathbf{B}_c^+ (\mathbf{B}_c \mathbf{B}_c^+)^{-1} \mathbf{B}_c] \mathbf{f} \tag{43}$$

The projection operator on the right-hand side of (43) annihilates the force components acting along the coordinates $q_1, \ldots, q_l$ to be constrained ($1 < 3N - 6$). $\mathbf{B}_c$ is the $\mathbf{B}$ matrix *only* for the coordinates $q_1, \ldots, q_l$. Infinitesimal changes along the reduced forces do not change the values of $q_1, \ldots, q_l$. In this way it is easy to fix all bond lengths which are reproduced rather poorly in semiempirical methods.

3.2. Force Constants

3.2.1. General Considerations and Diatomic Molecules

We define a force constant $F_{ijk\ldots}$ generally as

$$F_{ijk\ldots} = \partial\partial\partial \ldots E / \partial q_i \partial q_j \partial q_k \ldots \tag{44}$$

at the reference geometry. The experimental system of units is used: energy is measured in mdyn Å $= 10^{-18}$ J, bond lengths in Å $= 10^{-10}$ m, and bond angles in radians (1 hartree = 4.359828 mdyn Å, 1 bohr = 0.5291772 Å). Force constants are not directly accessible to measurements, only the spectroscopical observables are measurable, e.g., vibrational frequencies. At first it might seem

---

*Note, however, that the reaction path generally depends on the choice of the fixed coordinate.

more advantageous to compare the calculated spectroscopical observables with experiment. However, in cases when reliable experimental force constants are available it is better to carry out the comparison at the level of force constants. Spectroscopical observables usually depend on several physically different force constants and their error is determined by the largest error in the contributing force constants. Valuable information may be lost this way.

Calculation of the force constants from the gradient requires numerical differentiation of the forces. For harmonic force constants we can simply approximate the force curve linearly, i.e., use the difference quotient

$$F_{ij} = -\Delta\varphi_i/\Delta q_j \qquad (45)$$

In order to eliminate the effect of cubic anharmonicity, the points should lie symmetrically around the reference geometry at $q_j^0 \pm \frac{1}{2}\Delta q_j$. Note that from two evaluations of the gradient a whole row of the harmonic force constant matrix F_{ij}, $i = 1, \ldots, M$ is obtained.[8] To calculate all harmonic force constants. $2M$ gradient evaluations are necessary if no symmetry is present. The off-diagonal (coupling) elements of the force constant matrix are evaluated twice, as $-\Delta\varphi_i/\Delta q_j$ and as $-\Delta\varphi_j/\Delta q_i$; the agreement of the two values is a check for numerical stability. If all force constants are not required, a substantial savings in the number of gradient evaluations can be achieved, still yielding most of the coupling constants which, as will be shown below, are of primary interest. In symmetrical molecules one coordinate from each symmetry species may be changed simultaneously, also reducing the number of gradient evaluations. By adding a single calculation at the reference geometry we are able to evaluate all diagonal and semidiagonal cubic force constants, F_{iii} and F_{iij} from the above $2M$ points. As diagonal cubic constants are fairly well reproduced in Hartree–Fock calculations, this is worthwhile. Note that three energy and three force values at the points $q_j^0, q_j^0 \pm \frac{1}{2}\Delta q_j$ determine, in principle, the diagonal force constants up to quintic. However, terms higher than cubic are not determined by the forces alone and their evaluation is numerically unstable. As to the values of the finite distortions $\frac{1}{2}\Delta q_j$, the data in Table 1 indicate that ± 0.05 bohr (or perhaps a little less) is about the right value for bond stretchings; it is neither too small for numerical instabilities to appear, nor too large to render (45) inaccurate. For deformations, $\pm 2°$ seems to be a good value.

It is not the purpose of this chapter to deal with the force constants and related spectroscopic constants of diatomic molecules in detail, as the gradient method is of no particular value for diatomics and this question has been thoroughly discussed.[29,84–90] However, some general points are best illuminated in the case of diatomics.

The first question concerns the choice of reference geometry. The fully consistent procedure is to calculate the force constants at the theoretical energy minimum. As mentioned in Section 2.6, there is a certain amount of ambiguity in the definition of force constants at nonequilibrium geometries although this is usually not apparent when valence coordinates are used. On the other hand Schwendeman[90] argues that better values are obtained at the experimental geometry. Indeed, consider the difference of the theoretical and experimental quadratic force constants at their respective geometries:

$$F_2^{th}(R^{th}) - F_2^{exp}(R^{exp}) = [F_2^{th}(R^{exp}) - F_2^{exp}(R^{exp})]$$
$$+ (R^{th} - R^{exp})F_3^{th}(R^{exp}) + \cdots \tag{46}$$

Experience shows that the second term on the right-hand side of (46) usually exceeds the first. According to Schwendeman[90] this is understandable, considering the large compensation of the electronic and nuclear part of the force which causes R^{th} to behave like a pseudo first-order quantity. For bond stretchings, both terms on the right-hand side of (46) are usually positive: $F_2^{th}(R^{exp}) > F_2^{exp}(R^{exp})$, $R^{th} < R^{exp}$, and $F_3^{th} < 0$.

To illustrate the above point, the force constants of the hydrogen molecule from a near-Hartree–Fock wave function are compared to the very accurate values of Kolos and Wolniewicz[91] in Table 3. It is seen that the error of the harmonic force constant decreases from 8.2% to 3.1% when going from the theoretical to the exact energy minimum. It is also seen that the dominant

Table 3. *Force Constants of the Hydrogen Molecule from Hartree–Fock Calculations in Atomic Units*

Gaussian basis	R	E	dE/dR	d^2E/dR^2	d^3E/dR^3	d^4E/dR^4	R_e
3s	1.401	−1.122534	0.005150	0.3930	−1.3066	4.221	1.3882
5s	1.401	−1.127986	0.007249	0.3759	−1.2830	4.420	1.3823
s limit (15s)	1.401	−1.128525	0.006938	0.3766	−1.2593	4.289	1.3831
3s1p	1.401	−1.126758	0.000405	0.3991	−1.2813	4.224	1.4000
3s1b	1.401	−1.126822	0.000611	0.4023	−1.2870	4.173	1.3995
5s2p	1.401	−1.132919	0.004829	0.3854	−1.2807	4.304	1.3887
H–F limit[a] (7s4p2d/2b)	1.3865	−1.133501	−0.000003	0.4002	−1.3328	4.541	1.3865
H–F limit	1.401	−1.133460	0.005663	0.3814	−1.2686	4.309	—
Near exact[b]	1.401	−1.174475	−0.000031	0.3701	−1.2703	4.22$_4$	1.40108
Electronic part H–F limit	1.3865	−1.854742	0.520185	−0.3501	0.2908	−0.143	—
Electronic part H–F limit	1.401	−1.847236	0.515139	−0.3459	0.2888	−0.138	—
Electronic part Near exact	1.401	−1.888251	0.509445	−0.3572	0.2871	−0.22$_3$	—
Nuclear part	1.401	0.713776	−0.509476	0.7273	−1.5574	4.4465	—

[a] Calculated from the data of Table 1. The potential curve of Table 1 is only 0.00016 hartree above the very accurate SCF curve of Kolos and Roothaan[110] and the two curves are parallel to 4 cm^{-1} between 1.3 and 1.5 bohr. The deviation from parallellism is believed to be due to the deterioration of the Kolos–Roothaan results for large interatomic distances, and it is linear in R to the numerical accuracy of the Kolos–Roothaan curve. Because of its higher numerical accuracy the force curve of Table 1 was used.
[b] Calculated from a hexic fit to the *electronic* energy of the near-exact wave function of Kolos and Wolniewicz[91] at 9 points between 1.0 and 1.8 bohr. By splitting off the very anharmonic nuclear part a satisfactory fit was obtained, although the quartic term remains somewhat inaccurate.

contribution to the higher force constants comes from the nuclear part. At 1.401 bohr, correlation contributes -3.1%, $+0.13\%$, and $\sim -2\%$ to the quadratic, cubic, and quartic force constants, respectively.

Although the hydrogen molecule may be a somewhat fortunate example, the main features are the same for other molecules as well. For OH ($^2\Pi$), near-Hartree–Fock calculations[31] yield 16.8% error in the calculated harmonic force constant at the theoretical geometry (the experimental value is taken as 7.867 mdyn/Å,[92]). This error diminishes to 1.6% at the experimental geometry. It is interesting to evaluate the correlation contribution to the energy derivatives from a recent calculation yielding 89% of the total correlation energy.[92] At $R^{\text{exp}} = 1.3842$ bohr the Hartree–Fock values are

$$E' = 0.172 \text{ mdyn}, \qquad E'' = 7.993 \text{ mdyn/Å}$$

$$E''' = -54.94 \text{ mdyn/Å}^2, \qquad E^{iv} = 355 \text{ mdyn/Å}^3$$

whereas correlation contributes

$$-0.166, -0.107(-1.3\%), -0.21(+0.4\%), \text{ and } -4.5(-1.3\%)$$

respectively. Taking into account that calculations at the experimental geometry not only yield better results but also require less computational work, this is the recommended procedure.

Calculation of force constants at the experimental geometry is equivalent to adding an empirical linear term to the theoretical potential function which shifts the position of the minimum to the experimental value. Although empirical elements are introduced by this procedure, it can be easily justified. The Hartree–Fock model is expected to perform very well for the core–core repulsion interaction which dominates most force constants, especially the cubic and quartic ones. Because of its $1/R$ behavior, the core–core contribution to the force constants varies strongly with the internuclear distance, in contrast to the valence electron contribution which is a fairly smooth function of R. Therefore, the changes in the force constants with the molecular geometry come mainly from the core–core term and, of course, the latter is most accurate at the experimental r_e geometry. There is some difficulty in obtaining r_e geometries for polyatomic molecules; in general, however, the error in the Hartree–Fock interatomic distances is larger than the error of substituting r_0 or some other experimental structure.

The above procedure has only one drawback: in general, it does not leave the quadratic and higher force constants invariant. For example, if the function added to the Hartree–Fock surface is linear in curvilinear valence coordinates, it may contain quadratic terms when transformed to Cartesian coordinates. Because the largest force constants are associated with bond stretching coordinates, and the Hartree–Fock method gives substantial systematic errors for certain bond lengths, the added empirical linear function is dominated by bond stretching terms. These do not influence, e.g., deformational force constants if curvilinear coordinates are used. However, in Cartesian coordinates stretchings and deformations are not separated and the deformational force constants are significantly influenced. Using curvilinear valence coordinates we never experienced this trouble.

The question of what accuracy can be expected for force constants at the Hartree–Fock limit has been widely discussed. Gerratt and Mills[2] and Freed[93] show that the error is second order in the error parameter ε provided that $d\varepsilon/dR$ is of order ε. Gerratt and Mills therefore, expect an accuracy of 1%. However, taking into account the compensation of the electronic and nuclear part of the force,[88,90,93] a more realistic estimate is 10%, particularly for the smaller deformation and coupling force constants.

Table 3 also gives some indication about the variation in the force constants with the basis set. Contrary to former views this is not large if physically reasonable multicenter basis sets are used. The largest deviation in the harmonic force constant from the Hartree–Fock limit value is 5.5% for the wave functions of Table 3 at 1.401 bohr.

3.2.2. Harmonic Force Constants in Polyatomic Molecules

A systematic quantum chemical study of the harmonic force fields of polyatomic molecules has been carried out during the past five years, largely by the gradient method. Most of the previous investigations suffered either from unsatisfactory basis sets or from numerical errors. The recent results demonstrate that harmonic force fields from Hartree–Fock calculations can successfully compete with experimental determinations.

In comparing experimental and theoretical force constants it should be realized that the "experimental" values usually contain a number of implicit assumptions and systematic errors. The largest source of error is anharmonicity. Although it is possible to correct for anharmonicity on the basis of experimental data, this may introduce new errors. An example of this is the anharmonicity X_{11} of the A_1 "breathing" vibration of methane. It now seems sure[74] that the experimental value $X_{11} = 65$ cm^{-1} of Jones and McDowell[94] is severely in error, the correct value being about 13.6 cm^{-1}. It is interesting that the first indication to this was given by the surprisingly large error in the calculated CH–CH stretching–stretching coupling force constant.[24] Accepting 3143 cm^{-1} for ω_1, the harmonic breathing frequency of methane, the experimental value of this constant is +0.115 mdyn/Å, while a near Hartree–Fock wave function[24] yields +0.049. If, as it seems now, the harmonic frequency is near 3044 cm^{-1} [74] then the experimental value decreases to +0.03 ± 0.02 mdyn/Å, in good agreement with the calculation.

Standard deviations of experimental force constants do not include systematic errors such as anharmonicity effects. Therefore, as Strey and Mills[95] express it "the standard errors of force constants give an optimistic assessment

Table 4. Harmonic Force Constants of Formaldehyde[a]

Force constant	73/3/1[b]	95/4/1[b]	951/4/1[b]	GHFF[c]	GHFF[d]
$F_{11}(CO)$	13.905	13.646	13.663	12.903(62)	12.931(58)
$F_{12}(CO, CH)$	+0.676	+0.797	+0.791	+0.739(66)	+0.361(236)
$F_{13}(CO, \alpha)$	−0.414	−0.411	−0.402	−0.408(15)	−0.431(34)
$F_{22}(CH)$	4.999	4.885	—	4.963(34)	4.920(58)
$F_{23}(CH, \alpha)$	+0.106	+0.131	—	+0.077(45)	+0.308(134)
$F_{33}(\alpha)$	0.645	0.634	—	0.570(4)	0.589(22)
$F_{44}(CH)$	4.909	4.791	—	4.852(37)	4.869(25)
$F_{45}(CH, \beta)$	+0.157	+0.146	—	+0.171(37)	+0.203(28)
$F_{55}(\beta)$	0.946	0.907	—	0.833(5)	0.835(3)
$F_{66}(\gamma)$	0.514	0.501	—	0.403(2)	0.403(2)

[a] In units mdyn/Å, mdyn/rad, and mdyn Å/rad² for stretching–stretching, stretching–bending, and bending–bending force constants, respectively. For the definition of the coordinates see Ref. 75.
[b] Size of the Gaussian basis set.[75]
[c] Experimental general harmonic force field,[99] standard errors are given in parentheses.
[d] Results obtained without including the centrifugal distortion data on $H_2{}^{13}CO$ in the least-squares procedure.[100]

of the uncertainty; an assessment of 4 to 5 times the standard error looks more realistic."

Another type of error (in experimental force constants) which cannot be expressed by standard deviation is the uncertainty of assignment. For example, for ethylene two qualitatively different force fields were used in the A_{1g} species until 1971 when the question of assignment was unambiguously solved independently by experimental[96] and theoretical[97] means. Interestingly, the true solution turned out to be the set which was less preferred by Crawford et al.[98] on the basis of transferability considerations.

An example of successful force field prediction is the case of formaldehyde. Prior to 1973 the force constants of this simple compound were rather uncertainly known. An *ab initio*[75] and a new experimental[99] determination were started in 1973. The results are compared in Table 4. The fifth column is particularly interesting: it represents results where the new centrifugal distortion data on $H_2{}^{13}CO$ were not yet included.[100] In contrast to the final results (column 4) the results in column 5 (especially F_{12}) deviate markedly from the *ab initio* values. This was taken as an indication that new experimental values which are more sensitive to F_{12} have to be included in the least-squares analysis.

Near Hartree–Fock calculations have been performed for the force constants of water,[80–82] methane,[24] ammonia,[56] and formaldehyde.[75] The following can be inferred from these calculations (and from extrapolation of results using smaller basis sets).

(a) Coupling force constants are reproduced with remarkable accuracy, errors generally not exceeding ±0.05 mdyn/Å or 10% whichever is larger. Experimental values are often less accurate.

(b) Diagonal deformation force constants are consistently overestimated by about 10% except for the out-of-plane deformations of planar π systems, where the deviation is about 25%. The larger errors of the out-of-plane deformations are probably due to nondynamical (near degeneracy) effects caused by the low-lying π^* orbitals. Note that experimental deformation force constants are usually very accurate.

(c) Stretching force constants are generally overestimated by some few percent at the experimental r_e geometry, as in diatomic molecules. It is, however, difficult to compare stretching constants with experiment because they are very sensitive to the choice of reference geometry, and the experimental r_e distances in polyatomic molecules are generally not known to sufficient accuracy. Even in methane the value of Kuchitsu and Bartell,[101] $r_e = 1.085$ Å, is now in doubt, the true value probably being near 1.090 Å.[74] This difference of only 0.005 Å in r_e causes $\Delta F = F_{rrr} \Delta r = 0.16$ mdyn/Å $= 2.8\%$ change in F_{rr}.

It is seen from the above points that the calculation of diagonal force constants to experimental accuracy is not possible from Hartree–Fock wave functions (although empirical corrections for near-systematic errors significantly improve the agreement with experiment). On the other hand, theoretical coupling force constants can successfully contribute to the determination of molecular force fields.

Comparison of near Hartree–Fock calculations with results obtained using a modest $7s3p/3s/1b$ Gaussian basis set (not contracted in the valence shell and containing bond functions for polarization) shows that the latter are also

Table 5. *Comparison of Theoretical and Experimental Stretching–Stretching Type Force Constants*[a]

Molecule	References	Coordinates	Theoretical[b]	Experimental
H_2O	20	OH–OH	−0.156	−0.101 ± 0.001
NH_3	102	NH–NH	−0.063	+0.01 ± 0.18
CH_4	102	CH–CH	+0.036	+0.03 ± 0.02
C_2H_6	103	CC–sym. CH	+0.140	0 to 0.21
		sym. CH–sym. CH′	+0.007	−0.007 ± 0.08
		asym. CH–asym. CH′	−0.035	−0.050 ± 0.08
C_2H_4	97	CC–sym. CH	+0.136	0.1 to 0.26
		sym. CH–sym. CH′	+0.007	−0.032 to 0.040
		asym. CH–asym. CH′	−0.025	−0.03 to +0.082
C_2H_2	103	CC–CH	−0.134	−0.05 to −0.20
		CH–CH′	+0.007	−0.03 to +0.02
H_2CO	75	CO–sym. CH	+0.676	+0.739 ± 0.06
		CH–CH′	+0.045	+0.056 ± 0.037
ONF	104	ON–NF	+2.48	+2.00 to +2.50
HCN	105	HC–CN	−0.219	−0.200 ± 0.001
C_2N_2	105	CC–CN	+0.140	+0.42 ± 0.01
		CN–C′N′	−0.240	−0.256 ± 0.02

[a] In mdyn/Å. For the definition of the coordinates and the source of the experimental data see the references.
[b] Calculated using a $7s3p/3s/1b$ Gaussian basis set.

Table 6. Comparison of Theoretical and Experimental Stretching–Deformation Force Constants[a]

Molecule	Reference	Coordinates	Theoretical	Experimental
H_2O	20	OH–HOH	+0.280	+0.219±0.002
NH_3	102	sym. NH–sym. def	−0.607	+0.79±0.2
		asym. NH–asym. def	−0.207	−0.176±0.04
CH_4	102	S_3–S_4	+0.235	+0.225±0.05
C_2H_6	103	CC–sym. def	+0.390	+0.346±0.015
		sym. CH–sym. def	−0.123	−0.050[c]
		asym. CH–rocking	+0.206	+0.076[c]
		asym. CH–asym. def	−0.220	−0.076[c]
C_2H_4	97	CC–sym. def	+0.242	+0.222±0.007
		sym. CH–sym. def	−0.093	−0.018 to −0.100
		asym. CH–rocking	+0.161	+0.111 to +0.185
H_2CO	75	CO–HCH	+0.414	+0.408±0.015
		sym. CH–HCH	−0.106	−0.077±0.045
		asym. CH–rocking	+0.157	+0.171±0.037
ONF	104	ON–ONF	+0.49	+0.29 to +0.49
		NF–ONF	+0.21	+0.22 to +0.26

[a] In mdyn/rad. For the definition of the coordinates and the source of the experimental data see the references.
[b] Calculated using $7s3p/3s/1b$ Gaussian basis sets.
[c] Values probably too low by a factor of about 2, see Ref. 103.

surprisingly successful in predicting coupling constants, the errors usually being less than twice those at the Hartree–Fock limit. Tables 5–7 compare the theoretical ($7s3p/3s/1b$) and experimental stretching–stretching, stretching–deformation, and deformation–deformation force constants for H_2O,[20] CH_4 and NH_3,[102] C_2H_4,[97] C_2H_2 and C_2H_6,[103] H_2CO,[75] ONF,[104] HCN, and C_2N_2.[105] On the basis of the general agreement predictions have been made for the force fields of BH_4^-,[102] HCO,[76] H_2CNH,[77] FCN,[105] N_2O_2,[78] and HNO_2.[79]

Table 7. Comparison of Theoretical and Experimental Deformation–Deformation Force Constants[a]

Molecule	Reference	Coordinates	Theoretical[b]	Experimental
NH_3	102	HNH–HNH	−0.067	−0.045±0.023
CH_4	102	$(F_{22}-F_{44})$	+0.041	+0.034±0.017
C_2H_6	103	sym. def–sym. def′	+0.038	+0.033 to +0.038
		rocking–rocking′	+0.193	+0.139±0.02
		asym. def–asym. def′	−0.005	−0.005±0.015
C_2H_4	97	sym. def–sym. def′	+0.024	+0.014 to +0.018
		rocking–rocking′	+0.104	+0.07 to +0.096
		wagging–wagging′	−0.050	−0.039±0.001
C_2H_2	103	HCC′–CC′H′	−0.126	−0.091 to −0.100

[a] In mdyn Å/rad². For the definition of the coordinates and the source of the experimental data see the references.
[b] Calculated using $7s3p/3s/1b$ Gaussian basis sets.

As sufficient experimental information is available only for the simplest molecules, the best way to determine reliable force fields seems to be to combine experimental spectra and *ab initio* coupling force constants. This would enable us to predict the vibrational spectra of most organic molecules. The difficulty at present lies in the fact that calculations on molecular fragments exceeding the size of the benzene ring are needed to establish transferable force fields. Such calculations are very costly.

3.2.3. Higher Force Constants in Polyatomic Molecules

Higher force constants become increasingly important as a source of anharmonicity in spectroscopy and as means for vibrational energy transfer between normal modes. It is often stated that restricted Hartree–Fock wave functions cannot be expected to yield realistic higher force constants because of their wrong dissociation behavior. This statement should be taken with a grain of salt because, as the examples in Section 2.2.1 seem to indicate, the effect of correlation energy on the force constants is slight near r_e, i.e., the correlation energy is an almost linear function of the internuclear distance in the neighborhood of r_e.

The first near Hartree–Fock calculation of anharmonic force constants was that of Ermler and Kern.[80] This is a pointwise calculation but a large number of carefully selected points were used to maintain numerical accuracy. Recently, Meyer[82] has obtained more comprehensive data, including the quartic constants. The significance of his calculations lies in the fact that he compares the Hartree–Fock values to a highly correlated wave function. This allows a more realistic assessment of the Hartree-Fock method than in comparison to experimental values (which are quite uncertain for the higher coupling constants). A comparison of near Hartree–Fock force constants[80,82] with data evaluated from a $7s\,3p/3s/1b$ calculation[20] shows that the latter are also surprisingly good. The same conclusion emerges from the data on

Table 8. Some Quadratic and Cubic Force Constants of Hydrogen Cyanide[a]

| Constant | Theoretical[b] | | Experimental | | |
	I	II	Ref. 95	Ref. 106	Ref. 107
F_{rR}	-0.299	-0.221	-0.200 ± 0.001	-0.211 ± 0.006	-0.216 ± 0.08
F_{rrr}	-38.53	-35.63	-35.37 ± 0.48	-33.76 ± 0.42	-36.51 ± 0.90
F_{RRR}	-133.26	-130.30	-125.95 ± 1.35	-125.09 ± 0.96	-115.66 ± 2.16
F_r	-0.062	-0.178	-0.19 ± 0.12	-0.11 ± 0.06	-0.26 ± 0.14
F_R .	-0.585	-0.674	-0.65 ± 0.09	-0.66 ± 0.04	-0.54 ± 0.10
F_{rRR}	$+0.116$	$+0.063$	$+0.41\pm0.67$	$+0.09\pm0.44$	-2.66 ± 0.8
F_{rrR}	$+0.691$	$+0.570$	$+0.04\pm0.19$	-0.99 ± 0.22	-0.64 ± 0.54

[a] In experimental units (see text). r, R, and β refer to the change in the CH and CN bond lengths and in the HCN angle, respectively.
[b] I: $7s\,3p/3s$ basis set; II: $9s\,5p\,1d/5s\,1p$ basis set, at $R = 1.1532$ Å, $r = 1.0655$ Å. See also Refs. 105, 114.

hydrogen cyanide, shown in Table 8. Note that for HCN, a linear molecule, the experimental values are more reliable. Still, the theoretical values agree significantly better with the recent accurate data of Strey and Mills[95] than, e.g., the values of Suzuki et al.[107]

4. Analytic Calculation of Higher Energy Derivatives

An analytic expression for the exact second derivative of the SCF energy with respect to the nuclear coordinates was first given by Bratož[12] for closed shells. Gerratt and Mills[2] give formulas for the analytic derivative of the Hellmann–Feynman force, but their method can be easily extended to the exact second derivative. Moccia[108] has generalized these equations to the analytic third derivative and has given a lucid treatment of the first and second derivatives. The first (and so far only) program to calculate the analytic second derivative *with basis functions which depend on the nuclear coordinates* is that of Thomsen and Swanstrøm.[64,109] Instead of giving the detailed equations here, we only sketch the method and concentrate on the computational implications.

Remember that the efficiency of the gradient method is mainly due to the following characteristics:

(i) Calculation of the gradient is a simple summation, not requiring external data storage.

(ii) It is possible to calculate the forces for all nuclear coordinates by going through the basis integrals only once. This, in turn, leads to a substantial saving because the derivatives of the basis integrals with respect to different nuclear coordinates are closely related.

In comparing the analytical calculation of the second derivative with gradient evaluation and subsequent numerical differentiation these points should be kept in mind.

In contrast to the gradient, evaluation of the second derivative requires the derivatives of the orbital coefficients C_{ri}^a. These can be obtained from a linear system of equations,

$$\mathbf{G}\mathbf{C}^a = \mathbf{R}_a \qquad a = 1, \dots, 3N \qquad (47)$$

Bratož[12] and Gerratt and Mills[2] transform the matrices $\mathbf{C}^a$ to an orbital basis at the reference geometry; in this case elements of $\mathbf{G}$ are simple sums of some two-electron matrix elements. However, transformation of the two-electron integrals to an orbital basis may be a very expensive procedure, requiring in the best case $O(m^5)$ arithmetic operations. A partial transformation to an orbital basis is also implicit in the formulas of Moccia.[108] Storage of $\mathbf{G}$ and solution of Eq. (47) for $\mathbf{C}^a$ is also problematic for larger systems. There are $n(m-n)$ unknown elements C_{ri}^a in the Gerratt–Mills scheme, giving rise to $n^2(m-n)^2$ $\mathbf{G}$ elements and $O[n^3(m-n)^3]$ arithmetic operations in the course of solution. To take advantage of the fact that $\mathbf{G}$ does not depend on a, all $3N$ Eq. (47) should be solved simultaneously; this probably necessitates external data storage with the resulting loss of efficiency.

The final formula for the second derivative of the energy contains the first and second derivatives of the basis integrals, the above $\mathbf{C}^a$ and the derivatives of the orbital energies. The latter are easily calculated once $\mathbf{C}^a$ is known. In evaluating this formula one has the choice of two possibilities: (a) compute all first and second derivatives of the integrals together and store them on external medium, abandoning thus advantage (i) of the gradient method; or (b) calculate the integral derivatives for different nuclear coordinates independently, sacrificing advantage (ii).

Thomsen and Swanstrøm[109] have chosen the first possibility. However, as their computer timing data show (about 30 hr CDC 6400 time for water with a $(11s7p1d/6s1p)-[5s4p1d/3s1p]$ Gaussian basis set for only a part of the Cartesian second derivatives), analytic calculation of the second (and by the same token the third) derivative is not practical.* The same applies to the analytic calculation of expectation values, e.g., dipole moment derivatives. The numerical finite difference method seems to be simpler and cheaper. Moreover, no numerical problems arise for expectation values since we are usually interested in the first derivative.

ACKNOWLEDGMENTS

The author is very grateful to Professor W. Meyer who has considerably helped to shape this article by advice, critique, and original contributions, and to Dr. G. Fogarasi for reading the manuscript,

Addenda

In recent papers[111,112] Schlegel, Wolfe, Bernardi, and Mislow have carried out systematic gradient calculations on molecular geometries and force constants, using an extended version of the GAUSSIAN 70 computer program.[54] Results are reported for second- and third-period hydrides,[111] ethylene, silaethylene, and methylamine.[112] Both the STO-3G and the 4-31G basis sets were used. The latter provided reliable harmonic and cubic force constants which agree excellently with the $7s3p/3s$ results, but the STO-3G basis (essentially a minimum Slater basis) gave rather poor results.

---

*Note that the algorithm of Thomsen and Swanstrøm is not optimum because that part of the second derivative expression which contains the second derivatives of the basis integrals could be summed up without storing the derived integrals. However, it is still unlikely that this method can be made competitive with the gradient method, even if the enormous complications of programming are not considered.

Garton and Sutcliffe[113] have recently reviewed direct minimization methods in quantum chemistry, covering both gradient and nongradient methods. Although the discussion is centered on the determination of density matrices which minimize the total energy, geometry determination is also considered. In particular, the merits of various gradient optimization schemes, only briefly discussed in Section 3.1, are critically assessed in this paper. An important comparison of the Hartree–Fock and CI potential surfaces has been carried out recently for the water molecule.[115]

References

1. D. R. Hartree, *Numerical Analysis*, Oxford University Press, Oxford (1968).
2. J. Gerratt and I. M. Mills, Force constants and dipole-moment derivatives of molecules from perturbed Hartree–Fock calculations. I, II, *J. Chem. Phys.* **49**, 1719–1739 (1968).
3. W. L. Clinton, Forces in molecules. I. Application of the virial theorem, *J. Chem. Phys.* **33**, 1603–1606 (1960).
4. R. M. Stevens, Geometry optimization in the computation of barriers to internal rotation, *J. Chem. Phys.* **52**, 1397–1402 (1970).
5. A. Veillard, Distortional effects on the ethane internal rotation barrier and rotational barriers in borazane and methylsilane, *Chem. Phys. Lett.* **3**, 128–130 (1969).
6. H. J. Monkhorst, Geometrical changes during the internal rotation in ethane, *Chem. Phys. Lett.* **3**, 289–291 (1969).
7. T. H. Dunning and N. W. Winter, Hartree–Fock calculation of the barrier to internal rotation in hydrogen peroxide, *Chem. Phys. Lett.* **11**, 194–195 (1971).
8. P. Pulay, *Ab initio* calculation of force constants and equilibrium geometries in polyatomic molecules. I. Theory, *Mol. Phys.* **17**, 197–204 (1969).
9. J. Hinze, in: *Advances in Chemical Physics* (I. Prigogine and S. A. Rice, eds.), Vol. 26, pp. 213–263, John Wiley and Sons, New York (1974).
10. J. Hellmann, *Einführung in die Quantenchemie*, Deuticke & Co., Leipzig (1937).
11. R. P. Feynman, Forces in molecules, *Phys. Rev.* **56**, 340–343 (1939).
12. S. Bratož, Le calcul non empirique des constantes de force et des dérivées du moment dipolaire, *Colloq. Int. C.N.R.S.* **82**, 287–301 (1958).
13. S. Bratož and M. Allavena, Electronic calculation on NH_3. Harmonic force constants, infrared and ultraviolet spectra, *J. Chem. Phys.* **37**, 2138–2143 (1962).
14. M. Allavena and S. Bratož, Electronic calculation of force constants and infrared spectra of H_2O and D_2O, *J. Chim. Phys.* **60**, 1199–1202 (1963).
15. M. Allavena, Calculation of the force constants of the methane molecule with the aid of electron wave functions, *Theor. Chim. Acta* **5**, 21–28 (1966).
16. R. Moccia, Optimization of the basis functions in SCF MO calculations; optimized one-center SCF basis set for HCl, *Theor. Chim. Acta* **8**, 8–17 (1967).
17. W. L. Bloemer and B. L. Bruner, Optimization of variational trial functions, *J. Chem. Phys.* **58**, 3735–3744 (1973).
18. W. Meyer, private communication.
19. W. Meyer and P. Pulay, Generalization of the force method of open-shell wavefunctions, Proceedings of the Second Seminar on Computational Problems in Quantum Chemistry, Strasbourg, France, September, 1972, pp. 44–48.
20. P. Pulay, *Ab initio* calculation of force constants and equilibrium geometries in polyatomic molecules. II. Force constants of water, *Mol. Phys.* **18**, 473–480 (1970).
21. K. Klauss, Zwei Anwendungsmöglichkeiten von perturbed Hartree–Fock Rechungen, Arbeitsbericht des Institutes für Theoretische Physikalische Chemie der Universität Stuttgart, No. 14, pp. 71–84, (1971).
22. T. Vladimiroff, Computation of molecular equilibrium geometries using self-consistent perturbation theory, *J. Chem. Phys.* **54**, 2292 (1971).

23. R. M. Stevens, R. M. Pitzer, and W. N. Lipscomb, Perturbed Hartree–Fock calculations. I. Magnetic susceptibility and shielding in the LiH molecule, *J. Chem. Phys.* **38**, 550–560 (1963).

24. W. Meyer and P. Pulay, Near-Hartree–Fock calculations of the force constants and dipole moment derivatives in methane, *J. Chem. Phys.* **56**, 2109–2116 (1973).

25. A. C. Hurley, The electrostatic calculation of molecular energies, *Proc. R. Soc. London, Ser. A.* **226**, 170–192 (1954).

26. G. G. Hall, The stability of a wavefunction under a perturbation, *Philos. Mag.* **6**, 249–258 (1961).

27. A. A. Frost, Floating spherical gaussian orbital model of molecular structure. I. Computational procedure. LiH as an example. *J. Chem. Phys.* **47**, 3707–3714 (1967).

28. D. M. Bishop and M. Randić, *Ab initio* calculation of harmonic force constants, *J. Chem. Phys.* **44**, 2480–2487 (1966).

29. D. M. Bishop and A. Macias, *Ab initio* calculation of harmonic force constants. IV. Comparison of different methods, *J. Chem. Phys.* **53**, 3515–3521 (1970).

30. D. P. Chong, P. J. Gagnon, and J. Thorhallson, Virial scaling and diatomic force constants, *Can. J. Chem.* **49**, 1047–1052 (1971).

31. P. E. Cade and W. M. Huo, Electronic structure of diatomic molecules. VI.A. Hartree–Fock wavefunctions and energy quantities for the ground states of the first-row hydrides, AH, *J. Chem. Phys.* **47**, 614–648 (1967).

32. P. E. Cade, K. D. Sales, and A. C. Wahl, Electronic structure of diatomic molecules. III.A. Hartree–Fock wavefunctions and energy quantities for $N_2(X^1\Sigma_g^+)$ and $N_2^+(X^2\Sigma_g^+, A^2\Pi_u, B^2\Sigma_u^+)$ molecular ions, *J. Chem. Phys.* **44**, 1973–2003 (1966).

33. W. Meyer, Molecular spectroscopic constants by the coupled electron pair approach, Proceedings of the SRC atlas symposium No. 4, *in*: *Quantum Chemistry—the State of Art*, Chilton, Berkshire, England (1974).

34. L. Salem, Theoretical interpretation of force constants, *J. Chem. Phys.* **38**, 1227–1236 (1963).

35. P. Phillipson, Electronic bases of molecular vibrations. I. General theory for diatomic molecules, *J. Chem. Phys.* **39**, 3010–3016 (1963).

36. R. F. W. Bader and G. A. Jones, Electron-density distributions in hydride molecules. The ammonia molecule, *J. Chem. Phys.* **38**, 2791–2802 (1963).

37. H. J. Kim and R. G. Parr, Integral Hellmann–Feynman theorem, *J. Chem. Phys.* **41**, 2892–2897 (1964).

38. R. H. Schwendeman, Application of the Hellmann–Feynman ahd virial theorems to the theoretical calculation of molecular potential constants, *J. Chem. Phys.* **44**, 556–561 (1966).

39. J. Goodisman, Calculation of the barrier to internal rotation of ethane, *J. Chem. Phys.* **44**, 2085–2092 (1966).

40. R. F. W. Bader and A. D. Bandrauk, Relaxation of the molecular charge distribution and the vibrational force constants, *J. Chem. Phys.* **49**, 1666–1675 (1968).

41. V. V. Rossikhin and V. P. Morozov, O vychislenii silovych postoyannych molekuls primeneniem teoremy Hellmanna–Feynmana, *Teor. Eksp. Khim.* **5**, 32–37 (1969).

42. M. Cohen and A. Dalgarno, Stationary properties of the Hartree–Fock approximation, *Proc. Phys. Soc., London* **77**, 748–750 (1961).

43. R. E. Stanton, Hellmann–Feynman theorem and correlation energies, *J. Chem. Phys.* **36**, 1298–1300 (1962).

44. D. F. Tuan, Hellmann–Feynman theorem for multiconfiguration self-consistent field theory and correlation energy, *J. Chem. Phys.* **51**, 607–611 (1969).

45. C. A. Coulson, Brillouin's theorem and the Hellmann–Feynman theorem for Hartree–Fock wavefunctions, *Mol. Phys.* **20**, 687–694 (1971).

46. L. Salem and E. B. Wilson, Jr., Reliability of the Hellmann–Feynman theorem for approximate charge densities, *J. Chem. Phys.* **36**, 3421–3427 (1962).

47. L. Salem and M. Alexander, Numerical calculations by the Hellmann–Feynman prodecure, *J. Chem. Phys.* **39**, 2994–2996 (1963).

48. C. W. Kern and M. Karplus, Analysis of charge distributions: hydrogen fluoride, *J. Chem. Phys.* **40**, 1374–1389 (1964).

49. M. L. Bentson and B. Kirtman, Diatomic forces and force constants. I. Errors in the Hellmann–Feynman method, *J. Chem. Phys.* **44**, 119–129 (1966).
50. W. Fink and L. C. Allen, Numerical test of the integral Hellmann–Feynman theorem, *J. Chem. Phys* **46**, 3270–3271 (1967).
51. S. Rothenberg and H. F. Schaefer, Theoretical study of SO_2 molecular properties, *J. Chem. Phys.* **53**, 3014–3019 (1970).
52. A. G. Lazarev and I. F. Kovalev, Vychislenie silovych postoyannych molekuly ammiaka v valentno-silovoi sisteme koordinat s ispolzovaniem kvantovomechanicheskoi teoremy Hellmanna–Feynmana, *Opt. Spektrosk.* **30**, 660–663 (1971).
53. W. Meyer and P. Pulay, MOLPRO *Program Description*, München and Stuttgart, Germany (1969).
54. W. J. Hehre, W. A. Latham, R. Ditchfield, M. D. Newton, and J. A. Pople, GAUSSIAN 70, Quantum Chemistry Program Exchange, Indiana University, Bloomington, Indiana, Program No. 236.
55. J. A. Pople, J. W. McIver, and N. S. Ostlund, Self-consistent perturbation theory. I. Finite perturbation methods, *J. Chem. Phys.* **49**, 2960–2964 (1968).
56. P. Pulay and W. Meyer, Force constants and dipole moment derivatives of ammonia from Hartree–Fock calculations, *J. Chem. Phys.* **57**, 3337–3340 (1972).
57. J. W. McIver and A. Komornicki, Rapid geometry optimization for semi-empirical molecular orbital methods, *Chem. Phys. Lett.* **10**, 303–306 (1971).
58. D. Rinaldi and J. -L. Rivail, Recherche rapide de la géométrie d'une molecule à l'aide des methodes LCAO semi-empiriques, *C. R. Acad. Sci.* **274**, 1664–1667 (1972).
59. P. Pulay and F. Török, Calculation of molecular geometries and force constants from CNDO wavefunctions by the force method, *Mol. Phys.* **25**, 1153–1161 (1973).
60. J. Pancir̆, Optimization of the geometry of the molecule in the framework of a single calculation of the energy function, *Theor. Chim. Acta* **29**, 21–28 (1973).
61. M. Grimmer and D. Heidrich, Eine Variante der quantenchemischen Geometrieoptimierung über den Gradienten der Potentialenergie am Beispiel von Pyrrol, Furan und Cyclopentan, *Z. Chem.* **13**, 356–358 (1973).
62. E. B. Wilson, Jr., J. C. Decius, and P. C. Cross, *Molecular Vibrations*, McGraw-Hill, New York (1955).
63. P. Pulay, Gy. Borossay, and F. Török, A general method for the calculation of matrices depending on the equilibrium conformation of the molecule by computers, *J. Mol. Struct.* **2**, 336–340 (1968).
64. K. Thomsen and P. Swanstrøm, Calculation of molecular one-electron properties using coupled Hartree–Fock methods. II. The water molecule, *Mol. Phys.* **26**, 751–763 (1973).
65. P. Pulay and F. Török, Force constants, vibrational assignment and geometry of methyl amine from Hartree–Fock calculations, *J. Mol. Struct.* **29**, 239–246 (1975).
66. T. Nishikawa, T. Itoh, and K. Shimoda, Molecular structure of methylamine from its microwave spectrum, *J. Chem. Phys.* **23**, 1735–1736 (1955).
67. D. R. Lide, Structure of the methylamine molecule. I. Microwave spectrum of CD_3ND_2, *J. Chem. Phys.* **27**, 343–360 (1957).
68. W. Meyer, private communication.
69. R. Fletcher and M. J. D. Powell, A rapidly convergent descent method for minimization, *Comput. J.* **6**, 163–168 (1963).
70. B. A. Murtagh and R. W. H. Sargent, Computational experience with quadratically convergent minimization methods, *Comput. J.* **13**, 185–194 (1970).
71. W. Meyer and P. Pulay, unpublished.
72. A. D. Baker, C. Baker, C. R. Brundle, and D. W. Turner, The electronic structures of methane, ethane, ethylene and formaldehyde studied by high-resolution molecular photoelectron spectroscopy, *Int. J. Mass Spectrom. Ion Phys.* **1**, 285–301 (1968).
73. W. Meyer, Ionization energies of water from PNO-CI calculations, *Int. J. Quantum Chem.* **5**, 341–348 (1971).
74. W. Meyer, PNO–CI studies of electron correlation effects. I. Configuration expansion by means of nonorthogonal orbitals, and application to the ground state and ionized states of methane, *J. Chem. Phys.* **58**, 1017–1035 (1973).

75. W. Meyer and P. Pulay, Hartree–Fock calculation of the harmonic force constants and equilibrium geometry of formaldehyde, *Theor. Chim. Acta* **32**, 253–264 (1974).

76. P. Botschwina, Unrestricted Hartree–Fock calculation of force constants and vibrational frequencies of the HCO radical, *Chem. Phys. Lett.* **29**, 98–101 (1974).

77. P. Botschwina, An *ab initio* calculation of the force field and vibrational frequencies of H_2CNH, *Chem. Phys. Lett.* **29**, 580–583 (1974).

78. S. Skaarup and J. E. Boggs, *Ab initio* calculation of the structures and force fields of the isomers of $(NO)_2$, Proceedings of the Fifth Austin Symposium on Gas Phase Molecular Structure, March, 1974, pp. 69–72.

79. S. Skaarup and J. E. Boggs, An *ab initio* study of the conformational isomerism in HNO_2, *J. Mol. Struct.* **30**, 389–398 (1976).

80. W. C. Ermler and C. W. Kern, Zero-point vibrational corrections to one-electron properties of the water molecule in the near-Hartree–Fock limit, *J. Chem. Phys.* **55**, 4851–4860 (1971).

81. T. H. Dunning, R. M. Pitzer, and S. Aung, Near-Hartree–Fock calculations on the ground state of the water molecule: energies, ionization potentials, geometry, force constants and one-electron properties, *J. Chem. Phys.* **57**, 5044–5051 (1972).

82. W. Meyer, unpublished.

83. A. R. Hoy and P. R. Bunker, Effective rotational-bending Hamiltonian of the water molecule, *J. Mol. Spectrosc.* **52**, 439–456 (1974).

84. H. F. Schaefer, *The Electronic Structure of Atoms and Molecules*, Addison-Wesley, Reading, Massachusetts (1972).

85. A. D. McLean, Accuracy of computed spectroscopic constants from Hartree–Fock wavefunctions for diatomic molecules, *J. Chem. Phys.* **40**, 243–244 (1964).

86. A. C. Wahl, P. J. Bertoncini, G. Das, and T. L. Gilbert, Recent progress beyond the Hartree–Fock method for diatomic molecules: the method of optimized valence configurations, *Int. J. Quantum Chem., Symp.* **1967**, 123–152.

87. P. Swanstrøm, K. Thomsen, and P. B. Yde, Calculation of harmonic force constants from Hartree–Fock–Roothaan wave functions, *Mol. Phys.* **20**, 1135–1146 (1971).

88. P. B. Yde, K. Thomsen, and P. Swanstrøm, Analytical *ab initio* calculation of force constants and dipole moment derivatives: LiH, Li_2 and BH, *Mol. Phys.* **23**, 691–697 (1972).

89. C. J. H. Schutte, *Ab initio* calculation of molecular vibrational frequencies and force constants, *Struct. Bonding (Berlin)* **9**, 213–263 (1971).

90. R. H. Schwendeman, Comparison of experimentally derived and theoretically calculated derivatives of the energy, kinetic energy and potential energy for CO, *J. Chem. Phys.* **44**, 2115–2119 (1966).

91. W. Kolos and L. Wolniewicz, Improved theoretical ground-state energy of the hydrogen molecule, *J. Chem. Phys.* **49**, 404–410 (1968).

92. W. Meyer, PNO-CI and CEPA studies of electron correlation effects. II. Potential curves and dipole moment functions of the OH radical, *Theor. Chim. Acta* **35**, 277–292.(1974).

93. K. F. Freed, Force constants in Hartree–Fock theory, *J. Chem. Phys.* **52**, 253–257 (1970).

94. L. H. Jones and R. S. McDowell, Force constants of CH_4—infrared spectra and thermodynamic functions of isotopic methanes, *J. Mol. Spectrosc.* **3**, 632–653 (1959).

95. G. Strey and I. M. Mills, The anharmonic force field and equilibrium structure of HCN and HCP, *Mol. Phys.* **26**, 129–138 (1973).

96. D. C. McKean and J. L. Duncan, On isotropic substitution and the choice between alternative sets of force constants with special reference to the cases of ethylene, ketene, diazomethane and formaldehyde, *Spectrochim. Acta A* **27**, 1879–1891 (1971).

97. P. Pulay and W. Meyer, *Ab initio* calculation of the force field of ethylene, *J. Mol. Spectrosc.* **40**, 59–70 (1971).

98. B. L. Crawford, Jr., J. E. Lancaster, and R. G. Inskeep, The potential function of ethylene, *J. Chem. Phys.* **21**, 678–686 (1953).

99. J. L. Duncan and P. D. Mallinson, The general harmonic force field of formaldehyde, *Chem. Phys. Lett.* **23**, 597–599 (1973).

100. J. L. Duncan, private communication.

101. K. Kuchitsu and L. S. Bartell, Effect of anharmonic vibrations on the bond length of polyatomic molecules. II. Cubic constants ahd equilibrium bond lengths in methane, *J. Chem. Phys.* **36**, 2470–2481 (1962).

102. P. Pulay, *Ab initio* calculation of force constants and equilibrium geometries. III. Second-row hydrides. *Mol. Phys.* **21**, 329–339 (1971).

103. P. Pulay and W. Meyer, Comparison of the *ab initio* force constants of ethane, ethylene and acetylene, *Mol. Phys.* **27**, 473–490 (1974).

104. W. Sawodny and P. Pulay, *Ab initio* study of the force constants of inorganic molecules. ONF and NF$_3$, *J. Mol. Spectrosc.* **51**, 135–141 (1974).

105. P. Pulay, A. Ruoff, and W. Sawodny, *Ab initio* calculation of force constants for the linear molecules HCN, FCN, (CN)$_2$ and FN$_2^+$, *Mol. Phys.* **30**, 1123–1131 (1975).

106. T. Nakagawa and Y. Morino, Anharmonic potential constants and vibrational and rotational parameters for hydrogen cyanide, *J. Mol. Spectrosc.* **31**, 208–229 (1969).

107. J. Suzuki, M. A. Pariseau, and J. Overend, General quartic force field of HCN, *J. Chem. Phys.* **44**, 3561–3567 (1966).

108. R. Moccia, Variable bases in SCF MO calculations, *Chem. Phys. Lett.* **5**, 260–268 (1970).

109. K. Thomsen and P. Swanstrøm, Calculation of molecular one-electron properties using coupled Hartree–Fock methods. I. Computational scheme, *Mol. Phys.* **26**, 735–750 (1973).

110. W. Kolos and C. C. J. Roothaan, Accurate electronic wave functions for the hydrogen molecule, *Rev. Mod. Phys.* **32**, 219–232 (1960).

111. H. B. Schlegel, S. Wolfe, and F. Bernardi, *Ab initio* computation of force constants from Gaussian 70 wavefunctions. The second and third period hydrides, *J. Chem. Phys.* **63**, 3632–3638 (1975).

112. H. B. Schlegel, S. Wolfe, and K. Mislow, *Ab initio* molecular orbital calculations on silaethylene H$_2$Si=CH$_2$. The theoretical infrared spectum. *J. Chem. Soc. Chem. Comm.* **1975**, 246–247 (1975).

113. D. Garton and B. T. Sutcliffe, *in*: Specialist Periodical Report, Theoretical Chemistry (R. N. Dixon, senior reporter), Vol. 1, pp. 34–59, The Chemical Society, London (1974).

114. U. Wahlgren, J. Pacansky, and P. S. Bagus, *Ab initio* force constants for the HCN molecule: SCF and CI results, *J. Chem. Phys.* **63**, 2874–2881 (1975).

115. B. J. Rosenberg, W. C. Ermler, and I. Shavitt, *Ab initio* SCF and CI studies on the ground state of the water molecule. II. Potential energy and property surfaces. *J. Chem. Phys* **65**, 4072–4082 (1976).

5

Transition Metal Compounds

A. Veillard
and
J. Demuynck

1. Introduction

The field of transition metal compounds has always been in a special position in theoretical chemistry. For a long period, up to the sixties, the basic theory which governed this field was the crystal field theory[1] and its daughter the ligand field theory (born from the wedding of the crystal field theory, a physicist's approach, with the molecular orbital theory, a chemist's approach; see for instance Ref. 2). However, the reader is reminded that the first extended Hückel calculation dealt not with some hydrocarbons but with MnO_4^-.[3] The molecular orbital approach to the electronic structure of transition metal complexes flourished in the sixties through many semiempirical approximations and in 1969 the *ab initio* treatment of the NiF_6^{4-} cluster[4,5] paved the way for *ab initio* calculations of transition metal compounds. It is mostly computational limitations which have in the past more or less prevented a wide application of the *ab initio* techniques to the chemistry of transition metal compounds. However, with the technical developments which may be forecast for the next few years, this type of calculation will probably become much more common. In this vast field that is open to the quantum chemist (in an authoritative book of inorganic chemistry, more than half of the text is devoted

A. Veillard • C.N.R.S., Strasbourg, France, and *J. Demuynck* • Université L. Pasteur, Strasbourg, France

to the chemistry of the transition elements[6]), the most fruitful studies will probably correspond to some specific areas such as the study of conformations or the study of unstable species and transition states. We present here some results of the work carried out recently in Strasbourg for a number of transition metal complexes and organometallics.

2. The Technique of Ab Initio LCAO–MO–SCF Calculations

The reader will find that the technique of LCAO–MO–SCF calculations is described thoroughly in Volume 3 of this series (Chapters 1 and 4). However, we shall briefly review some of the features in the LCAO–MO–SCF treatment which are specific for transition metal compounds, namely the choice of the basis set and the importance of molecular symmetry.

2.1. The Choice of the Basis Set

In the LCAO–MO–SCF method, the molecular orbitals ϕ_i are expanded in terms of a given basis set χ

$$\phi_i = \sum_p C_{ip}\chi_p \tag{1}$$

with the expansion coefficients C given as the solutions of the Roothaan SCF equations[7]

$$FC = \varepsilon SC \tag{2}$$

All the present *ab initio* calculations for transition metal compounds have been done with Gaussian functions[8]

$$Nr^{n-1}e^{-\zeta r^2}Y_{lm}(\theta, \phi)$$

centered on the nuclei. For the first and second row atoms of the ligands, many Gaussian basis sets are available in the literature (references may be found in Ref. 9). For the first-transition series, three different Gaussian basis sets may be found in the literature.[10–12] Two are large basis sets, respectively $(15, 8, 5)$[10] and $(14, 9, 5)$[11] with the standard notation,* leading to very accurate wave functions.[13–15] However, their use for relatively large systems is rather prohibitive and one has then to resort to a smaller basis set such as the $(12, 6, 4)$ set.[12] This basis set has been optimized for the neutral atom with the $3d^n4s^2$ configuration, which usually corresponds to the ground state (except for Cr and Cu). As such it is not directly usable for molecular calculations. Of the 12 s

---

*The notation (a, b, c) for a Gaussian basis set means a 1s functions, b 2p functions (namely, for each subset $2p_x$, $2p_y$, $2p_z$) and c 3d functions (for each subset of 3d).

functions, two are diffuse (with low exponents less than 0.15) and correspond to the description of the $4s$ atomic orbital. However, experience has shown that for ground-state molecular calculations with the atoms Fe–Cu it is usually sufficient to work with only one Gaussian function to describe the $4s$ "atomic orbital" provided that the exponent corresponds to a less diffuse function (an appropriate choice for the exponent is such that the Gaussian function has a maximum of radial density about midway of the metal–ligand bond). On the contrary, there is no provision for a $4p$ atomic orbital in the basis set optimized with the configuration $3d^n4s^2$, so that one p function with an appropriate exponent (usually close to the one used for the $4s$ orbital) has to be added to the atomic set. Finally, there is a tendency of the $3d$ atomic orbitals to sink into the core as one goes along the series from Sc to Cu. Consequently, when a medium-size basis set is used, the $3d$ functions become too contracted at the end of the series to describe correctly the metal–ligand bonding, and one additional $3d$ function with a low exponent (usually of the order of 0.2) is needed in order to achieve a correct description of the bonding in the complex. The basis set then used for the metal in many studies is a (11, 7, 5) set. Although no systematic basis set has been reported so far for the second and third periods, basis sets for the elements Tc and Ag may be found in the literature.[16,17]

In order to keep a tractable number of two-electron integrals to be handled during the SCF calculation, the basis set of Gaussian functions has to be replaced with a basis set of contracted functions[18] which are linear combinations of Gaussian functions. Common choices for the contracted basis sets are either the minimal set or the "double-zeta" set corresponding, respectively, to one or two contracted functions for each atomic subshell. For the largest systems such as some organometallics[19] one is usually limited to the use of a minimal basis set except for the metal $3d$ orbitals which are described by split functions (it has been known for a long time that the description of the metal $3d$ orbitals with only one function is rather poor[20]). It is only for the smallest systems such as VF_5 that a double-zeta basis set can, at present, be used.[21] For these reasons a popular and convenient approximation,[22] based on the fact that inner shells are not very different in the atoms and the molecules, is the use of a minimal basis set for the inner-shell orbitals with a set of "double-zeta" quality for the valence-shell orbitals of both the metal and the ligands.[23,24] We have reported in Table 1 the basis sets used for a number of complexes and organometallics, together with the total number of Gaussian and contracted functions.

2.2. The Use of Molecular Symmetry

Many transition metal compounds have a relatively high symmetry such as C_{4v}, D_{5h} or O_h (these are extremely common symmetries in four, five, or six

Table 1. Basis Sets, Symmetry, and Number of Two-Electron Integrals

System	Basis set[a]	Contracted basis set	Number of Gaussian functions	Number of contracted functions	Molecular point group[b]	Number of two-electron integrals $(\times 10^6)$[c]	Reference
VF_5	11, 7, 4/8, 4	7, 5, 2/4, 2	156	84	D_{3h}	6.374 (1.416)	21
$Fe(CO)_5$	11, 7, 5/8, 4	4, 3, 2/3, 2	262	115	D_{3h}	22.247 (5.520)	21
$Fe(C_5H_5)_2$	12, 7, 5/8, 4/3	4, 3, 2/2, 1/1	293	85	D_{5h}	6.681	25
$Ni(C_3H_5)_2$	11, 7, 5/8, 4/4	4, 3, 2/3, 2/2	222	99	C_{2h}	12.253	24
$Co(acacen)O_2Im$[d]	10, 6, 4/7, 3/3	4, 3, 2/2, 1/1	360	122	C_1	28.151 (22.019)	19

[a]In the notation $a, b, c/d, e/f$: (a, b, c) represent the Gaussian basis set for the metal atom, (d, e) the basis set for the ligand first row atoms, and f the basis set for hydrogen.
[b]The point group used in the integral evaluation is usually different from the molecular point group.
[c]The number between parentheses is the number of unique nonzero integrals to be computed.
[d]Im is imidazole.

coordination). This results in a large number of one- and two-electron integrals being zero by symmetry or being equal to within a sign. Since with the large basis sets such as the ones of Table 1 a sizable computer time is needed for the calculation of the integrals, one obvious way to cut down this time is to take advantage of the symmetry at the level of integral evaluation. Two slightly different procedures have been put to use. Both are based on the idea of grouping together those integrals that are equal to within a sign so that only one member of the group needs evaluation. The first one, which was originally proposed and implemented in the POLYATOM program[26] is based on the use of a nonredundant list of integrals. That list has no integrals in it which are zero by symmetry and will group together those integrals which are equal. When the list is processed during the integral evaluation phase, only the first member of each group will need evaluation. One obvious advantage is that the same list may be used for different calculations. In the second procedure,[27,28] each integral is processed in turn during the integral evaluation phase and at that stage a list of integrals which are equal to the processed one is generated. If this list includes an integral which has been processed previously, the integral under evaluation is dropped. Otherwise it is evaluated and stored together with the other members of the list. Obviously this procedure may be less economical than the previous one since it has to be repeated for each calculation. However it can be used in connection with the grouping of integrals which have certain terms in common, a procedure which has been found extremely efficient for *sp* basis sets.[29,30] We have reported in Table 1, for a number of calculations, the number of two-electron integrals together with the number of unique nonzero integrals to be computed. The use of molecular symmetry during the SCF steps has been recently reviewed.[31]

3. Bonding in Transition Metal Compounds

There is not an unique mode of bonding in transition metal complexes and organometallics. Different types of bonding have been postulated to account for the electronic structure of complexes of the type that Werner dealt with (such as ammine complexes) and of the organometallics such as the metal-locenes. A classification which, although necessarily imperfect and arbitrary, has been used for want of a better one[32] distinguishes between the following:

The "classical" complexes with metal–ligand bonding insured by electron donation from a σ orbital of the ligand (by σ orbital one means an orbital of symmetry σ with respect to the metal–ligand axis) to the metal, with the metal considered to be in a high oxidation number (+2 or higher).

The complexes of π-acceptor ligands (such as CO) where back donation from the metal to vacant π orbitals of the ligands is superimposed to the above σ donation from the ligand to the metal. The covalent character of metal–ligand bonding is comparatively larger and the metal atom is usually in a low (+1 or below) oxidation number.

Organometallic complexes where a synergic bonding is based again on donation and back acceptance by the ligand but now through the use of the ligand π orbitals.

This chapter deals successively with the bonding of compounds which are representative of the above classes: $CuCl_4^{2-}$ for the "classical" complexes; $Fe(CO)_5$ for the complexes of π-acceptor ligands; and $Fe(CO)_4C_2H_4$, bis-(π-allyl)nickel and ferrocene for the organometallic compounds.

3.1. Bonding in "Classical" Complexes: $CuCl_4^{2-}$

A bonding scheme for the $CuCl_4^{2-}$ complex (square-planar with the ligands on the x and y axis) is represented in Fig. 1 based on the results (orbital energies, molecular orbitals, and population analysis) of Ref. 23. The most important interactions between the valence orbitals $3d$, $4s$, $4p$ of the metal and $3s$, $3p$ of the ligands correspond to the molecular orbitals of symmetry a_{1g}, b_{1g},

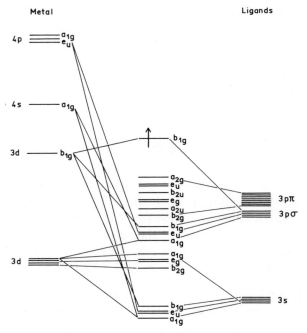

Fig. 1. Interaction diagram for the bonding in $CuCl_4^{2-}$ according to the wave function and orbital energies of Ref. 23 (the molecular orbitals are doubly occupied except for the highest one).

Table 2. *Population Analysis for the Valence Molecular Orbitals of Symmetry a_{1g}, b_{1g}, and e_u in CuCl$_4^{2-}$ (D_{4h} Symmetry)*

Molecular orbital	Gross orbital population						Cu–Cl overlap population	Character	
	3d	4s	4p	3s$_{Cl}$	3pσ$_{Cl}$	3pπ$_{Cl}$		Bonding	Antibonding
9a$_{1g}$	0.03	0.06	—	0.01×4	0.46×4	—	+0.006	3pσ–4s	3pσ–3d
8a$_{1g}$	1.92	0.00	—	0.01×4	0.005×4	—	0.000	3pσ–3d	3s–3d
7a$_{1g}$	0.04	0.02	—	0.47×4	0.01×4	—	+0.012	3s–3d	—
6b$_{1g}$	0.91	—	—	—	0.02×4	—	−0.021	—	3pσ–3d
5b$_{1g}$	0.15	—	—	—	0.46×4	—	+0.022	3pσ–3d	(3s–3d)
4b$_{1g}$	0.02	—	—	0.49×4	—	—	+0.004	3s–3d	—
9e$_u$	—	—	0.01	—	0.23×2	0.76×2	+0.007	3p–4p	—
8e$_u$	—	—	0.08	0.01×2	0.72×2	0.22×2	+0.056	3p–4p	3s–4p
7e$_u$	—	—	0.04	0.97×2	—	—	+0.028	3s–4p	—

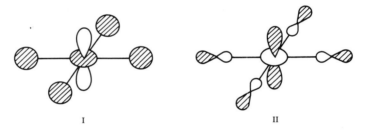

Fig. 2. Effect of the $3d_{z^2}$–$4s$ hybridization on the bonding or antibonding character of the $7a_{1g}$ (upper) and $9a_{1g}$ (lower) orbitals of $CuCl_4^{2-}$.

and e_u.* A key to an analysis of these interactions is given by the population analysis for the corresponding molecular orbitals which is reported in Table 2. The $7a_{1g}$ orbital is mostly made up of chlorine $3s$ orbitals with a small admixture of metal $3d_{z^2}$, the combination being bonding as shown in I. The $8a_{1g}$ orbital is mostly a metal $3d_{z^2}$ orbital with a small admixture of ligand $3s$ and $3p\sigma$ orbitals as in II, the combinations being antibonding with $3s$ and

I   II

bonding with $3p\sigma$ so that this molecular orbital is, on the whole, nonbonding. It is notable that hybridization of the $3s$ and $3p\sigma$ orbitals of the chlorine atoms tends to reduce the antibonding character of the $8a_{1g}$ orbital by producing an hybrid orbital that points away from the metal (Hoffmann has emphasized that metal $3d$–$4p$ hybridization in pentacoordinate complexes results in an increased bonding character of the metal–ligand bonding combination and in a decreased antibonding character of the metal–ligand antibonding combination[33]). The $9a_{1g}$ is mostly made of chlorine $3p\sigma$ orbitals with a small admixture of metal $3d_{z^2}$ and $4s$ orbitals, the combination being metal-ligand bonding. The same conclusion as that reached by Hoffmann emerges about the effect of the $3d_{z^2}$–$4s$ hybridization on the bonding in the $7a_{1g}$ and $9a_{1g}$ orbitals (Fig. 2). Hybridization of the $3d_{z^2}$ and $4s$ orbitals increases the bonding

*In the scheme of Fig. 1 the $3d$ orbitals of the *metal* are split with the $3d_{x^2-y^2}$ orbital of b_{1g} symmetry at higher energy. This turns to be a requisite to the construction of this interaction diagram. Although part of this difficulty may be due to the fact that this orbital corresponds to the open shell, it will be found that it is a general feature of the interaction diagrams from *ab initio* wave functions.

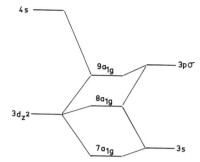

Fig. 3. Interaction diagram between the metal $3d_{z^2}$ and $4s$ orbitals and the chlorine $3s$ and $3p\sigma$ orbitals within the a_{1g} representation for $CuCl_4^{2-}$.

character of the combination $3d_{z^2}-3s$ in the $7a_{1g}$ orbital and decreases the antibonding character of the combination $3d_{z^2}-3p\sigma$ in the $9a_{1g}$ orbital. The interaction diagram between the metal and ligands orbitals within the a_{1g} representation, as found in Fig. 1, is shown in more detail in Fig. 3.

The $4b_{1g}$ and $5b_{1g}$ (III) orbitals are, respectively, made of chlorine $3s$ and $3p\sigma$ orbitals with some metal–ligand bonding character (small for the $4b_{1g}$) due to some admixture of metal $3d_{x^2-y^2}$ orbital. The $6b_{1g}$ (IV) open-shell orbital, mostly metal $3d_{x^2-y^2}$, represents the antibonding combination $3d_{x^2-y^2}-3p\sigma$. The $7e_u$ (V), $8e_u$ (VI) and $9e_u$ orbitals are ligand $3s$, $3p\sigma$, and $3p\pi$ orbitals with some bonding character due to some admixture of metal $4p$ orbitals, mostly with the $3s$ and $3p\sigma$ orbitals as in V and VI.

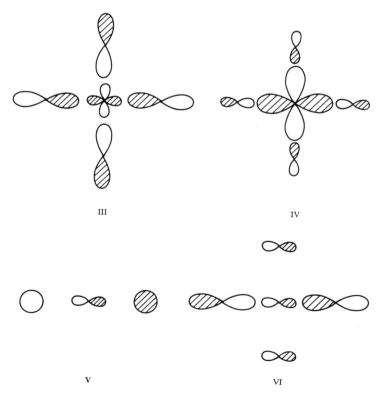

III IV

V VI

3.2. Bonding in Complexes of π-Acceptor Ligands: Fe(CO)$_5$

This type of bonding is usually represented in terms of σ donation from each carbon atom to the metal through the lone pair of the C atom (corresponding to the 5σ molecular orbital of the ligand CO) together with some backbonding from a filled $3d$ orbital of the metal toward the π^* antibonding empty orbital of the ligand. The ligand-to-metal σ bonding is similar in nature to the σ bonding in CuCl$_4^{2-}$ (although CO has less donor character than Cl$^-$), so that the discussion may focus on the π bonding. Iron pentacarbonyl has a trigonal bipyramid structure VII (molecular point group D_{3h}). The gross atomic and

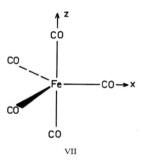

VII

orbital populations for Fe(CO)$_5$ from the population analysis are reported in Table 3. We focus first on the bonding of the equatorial carbonyl ligand along the x axis. The σ transfer from the ligand to the metal is evidenced through the decrease in the population of the $2s$ orbital of the carbon atom (the 5σ molecular orbital of the free ligand is composed mostly of the $2s$ atomic orbital of carbon). The π backbonding transfer is evidenced in the increased populations of the $2p_y$ orbital ($+0.18e$) and $2p_z$ orbital ($+0.09e$) of the carbon atom. The transfer from the metal-to-the-ligand $2p_y$ orbitals ($0.24e$ for the $2p_y$ orbitals of carbon and oxygen) is larger than the transfer to the ligand $2p_z$

Table 3. Population Analysis and Electronic Transfers in Fe(CO)$_5$ and CO

Atomic orbital	Fe(CO)$_5$					CO	
	Fe	C_{eq}[a]	O_{eq}[a]	C_{axial}	O_{axial}	C	O
$4s/2s$	0.04	1.52	1.96	1.41	1.96	1.72	1.94
$4p_x/2p_x$	0.06	0.90	1.49	0.61	1.52	0.85	1.50
$4p_y/2p_y$	0.06	0.67	1.57	0.61	1.52	0.49	1.51
$4p_z/2p_z$	0.06	0.58	1.49	0.88	1.48	0.49	1.51
$3d_{xy}, 3d_{x^2-y^2}$	1.60	—	—	—	—	—	—
$3d_{xz}, 3d_{yz}$	1.76	—	—	—	—	—	—
$3d_{z^2}$	0.50	—	—	—	—	—	—
Formal charge	+0.56	+0.32	−0.51	+0.48	−0.49	+0.46	−0.46

[a] On the Ox axis.

orbitals ($0.07e$). This is in agreement with the prediction by Hoffmann that the $eq_\perp$ interaction VIII should be larger than the $eq_\parallel$ interaction IX as a consequence of hybridization of the d_{xy} orbital with the $4p_y$ orbital as in X (hybridization reduces the antibonding interactions with the σ lone pairs of the two other equatorial ligands[33]).

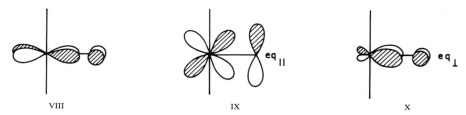

VIII IX X

We have also compared, in Table 3, the σ and π electronic transfers between the metal and an equatorial or axial ligand. We find that the ligand-to-metal σ charge transfer is larger for the axial ligand, a consequence of the fact that the orbitals e' and a_1', which are mostly metal $3d_{xy}$, $3d_{x^2-y^2}$ and $3d_{z^2}$ are metal–ligand σ antibonding and that only the e' orbitals (which correspond to the antibonding interaction with the equatorial ligands) are filled for a d^8 complex.[33] The extent of metal–ligand π interactions such as IX–XII follows the order $eq_\perp > ax > eq_\parallel$, a conclusion which is not very different from the one reached by Hoffmann[33]: $eq_\perp > ax = eq_\parallel$.

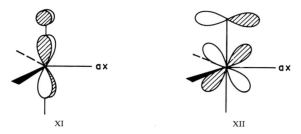

XI XII

3.3. Bonding in Some Organometallics

Although extremely different in nature, molecules such as $Fe(CO)_4(C_2H_4)$ (iron pentacarbonyl substituted with an ethylene ligand), bis(π-allyl)nickel $Ni(C_3H_5)_2$, and ferrocene $Fe(C_5H_5)_2$ share some common features in their electronic structure. It is easier to proceed from the relatively simple problem of the bonding between an ethylene ligand and a metal atom to the more complex situation found in bis(π-allyl)nickel and ferrocene.

The generally accepted description of the bonding of an ethylene ligand to the metal (with the ethylene ligand in a plane perpendicular to the plane containing the metal and the C=C bond) has been given by Dewar and Chatt.[34,35] It is a synergic process with donation of electrons from the filled π

d_z2
π
d_xz
π*
s
π
L→M
M→L

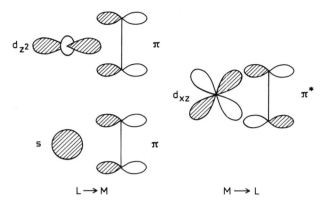

Fig. 4. Ligand-to-metal donation (left) and metal-to-ligand back-donation (right) according to the model of Dewar and Chatt for the ethylene–metal bonding.

orbital of ethylene (which is of symmetry σ with respect to the metal–olefin axis, see Fig. 4) into an empty orbital of the metal $[nd, (n+1)s$ or $(n+1)p]$ of appropriate symmetry, together with back donation of electrons from a *nd* orbital of the metal (which is of symmetry π with respect to the metal–olefin axis) into the π^* antibonding empty orbital of ethylene (Fig. 4).

The picture which emerges from an examination of the bonding in $Fe(CO)_4C_2H_4$ is in agreement with the above description. The structure of the molecule (C_{2v} symmetry) is such that the C–C bond is in the equatorial plane (Fig. 5).[36] The energy levels of the five highest molecular orbitals, corresponding to four occupied orbitals which are primarily metal 3*d* orbitals together with the π orbital of ethylene, are represented in Fig. 6. The $12b_1$ orbital is a

z
CO
y
CO
x
CO
CO

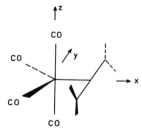

Fig. 5. Structure of the $Fe(CO)_4C_2H_4$ molecule.

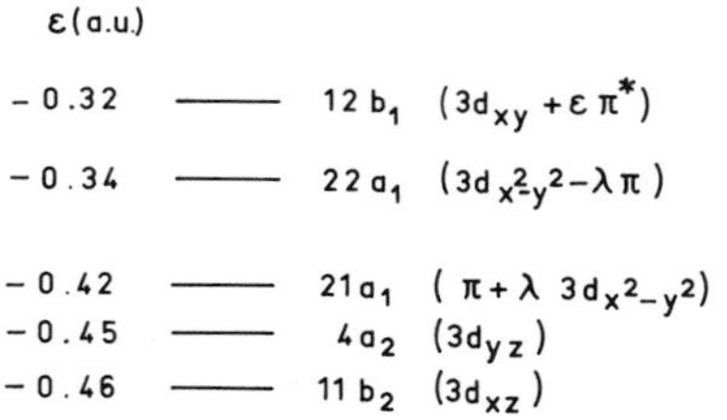

ε(a.u.)
−0.32 12 b1 (3d_xy + ε π*)
−0.34 22 a1 (3d_x²−y² − λπ)
−0.42 21a1 (π + λ 3d_x²−y²)
−0.45 4a2 (3d_yz)
−0.46 11 b2 (3d_xz)

Fig. 6. Upper energy levels of $Fe(CO)_4C_2H_4$ according to the *ab initio* calculation of Ref. 21.

bonding combination of primarily metal $3d_{xy}$ orbital with a small admixture of ethylene π^* antibonding orbital. The $21a_1$ and $22a_1$ are mostly bonding and antibonding combinations of metal $3d_{x^2-y^2}$ with the π orbital of ethylene ("four-electron destabilizing interaction"[37]). The population of the $2p_x$ orbital for each carbon of the ethylene ligand increases from 1.0 for the free ligand to 1.06 in the complex, thus pointing to a dominant metal-to-ligand back donation, with the C–C overlap population decreasing from 1.14 for the free ligand to 0.82 in the complex (as a consequence of populating the π^* antibonding orbital). The relative importance of the $\pi\, M \to L$ and $\sigma\, L \to M$ charge transfers may be assessed, from the details of the population analysis, to about $0.36e$ for the metal-to-ligand donation (this is the population of the $2p\pi$ orbitals of the carbon atoms for the $12b_1$ MO) versus $0.22e$ for the ligand-to-metal donation, leaving a net gain of about $0.14e$ for the ethylene ligand (the reader is reminded that such numbers obtained from a population analysis *have no physical meaning* and should be taken only as indicative). Nevertheless, they are in agreement with the idea of a metal–olefin bond essentially electroneutral, with donation and back acceptance approximately balanced.[38]

On the basis of an interaction diagram, Hoffmann has rationalized the fact that the ethylene ligand prefers the in-plane orientation XIII rather than the perpendicular orientation XIV. The same conclusion is reached from the *ab initio* calculation with conformation XIII found more stable than XIV by

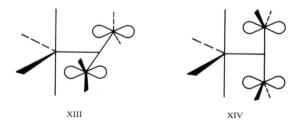

XIII   XIV

31 kcal/mole,[21] a conclusion which is in agreement with the electron diffraction structure.[36] The reduced extent of π backbonding for XIV as compared with XIII is apparent from the results of the population analysis. In the perpendicular orientation the population of the $2p_x$ orbital for a carbon of the ethylene ligand amounts to 0.96 (instead of 1.06 for the in-plane orientation), a consequence of a much decreased backdonation with a $\pi\, M \to L$ charge transfer of about $0.14e$ together with a nearly unchanged donation from the ligand to the metal (about $0.20e$).

The description of the bonding in bis(π-allyl)nickel $Ni(C_3H_5)_2$ is intermediate between the one for an ethylene ligand in $Fe(CO)_4(C_2H_4)$ and the one for a cyclopentadienyl ligand in ferrocene. The assumed structure of bis(π-allyl)nickel is shown in Fig. 7, in analogy with the experimental structure of bis(π-methylallyl)nickel,[40] the molecular point group being C_{2h}. Three

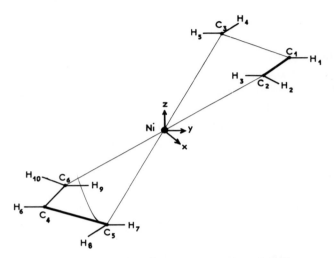

Fig. 7. Structure and choice of axis for bis(π-allyl)nickel.

interactions are usually considered between the metal orbitals and the π orbitals of the allyl group.[41]

(a)   One of these interactions is between the bonding π orbital of the allyl group and the orbitals $3d_{z^2}$, $4s$, and $4p_z$ of the metal as in XV. This interaction is similar to the σ interaction of Fig. 4 for the ethylene ligand, and should also correspond to a ligand-to-metal charge transfer. If the $3d_{z^2}$ orbital of the metal is doubly occupied, it will be effective only through the participation of the $4s$ and $4p_z$ orbitals.

(b)   Another interaction is between the nonbonding π orbital of the allyl group and the $3d_{xz}$ orbital of the metal as in XVI. This interaction is similar to the π interaction of Fig. 4 with the ethylene ligand. The electronic transfer may be either from the ligand to the metal or from the metal to the ligand, depending on the respective occupancy of these two orbitals.

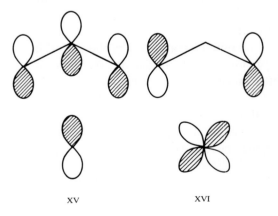

XV   XVI

(c)   Finally a possible interaction is between the antibonding π orbital of the allyl group and the $3d_{yz}$ orbital of the metal as in XVII. This interaction would be of the backbonding type, provided that the $3d_{yz}$ orbital of the metal is occupied, but should be relatively unfavorable on the basis of the large energy gap expected between the two interacting orbitals. In this commonly accepted[42] qualitative description the metal $3d_{xy}$ and $3d_{x^2-y^2}$ orbitals are considered to be nonbonding.

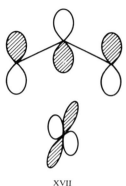

XVII

It is immaterial whether one considers the allyl group in bis(π-allyl)nickel as a radical or an anion, since the number of electrons supplied by the allyl group to the molecule can be formally compensated for by a change in the valency of the metal atom.[43] *For the sake of convenience* (since it is easier to deal with doubly occupied orbitals than with singly occupied orbitals) one may consider that the molecular orbitals of bis(π-allyl)nickel are described in terms of an interaction between two allyl anions and a Ni(II)$3d^8$ metal atom. The allyl anion has its bonding and nonbonding orbitals doubly occupied. The molecular orbital diagram of Fig. 8 corresponds to a wave function which is of double-zeta quality at the level of the valence orbitals.[24] The corresponding population analysis is given in Table 4. According to the diagram of Fig. 8 and to the contents of Table 4, four $3d$ orbitals of the Ni atom appear as doubly occupied

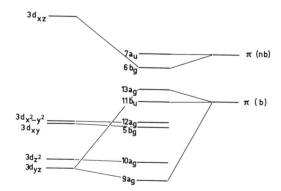

Fig. 8. A molecular orbital diagram for bis(π-allyl)nickel (reproduced from Ref. 24).

Table 4. Gross Atomic and Orbital Populations for
Bis(π-allyl)nickel[a]

Atomic orbital	Ni	$C_{(1)}$	$C_{(2)}$
$2s$ or $4s$	0.11	1.48	1.30
$2p_x$ or $4p_x$	0.11	0.93	0.86
$2p_y$ or $4p_y$	0.05	1.05	0.97
$2p_z$ or $4p_z$	0.12	0.88	1.25
$3d_{x^2-y^2}$	1.93	—	—
$3d_{xy}$	1.86	—	—
$3d_{xz}$	1.15	—	—
$3d_{yz}$	1.99	—	—
$3d_{z^2}$	2.03	—	—
Formal charge	+0.65	−0.33	−0.38

[a] From Ref. 24.

($3d_{yz}$, $3d_{z^2}$, $3d_{xy}$, and $3d_{x^2-y^2}$ in the order of increasing energy), while $3d_{xz}$ would be empty to a first approximation (if it were not for the covalent bonding). Then, one may infer that: interaction (a) between the two filled orbitals π and $3d_{z^2}$ is unimportant but that there is a limited amount of charge transfer to the $4s$ and $4p_z$ orbitals according to XV; interaction (b) between the filled nonbonding orbital of the allyl ligand and the empty $3d_{xz}$ orbital of the Ni atom as in XVI is largely responsible for the bonding (the largest metal–ligand bonding contribution comes from molecular orbital $6b_g$ which is mostly a ligand π orbital; an additional interaction XVIII which has not been considered previously corresponds to a bonding–antibonding interaction ("closed-shell four-electron interaction")[37] between the π bonding orbital of the allyl ligand and the $3d_{yz}$ orbital and is found in the molecular orbitals $9a_g$ and $13a_g$; and backbonding from the metal to the ligand is not operative through a type (c) interaction XVII as commonly postulated[41,42] (this may be seen from the population of 1.99 of the $3d_{yz}$ orbital) but rather results from interactions such as XIX between the $3d_{xy}$ and $3d_{x^2-y^2}$ orbitals and the σ and π orbitals of the ligands at the level of the $12a_g$ and $5b_g$ molecular orbitals.

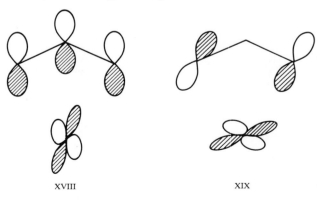

XVIII   XIX

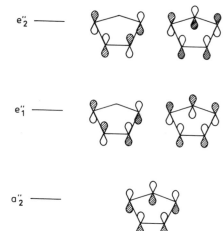

Fig. 9. The π molecular orbitals for the anion $C_5H_5^-$ (point group D_{5h}) (reproduced from Ref. 44).

The construction of a qualitative molecular orbital diagram for the ferrocene molecule has been illustrated recently by Evans *et al.*[44] These authors start with the π energy level manifold in the anion cyclopentadienyl as shown in Fig. 9 and next consider the "nonbonded" interactions between the two cyclopentadienyl ligands as in Fig. 10. Finally, they turn to the interaction between the orbitals of the ligand framework and the metal orbitals. We

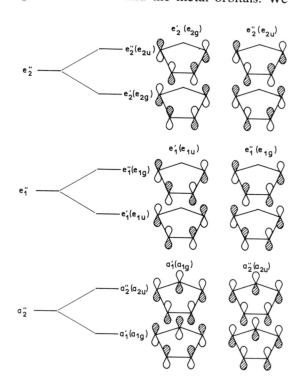

Fig. 10. The π symmetry orbitals for the $(C_5H_5)_2$ ligand framework of ferrocene. The eclipsed conformation (D_{5h} point group) is assumed (adapted from Ref. 44).

proceed in a similar way. To conform with earlier usage, we may use in the following discussion the notation of point group D_{5d} for a staggered conformation of the two cyclopentadienyl rings *although the calculation has been carried out within the D_{5h} point group*[25] *for an eclipsed conformation*[45] *and the symmetry properties correspond to the D_{5h} point group.* The relationship between the irreducible representations of the groups D_{5d} and D_{5h} may be found in Tables 5 and 6 of Ref. 44.

The energy levels corresponding to the upper valence orbitals of the cyclopentadienyl anion $C_5H_5^-$, the $(C_5H_5)_2^{2-}$ ligand framework and the ferrocene molecule as given from the *ab initio* calculation,[25] are reported in Table 5. In analogy with the case of benzene[46] and as suspected by Evans *et al.*,[44] the two uppermost σ levels in $C_5H_5^-(e_2')$ are nearly degenerate with the lowest π level (a_2''). The interaction between the two cyclopentadienyl groups in the $(C_5H_5)_2^{2-}$ ligand framework splits each level of C_5H_5 into two levels corresponding to the symmetric and antisymmetric combinations (the words symmetric and antisymmetric refer to the plane of symmetry σ_h of D_{5h}). This interaction between the two cyclopentadienyl fragments is such that the symmetric combination $(a_1', e_1', e_2',$ or $a_{1g}, e_{1u}, e_{2g})$ of the two ligand π orbitals is stabilized, and its antisymmetric partner $(a_2'', e_1'', e_2''$ or $a_{2u}, e_{1g}, e_{2u})$ destabilized with respect to the π orbitals of the noninteracting $C_5H_5^-$ systems (there is an additional shift to higher energies for the energy levels of the $(C_5H_5)_2^{2-}$ system compared to the ones of the $C_5H_5^-$ anion, since each ring feels now the field associated with the other anion; we reach a different conclusion than that of Ref. 44 regarding the order of the e_{2g} and e_{2u} levels). As expected, the interactions between the σ molecular orbitals of the two rings are more than one order of magnitude smaller (with a splitting of about 0.004 a.u.) than the interactions between the π orbitals (with a splitting of about 0.4 a.u.).

Table 5. Orbital Energies (in a.u.) for the Upper Valence Levels of $C_5H_5^-$, $(C_5H_5)_2^{2-}$, and the Ferrocene Molecule[a]

$C_5H_5^-$		$(C_5H_5)_2^{2-}$		$Fe(C_5H_5)_2$	
		0.594	$e_{2u}(\pi)$	0.180	$e_{2u}(\pi)$
				0.174	$e_{1g}(3d_{xz}, 3d_{yz})$
0.438	$e_2''(\pi)$			0.148	$e_{2g}(\pi)$
		0.557	$e_{2g}(\pi)$		
		0.035	$e_{1g}(\pi)$		
−0.124	$e_1''(\pi)$	−0.006	$e_{1u}(\pi)$	−0.429	$e_{1u}(\pi)$
				−0.437	$e_{1g}(\pi)$
				−0.530	$e_{2g}(3d_{xy}, 3d_{x^2-y^2})$
		−0.168	$a_{2u}(\pi)$	−0.585	$e_{2u}(\sigma)$
−0.330	$e_2'(\sigma)$	−0.197	$e_{2u}(\sigma)$	−0.589	$a_{2u}(\pi)$
−0.331	$a_2''(\pi)$	−0.200	$e_{2g}(\sigma)$	−0.594	$e_{2g}(\sigma)$
		−0.212	$a_{1g}(\pi)$	−0.609	$a_{1g}(d_{z^2})$
−0.353	$e_2'(\sigma)$	−0.220	$e_{2u}(\sigma)$	−0.622	$e_{2g}(\sigma)$
		−0.224	$e_{2g}(\sigma)$	−0.627	$e_{2u}(\sigma)$
				−0.653	$a_{1g}(\pi)$

[a] From Ref. 25.

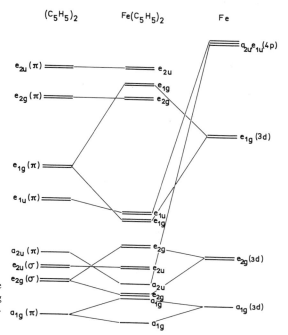

$(C_5H_5)_2$ $Fe(C_5H_5)_2$ Fe

Fig. 11. A molecular orbital scheme for the ferrocene molecule, according to the results of the *ab initio* calculation.[25]

If one turns to the energy levels of ferrocene, the results of Table 5 and Fig. 11 point to a *preferential* stabilization of four energy levels $e_{1g}(\pi)$, $a_{2u}(\pi)$, $e_{2g}(\sigma)$, and $a_{1g}(\pi)$, the stabilization being larger for the π levels than for the σ level. This stabilization corresponds to the bonding interactions with the empty valence orbitals $3d_{xz}$ and $3d_{yz}(e_{1g})$ as in XX and $4p_z(a_{2u})$ of the metal as in XXI together with the bonding–antibonding interactions with the filled valence orbitals $3d_{z^2}(a_{1g})$, $3d_{xy}$, and $3d_{x^2-y^2}(e_{2g})$. The bonding interactions XX and XXI in ferrocene are similar to the interactions XVI and XV of bis(π-allyl)nickel so that the bonding is very similar in nature in these two

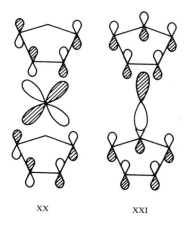

XX   XXI

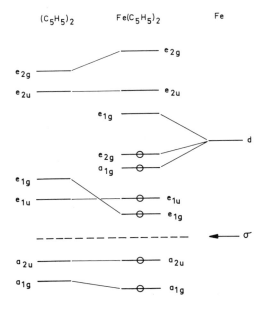

Fig. 12. The qualitative molecular orbital scheme for $Fe(C_5H_5)_2$ from Ref. 44 (the staggered D_{5d} configuration is assumed).

organometallics. This is substantiated by the existence of a backbonding transfer from the $3d_{xy}$ and $3d_{x^2-y^2}$ orbitals of the metal to the orbitals of the cyclopentadienyl rings,[25] which results from interactions similar in nature to XIX in bis(π-allyl)nickel.

It is interesting to compare the molecular orbital scheme of Fig. 11 as given by the *ab initio* calculation with the qualitative scheme proposed in Ref. 44 to interpret the photoelectron spectrum of ferrocene and reproduced in Fig. 12. The two schemes are in complete disagreement on the relative position of the orbitals e_{2g} and a_{1g} (which are predominantly metal $3d$ orbitals) with respect to the other occupied orbitals (which are mostly ligand orbitals). In the qualitative scheme of Ref. 44, these two orbitals e_{2g} and a_{1g} are the highest occupied, whereas in the *ab initio* calculation they are found below some π and σ orbitals of the ligands. This is due to the fact that the qualitative scheme of Ref. 44 relies on the validity of Koopmans' theorem,[47] whereas it has been shown that this theorem is not valid at the level of the *ab initio* calculation[25] (see below for a discussion of this problem in bis(π-allyl)nickel). The scheme of Ref. 44 ignores the stabilization of the a_{2u} and e_{1u} orbitals. It also ignores the σ orbitals of the ligands which are certainly affected by the bonding, although at a smaller degree than the π orbitals. Both schemes agree on the small degree (if any) of the interactions with the metal $4s$ orbital. A further noteworthy point is the fact that, in the diagram of Fig. 11, once more the metal $3d$ orbitals have been given different orbital energies on the right-hand side, whereas it is customary to represent them as degenerate since this is the situation in the atom. However, it turns that this is the only way to set up a correlation diagram at the *ab initio* level. This should be traced to the fact that the sum of the orbital energies in the *ab initio* calculation does not represent the total energy but includes some additional term corresponding to the electron repulsion.[39]

4. The Concept of Orbital Energy and the Interpretation of Electronic and Photoelectron Spectra

4.1. Photoelectron Spectra

There is currently a strong interest in the photoelectron spectroscopy of organometallics such as the metallocenes,[44] bis(π-allyl)nickel,[48] and many others (a limited list of references may be found in Ref. 24). Photoelectron spectroscopy has a strong relationship to theoretical studies[44] and one early incitement to the photoelectron spectroscopy of organometallics has been the hope of deriving the sequence of energy levels in the molecular ground state from the experimental sequence of ionization potentials. This idea was based on the use of Koopman's theorem,[47] which states that, for a closed-shell system, the ionization potential is equal to the negative value of the orbital energy (this theorem neglects both the electronic relaxation in the ionized system at the SCF level and the change in correlation energy between the molecule and the ion). On the basis of Koopmans' theorem, one would predict, from the sequence of molecular levels shown in Figs. 8 and 11 for bis(π-allyl)nickel and ferrocene, that the lowest ionization potentials correspond to molecular orbitals which are mostly π orbitals of the ligands [namely the highest occupied orbitals $e_{1u}(\pi)$ and $e_{1g}(\pi)$ of ferrocene or the π orbitals from $7a_u$ to $11b_u$ of bis(π-allyl)nickel]. This is at variance with the assignments found in the literature of the lowest two ionization potentials of ferrocene to the $e_{2g}(3d)$ and $a_{1g}(3d)$ orbitals[44,49,50] and of the lowest four ionization potentials of bis(π-allyl)nickel to orbitals which are mostly metal $3d$ in character.[48] Such a disagreement is rather common for metal complexes. The highest occupied orbitals of the $Ni(CN)_4^{2-}$ ion are ligand π orbitals,[51] whereas the peak at the lowest energies in the photoelectron spectrum probably corresponds to the Ni $3d$ orbitals.[52] The highest occupied orbitals of $CuCl_4^{2-}$ are nearly pure Cl $3p$ orbitals (Fig. 1) (the extension of Koopmans' theorem to the open-shell problem of $CuCl_4^{2-}$ has been discussed in Ref. 23), whereas in the related ion $PtCl_4^{2-}$ the lowest ionization energy in the x-ray photoelectron spectrum has been assigned to the removal of metal $5d$ electrons.[53]

However, Koopmans' theorem is based on the approximation of the neglect of the electronic reorganization upon ionization. The consistent way to compute ionization potentials is to work at the same level of approximation for the molecule and the ion, namely to carry independent SCF calculations for each electronic state of the ion. The successive ionization potentials are then given as the differences of the successive energies of the ion and of the ground-state energy of the molecule. This is sometimes called the ΔSCF method.[54]

We present here an application of this method to the molecule of bis(π-allyl)nickel.[24,55] The calculations used a minimal basis set of contracted orbitals (except for the metal $3d$ orbitals which are represented by split functions) but the essential conclusions are not changed when the basis set is improved to the double-zeta quality.[24] Figure 13 and Table 6 show the electronic states of $Ni(C_3H_5)_2^+$ and the ionization potentials of $Ni(C_3H_5)_2$ according to both Koopmans' theorem and the ΔSCF method. The dramatic effect of the electronic reorganization upon the energetics of ionization may be seen for the four orbitals $3b_g$, $9a_g$, $10a_g$, and $11a_g$. These are mixed metal $3d$ and ligand orbitals for the neutral molecule (Table 6) with relatively high orbital energies between 16 and 19 eV. However, the corresponding ionization potentials according to the ΔSCF method are close to 8 eV, in relationship with the fact that these orbitals undergo a marked electronic rearrangement and become nearly pure metal $3d$ orbitals in the ion, as indicated by the population analysis of Table 6. This points to a difference of more than 10 eV between the

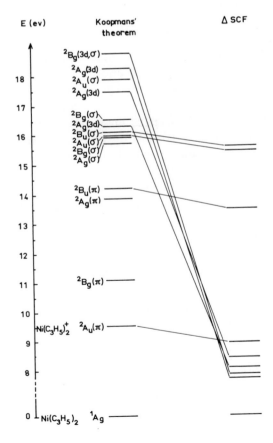

Fig. 13. The electronic states of $Ni(C_3H_5)_2^+$ according to Koopmans' theorem and to the ΔSCF method.

Table 6. Orbital Energies, Ionization Potentials by the ΔSCF Method, and Population Analysis for Bis(π-allyl)nickel and Its Positive Ion[a]

Orbital	Orbital energy, eV^b	Computed ionization potential, eV	MO composition in $Ni(C_3H_5)_2^c$			Open-shell MO composition in $Ni(C_3H_5)_2^{+c}$		
			Ni $3d$	Ligand π	Ligand σ	Ni $3d$	Ligand π	Ligand σ
$7a_u$	9.5	8.9	—	94	6	—	94	6
$6b_g$	11.1	—	34	56	10			
$13a_g$	13.8	—	27	53	20			
$11b_u$	14.2	13.6	—	93	7	—	94	6
$12a_g$	15.7	—	8	26	66			
$5b_g$	16.0	—	42	1	57			
$6a_u$	16.0	15.7	—	—	100	—	—	100
$10b_u$	16.0	15.8	—	3	97	—	4	96
$11a_g$	16.3	8.2	83	4	13	98	—	2
$4b_g$	16.5	—	25	—	75			
$10a_g$	17.5	8.5	69	2	29	99	1	—
$5a_u$	17.9	—	—	—	100			
$9a_g$	18.3	7.9	62	12	26	96	2	2
$3b_g$	18.8	8.0	38	—	62	99	—	1

[a] From Ref. 55.
[b] Negative of orbital energy, from the SCF calculation for the ground state of the neutral molecule.
[c] Percentages.

ionization potential according to Koopmans' theorem and the one given by the ΔSCF method for this type of orbital. It turns out that Koopmans' theorem is approximately valid for molecular orbitals which are nearly pure ligand orbitals (molecular orbitals $7a_u$ and $11b_u$ which are π-allyl orbitals and $6a_u$ and $10b_u$ which are σ-allyl orbitals), i.e., the ionization potentials computed by the ΔSCF method are not very different from the corresponding orbital energies. This is a consequence of the fact that these molecular orbitals show little electronic relaxation upon ionization (Table 6).

Similar conclusions regarding the dramatic effect of electron reorganization upon the energetics of ionization from a metal $3d$ orbital together with the near validity of Koopmans' theorem for the ligand orbitals have been reached for a number of organometallics: ferrocene,[25] dibenzenechromium,[56] and metal carbonyls,[56] together with some complexes such as $Ni(CN)_4^{2-}$ [51] and $CuCl_4^{2-}$.[23]

For the anions $Ni(CN)_4^{2-}$ and $CuCl_4^{2-}$, an unrealistic conclusion is that the computed ionization potentials are either very low (less than 2 eV for the ion $CuCl_4^{2-}$)[23] or even that they turn out to be negative, with the $Ni(CN)_4^-$ ion more stable at the SCF level than the $Ni(CN)_4^{2-}$ ion in its ground state.[51] This is probably an artifact due to the neglect of the stabilizing effect exerted by the surrounding crystal through its positive charge. It may be corrected by including the electrostatic potential of the crystal in the SCF calculation. For the $CuCl_4^{2-}$ ion the potential was evaluated following the method proposed by

Table 7. Orbital Energies and Computed Ionization Potentials (in a.u.) for $CuCl_4^{2-}$ (D_{4h} Configuration) either as a Free Ion or as an Ion in a Crystal[a]

Electronic configuration of the $CuCl_4^-$ ion[b]	Electronic state	Free ion		Ion with lattice	
		Orbital energy[c]	Ionization potential[d]	Orbital energy[c]	Ionization potential[d]
$6b_{1g}^0(3d)$	$^1A_{1g}$	0.476	0.110	0.794	0.437
$2a_{2g}^1$	$^1B_{2g}$	0.065	0.056	0.392	0.381
$9e_u^3$	^1E_u	0.091	—	0.418	—
$2b_{2u}^1$	$^1A_{2u}$	0.099	0.091	0.424	0.416
$3e_g^3$	^1E_g	0.112	—	0.437	—
$4a_{2u}^1$	$^1B_{2u}$	0.139	0.132	0.461	0.453
$3b_{2g}^1$	$^1A_{2g}$	0.150	0.143	0.478	0.470
$5b_{1g}^1$	$^1A_{1g}$	0.163	—	0.480	—
$8e_u^3$	^1E_u	0.165	—	0.488	—
$9a_{1g}^1$	$^1B_{1g}$	0.197	0.197	0.515	0.514
$8a_{1g}^1(3d)$	$^1B_{1g}$	0.465	0.110	0.769	0.422
$2e_g^3(3d)$	^1E_g	0.482	0.108	0.784	0.425
$2b_{2g}^1(3d)$	$^1A_{2g}$	0.522	0.103	0.829	0.424

[a] From Ref. 23.
[b] The molecular orbitals which are mostly metal $3d$ orbitals are indicated, the other ones are chlorine $3p$ orbitals.
[c] Negative of orbital energies, close to the ionization potentials computed according to an extension of Koopmans' theorem[23].
[d] By the ΔSCF method.

Ewald[57] for 8000 points within the van der Waals radius of the ion and then simulated with 44 point charges, the size of which were found by a least-squares procedure. These point charges were included in the SCF calculations and the corresponding results are reported in Table 7, under the heading "ion with lattice."[23] The inclusion of the crystal potential results in a stabilization of the energy levels since both the negative of the orbital energies and the ionization potentials given by the ΔSCF method are increased by about 0.30–0.33 a.u. (about 8 eV, Table 7), this increase reflecting the potential in the region of the ion with the corresponding values in the range 0.29 a.u. (at the site of the Cu atom) to 0.33 a.u. (at the boundaries of the ion). However, the inclusion of the crystal potential does not affect the sequence of computed ionization potentials.

4.2. Electronic Spectra

For more than a decade, the interpretation of the electronic spectra of transition metal complexes and organometallics has been based on the implicit assumption that the sequence of electronic excitations should reflect the sequence of energy levels. This was a consequence of the fact that, in the semiempirical methods which do not introduce *explicitly* the electron repulsion, excitation energies are written as the difference of the energies of orbitals i and j involved in the excitation,

$$\Delta E_{i \to j} = \varepsilon_j - \varepsilon_i \qquad (3)$$

Conversely, a sequence of energy levels was often proposed for the ground state of the molecule or the ion on the basis that it does account for the sequence of electronic transitions.[58,59] Since the transitions at lowest energy in the electronic spectrum are usually the $d-d$ or "ligand–field" transitions, the highest occupied and lowest empty orbitals were assigned as metal $3d$ orbitals.

However, the rigorous way to calculate an excitation energy $\Delta E_{i \to j}$ is to achieve a separate minimization of the energies E_i and E_j of the ground state and the excited state, with the transition energy given as the difference,

$$\Delta E_{i \to j} = E_j - E_i \qquad (4)$$

A common approximation has been the use of the occupied and virtual orbitals from the ground-state calculation in describing the excited states, with the excitation energies given through[7]

$$^1E(i \to j) = \varepsilon_j - \varepsilon_i - J_{ij} + 2K_{ij}$$

$$^3E(i \to j) = \varepsilon_j - \varepsilon_i - J_{ij} \qquad (5)$$

The sequence of orbital energies corresponding to the ground-state calculation for the $Ni(CN)_4^{2-}$ ion is shown in Table 8 (however, the concept of orbital energy is not uniquely defined, see for instance Ref. 60). Each excited state corresponding to a single excitation from an occupied orbital to an empty orbital of Table 8 has been the subject of an independent SCF calculation in the restricted Hartree–Fock scheme, and the computed transition energies for the singlet states are reported in Table 9. From a comparison of Tables 8 and 9, it is

Table 8. Highest Occupied and Lowest Empty Orbitals of $Ni(CN)_4^{2-}$ (from the Ground-State Calculation)^a

Orbital	Orbital energy^b	Nature	Orbital	Orbital energy^b	Nature
$3e_g$	0.589	π^*	$2b_{2g}$	−0.156	π
$6b_{1g}$	0.587	$d_{x^2-y^2}$	$3a_{2u}$	−0.160	π
$9e_u$	0.576	$4p_x, 4p_y$	$9a_{1g}$	−0.182	d_{z^2}
$10a_{1g}$	0.550	$4s$	$5b_{1g}$	−0.182	5σ
$3b_{2g}$	0.527	π^*	$7e_u$	−0.195	5σ
$4a_{2u}$	0.402	π^*	$8a_{1g}$	−0.231	4σ
$8e_u$	−0.121	π	$4b_{1g}$	−0.263	4σ
$1a_{2g}$	−0.123	π	$1e_g$	−0.267	$d_{xz,yz}$
$2e_g$	−0.132	π	$6e_u$	−0.278	4σ
$1b_{2u}$	−0.136	π	$1b_{2g}$	−0.335	d_{xy}

^{a} From Ref. 51.
^{b} In a.u.

Table 9. Computed Transition Energies^a (in cm^{-1}) for the Singlet States of Ni(CN)$_4^{2-}$

Transition	Excited state	Transition energy	Nature of the transition
$9a_{1g} \to 6b_{1g}$	$^1B_{1g}$	20,600	$d_{z^2} \to d_{x^2-y^2}$
$1e_g \to 6b_{1g}$	^1E_g	21,900	$d_{xz}, d_{yz} \to d_{x^2-y^2}$
$1b_{2g} \to 6b_{1g}$	$^1A_{2g}$	22,500	$d_{xy} \to d_{x^2-y^2}$
$9a_{1g} \to 4a_{2u}$	$^1A_{2u}$	33,900	$d_{z^2} \to \pi^*$
$1e_g \to 4a_{2u}$	^1E_u	37,900	$d_{xz}, d_{yz} \to \pi^*$
$1b_{2g} \to 4a_{2u}$	$^1B_{1u}$	46,900	$d_{xy} \to \pi^*$
$9a_{1g} \to 3b_{2g}$	$^1B_{2g}$	58,500	$d_{z^2} \to \pi^*$
$1e_g \to 3b_{2g}$	^1E_g	62,300	$d_{xz}, d_{yz} \to \pi^*$
$8e_u \to 6b_{1g}$	^1E_u	68,600	$\pi \to d_{x^2-y^2}$
$1a_{2g} \to 6b_{1g}$	$^1B_{2g}$	69,700	$\pi \to d_{x^2-y^2}$

^{a} From Ref. 51.

obvious that the sequence of excitation energies has no relationship with the sequence of orbital energies. On the basis of the orbital energies one would expect that the lowest energy transitions would correspond to the $\pi \to \pi^*$ excitations whereas they correspond to the $d \to d$ excitations, in agreement with the previous interpretation of the electronic spectrum.[61] The reason for this inversion is to be found both in the fact that the excitation energy is not merely a difference of orbital energies (since, with the assumption of using for the excited state the occupied and virtual orbitals of the ground state, the excitation energy includes a Coulomb and an exchange term) and in the occurence of electronic reorganization upon excitation (namely the orbitals of the excited state are different from the ones for the ground state). The relative importance of both factors may be assessed from Table 10 which includes the values of $\varepsilon_j - \varepsilon_i$, J_{ij}, K_{ij}, and the excitation energies given by Eqs. (4) and (5). One may see that, if the excitation energies given by (5) parallel the difference

Table 10. Transition Energies Computed as the Difference of the Energies of the Two States and Estimated from Orbital Energies, Coulomb and Exchange Integrals^a

Transition $i \to j$	$\varepsilon_j - \varepsilon_i$, a.u.^{b}	J_{ij}, a.u.^{b}	K_{ij}, a.u.^{b}	Transition energy,^{c} cm^{-1}	Transition energy,^{d} cm^{-1}
$1a_{2g} \to 4a_{2u}$	0.525	0.185	0.001	75,170	71,700
$9a_{1g} \to 4a_{2u}$	0.584	0.248	0.015	80,330	33,900
$1a_{2g} \to 3b_{2g}$	0.650	0.226	0.034	108,000	106,000
$9a_{1g} \to 3b_{2g}$	0.709	0.247	0.005	103,600	58,500
$1a_{2g} \to 6b_{1g}$	0.710	0.209	0.002	111,000	69,700
$9a_{1g} \to 6b_{1g}$	0.769	0.529	0.023	62,770	20,600

^{a} From Ref. 51.
^{b} 1 a.u. = 219,470 cm^{-1}
^{c} Computed according to $^1E(i \to j) = \varepsilon_j - \varepsilon_i - J_{ij} + 2K_{ij}$.
^{d} Computed as the difference of the energy for the two states.

$\varepsilon_j - \varepsilon_i$ for five excitations out of six, this correlation breaks down for the d–d excitation, a consequence of the large value for the corresponding integral J_{ij}. A comparison of the excitation energies given by the formulas (4) and (5) shows that formula (5) (based on the use of the ground-state orbitals) represents a poor approximation whenever the $i \to j$ transition involves one metal orbital such as i or j, with an error which may be as large as 45,000 cm^{-1}! This may be traced to the fact that the orbitals which are mostly metal $3d$ are affected by the electronic reorganization much more than the ligand orbitals.[51]

The ferrocene molecule has its two highest occupied orbitals and the lowest virtual orbital as predominantly π ligand orbitals (Fig. 11),[25] whereas the lowest energy bands in the electronic spectrum have been assigned to the d–d excitations.[59] The excitation energies of Table 11 were computed according to (4) through an independent SCF calculation for each excited state.[62] The lowest transition energies correspond to the d–d excitations, a result unexpected on the basis of the energy levels of the ground state. Although the agreement between the computed and experimental transition energies is not very satisfactory, owing to the use of a near minimal basis set together with the neglect of correlation energy, the results of Table 11 account for a number of features in the experimental spectrum (the band positions in the experimental spectrum are reported in Table 11). According to the calculation, the lowest d–d transition energies correspond to the $e_{2g} \to e_{1g}$ excitation, while the previous assignment based on ligand–field theory considered the lowest d–d transition to be due to the $a_{1g} \to e_{1g}$ excitation.[59] According to the computed energies of Table 11 the splitting between the two excited states $^1E_{2g}$ and $^1E_{1g}$

Table 11. *Excitation Energies of Ferrocene, According to SCF Computations*[a]

Excitation	Excited state	Excitation energy,[b] cm^{-1}	Nature of the excitation		Band position,[c] cm^{-1}
$e_{2g} \to e_{1g}$	$^1E_{2g}$	13,300	$d \to d$		21800 (IIa)
	$^1E_{1g}$	14,200	$d \to d$		24000 (IIb)
$a_{1g} \to e_{1g}$	$^1E_{1g}$	21,800	$d \to d$		30800 (III)
$e_{1u} \to e_{1g}$	$^1A_{1u}$	44,900	$L \to M$ C.T.	$\pi \to d$	37700 (IV)–41700 (V)
	$^1A_{2u}$	47,700	$L \to M$ C.T.	$\pi \to d$	50000 (VI)
	$^1E_{2u}$	46,300	$L \to M$ C.T.	$\pi \to d$	—
$e_{2g} \to e_{2u}$	$^1A_{1u}$	60,700	$M \to L$ C.T.	$d \to \pi^*$	—
	$^1A_{2u}$	62,000	$M \to L$ C.T.	$d \to \pi^*$	—
	$^1E_{1u}$	61,400	$M \to L$ C.T.	$d \to \pi^*$	—
$a_{1g} \to e_{2u}$	$^1E_{2u}$	75,500	$M \to L$ C.T.	$d \to \pi^*$	—
$e_{1g} \to e_{2g}$	$^1E_{1g}$	85,800	Intraligand	$\pi \to \pi^*$	—
	$^1E_{2g}$	84,700	Intraligand	$\pi \to \pi^*$	—
$a_{2u} \to e_{2g}$	$^1E_{2u}$	117,000	Intraligand	$\pi \to \pi^*$	—

[a] From Ref. 62.
[b] Computed.
[c] In the experimental spectrum (Ref. 59).

corresponding to the $e_{2g} \rightarrow e_{1g}$ excitation should be small, of the order of 900 cm^{-1}. The experimental splitting between bands (IIa) and (IIb) of about 2000 cm^{-1} is much smaller than the one between bands (IIb) and (III) of about 7000 cm^{-1}. This supports the assignment of bands (IIa) and (IIb) to the $e_{2g} \rightarrow e_{1g}$ excitation.[62] According to Table 11, the transitions next to the $d-d$ excitations should be the ligand-to-metal ($L \rightarrow M$) charge transfer transitions with computed energies in the range 45,000–48,000 cm^{-1}. This is in agreement with the assignment of band (VI) as the $L \rightarrow M$ $^1A_{1g} \rightarrow {}^1A_{2u}$ transition. Similarly bands (IV) and (V) which had not been assigned previously may be assigned on the basis of the SCF calculations to $L \rightarrow M$ charge-transfer excitations.[62]

5. Electronic Structure and Stereochemistry of Dioxygen Adducts of Cobalt–Schiff-Base Complexes

We deal in this section with one specific application, namely the electronic aspects and the stereochemistry of dioxygen binding to cobalt–Schiff-base complexes. Such complexes [like Co(acacen)L(acacen=N,N'-ethylenebis-(acetylacetoneiminato) with L a fifth axial ligand] reversibly bind molecular oxygen according to

$$Co(acacen)L + O_2 \rightleftharpoons Co(acacen)LO_2$$

to give an adduct such as the one represented in Fig. 14.[63] Interest in these synthetic oxygen carrier complexes arose from the fact that they are sometimes considered as possible models of the hemoglobin and myoglobin molecules which carry reversible oxygenation in life processes, although they are not isoelectronic with the iron(II)dioxygen moiety of hemoproteins [it is only recently that synthetic oxygen carriers of iron(II) have been synthetized[64]].

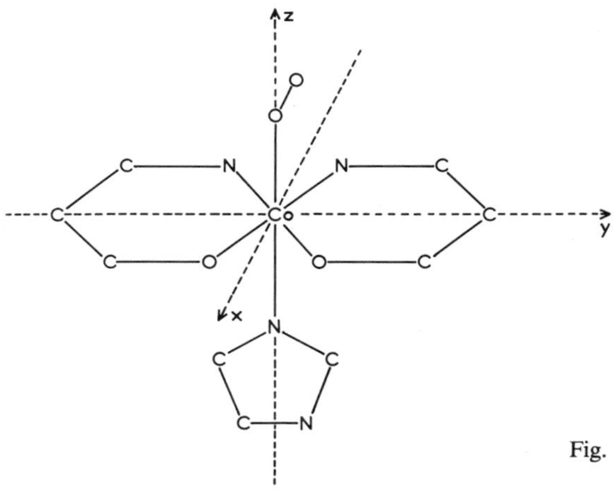

Fig. 14. The Co(acacen)ImO$_2$ molecule (reproduced from Ref. 19).

More specifically, we shall consider the problems of the electronic structure of these adducts and of the stereochemistry of dioxygen binding. ESR studies of oxygen adducts of a wide variety of Co(II) complexes such as acacen, salen, porphyrin, and vitamin B_{12r} (references may be found in Ref. 19) have indicated that dioxygen binding is accompanied by an electron transfer from cobalt to oxygen, namely the unpaired electron is largely delocalized on the oxygen atom [the Co(acacen)LO_2 complexes are experimentally known to have only one unpaired electron[63]]. Then these adducts should rather be formulated as Co(III)–O_2^- complexes.[63] Although this conclusion seems now firmly established,[65] no rationale has been given in terms of the bonding and the electronic structure of these adducts.

Different structural models have been proposed for dioxygen binding in the oxygen carriers, including a linear M–O–O unit XXII[66] or an "end-on" angular bond XXIII[67–69] or a sideways triangular structure XXIV.[70] In fact the dioxygen complexes of both Co(II) and Fe(II) have been structurally characterized as systems with M–O–O bent bonds, the corresponding angle being in the range 124°–136°.[64,71] However, there is still the possibility that for a given complex the energy difference between the three structures XXII–XXIV could be small, with the environmental factors influencing which of the forms is the more stable. EPR of cobalt oxygen carriers labeled with ^{17}O shows a complex containing magnetically equivalent oxygen atoms, a result which is consistent either with a triangular structure or with a rapid flipping of the O–O group between two bent positions as in XXV,[72] with the sideways structure representing probably a transition state.

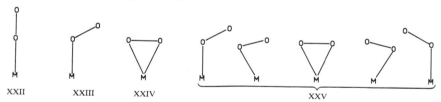

| XXII | XXIII | XXIV | XXV |

The bonding in a bent dioxygen adduct Co(acacen)LO_2 may be described essentially in terms of the interactions between the metal $3d$ orbitals and the dioxygen π_g antibonding orbitals. In the complex the degenerate π_g and $\overline{\pi}_g$ orbitals of the dioxygen molecule are split due to the lower symmetry around the O–O axis.[70] If one considers the case of an axially symmetric ligand L such as CN^- or CO, the plane xOz which includes the dioxygen ligand in Fig. 14 is a plane of symmetry for the molecule. We denote π_g^a the antibonding orbital π_g of O_2 which is symmetrical with respect to xOz and π_g^b the one which is antisymmetrical (π_g^b is made from the $2p_y$ orbitals of the oxygen atoms and is represented in XXVI). The most important interaction is the one of π_g^a with the $3d_{z^2}$ orbital of Co, which is represented in XXVII (for a more detailed discussion of the interactions between a metal atom and a diatomic ligand, we refer the reader to the work of Hoffmann *et al.* on the coordination of the nitrosyl ligand[73]). Other possible interactions include the ones represented in **XXVIII** and **XXIX** between π_g^a or π_g^g and the $3d_{xz}$ or $3d_{yz}$ orbitals (the

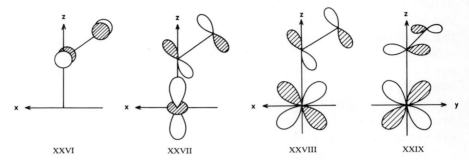

XXVI   XXVII   XXVIII   XXIX

interaction $3d_{yz}-\pi_g^b$ is of the $d\pi-p\pi$ type, the interaction $3d_{xz}-\pi_g^a$ would be of the $d\pi-p\pi$ type for linear coordination). Without a precise knowledge of the relative importance of these different interactions, there are *a priori* four possible electronic configurations for the ground state of the dioxygen adduct, which are represented in Fig. 15. They correspond formally to the different possibilities regarding the relative order of π_g^a and π_g^b in the complex and the relative location of the $3d_{z^2}$ orbital with respect to π_g^a and π_g^b. Of these four configurations, three, denoted S_1, S_2, and S_3, belong to the symmetric representation and one, denoted A, belongs to the antisymmetric one. The configurations S_1 and S_2 correspond to a formal oxidation number of II for Co, hence to a description of the bonding in terms of a structure Co(II)–O$_2$, whereas A and S_3 correspond to a charge-transfer configuration Co(III)–O$_2^-$.

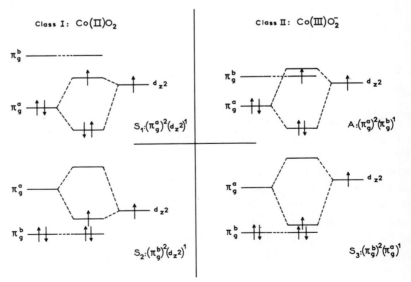

Fig. 15. Possible electronic configurations for a dioxygen adduct Co(acacen)LO$_2$ (reproduced from Ref. 19).

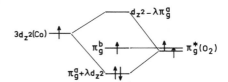

Fig. 16. A simplified interaction diagram and the ground-state configuration $(\pi_g^a)^2(\pi_g^b)^1$ for the adduct Co(acacen)LO$_2$.

According to the *ab initio* calculations, the A configuration is the most stable for a variety of fifth ligand L (L = none, H$_2$O, imidazole, CN$^-$, and CO) and this provides a rationale for the formulation as Co(III)–O$_2^-$ complexes.[19] The greater stability of the A configuration $(\pi_g^a)^2(\pi_g^b)^1$ may be understood on the basis of interaction XXVII which stabilizes π_g^a and destabilizes $3d_{z^2}$ as in Fig. 16, whereas the other interactions such as XXVIII and XXIX are found unimportant, a conclusion from *ab initio* calculations.[74] If the nature of the interaction between Fe and O$_2$ in hemoglobin is similar, then one may see that by removing one electron from the diagram of Fig. 16 the bond in hemoglobin should be formulated as Fe(II)–O$_2$, a conclusion which is in agreement with the one reached by Wayland *et al.* on the basis of a qualitative molecular orbital diagram,[75] together with the value found for the O$_2$ infrared frequency in a "picket fence" porphyrin.[76]

So far we have considered only the case of a Co–O–O bent conformation. The relative stabilities of the linear, bent and perpendicular structures XXII–XXIV for the low-spin adducts Co(acacen)LO$_2$ are reported in Table 12.[77] Although the results depend somewhat on the nature of the fifth ligand, the linear structure is found somewhat less stable (by 4–26 kcal/mole) than the bent one, with the perpendicular structure considerably less stable (by 70–80 kcal/mole). Possible interaction diagrams for these three structures which may account for their relative stabilities are represented in Fig. 17 (only the σ-type interaction is considered in Fig. 17, we have already mentioned that $d\pi$–$p\pi$ backbonding is unimportant in the bent structure). The reader is reminded that, according to perturbation theory, the degree of interaction between two orbitals depends on both their overlap and the corresponding energy gap denominator.[78] In the bent and linear structures, the overlap terms are probably comparable but the energy denominator should favor the bent structure since the $1\pi_g$ orbital of the oxygen molecule is well above the $3\sigma_g$

Table 12. Relative Stabilities (in kcal/mole) of the Bent, Linear, and Perpendicular Structures for Some Adducts Co(acacen)LO$_2$[a]

L	Bent	Linear	Perpendicular
None	0	4	71
CN$^-$	0	20	71
CO	0	26	80

[a] Ref. 77.

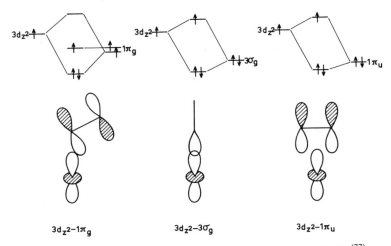

Fig. 17. Possible interaction diagrams for the three structures Co–O$_2$.[77]

orbital. Now, according to Fig. 17, the interaction is stabilizing for two electrons in the bent structure, whereas it appears stabilizing for two electrons and destabilizing for one electron in the linear structure. Thus, the bent structure will be slightly favored over the linear one, in agreement with the results of Table 12. Turning now to a comparison of the linear and perpendicular structures, both the overlap term and the energy denominator are comparable (the $3\sigma_g$ and $1\pi_u$ orbitals are nearly degenerate in the oxygen molecule). Since for both structures the interaction is stabilizing for two electrons and destabilizing for one electron (Fig. 17), their stability should be comparable on the basis of the interactions represented in Fig. 17. However, interaction of the $d\pi$–$p\pi$ type is comparatively more important in the perpendicular structure than in the linear one, with the overlap term larger in XXX than in XXXI. Since this is a four-electron destabilizing interaction, the perpendicular structure should be destabilized with respect to the linear one, in agreement with the results of the calculation.*

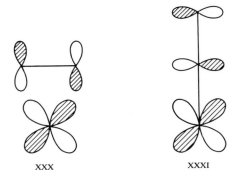

XXX   XXXI

*See Note Added in Proof, page 222.

References

1. H. Bethe, Termaufspaltung in Kristallen, *Ann. Phys.* **3**, 133–208 (1929).
2. C. J. Ballhausen, *Introduction to Ligand-Field Theory*, McGraw-Hill, New York (1962).
3. M. Wolfsberg and L. Helmholz, The spectra and electronic structure of the tetrahedral ions MnO_4^-, CrO_4^{2-} and ClO_4^-, *J. Chem. Phys.* **20**, 837–843 (1952).
4. H. M. Gladney and A. Veillard, Limited basis set Hartree–Fock theory of NiF_6^{4-}, *Phys. Rev.* **180**, 385–395 (1969).
5. J. W. Moskowitz, C. Hollister, C. J. Hornback, and H. Basch, Self-consistent field study of the cluster model in ionic salts. I. NiF_6^{4-}, *J. Chem. Phys.* **53**, 2570–2580 (1970).
6. F. A. Cotton and G. Wilkinson, *Advanced Inorganic Chemistry*, Interscience Publishers, New York (1972).
7. C. C. J. Roothaan, New developments in molecular orbital theory, *Rev. Mod. Phys.* **23**, 69–89 (1951).
8. S. F. Boys, Electronic wave functions. I. A general method of calculation for the stationary states of any molecular system, *Proc. R. Soc. London*, Ser. A. **200**, 542–554 (1950).
9. B. Roos and P. Siegbahn, Gaussian basis sets for the first and second row atoms, *Theor. Chim. Acta* **17**, 209–215 (1970).
10. H. Basch, C. J. Hornback, and J. W. Moskowitz, Gaussian-orbital basis sets for the first-row transition-metal atoms, *J. Chem. Phys* **51**, 1311–1318 (1969).
11. A. J. H. Wachters, Gaussian basis set for molecular wave functions containing third-row atoms, *J. Chem. Phys.* **52**, 1033–1036 (1970).
12. B. Roos, A. Veillard, and G. Vinot, Gaussian basis sets for molecular wavefunctions containing third-row atoms, *Theor. Chim. Acta* **20**, 1–11 (1971).
13. A. J. H. Wachters, Ligand field splitting and magnetic exchange interaction in $KNiF_3$, Ph.D. thesis, Rijksuniversiteit te Groningen, 1971.
14. H. Johansen, SCF LCAO MO calculations for MnO_4^-, *Chem. Phys. Lett.* **17**, 569–573 (1972).
15. A. P. Mortola, H. Basch, and J. W. Moskowitz, An *ab initio* study of the permanganate ion, *Int. J. Quantum Chem.* **7**, 725–737 (1973).
16. H. Basch and A. P. Ginsberg, A molecular orbital description of TcH_9^{2-}, *J. Phys. Chem.* **73**, 854–857 (1969).
17. H. Basch, Electronic structure of the silver (1+)-ethylene complex, *J. Chem. Phys.* **56**, 441–450 (1972).
18. E. Clementi and D. R. Davis, Electronic structure of large molecular systems, *J. Comput. Phys.* **1**, 223–244 (1966).
19. A. Dedieu and A. Veillard, Electronic aspects of dioxygen binding to cobalt-Schiff-base-complexes, *Theor. Chim. Acta* **36**, 231–235 (1975).
20. J. W. Richardson, W. C. Nieuwpoort, R. R. Powell, and W. F. Edgell, Approximate radial functions for first-row transition metal atoms and ions. I. Inner-shell, $3d$ and $4s$ atomic orbitals, *J. Chem. Phys.* **36**, 1057–1061 (1962).
21. A. Strich, J. Demuynck, and A. Veillard, Intramolecular rearrangements in transition metal complexes. *Ab initio* calculations, *Nouveau J. Chim.* **1** (3) (1977).
22. C. R. Claydon and K. D. Carlson, Ground states, configurations and truncated orbital bases of the iron-series atoms, *J. Chem. Phys.* **49**, 1331–1339 (1968).
23. J. Demuynck, A. Veillard, and U. Wahlgren, Bonding, spectra and geometry of the tetrachlorocuprate ion $CuCl_4^{2-}$. An *ab initio* LCAO-MO-SCF calculation, *J. Am. Chem. Soc.* **95**, 5563–5574 (1973).
24. M-M. Rohmer, J. Demuynck, and A. Veillard, A double-zeta type wave function for an organometallic: bis-(π-allyl)nickel, *Theor. Chim. Acta* **36**, 93–102 (1974).
25. M-M. Coutière, J. Demuynck, and A. Veillard, Ionization potentials of ferrocene and Koopmans' theorem. An *ab initio* LCAO-MO-SCF calculation, *Theor. Chim. Acta* **27**, 281–287 (1972).
26. D. B. Neumann, H. Basch, R. L. Kornegay, L. C. Snyder, J. W. Moskowitz, C. Hornback, and S. P. Liebmann, The POLYATOM (version 2) System of Programs, Program No. 199, Quantum Chemistry Program Exchange, Indiana University, Bloomington, Indiana, 47401.

27. M. Benard, A. Dedieu, J. Demuynck, A. Strich, and A. Veillard, Asterix: a system of programs for the Univac 1110, unpublished.
28. V. Saunders, the ATMOL program, private communication.
29. W. J. Hehre, W. A. Lathan, R. Ditchfield, M. D. Newton, and J. A. Pople, Gaussian 70, Program No. 236, Quantum Chemistry Program Exchange, Indiana University, Bloomington, Indiana 47401, U.S.A.
30. G. H. F. Diercksen and W. P. Kraemer, Munich, molecular program system, Special Technical Report, Max-Planck Institut für Physik und Astrophysik, Munich, Germany.
31. A. Veillard, in: Computational Techniques in Quantum Chemistry and Molecular Physics (G. Diercksen, B. T. Sutcliffe, and A. Veillard, eds.), NATO ASI Series, D. Reidel, Dordrecht, Holland (1975).
32. Ref. 6, p. 620 and following.
33. A. R. Rossi and R. Hoffmann, Transition metal pentacoordination, Inorg. Chem. 14, 365–374 (1975).
34. M. J. S. Dewar, A review of the π-complex theory, Bull. Soc. Chim. Fr. 18, C71–C79 (1951).
35. J. Chatt and L. A. Duncanson, Olefin coordination compounds. Part III. Infra-red spectra and structure: Attempted preparation of acetylene complexes, J. Chem. Soc. 1953, 2939–2947.
36. M. I. Davis and C. S. Speed, Gas-phase electron diffraction studies of some iron carbonyl complexes, J. Organometal. Chem. 21, 401–413 (1970).
37. L. Salem, Forces between polyatomic molecules. II. Short-range repulsive forces, Proc. R. Soc. London, Ser. A 264, 379–391 (1961).
38. Ref. 6, p. 731.
39. R. J. Buenker and S. D. Peyerimhoff, Molecular geometry and the Mulliken–Walsh molecular orbital model. An ab initio study, Chem. Rev. 74, 127–188 (1974).
40. R. Uttech and H. Dietrich, Kristall- und Molekülstruktur von bis-methallylnickel Ni $[(CH_2)_2C\cdot CH_3]_2$, Z. Kristallogr. 122, 60–72 (1965).
41. K. Vrieze, C. MacLean, P. Cossee, and C. W. Hilbers, Nuclear magnetic resonance studies in coordination chemistry I. Structure and conformational rearrangements of π-allyl complexes containing group-V donor ligands, Rec. Trav. Chim. Pays-Bas 85, 1077–1098 (1966).
42. P. W. Jolly and G. Wilke, The Organic Chemistry of Nickel, p. 329, Academic Press, New York (1974).
43. E. O. Fischer and H. Werner, Metal π-Complexes, Vol. 1, p. 176, Elsevier, Amsterdam (1966).
44. S. Evans, M. L. H. Green, B. Jewitt, A. F. Orchard, and C. Pygall, Electronic structure of metal complexes containing π-cyclopentadienyl and related ligands, J. Chem. Soc., Faraday Trans. 2 68, 1847–1865 (1972).
45. R. K. Bohn and A. Haaland, On the molecular structure of ferrocene, J. Organomet. Chem. 5, 470–476 (1966).
46. M. D. Newton, F. P. Boer, and W. N. Lipscomb, Molecular orbitals for organic systems parametrized from SCF model calculations, J. Am. Chem. Soc. 88, 2367–2384 (1966).
47. T. Koopmans, Über die Zuordnung von Wellenfunktionen und eigenwerten zu den einzelnen elektronen eines atoms, Phys. Fenn. 1, 104–113 (1934).
48. D. R. Lloyd and N. Lynaugh, in: Electron Spectroscopy, Proceedings of an International Conference held at Asilomar, California, September 7–10, 1971 (D. E. Shirley ed.), North-Holland, Amsterdam (1972).
49. D. W. Turner, in: Physical Methods in Advanced Inorganic Chemistry (H. A. O. Hill and P. Day eds.), Interscience Publishers, New York (1968).
50. J. W. Rabalais, L. O. Werme, T. Bergmark, L. Karlsson, M. Hussain, and K. Siegbahn, Electron spectroscopy of open-shell systems: spectra of $Ni(C_5H_5)_2$, $Fe(C_5H_5)_2$, $Mn(C_5H_5)_2$ and $Cr(C_5H_5)_2$, J. Chem. Phys. 57, 1185–1192 (1972).
51. J. Demuynck and A. Veillard, Electronic structure of the nickel tetracyanonickelate $Ni(CN)_4^{2-}$ and nickel carbonyl $Ni(CO)_4$. An ab initio LCAO–MO–SCF calculation, Theor. Chim. Acta 28, 241–265 (1973).
52. K. Jorgensen, private communication.
53. P. Biloen and R. Prins, Level ordering in transition halide complexes. An X-ray photoelectron spectroscopy study, Chem. Phys. Lett. 16, 611–613 (1972).

54. P. S. Bagus, Self-consistent-field wave functions for hole states of some Ne-like and Ar-like ions, *Phys. Rev. A*, **139**, 619–634 (1965).

55. M-M. Rohmer and A. Veillard, Photoelectron spectrum of bis-(π-allyl) nickel, *J. Chem. Soc. D*. **1973**, 250–251 (1973).

56. M. F. Guest, I. H. Hillier, B. R. Higginson, and D. R. Lloyd, The electronic structure of transition metal complexes containing organic ligands. II. Low energy photoelectron spectra and *ab initio* SCF MO calculations of dibenzenechromium and benzenechromiumtricarbonyl, *Mol. Phys.* **29**, 113–128 (1975).

57. P. P. Ewald, Die Berechnung optischer und elecktrostatischer Gitterpotentiale, *Ann. Phys. (Leipzig)* **64**, 253–287 (1921).

58. H. B. Gray and C. J. Ballhausen, A molecular orbital theory for square planar metal complexes, *J. Am. Chem. Soc.* **85**, 260–265 (1963).

59. Y. S. Sohn, D. N. Hendrickson, and H. B. Gray, Electronic structure of metallocenes, *J. Am. Chem. Soc.* **93**, 3603–3612 (1971).

60. E. R. Davidson, Selection of the proper canonical Roothaan–Hartree–Fock orbitals for particular applications. I. Theory, *J. Chem. Phys.* **57**, 1999–2005 (1972).

61. S. B. Piepho, P. N. Schatz, and A. J. McCaffery, Ultraviolet spectral assignment in the tetracyanocomplexes of platinum, palladium and nickel from magnetic circular dichroism, *J. Am. Chem. Soc.* **91**, 5994–6001 (1969).

62. M-M. Rohmer, A. Veillard, and M. H. Wood, Excited states and electronic spectrum of ferrocene, *Chem. Phys. Lett.* **29**, 466–468 (1974).

63. B. M. Hoffman, D. Diemente, and F. Basolo, Electron paramagnetic resonance studies of some cobalt(II) Schiff base compounds and their monomeric oxygen adducts, *J. Am. Chem. Soc.* **92**, 61–65 (1970).

64. J. P. Collman, R. R. Gagne, C. A. Reed, T. R. Halbert, G. Lang, and W. T. Robinson, Picket fence porphyrins. Synthetic models for oxygen binding hemoproteins, *J. Am. Chem. Soc.* **97**, 1427–1439 (1975).

65. B. M. Hoffman, T. Szymanski, and F. Basolo, Consideration of a report on the formulation of monomeric cobalt-dioxygen adducts. Continued support for Co(III)-O_2^-, *J. Am. Chem. Soc.* **97**, 673–674 (1975).

66. L. Pauling and C. D. Coryell, The magnetic properties and structure of hemoglobin, oxyhemoglobin and carbonmonoxyhemoglobin, *Proc. Natl. Acad. Sci. USA* **22**, 210–216 (1936).

67. L. Pauling, *Haemoglobin*, p. 57, Butterworth, London, (1949).

68. L. Pauling, Nature of the iron-oxygen bond in oxyhaemoglobin, *Nature (London)* **203**, 182–183 (1964).

69. J. J. Weiss, Nature of the iron–oxygen bond in oxyhaemoglobin, *Nature (London)* **202**, 83–84 (1964).

70. J. S. Griffith, On the magnetic properties of some haemoglobin complexes, *Proc. R. Soc. London, Ser. A* **235**, 23–36 (1956).

71. G. A. Rodley and W. T. Robinson, Structure of a monomeric oxygen-carrying complex, *Nature (London)* **235**, 438–439 (1972).

72. E. Melamud, B. L. Silver, and Z. Dori, Electron paramagnetic resonance of mononuclear cobalt oxygen carriers labeled with oxygen-17, *J. Am. Chem. Soc.* **96**, 4689–4690 (1974).

73. R. Hoffmann, M. M. L. Chen, M. Elian, A. R. Rossi, and D. M. P. Mingos, Pentacoordinate nitrosyls, *Inorg. Chem.* **13**, 2666–2675 (1974).

74. A. Dedieu, Thèse de Doctorat d'Etat, Strasbourg, France, 1975.

75. B. B. Wayland, J. V. Minkiewicz, and M. E. Abd-Elmageed, Spectroscopic studies for tetraphenylporphyrincobalt(II)complexes of CO, NO, O_2, RNC and $(RO)_3P$, and a bonding model for complexes of CO, NO and O_2 with cobalt(II) and iron(II)porphyrins, *J. Am. Chem. Soc.* **96**, 2795–2801 (1974).

76. J. P. Collman, R. R. Gagne, H. B. Gray, and J. W. Hare, A low temperature infrared spectral study of iron(II)dioxygen complexes derived from a "picket fence" porphyrin, *J. Am. Chem. Soc.* **96**, 6522–6524 (1974).

77. A. Dedieu, M.-M. Rohmer, and A. Veillard, Binding of dioxygen to metal complexes. The oxygen adduct of Co(acacen), *J. Am. Chem. Soc.* **98**, 5789 (1976).

78. L. Salem, Intermolecular orbital theory of the interaction between conjugated systems. I. General theory, *J. Am. Chem. Soc.* **90**, 543–552 (1968).

Note Added in Proof

Since completion of this manuscript, the field of *ab initio* calculations for transition metal complexes and organometallics has expanded rapidly, largely due to the availability of very efficient programs.[79] A (13, 9, 7) Gaussian basis set is now available for the second transition series.[80] An *ab initio* calculation for $Pd(CO)_4$[81] invalidates the conclusions of molecular pseudopotential calculations regarding the sequence of the outermost occupied orbitals,[82] and the bonding schemes of $Ni(CO)_4$ and $Pd(CO)_4$ have been compared.[81] Calculated excitation energies for the $PdCl_4^{2-}$ anion, with a double-zeta-type basis set for the valence shells, are in excellent agreement with the data (assignments and energy separations) derived from the polarized crystal spectrum.[83] Calculations for the ground and excited states have been reported for MnO_4^-,[84] $TiCl_4$ and VCl_4,[85] $CrOF_5$,[86] $CrOCl_4^-$,[87] $CoCl_4^{2-}$.[88] The nature of the metal–metal interaction in binuclear complexes of Cr and Mo has been discussed in light of SCF and CI calculations.[89] The CI calculations for the binuclear complexes of Cr do not support a previous description in terms of a *nonbonding* configuration for the ground state of tetra-μ-carboxylatochromium(II) compounds.[90] Electronic and structural aspects of dioxygen binding to iron porphyrins considered as heme models have been discussed recently.[91–93]

79. M. Benard, Efficient computing of two-electron integrals for gaussian d-type orbitals, *J. Chim. Phys.* **1976**, 413 (1976).

80. M. Benard and J. Demuynck, unpublished.

81. J. Demuynck, Failure of a pseudopotential calculation for $Pd(CO)_4$, *Chem. Phys. Lett.* **45**, 74 (1977).

82. R. Osman, C. S. Ewig, and J. R. Van Wazer, Molecular pseudopotential calculations on transition metal complexes: $Ni(CO)_4$, $Pd(CO)_4$, $Pt(CO)_4$, *Chem. Phys. Lett.* **39**, 27 (1976).

83. M. Benard, J. Demuynck, M-M. Rohmer, and A. Veillard, Ground and excited states of transition metal complexes, *in*: *Spectroscopie des éléments de transition et des éléments lourds dans les solides* (F. Gaume, ed.), Editions du C.N.R.S., Paris, in press.

84. H. Hsu, C. Peterson, and R. M. Pitzer, Calculations on the permanganate ion in the ground and excited states, *J. Chem. Phys.* **64**, 791 (1976).

85. I. H. Hillier and J. Kendrick, *Ab initio* calculations of the ground, excited, and ionic states of titanium and vanadium tetrachlorides, *Inorg. Chem.* **15**, 520 (1976).

86. C. D. Garner, I. H. Hillier, F. E. Mabbs, and M. F. Guest, The electronic structure of CrO^{3+} and MoO^{3+} complexes, *Chem. Phys. Lett.* **32**, 224 (1976).

87. C. D. Garner, J. Kendrick, P. Lambert, F. E. Mabbs, and I. H. Hillier, Single-crystal electronic spectrum of tetraphenylarsonium oxotetrachlorochromate(V) and an *ab initio* calculation of the bonding and excited states of oxotetrachlorochromate(V), *Inorg. Chem.* **15**, 1287 (1976).

88. I. H. Hillier, J. Kendrick, F. E. Mabbs, and C. D. Garner, An *ab initio* calculation of the bonding, excited states and g value of tetrachlorocobaltate(II) $CoCl_4^{2-}$, *J. Am. Chem. Soc.* **98**, 395 (1976).

89. M. Benard and A. Veillard, Nature of the metal–metal interaction in binuclear complexes of Cr and Mo. *Nouveau J. Chim.* **1**, 97 (1977).

90. C. D. Garner, I. H. Hiller, M. F. Guest, J. C. Green, and A. W. Coleman, The nature of the metal–metal interaction in tetra-μ-carboxylatochromium(II) systems, *Chem Phys. Lett.* **41**, 91 (1976).

91. A. Dedieu, M.-M. Rohmer, M. Benard, and A. Veillard, Oxygen binding to iron porphyrins. An *ab initio* calculation, *J. Am. Chem. Soc.* **98**, 3717 (1976).

92. A. Dedieu, M.-M. Rohmer, and A. Veillard, Oxygen binding to iron porphyrins. *Ab initio* calculations, *in*: Metal-Ligand Interactions in Organic Chemistry and Biochemistry (B. Pullman and N. Goldblum, eds.), D. Reidel, Holland, in press.

93. A. Dedieu, M.-M. Rohmer, H. Veillard, and A. Veillard, The nature of oxygen binding in heme models, *Bull. Soc. Chim. Belge* **85**, 953 (1976).

6

Strained Organic Molecules

Marshall D. Newton

1. Introduction

In its spectacular evolution through the past decades, the orbital theory of molecular structure has steadily progressed from simple semiempirical models of limited application (e.g., π-electron theories for planar systems) to elaborate *ab initio* models which include all electrons and are applicable to general molecular geometries.[1-6] Furthermore, the power of present generation computers means that these *ab initio* techniques can now be feasibly applied to rather large molecular systems (i.e., ~50–75 electrons). As a result, the past few years have seen major progress in our understanding of a fundamentally important aspect of organic chemistry—the nature of strained organic molecules.[7,8] Although current semiempirical techniques provide much important information about strained molecules,[2,9,10] *ab initio* studies have been imperative because the strained molecules transcend our usual concepts about the nature of bonding and thus, among other things, play a vital role in helping extend and recalibrate semiempirical methods. *Ab initio* capability for treating *large* molecules is very important in this connection, since many of the most intriguing aspects of strained systems only appear in rather large systems (i.e., ~5–10 heavy atoms), especially the polycyclic variety. Fortunately, a large base of *ab initio* calculations for both strained and unstrained systems currently exists,[1,4-6,8,11-45]* thus allowing reasonable estimates of the

---

*These references provide examples which fall within the scope of this presentation and by no means consitute a definitive listing of calculations for strained molecules.

Marshall D. Newton • Department of Chemistry, Brookhaven National Laboratory, Upton, New York

predictive accuracy of *ab initio* results for experimentally unknown properties.

This chapter does not attempt to review thoroughly the recent *ab initio* literature for calculations on strained molecules, but rather to illustrate with carefully chosen examples the ability of *ab initio* calculations to elucidate some of the basic features and consequences of strain in organic molecules. Our examples will be limited to hydrocarbons. Calculations for heterocyclic 3-membered rings have recently been reviewed.[8] In addition to the traditional focus on the static properties of molecules, much recent attention has been given to the energetics and dynamics of bond-breaking processes (including thermal isomerization[29,35,39–45] and attack by electrophilic reagents[12,23,24]) in strained rings, whose rigid structure is ideal for studying stereospecificity. Although some reference will be made to this important development, it will largely be considered beyond the scope of the present work. We finally note that while primary emphasis will be on ground-state properties, the recent availability of high-resolution electronic spectra for bicyclo[1.1.0]butane[46] presages the arrival of major new challenges to theory in the realm of strained systems.

2. The Nature of Strained Organic Molecules

The term "strain" has been repeatedly used in the preceding paragraphs. It is now appropriate to provide some operational definitions of "strain," and to state more specifically the various challenges which strained molecules pose for theory.

2.1. Definition of Strain

Structural Definition. Perhaps the simplest definition of strain is the structural or geometrical one. Carbon atoms whose bond angles (as determined from interatomic vectors) depart markedly from the standard bond angles (109° 28', 120°, or 180°) are said to be strained. This is the classical definition of strain introduced by Baeyer a century ago.[47] Strain also accompanies torsional eclipsing. It should be noted that angle strain may show up either as ring strain or as the result of a compromise necessitated by steric crowding (tri-*t*-butyl methane provides a simple but dramatic example of the latter effect[48]). We shall deal only with cases of ring strain.

Thermochemical Definition. The other familiar definition of strain arises from deviation of heats of formation from those expected on the basis of additivity relationships which hold for "standard" or "unstrained" (by definition) molecules.[49,50] Strain energies are typically an appreciable fraction of an electron volt (on a per CC bond basis). Schleyer *et al.*[50] have recently discussed various definitions of strain and have compared different methods[49] of estimating strain energy, suggesting new "strain-free" group equivalents which are preferable to the original Franklin values.[49a]

Some Problems in Defining Strained Bonds. Although the above definitions are rather unequivocal, as far as they go, complications arise when one attempts to talk about a strained bond. Thus, while it may be clear from thermochemistry that a given molecule as a whole is strained, and from geometry that certain atoms are in a strained environment, it is not so obvious to what extent a "bond" exists between a given pair of atoms in such a molecule, using the familiar measures of bond length, force constant, and bond energy. This situation arises because atoms in strained rings are invariably multiply connected; e.g., in cyclopropane there is a 1-bond and 2-bond link between each pair of carbon atoms. Furthermore, we cannot break the direct link between two atoms in cyclopropane without also affecting the other linkage (e.g., changing a CCC bond angle). This situation is concisely summarized in the following redundancy condition for the stretching of one CC bond in cyclopropane (correct through first order),

$$(3^{1/2}/2)R\,\Delta\theta_i - \Delta R_i \equiv 0 \tag{1}$$

where R is the CC bond length, and ΔR_i and $\Delta\theta_i$ are, respectively, the distance and angle increments (θ_i is opposite bond i). Thus, the significance of the stretching force constant for a *single* CC bond in cyclopropane is much less clear than in the case of, say, n-propane. For similar reasons, the distance between two atoms in a strained ring gives no direct indication of the strength of the bond between them, if indeed one exists at all. For example, a CC bond in cyclopropane is not stronger in any sense than the *longer* bond in propane. The multiple connectivity referred to above often leads to situations in polycyclic molecules where dominant nonbonded interactions force other atoms into unusually short but still nonbonded contacts. Thus, the use of the terms bonded and nonbonded requires care, and when ambiguity arises in the following sections, the question will often be resolved by direct appeal to the electron density between the atoms in question.

2.2. Challenges to Theory

A successful theory of strained organic molecules must give meaning to the various novel concepts and properties which arise in the course of describing such molecules. Some examples pertinent to strained *saturated* hydrocarbons are the following:

(1) Bent bonds (by up to ~30°) and partial π character in single bonds. Combined use of high-precision x-ray and neutron diffraction data now

provides a direct probe of bonding densities, which are available for comparison with theory.[51]

(2)   Strange atomic hybridization (from $\sim sp^{1.5}$ to nearly pure $2p$ orbitals) and related effects in directly bonded spin–spin coupling constants.

(3)   Unusual bond lengths (~ 1.45–1.58 Å) and bond angles ($\angle HCC$ as large as $130°$).

(4)   Large dipole moments ($\leqslant 1$ D).

(5)   Strain energies.

In anticipation of the discussion in later sections, we comment briefly on the set of molecules represented by computer-generated drawings and designated by the roman numerals I–XX.* The hydrogen atoms have also been included so as to give a better feeling for steric effects. The prototype strained rings cyclopropane(I) and cyclobutane(II) and their olefinic counterparts, cyclopropene(III) and cyclobutene(IV) are displayed first. The original valence

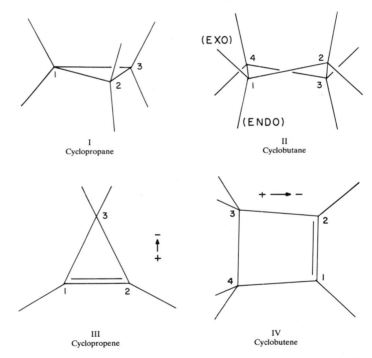

I
Cyclopropane

II
Cyclobutane

III
Cyclopropene

IV
Cyclobutene

bond and molecular orbital studies of (I) by Coulson and Moffitt[52] and Walsh[53] have now been extended by numerous *ab initio* calculations which confirm the presence of bent bonds,[12–17] in agreement with diffraction experiments on related species.[54–57] In Section 4 we shall also examine the electron

---

*The carbon atoms are numbered and the hydrogen atoms are in general unlabeled, although some *endo* and *exo* hydrogens are distinguished. Dipole moment polarities are indicated by arrows, except for Dewar benzene (IX) where the positive end is *exo* to the framework.

density in the middle of the cyclopropane ring, as revealed by theory and experiment. An obvious question raised by II is the nature of the departure from D_{4h} symmetry. Structures V–XV illustrate a variety of fused 3- and 4-membered rings, and our major goal will be to describe the characteristics of the bonding between the bridgehead atoms at the points of ring fusion. A key

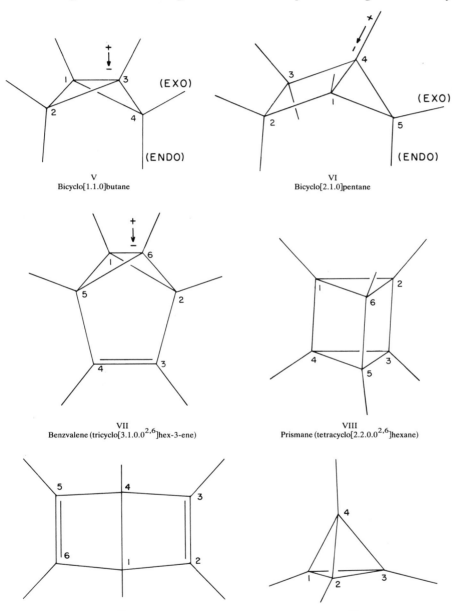

V
Bicyclo[1.1.0]butane

VI
Bicyclo[2.1.0]pentane

VII
Benzvalene (tricyclo[3.1.0.0^{2,6}]hex-3-ene)

VIII
Prismane (tetracyclo[2.2.0.0^{2,6}]hexane)

IX
Dewar benzene (bicyclo[2.2.0]hexa-2,5-diene)

X
Tetrahedrane (tricyclo[1.1.0.0^{2,4}]butane)

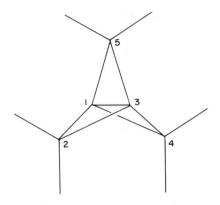

XI
[1.1.1]Propellane (tricyclo[1.1.1.0^{1,3}]pentane)

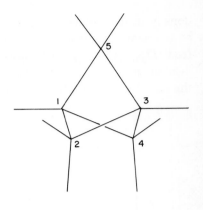

XII
Bicyclo[1.1.1]pentane

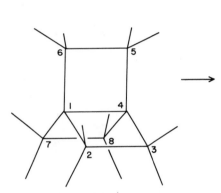

XIII
[2.2.2]Propellane (tricyclo[2.2.2.0^{1,4}]octane)

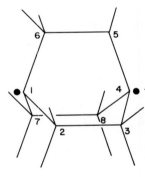

XIV
Bicyclo[2.2.2]octanyl diradical

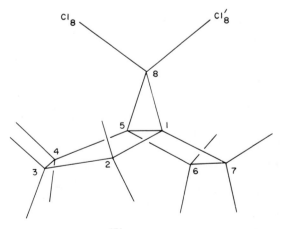

XV
8,8-dichloro[3.2.1]propellane(8,8-dichloro-tricyclo[3.2.1.0^{1,5}]octane)

molecule in this group is the prototype fused system, bicyclo[1.1.0]butane (V), whose "olefinic" and "acetylenic" properties have recently been explained by an *ab initio* model due originally to Schulman and Fisanick,[25] which postulates nearly pure $2p$–$2p$ bonding between atoms 1 and 3. Special attention should also be directed to two molecules which have not yet been observed: (1) tetrahedrane (X), whose temporary existence has been inferred,[58,59] and for which a potential energy minimum has been calculated[35]; and (2) the prototype propellane (XI), which is not thought to be stable but which provides an interesting comparison with the related known system, bicyclo[1.1.1]pentane (XII)[37]; both systems exhibit a delicate balance of nonbonded interactions. Propellanes are those molecules composed of three saturated rings with a common bond.[60] The [2.2.2]propellane (XIII) offers an example of bond-stretch isomerism[61]; i.e., XIII and XIV are separated by a potential barrier whose calculated magnitude[39] was in accord with subsequent experimental data.[62] While a bridgehead atom in XIII has three CC bonds in a plane essentially perpendicular to the fourth bond, [3.2.1]propellane (XV) finally attains the condition denoted by Wiberg *et al.*[63] as "inverted" tetrahedral geometry, where all four bonds from a carbon atom lie on one side of a plane containing the atom (Fig. 1). This situation sharply accentuates the difference between interatomic vectors and the actual directions of "bond" densities. Although no *ab initio* studies of (XV) have been made, its unusual bonding has been discussed in terms of CNDO calculations.[63]

Structures XVI and XVII schematically depict the formation of orthobenzyne from benzene. Theoretical studies have addressed the question as to whether *o*-benzyne prefers to exist as the diradical XVII with an unstrained aromatic system, or chooses to form additional bonding between atoms 1 and 2 (XVIII, a and b), at the expense of strain in the ring.[37,38,64,65] It is also of interest to know whether a strong C_1C_2 bond would lead to an observable infrared spectrum, since isolation of *o*-benzyne has been reported in a matrix.[66]

Finally in (XIX) and (XX) we have the traditionally assumed structures of the triplet and singlet states of cyclobutadiene, long the subject of theoretical

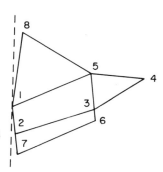

Fig. 1. View of the inverted[63] carbon atom in 8,8-dichloro[3.2.1]propellane (XV). The four CC bonds emanating from carbon atom 1 are on one side of the plane perpendicular to the figure plane (dotted line). Note that the C_1C_2 and C_1C_7 bonds are partially eclipsed.

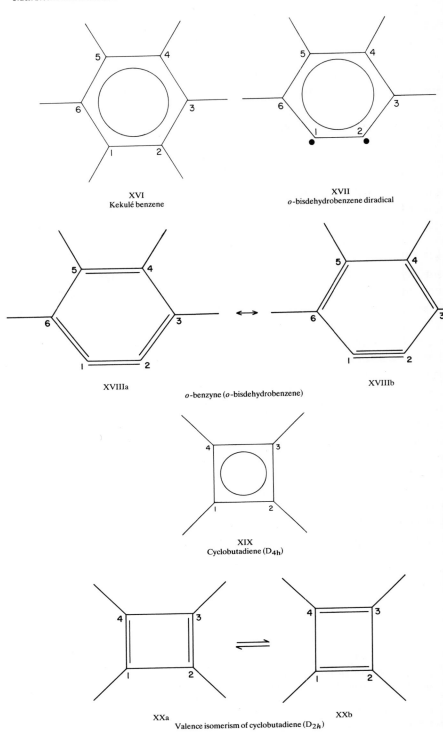

XVI
Kekulé benzene

XVII
o-bisdehydrobenzene diradical

XVIIIa

o-benzyne (*o*-bisdehydrobenzene)

XVIIIb

XIX
Cyclobutadiene (D$_{4h}$)

XXa

XXb

Valence isomerism of cyclobutadiene (D$_{2h}$)

and experimental investigations[67] and recently isolated in a matrix.[69–72] It still remains to be demonstrated which species (singlet or triplet) is more stable and which species has been seen experimentally. A related question involves the magnitude of the barrier separating the two singlet structures (XXa and b) and possible barriers separating the singlet and triplet states. Finally, it is important to ascertain whether the assumed D_{4h} and D_{2h} geometries do in fact correspond to local potential minima.

Table 1 lists the empirical strain energies for the molecules under discussion. These are based on the standard group equivalents[49a] given in Table 2.

Table 1. Empirical Strain Energies

Molecule	Structure	Strain energy, kcal/mole[a]	
		Total	Per CC bond
Alkanes			
Cyclobutane	II	27	6.8
Bicyclo[2.1.0]pentane	VI	55	9.1
Cyclopropane	I	28	9.3
Bicyclo[1.1.1]pentane	XII	(62)	(10)
Prismane	VIII	(114)	(12.7)
Cubane		155	12.9
Bicyclo[1.1.0]butane	V	63	13
Tetrahedrane	X	(127)	(21.2)
Unsaturated systems			
Unconjugated			
Dewar benzene	IX	(43)	(6.1)
Cyclobutene	IV	29	7.1
Benzvalene	VII	(69)	(8.6)
Cyclopropene	III	52	17
Conjugated			
Kekulé benzene	XVI	-37^{b}	-6.2^{b}
o-Benzyne	XVIII	$37^{c,d}$	$4.5^{c,d}$
Cyclobutadiene	XIX	$(55)^{c,e}$	$(14)^{c,e}$

[a] Based on the ΔH_f^o (0 K) values given in Table 4 and the data of Table 2, as described in the text. For cubane, ΔH_f^o was taken from B. D. Kybett *et al.* Thermodynamic properties of cubane, *J. Am. Chem. Soc.* **88**, 626 (1966), and adjusted from 298 to 0 K, as described in Ref. 98. Values in parentheses are based on estimated ΔH_f^o values (Table 4).
[b] The negative quantity corresponds to a *positive* π-electron resonance energy (based on a Kekulé valence structure).
[c] Resultant of σ-electron ring strain and π-electron resonance energy. For cyclobutadiene the resonance energy is negative due to antiaromaticity.[125]
[d] Based on valence structure XVIIIb.
[e] Based on a Kekulé valence structure (XXa or b).

*Table 2. Franklin's Group Equivalents for
Strain-Free Molecules*[a]

Group[b]	ΔH_f (0 K) contribution, kcal/mole
—CH$_3$	−8.26
CH$_2$ (with two bonds shown)	−3.67
—CH (with three bonds shown)	+0.18
C (with four bonds shown)	+1.74
C=C (with H, H shown)	+20.31
—C≡	+27.12

[a] Reference 49a.
[b] Straight-line segments without the terminal H indicate that the
carbon atoms are singly bonded to adjacent carbon atoms (or
triply bonded in the case of C≡).

In summary, the set of molecules chosen for the current survey illustrates a
great variety of interesting structural and energetic features of strained
molecules. Before the detailed theoretical discussion of these points in Section
4, we summarize the various theoretical methods available for studying
strained systems and their accuracy for various properties in Section 3.

3. Theoretical Methods for Strained Organic Systems

3.1. Empirical and Semiempirical Methods

A variety of theoretical techniques have been applied to strained systems,
from the purely empirical molecular mechanics (MM) approach,[48,73] to the
semiempirical maximum overlap method (MOA),[9] to LCAO methods such as
the extended Hückel theory (EHT)[74] and the self-consistent field CNDO,[75]
INDO,[76] and MINDO techniques.[2,10] The classic studies of Coulson and
Moffitt,[52] using largely nonempirical valence bond theory, have already been
referred to. The MM approach has been used to obtain geometries and strain
energies, but is not reliable for three-membered rings and nonadditive strain
energy effects in some polycyclic systems.[73] The hybrid orbitals obtained from
the MOA method have been correlated with many properties of molecular and
electronic structure.[77] The EHT theory as employed by Hoffmann and

co-workers has provided a wealth of qualitative predictions regarding the electronic structure of both simple rings[78] and complex polycyclic systems (e.g., the XIII → XIV sequence).[61] The CNDO and INDO methods have proved very useful in correlating atomic hybridization with spin–spin coupling constants.[26,27,32,79] One of the most versatile semiempirical techniques for strained systems is the MINDO approach,[2] which at its current stage of development (MINDO/3[10,80]) seems to yield generally reliable equilibrium geometries and energies, including activation energies for various bond-breaking processes.[10] Typically, MINDO/3 heats of formation (ΔH_f) are in error by ~5 kcal/mole, while, for small rings, strain energy is systematically *underestimated* by 5–10 kcal/mole.[10]

3.2. *Ab Initio* Methods and Basis Sets

Molecular orbital SCF calculations for strained systems have been carried out with a number of basis sets, principally including the simple basis of floating spherical Gaussian orbitals (FSGO),[20,81] the minimal STO basis, [16,17] and a variety of contracted[82] Gaussian orbital (GTO) basis sets, especially those developed by Basch, Robin, and Kuebler,[83] Pople and co-workers (STO-3G,[84] 4-31G,[85] and 6-31G*[86]),† and Lehn and Wipff.[22] These contracted GTO sets range from the minimal basis (both STO-type[84] and SCF-AO type[22,87–89]) to the split valence or "double-zeta" basis plus 3d functions for carbon.[86]

In cases where a single configuration wave function is clearly inadequate, [such as ring-opening processes[29,39–45] and the interconversions of rectangular cyclobutadiene (XX)[33,72]], CI calculations have been carried out with the above basis sets, using either simple 2-configuration wave functions when physically justifiable (analogous to σ_g^2 and σ_u^2 for H_2) or larger CI wave functions when necessary.

For purposes of facile intercomparisons it is, of course, desirable that calculations for the various strained molecules of interest be carried out at a uniform level of quality. Since most calculations for large strained systems have, in the past, employed the minimal STO-3G basis or the split valence 4-31G basis, using standard molecular scale factors,[84,85] the majority of the results discussed in the following sections will involve these basis sets, even though other calculations using comparable or superior basis sets have also been carried out. We note that Wipff and Lehn[24] have very recently completed an extensive and largely unpublished set of calculations for about twenty-five polycyclic systems, using other minimal and extended basis sets.

---

†In the standard notation of contracted GTO basis sets,[82] the STO-3G, 4-31G, 6-31G, and 6-31G* sets would be designated as: (63/3) → [21/1], (84/4) → [32/2], (10 4/4) → [32/2], and (10 4 1/4) → [321/1], respectively. The 4-31G and 6-31G basis sets give very similar results except, of course, for total molecular energies.

Their results are in general agreement with the calculations considered in the present survey, but they came to our attention too late to be presented here in any great detail.

Results not otherwise referenced in the following discussion are presented here for the first time and were obtained with the GAUSSIAN 70 series of computer programs.[90]

3.3. Localized Molecular Orbitals

In elucidating electronic structure of molecules it is often useful to localize the occupied molecular orbitals (MO) by subjecting them to a unitary transformation uniquely determined by an appropriate physical criterion. The two most common criteria are: (1) the Edmiston–Ruedenberg (ER) procedure,[91] based on an original suggestion of Pople and Lennard-Jones,[92] which postulates that localized molecular orbitals (LMO) are defined by minimizing the sum of exchange (or Coulomb) interaction energies between occupied MOs (the results of this approach will be presented below); and (2) the Foster–Boys (FB) procedure,[93] according to which the sum of squares of the centroids of occupied MOs is maximized.

The ER and FB methods usually lead to similar LMOs.[94] In any case, once the LMOs are determined, truncation of the usually small delocalized "tails" (necessary to preserve orthogonality), followed by renormalization, leads to 2-center MOs from which hydrid atomic orbitals can be extracted.[17a]

3.4. Reliability of *Ab Initio* Methods

The accuracy of *ab initio* MO theory for calculating standard molecular properties has been the subject of recent reviews.[3b,4,5,8] We shall focus on hydrocarbons, where minimal basis sets have been shown[3b,8,19,32,43,95–97] to yield reliable equilibrium geometries (to within ≤ 0.03 Å and $\leq 5°$ for bond lengths and angles, respectively) and dipole moments. Split-valence basis sets generally give more accurate geometries and force constants,[3b,8,96,97] and can also give an excellent account of ionization energies.[15,32] Energies for appropriately balanced chemical reactions can be reproduced to within ≤ 5–10 kcal/mole, depending on basis set and type of reaction.[9,19,26,86,98] For example, in the case of so-called isodesmic reactions,[98] where the number of bonds of a given formal type is conserved, the 4-31G basis typically gives ± 5 kcal/mole accuracy if unstrained species are involved. Carbon atom polarization functions (6-31G*) are required if strained systems are to be handled with similar accuracy.[86] At the 4-31G level, energies to within ~ 10 kcal/mole can be obtained for reactions including highly strained species only if similar numbers of strained ring types (3- or 4-membered rings) appear on both sides of the equation. Hydrogenation reactions are handled somewhat less accurately,[98] especially when hydrogenation changes the number of rings (e.g., cyclopropane $+ H_2 \rightarrow n$-propane).

These points are illustrated in Table 3, where reaction energies, including zero-point corrections, have been calculated in terms of the basic energy data

Table 3. Energetics of Some Hydrogenation and Isodesmic Reactions for Strained Systems

Reaction	$\Delta H°$ (0 K), kcal/mole Calculated,[a] 4-31G(6-31G*)	Experimental[b]
Hydrogenation		
$C_2H_4 + H_2 \longrightarrow C_2H_6$	$-33.9\ (-36.1)$	-31.0
[□ with double bond] $+ H_2 \longrightarrow$ [□]	$-36.8\ (-34.6)$	-29
[△] $+ H_2 \longrightarrow$ [△]	$-63.2\ (-59.4)$	-52
[□ with double bond] $+ H_2 \longrightarrow$ [□]	$-74.5\ (-74.4)^c$	(-62)
[benzene ring] $+ H_2 \longrightarrow$ [cyclohexane ring]	-112.3	-98 ± 5
Isodesmic[d,e]		
[▷] $+ C_2H_6 \longrightarrow$ [□] $+ CH_4$	$-4.8\ (-2.1)$	-3.8
[◇ with line] $+ C_2H_6 \longrightarrow 2$ [▷]	$-15.7\ (-10.0)$	-5.9
[□▷ fused] $+ C_2H_6 \longrightarrow$ [□] $+$ [▷]	-1.2	1.8
[diamond/bicyclic] $+$ { $2C_2H_6 \longrightarrow 2$ [□] $+ CH_4$	-19.8	(-11)
	$CH_4 \longrightarrow$ [◇ with line] $+ C_2H_6$	5.4	(4.5)
[tetrahedral cage] $+$ [□] $\longrightarrow 2$ [◇ with line]	-25.3	(-28.5)

[a] Based on the *ab initio* total energies and experimental zero-point energies given in Table 4. Results at the 6-31G* level are in parentheses.
[b] Based on the $\Delta H_f°$ (0 K) values of Table 4. Quantities in parentheses involve the use of estimates based on semiempirical calculations.
[c] Based on the RHF $^1A_{1g}$ energy.
[d] Reference 98.
[e] The polycyclic molecules correspond to structures V, VI, X, and XII.

Table 4. Basic Energy Data for Some Prototype Strained and Unstrained Systems

| Molecule | Structure | Calculated total energy,[a] a.u. | | Experimental data | |
		4-31G	6-31G*	E_{zp},[b] a.u.	ΔH_f° (0 K),[b] kcal/mole
Hydrogen		−1.1268	−1.1268	0.0100	0.0
Methane		−40.1398	−40.1952	0.0432	−16.0
Ethane		−79.1159	−79.2287	0.0721	−16.5
Ethylene		−77.9222	−78.0315	0.0492	14.5
Cyclopropane	I	−116.8835	−117.0587	0.0788	16.8
Cyclopropene	III	−115.6417	−115.8229	0.0545	68.7
Cyclobutane	II	−155.8663[c]	(−156.0947)	0.1068[d]	12.5[e]
Cyclobutene	IV	−154.6673	−154.8993	0.0833[f]	41.5[g]
Cyclobutadiene					
($^3A_{2g}$, RHF)	XIX	−153.4130[h]	—	(0.060)[i]	102[j]
($^1A_{1g}$, RHF)	XX	−153.4085[h]	−153.6408	(0.060)[i]	96[j]
($^1A_{1g}$, 2×2 CI)	XX	−153.4215[h]	—	(0.060)[i]	
Bicyclo[1.1.0]butane	V	−154.6237	−154.8705	0.0831[k]	56[g]
Tetrahedrane	X	−153.3414[l]	−153.5974	(0.060)[i]	(128)[j]
Bicyclo[1.1.1]pentane	XII	−193.6085[m]	—	0.1121[n]	(51)[o]
Bicyclo[2.1.0]pentane	VI	−193.6306[p]	—	0.1141[q]	44[o]
[1.1.1]Propellane	XI	−192.3601[r]	—	—	—
Kekulé benzene	XVI	−230.3774[s]	—	0.0975	24
Dewar benzene	IX	−230.2145[s]	—	(0.0975)[t]	(84)[u]
Benzvalene	VII	−230.2108[s]	—	(0.0975)[t]	(90)[u]
Prismane	VIII	−230.1383[s]	—	(0.0975)[t]	(115)[u]
o-Benzene	XVIII	−229.0575[v]	—	(0.0733)[w]	122 ± 5[x]

[a]Based on the STO-3G or 4-31G equilibrium geometries tabulated in Refs. 19 and 96, unless otherwise noted. Quantities in parentheses are estimated.[(19)]

[b]Taken from tabulation in Ref. 98, unless otherwise noted. Data for 298 K have been corrected to 0 K as described in Ref. 98, using available or estimated vibrational frequencies.

[c]Based on $R_{CC} = 1.548$ Å, $R_{CH} = 1.09$ Å, and calculated (4-31G) dihedral (24°) and CH bond angles (Table 15). This total energy supersedes a previously reported energy (Ref. 27) which was subsequently found to be based on an incorrect geometry.

[d]R. C. Lord and I. Nakagawa, Normal Vibrations, Potential constants, and vibration–rotation interaction constants in cyclobutane and cyclobutane-d_8, J. Chem. Phys. 39, 2951–2965 (1963).

[e]Reference 50.

[f]R. C. Lord and D. G. Rea, The vibrational spectra and structure of cyclobutene and cyclobutene-d_6, J. Am. Chem. Soc. 79, 2401–2406 (1957).

[g]K. B. Wiberg and R. A. Fenoglio, Heats of formation of C_4H_6 hydrocarbons, J. Am. Chem. Soc. 90, 3395–3397 (1968).

[h]The RHF and 2-configuration wave functions were all obtained self-consistently, using the formalism and computer programs developed by W. J. Hunt et al. (Ref. 127). The equilibrium geometries (4-31G) for $^3A_{2g}$ (RHF) and $^1A_{2g}$ (CI) are given in Table 21. For $^1A_{1g}$ (RHF), $R_{CC} = 1.323, 1.581$ Å. All calculations employed $R_{CH} = 1.075$ Å. The unrestricted Hartree–Fock[(129)] energy of $^3A_{2g}$ is −153.4176 ($R_{CC} = 1.434$ Å) at the 4-31G level; the corresponding 6-31G* value[(19)] is −153.6501 ($R_{CC} = 1.431$ Å).

[i]Estimated by extrapolation of known values.

[j]Semiempirical (MINDO/3) estimate (Ref. 101).

[k]I. Haller and R. Srinivasan, Vibrational spectra and molecular structure of bicyclo[1.1.0] butane, J. Chem. Phys. 41 2745–2752 (1964).

[l]Reference 35.

[m]Reference 27.

[n]K. B. Wiberg, D. Sturmer, T. P. Lewis, and I. W. Levin, Vibrational spectrum of bicyclo[1.1.1]pentane, Spectrochim. Acta 31A, 57–73 (1975).

[o]Based on the experimental value for the [2.1.0] isomer [R. B. Turner, P. Goebel, B. J. Mallon, W. von E. Doering, J. F. Coburn, Jr., and M. Pomerantz, Heats of hydrogenation. VIII. Compounds with three- and four-membered rings, J. Am. Chem. Soc. 90, 4315–4322 (1968)] and the calculated (MINDO/1) difference between the [1.1.1] and [2.1.0] isomers (Ref. 100).

listed in Table 4. In cases where experimental energies are not available, estimates have been made on the basis of recent MINDO calculations.[99–101] As was noted above, the results for isodesmic reactions are usually superior to those for hydrogenation reactions. It is clear from Table 3 that the 4-31G level allows useful rough energy comparisons to be made between a variety of highly strained molecules. The 4-31G and MINDO calculations yield similar results for those reactions containing species of unknown enthalpy, with the largest deviation occurring for the hydrogenation of cyclobutadiene. It should be noted that the 4-31G level systematically exaggerates the increase in strain energy as 3- and 4-membered rings are fused to each other. For example, the nonadditivity in the strain energy of bicyclo[1.1.0]butane (V) as compared to that for two cyclopropanes is too large by ~10 kcal/mole. At the 6-31G* level, this discrepancy is reduced to ~4 kcal/mole (Table 4 and Ref. 19b). The energies of systems with vastly differing degrees of strain cannot be meaningfully compared at the 4-31G level. The data of Table 3 reveal that while the correct energy ordering of the benzene valence isomers is obtained (VII, VIII, IX, and XVI), the differences are much too large. The MINDO/3 method also has some difficulties with the relative energies of benzene isomers but the errors are smaller.[80]† It is anticipated that, in general, the use of polarization functions (e.g., the 6-31G* basis) will markedly improve the accuracy of *ab initio* energies for isodesmic reactions of the type illustrated in Table 3.

   The geometries calculated for bridgehead carbon atoms at the STO-3G level (see Table 5) are in very good accord with experiment, with the major discrepancies being a slight underestimation of the C_1C_3 bond lengths in V and VI and an exaggeration of the large HCC bridgehead angles in V, VI, and

---

†Based on experimental data cited in footnote *u* of Table 4 for derivatives of the benzene valence isomers.

---

*p*Based on the microwave geometry for the carbon framework [R. D. Suenram and M. D. Harmony, Microwave spectrum, structure, and dipole moment of bicyclo[2.1.0]pentane, *J. Chem. Phys.* **56**, 3837–3842 (1972)] and the calculated (STO-3G) equilibrium CH bond angles, with R_{CH} kept at 1.09 Å (see Tables 5 and 15).

*q*J. Bragin and D. Guthals, Vibrational spectra and structure of bicyclo[2.1.0]pentane *J. Phys. Chem.* **79**, 2139–2144 (1975).

*r*Reference 27.

*s*Based on the geometries given in Ref. 32.

*t*Approximated by the value of Kekulé benzene.

*u*Based on the relative energies of the hexamethyl and hexatrifluoromethyl derivatives of the benzene valence isomers as determined, respectively, by J. M. Oth, The kinetics and thermochemistry of the thermal rearrangement of hexamethylbicyclo[2.2.0]hexa-2,5-diene(hexamethyldewarbenzene) and of hexamethyltetracyclo[2.2.0,0²,⁶.0³,⁵]hexane (hexamethylprismane), *Rec. Trav. Chim. Pays-Bas* **87**, 1185–1195 (1968) and D. M. Lemal and L. H. Dunlap, Jr., Kinetics and thermodynamics of (CCF₃)₆ valence isomer interconversions, *J. Am. Chem. Soc.* **94**, 6562–6564 (1972).

*v*Based on optimal values of $R_{C_1C_2}$ (1.226 Å) and $<C_3C_4C_5$ (122.3°) at the 4-31G level, with other parameters as given in Ref. 38.

*x*H.-F. Grützmacher and J. Lohmann, Ionizationspotential und Bildungsenthalpie von Dehydrobenzol, *Liebsig Ann. Chem.* **705**, 81–90 (1967). A somewhat lower value (~100 kcal/mole) has recently been suggested by H. Rosenstock, J. T. Larkins, and J. A. Walker, Interpretation of photoionization thresholds: quasi-equilibrium theory and the fragmentation of benzene, *Int. J. Mass Spectrom. Ion Phys.* **11**, 309–328 (1973). The MINDO/3 estimate (adjusted to 0 K) is 122 kcal/mole (Ref. 65).

Table 5. Calculated (STO-3G) Equilibrium Geometries in Strained Hydrocarbons[a]

Molecule[b]	Structure	R_{CC}, Å	∠HCX, degrees	Framework dihedral angle, degrees
Reference monocycles, X≡H				
Cyclopropane[c]	I	1.502 (1.510)	113 (115.1)	—
Cyclobutane[d]	II	1.550 (1.548)	108.2 (110)	15 (27,[e] 35[f])
Bridgehead (C_{br}) in polycyclic systems, CX≡C_{br},C_{br}				
Benzvalene (C_1C_6)[g]	VII	1.45 (1.45)	139 (134)	107 (106)
Bicyclo[1.1.0]butane[h]	V	1.469 (1.497)	135 (128.3)	117 (121.7)
Tetrahedrane[i]	X	1.472	144.7 (144.7)[j]	70.5 (70.5)[j]
Bicyclo[2.1.0]pentane[g]	VI	1.50 (1,536)	128 (124)[k]	109 (112.7)
Prismane (C_1–C_2)[g,l]	VIII	(1.51 (1.500)	130 —	90 (90)[j]
Prismane (C_1–C_4)[g,l]	VIII	1.56 (1.585)	132 —	60 (60)[j]
Dewar benzene[g,l]	IX	1.58 (1.58, 1.597, 1.63)	121 (120.1)	117 (117.7)

[a] Experimental values are given in parentheses.
[b] Footnotes in this column are to theoretical papers, which include references to pertinent experimental data.
[c] Reference 96b. [d] Reference 43. [e] Reference 102. [f] Reference 103. [g] Reference 32. [h] Reference 19. [i] Reference 35.
[j] Determined by symmetry.
[k] Consistent with experimental microwave data (footnote p, Table 4), but not uniquely determined.
[l] Experimental geometries taken from hexamethyl and hexafluoro derivatives (see Ref. 32).

VII.[19,32] The striking variation in bond lengths for single CC bonds, from 1.45 Å in VII to ~1.58 in VIII and IX is accurately reproduced. The dihedral angles between fused rings are accurate to within a few degrees, while the ring puckering angle of cyclobutane is too small by ~20° at the STO-3G level. The 4-31G value (~24°) is almost within the range of experimental uncertainty,[102,103] but the calculated barrier for ring inversion (≲0.4 kcal/mole) is appreciably less than the empirical value of 1.44 kcal/mole.[102] The tilt of the methylene group will be discussed below. The simple FSGO method, with appropriate variational determination of the locations of the "bent bond" FSGOs also gives a good account of the cyclobutane structure.[20]

Force constants for symmetric stretching of some single and multiple bonds in strained and unstrained molecules are given in Tables 6 and 7. In the case of totally symmetric modes, redundancies between CC bond lengths and bond angles are avoided [cf. Eq. (1)]. The 4-31G level quite uniformly reproduces experimental trends for both saturated (≲11% exaggeration) and unsaturated (≲28% exaggeration) systems. Even the STO-3G level is shown to be reliable for qualitative trends. The symmetric stretching modes for the strained bonds are all found to have smaller force constants than their unstrained counterparts. We finally note that as expected from the discussion in Section 1, the delicate balance between various nonbonded repulsions in

Table 6. Force Constants for Some Saturated Systems[a]

Molecule	Calculated		Experimental
	STO-3G	4-31G	
Symmetric CC stretching			
Ethane	6.5^b	4.9^c	4.5^d
Propane	6.5^e	5.0^e	4.5^f
Cyclopropane (I)	6.2^e	4.4^e	4.0^g
Cyclobutane (II)	6.1^h	4.4^h	4.2^i
Tetrahedrane (X)	5.8^i	4.6^i	—
Symmetric CCC bending			
Propane	1.5^e	1.4^e	0.90^f
Bicyclo[1.1.1]pentane (XII) $\angle C_1C_2C_3$	—	4.4^k	2.7^l

[a] Obtained by parabolic interpolation, using a grid of 0.005–0.01 Å for stretching and 1°–2° for bending.
[b] Reference 95.
[c] Based on calculated equilibrium geometry given in Ref. 96a.
[d] Tabulated in Ref. 95.
[e] Based on calculated equilibrium geometry given in Ref. 96b.
[f] J. H. Schactschneider and R. G. Snyder, Vibrational analysis of the *n*-paraffins II. Normal co-ordinate calculations, *Spectrochim. Acta* **19**, 117–168 (1963) (Calculation V).
[g] J. L. Duncan, The force field and normal coordinates of cyclopropane-H_6 and -D_6. *J. Mol. Spectrosc.* **30**, 253–265 (1969).
[h] Based on $R_{CH} = 1.09$ Å and calculated (4-31G) equilibrium angles (see footnote c, Table 4).
[i] See footnote d in Table 4.
[j] Reference 35.
[k] Obtained from data given in Table 1 of Ref. 27. The total energy for $R_{13} = 1.865$ Å should be corrected to read -193.6039 a.u.
[l] Obtained by appropriate scaling of the value given by Wiberg *et al.* (footnote *n*, Table 4), which employed a different definition of the symmetry coordinate.

Table 7. Symmetric CC Stretching Force Constants for Some Unsaturated Systems[a]

Molecule	Calculated		Experimental
	STO-3G	4-31G	
Cyclobutadiene [$^3A_{2g}(D_{4h})$]	—	$7.3^{b,c}$	—
Benzene	10.5^d	9.1^c	7.1^e
Cyclobutadiene [$^1A_{1g}(D_{2h})$]	—	11.6^f	—
Ethylene	14.4^d	11.8^g	9.3^e
o-Benzyne	17.6^h	16.9^h	—
Acetylene	25.8^d	20.3^g	15.9^e

[a] Obtained by parabolic interpolation, using a grid size of 0.01 Å.
[b] The same value is obtained with both the RHF and UHF formalisms.
[c] Reported in Ref. 72.
[d] Reference 95.
[e] Tabulated in Ref. 95.
[f] Based on RHF calculation (see footnote h of Table 4). The force constant pertains to the double bonds.
[g] Based on calculated equilibrium geometry given in Ref. 96a.
[h] Calculated for C_1C_2 stretching, with the $\angle C_3C_4C_5$ angle frozen at its equilibrium angle (Ref. 38 and footnote *v* of Table 4).

bicyclo[1.1.1]pentane (XII) leads to a calculated bending force constant about three times that in propane, in accord with experimental determinations.

An important probe of electronic structure is provided by photoionization spectra. Basch *et al.*[15a] have shown that appropriately scaled 1-electron SCF energies (ε) give remarkable agreement with vertical ionization energies (IPs) and thus prove extremely useful in assigning electron levels. Table 8 exhibits results for bicyclobutane (C_{2v} geometry).[32] Typical discrepancies between -0.92ε and experimental IPs[104] are ≤ 0.5 eV, suggesting that this approach will be as useful for larger polycyclic systems as it has been for simpler ring systems.

An unusual feature of strained hydrocarbons is that even the totally saturated ones often have appreciable dipole moments. Table 9 gives typical results and illustrates the generally excellent quantitative agreement between theory and experiment. The 4-31G level appears to be the most reliable of the three theories represented. Dipole moment polarities are indicated in the structural diagrams. The same direction is predicted for cyclopropene $(+\boxed{\triangleright}-)$ by all of the theoretical calculations, and this prediction is strengthened by applying additivity relations in comparing cyclopropene with its 1-methyl derivative.[19] Interestingly, the opposite polarity has been inferred from experiment.[105]† Inclusion of carbon $3d$ orbitals (6-31G*) has been shown to have virtually no effect on the calculated dipole moment.[8] The dipole moment directions predicted (4-31G) for bicyclo[1.1.0]butane and bicyclo[2.1.0]pentane have been confirmed by experiment (see Table 9).

Even though atomic hybridization is not a directly observable property, it has traditionally played a pivotal conceptual role in theories of bond strain.[52,53] Accordingly, we display the atomic hybrids and related properties obtained from localized *ab initio* MOs for a representative set of saturated strained ring systems in Table 10. Large departures from the hybridization of ethane are observed, generally corresponding to $2s$-rich CH bonds and $2p$-rich CC bonds, including a case of virtually pure $2p$ hybrids. Decreasing fractional s-character (f_s), defined as

$$f_s = 1/(1+\lambda) \qquad (2)$$

for the hybrid $h = 2s + \lambda^{1/2}2p$, is associated with increasingly large deviations of hybrid directions from interatomic vectors. Table 11 demonstrates the essentially monotonic relationship between s character and directly bonded CC and CH spin–spin coupling constants. In fact good linear least-squares relations can be obtained, as expected from simple perturbation theory.[26,32,79] An important consequence of the above results is the prediction that J_{CC} values approach zero (or even become negative) in molecules like bicyclo[1.1.0]butane (V) or benzvalene (VII).[26,32,79] This prediction has recently been confirmed for V (see Table 11).

†See note 1 of Notes Added in Proof on page 275.

Table 8. *Comparison of Vertical Photoionization Energies (IP) and Scaled Orbital Energies for Bicyclo[1.1.0]butane Valence Electrons*

Orbital[a]	$-0.92\varepsilon^b$	IP[c]	Orbital[a]	$-0.92\varepsilon^b$	IP[c]
$7a_1$	8.67	9.14	$5a_1$	17.42	16.86
$1a_2$	10.91⎱		$2b_1$	19.62⎱	
$3b_1$	11.58⎰	11.23	$4a_1$	19.68⎰	18.91
$4b_2$	12.90	12.87	$2b_2$	23.31	22.1
$6a_1$	14.52	14.51	$3a_1$	30.82	—
$3b_2$	14.60	14.85			

[a] Classified according to irreducible representation in C_{2v} symmetry. b_1 and b_2 are, respectively, symmetric and antisymmetric in the plane of the bridgehead CH groups.
[b] The ε are the 4-31G values given in Ref. 26. Similar results, based on a scale factor of 0.9, were presented in Ref. 32.
[c] From Ref. 104.

Table 9. *Calculated and Experimental Dipole Moments (in Debyes)*

		Calculated			
Molecule[a]	Structure	STO-3G	4-31G or 6-31G	INDO	Experimental
n-Propane[b]		0.02[c]	0.06	(0.00)[d]	0.083
Methylcyclopropane[e]		—	0.11	—	0.139
Methylenecyclopropane[e]		—	0.35	—	0.402
Cyclopropene[b]	III	0.51[c]	0.54	0.92[c]	0.45
1-Methylcylopropene[e]		—	0.90	—	0.84
Cyclobutene[e]	IV	0.054[c]	0.07	0.095[c]	0.132
Bicyclo[1.1.0]butane[f]	V	0.66	0.88	1.19	0.68
Bicyclo[2.1.0]pentane[f]	VI	0.18	0.28	0.44	0.26
Dewar benzene[f]	IX	0.02	0.04	0.27	0.044[g]
Benzvalene[f]	VII	0.74	0.95	1.36	0.88
o-Benzyne[h]	XVIII	0.87	1.78	0.75	—

[a] The footnotes in this column give reference to all pertinent calculated and experimental data, except as noted.
[b] Reference 96b.
[c] Unpublished data, based on experimental geometries.
[d] Based on all tetrahedral bond angles; J. A. Pople and M. Gordon, Molecular orbital theory of the electronic structure of organic compounds. I. Substituent effects and dipole moments, *J. Am. Chem. Soc.* **89**, 4253–4261 (1967).
[e] Reference 19.
[f] Reference 32. The predicted directions of the dipole moments for V and VI are in agreement with subsequently obtained experimental data: Ref. 111 and M. D. Harmony, C. S. Wang, K. B. Wiberg, and K. C. Bishop III, Microwave spectra of *endo*- and *exo*-methylbicyclo[2.1.0]pentane. Methyl group polarity and the sign of the dipole moment in bicyclo[2.1.0]pentane, *J. Chem. Phys.* **63**, 3312–3316 (1975).
[g] D. W. T. Griffith and J. E. Kent, Microwave spectrum and dipole moment of dewar benzene, *Chem. Phys. Lett.* **25**, 290–292 (1974).
[h] Unpublished data based on calculated (STO-3G) geometry (Ref. 38).

Table 10. Hybrid Atomic Orbitals from Localized Molecular Orbitals: $LMO = \alpha h_A + \beta h_B$; $h(carbon) = 2s + \lambda^{1/2}p$; $h(hydrogen) = 1s$

Molecule[a]	Bond AB	Hybridization		Angle of deviation[b]		Polarity	
		λ_A	λ_B	θ_A	θ_B	α	β
Ethane	C_1C_2	2.11	2.11	0.0	0.0	0.554	0.554
	C_1H_1	2.52	—	1.6	—	0.543	0.559
Cyclobutane[c] (II)	C_1C_2	2.44	2.44	10.7	10.7	0.558	0.558
	C_1H (endo)	2.54	—	1.2	—	0.565	0.534
	C_1H (exo)	2.35	—	2.1	—	0.564	0.533
Cyclopropane (I)	C_1C_2	3.38	3.38	27.5	27.5	0.571	0.571
	C_1H_1	1.86	—	1.0	—	0.563	0.528
Bicyclo[1.1.0]butane[d] (V)	C_1C_2	2.08	3.36	34.8	27.8	0.581	0.551
	C_1C_3	14.7	14.7	32.7	32.7	0.592	0.551
	C_1H_1	1.37	—	3.1	—	0.576	0.513
	C_2H_2 (endo)	1.90	—	1.4	—	0.571	0.525
	C_2H_2 (exo)	2.00	—	0.8	—	0.573	0.522
Bicyclo[2.1.0]pentane[e] (VI)	C_1C_2	1.82	2.48	12.0	10.0	0.562	0.548
	C_1C_4	4.54	4.54	30.2	30.2	0.576	0.576
	C_1C_5	3.15	3.39	30.3	27.7	0.568	0.567
	C_1H_1	1.67	—	0.6	—	0.570	0.524
	C_5H_5 (endo)	1.94	—	1.4	—	0.569	0.526
	C_5H_5 (exo)	2.06	—	0.7	—	0.571	0.524

[a] Calculations are based on experimental geometries, except as noted.
[b] From AB vector.
[c] Based on calculated (4-31G) dihedral and CH bond angles (see footnote c of Tables 4 and 15).
[d] Based on calculated (STO-3G) methylene bond angles (see Table 15).
[e] Based on calculated (STO-3G) CH bond angles (see Tables 5 and 15).

Table 11. Fractional s Character and Nuclear Spin–Spin Coupling Constants in Some Strained Systems

Molecule[a]	Structure	Fractional s character[b]	$^1J_{CH}$, Hz[c]
CH bonds			
Ethane		0.28	125
Cyclobutane			
CH (endo) ⎫	II	⎰0.28⎱	136[d]
CH (exo) ⎭		⎱0.30⎰	
Cyclopropane	I	0.35	162
Bicyclobutane			
C_2H (exo) ⎫	V	⎰0.33⎱	153
C_2H (endo) ⎭		⎱0.35⎰	169
Bicyclopentane			
C_1H_1	VI	0.37	178
Bicyclobutane			
C_1H_1	V	0.42	205

Molecule[a]	Structure	Product of fractional s characters[b]	$^1J_{CC}$, Hz[c]
CC bonds			
Cyclopropane			
C_1C_2	I	0.052	10
Bicyclopentane			
C_1C_5	VI	0.055	16
Bicyclobutane			
C_1C_2	V	0.0074	21
Cyclobutane			
C_1C_2	II	0.085	30
Bicyclopentane			
C_1C_2	VI	0.102	37
Ethane			
C_1C_2		0.103	35

[a] Based on geometries documented in Table 10.
[b] Defined in Eq. (2).
[c] The experimental literature is cited in Refs. 26, 32, and 79. The small *negative* value predicted for the central bond of V in the above references, has been confirmed by M. Pomerantz, R. Fink, and G. A. Gray, The sign of the bridgehead–bridgehead ^{13}C–^{13}C coupling constant in a bicyclobutane, *J. Am. Chem. Soc.* **98**, 291–292 (1976).
[d] Only the average of the *endo* and *exo* constants was observed.

It should be noted that atomic hybrids of the type defined above are somewhat basis-set or method-dependent.* These variations, however, do not usually affect qualitative correlations based on the hybrids.[77] †

The data presented in this section document the claim that currently feasible *ab initio* techniques give a remarkably consistent and generally quite

---

*Thus in general the only meaningful comparisons are between hybrids obtained by the same methods and for the same basis set.
†The reader is reminded that hybrids discussed here are not constrained to be orthogonal on a given atom.

accurate account of many important molecular properties. The techniques appear to be as reliable for large strained systems as they are for simpler unstrained systems, with the proviso that care must be used in comparing the energies of highly strained molecules, as discussed above in reference to Table 3.

4. Discussion of Ab Initio Results

Having demonstrated the ability of *ab initio* molecular orbital theory to give an adequate account of the structural and energetic aspects of strained organic molecules, we now consider the various ways in which the theory gives insight into the nature of strained bonding in more detail.

4.1. Distorted Methane as a Model for Strained Hydrocarbons

If the strain in saturated hydrocarbons is associated with departures from tetrahedral geometry, then suitably distorted methane molecules may be helpful in accounting for part of the strain energy, since the CH_4 geometry can be made to conform to those geometries which arise in strained situations.[106] Such models turn out to be of little use for three-membered rings, since the nonbonded contact between the two hydrogens forming a 60° HCH angle is as short as the CH bond itself. (For related reasons, the molecular mechanics approach is not very useful for three-membered rings.[73]) Distorted methanes provide information regarding the intrinsic energy of hybrid angle deformations and the so-called 1–3 nonbonded interactions. Since carbon hybrid orbitals tend to retain tetrahedral angles in spite of strained atomic geometries (Table 10), nonbonded repulsions are thought to be the more important energetic factor.[107] Methane, of course, involves only H···H 1–3 interactions. However, they should be similar in magnitude to the corresponding C···C 1–3 interactions in alkanes.[107] Hence, distorted methanes can yield rough estimates of the contributions which these interactions make to the total strain energy of saturated hydrocarbons.* In Table 12 we have collected some data pertinent to various fused four-membered ring systems. The quantities in the first three columns are the calculated energies of an appropriate number of distorted methane molecules relative to the energy of the same number of tetrahedral methanes. The CH bond angles of the distorted methanes are set equal to the strained framework angles in the molecule of interest, with any additional bond angles determined by energy minimization. The two 1–3 C···C interactions in cyclobutane(II) are simulated by two C_{2v} methanes, while four C_{3v} methanes are used to represent the twelve 1–3 framework interactions in

*We are indebted to Professor Wiberg for helpful comments on these points.

Table 12. *Strain Energy Data for Some Fused Four-Membered Ring Systems*

Molecule		Partial strain energy from *ab initio* distorted methane calculations[a]			Total strain energy from empirical or semiempirical data
		4-31G SCF	6-31G* SCF	6-31G* CI	
Cyclobutane[b] (II)		18.5 (14.1)	17.8 (13.6)	16.4 (12.6)	27[c]
Bicyclo[2.2.0]hexane[d]		36.0	—	—	(52)[e]
[2.2.2]Propellane[f] (XIII)		57.3	53.4	48.7	(73)[g]
Cubane[h]		96.5	93.0	86.5	155[i]

[a] With the exception of the bicyclo[2.2.0]butane bridgehead atoms, the nonbonded 1–3 C⋅⋅⋅C interactions are simulated by C_{2v} or C_{3v} methane with appropriate HCH angles (equal to the corresponding CCC angles) and optimized CH bond lengths. The extra HCH angle in C_{2v} methane was also optimized. All the C_{2v} and C_{3v} methane data were taken from Ref. 106b.

[b] Based on two C_{2v} methanes (i.e., methylene tilting is ignored) and a framework dihedral angle of 35° (ω in Fig. 5); i.e., $\angle CCC = 87.3°$. The values in parentheses are for a planar framework.

[c] Table 1.

[d] Based on two C_s methanes assigned the electron diffraction CCC bridgehead angles (the additional bond angle was optimized by energy minimization) of B. Andersen and R. Srinivasan, An electron diffraction investigation of the molecular structure of bicyclo[2.2.0] hexane in the vapour phase, *Acta Chem. Scand.* **26**, 3468–3474 (1972), with CC bond lengths of 1.071 Å, 1.081 Å, and 1.091 Å, corresponding, respoectively, to the CH, CC side bonds, and CC central bond.

[e] Based on the experimental value for the [3.1.0] isomer and the calculated (MINDO/1) difference between the [2.2.0] and [3.1.0] isomers (Ref. 100). A similar value (51 kcal/mole) is reported in Ref. 73.

[f] Based on two C_{3v} methanes (CCC angles of 90° and 120°).

[g] Obtained from the empirical calculations reported in Ref. 73.

[h] Based on four C_{3v} methanes ($\angle CCC = 90°$).

[i] Based on the data of Table 1; the ΔH_f value was taken from Kybett *et al.* (footnote h, Table 1). A somewhat larger value (166 kcal/mole) is reported in Ref. 73.

cubane. The 1–3 C⋅⋅⋅C interactions in bicyclo[2.2.0]hexane and [2.2.2]propellane(XIII) are represented, respectively, by two C_s and two C_{3v} methanes corresponding to the bridgehead geometries. For cyclobutane, the numbers in parentheses refer to a square planar geometry, for comparison with the equilibrium structure which has CCC angles of 87.3°. The last column of Table 12 contains the total strain energies estimated from various empirical or semiempirical data. The calculated distortion energies are seen to decrease significantly as the quality of the calculation improves, with polarization functions and CI making comparable contributions. At the 6-31G* CI level, the models based on distorted methanes account for ~60±5% of the total strain energies. Most of the remainder of the strain energy arises from 1–4 interactions such as H···H eclipsing, and from other long-range interactions. At the 4-31G level (the only level available for bicyclo[2.2.0]hexane) the ratio of *ab initio* distortion energies for cyclobutane and bicyclo[2.2.0]hexane is the same as the empirical ratio (i.e., ~1:2). Finally, we note that the ratio of the calculated distortion energies for cubane and cyclobutane (based on a 90° bond angle) is greater than 6:1 (i.e., 6.85 ± 0.01), indicating a significant positive nonadditivity of 1–3 C···C repulsions.

Most of the data of Table 12 was obtained by Palalikit, Hariharan, and Shavitt in a systematic study of various distorted methanes.[106b] Related studies primarily at the STO-3G level have been reported by Wiberg and Ellison.[106a]

4.2. Cyclopropane and Cyclobutane

The last few years have seen the growth of an extensive literature on *ab initio* studies of the electronic structure of cyclopropane,[8,11–17] cyclobutane[19,20,43] and related heterocycles and unsaturated analogs.[8,11,14–16,18] Among the many studies of cyclopropane we note especially: (1) the "double-zeta" study of the electron density by Kochanski and Lehn,[14] which indicated appreciably bent bonds and density depletion at the center of the ring; (2) the studies of electron-density *differences* maps obtained by Stevens *et al.*[17b] (through the use of spherically averaged unbonded reference atoms), which confirmed and accentuated the effects inferred from the density itself; and (3) the analysis of optical and photoionization spectra by Basch *et al.*[15a] in terms of molecular orbitals. Total and difference densities [based on spherically averaged $C(^3P)$ atoms] at the 4-31G level are illustrated in Figs. 2 and 3. Note that the density units used here are $e/\text{a.u.}^3$ instead of the $e/\text{Å}^3$ units used in some of the literature. The degree of bond bending inferred from maxima in the difference density (~24°) is similar to earlier estimates based on theory ($22°^{[52]}$ and $28°^{[17b]}$) and diffraction studies on various derivatives and isoelectronic analogs of cyclopropane ($20°,^{[54]}$ $22°,^{[55]}$ and $24°^{[57]}$). Even though the use of difference densities implies a somewhat arbitrary choice of reference atoms, it still appears to provide the most revealing link with experiment, provided that reference atoms are consistently chosen.[17b,51] Recent studies[108] suggest that, in general, a basis set level of "double zeta" or better should be used in generating quantitatively (or even qualitatively) useful difference densities. The CC bonding hybrids for cyclopropane presented in Table 10 are directed outward by 28°.

Scaled 1-electron energies (cf. Table 8) for cyclopropane obtained from SCF double-zeta calculations[11,15] are presented in Table 13, along with the vertical ionization energies[15a] and calculated CC overlap populations.[11] The correlation of calculated and measured energies is excellent, provided that the lowest-energy experimental doublet is identified with Jahn–Teller splitting of the ionic $^2E'$ state. The familiar 3-center CC bonding orbitals of Walsh[53] are most closely identified with the $2a_1'$ and $3\varepsilon'$ orbitals. Although the highest lying $3\varepsilon'$ level is found to be composed primarily of tangential $2p$ orbitals as expected, the $2a_1'$ level has very little carbon $2p$ character, whereas Walsh had suggested an orbital composed of sp^2 hybrids.[53] Furthermore, other orbitals are seen to have appreciable CC bonding character. The main success of Walsh's model, then, appears to be the prediction of the $2p$ character of the highest occupied level ($3\varepsilon'$). As has been pointed out many times, in terms of the total wave function there is no difference in principle between the models of Walsh[53] (delocalized MOs) and Coulson–Moffitt (localized MOs; see the appendix of Ref. 52), although the Walsh model is often more convenient for amalyzing electronic effects.

A Walsh-type analysis of the delocalized MOs in cyclobutane has been discussed by Wright and Salem[43] and also by Hoffmann and Davidson.[78]

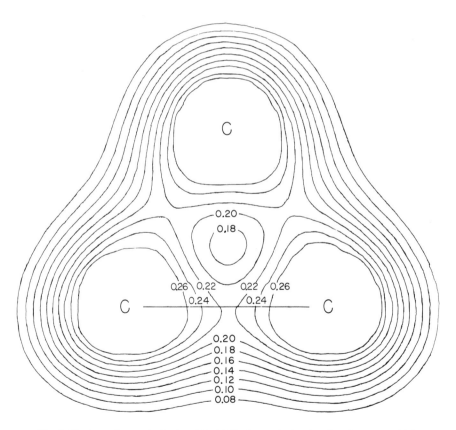

Fig. 2. Contours for total calculated (4-31G) electronic density of cyclopropane (I) in carbon atom plane (e/a.u.^{3}).

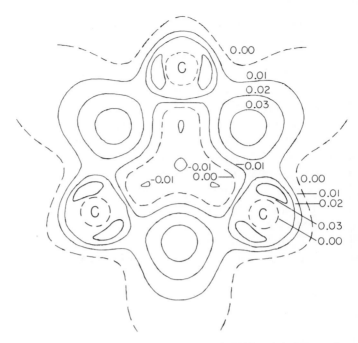

Fig. 3. Calculated (4-31G) difference density for cyclopropane in carbon atom plane (e/a.u.³): molecular density minus superposed atomic densities [C(³P), spherically averaged, and H(²S)]. The zero contour is represented by the dashed line.

Once again a degenerate highest-lying occupied level is expected to involve carbon–carbon bonding via tangential $2p$ orbitals. Calculated (scaled) one-electron energies are compared with experiment in Table 14, and good quantitative correlation is obtained, subject to the assumption of Jahn–Teller

Table 13. Valence Molecular Orbitals of Cyclopropane

| Symmetry type (D_{3h}) | Ionization energy, eV | | Bonding characteristics[a] | |
	-0.92ε [a]	Experiment[b]	Principal bonding	CC overlap population
$3\varepsilon'$	10.5	10.5, 11.3	CC($2p$)	0.115
$1\varepsilon''$	12.8	13.2	CH	−0.163
$3a_1'$	15.6	15.7	CC, CH	0.113
$1a_2''$	16.8	16.5	CC, CH	0.097
$2\varepsilon'$	20.4	—	CH	−0.115
$2a_1'$	28.5	—	CC ($2s$)	0.267

[a] Theoretical quantities (double-zeta level) taken from Ref. 11. The orbital energies ε were scaled as suggested in Ref. 15.
[b] Reference 15.

Table 14. Valence Molecular Orbitals of Cyclobutane

Symmetry type		Ionization energy, eV		
D_{2d}	D_{4h}	Calculated[a,b]	Experiment[c]	Principal bonding
4ε	$3\varepsilon_u$	10.8 (11.1)	10.7, 11.3	CC (2p)
$4a_1$	$1b_{1u}$	11.3 (11.6)	11.7	$CH_2 (\pi)$
$1b_1$	$1b_{1g}$	12.0 (12.7)	12.5	CC
3ε	$1\varepsilon_g$	13.4 (14.1)	13.4, 13.6	$CH_2 (\pi)$
$3a_1$	$3a_{1g}$	16.0 (16.7) ⎫		$CH_2 (\sigma)$
$3b_2$	$1a_{2u}$	16.3 (17.3) ⎬	15.9	$CH_2 (\pi)$
$2b_2$	$2b_{2g}$	18.8 (19.5)	— ⎫	
2ε	$2\varepsilon_u$	22.3 (23.7)	— ⎬	CC, CH
$2a_1$	$2a_{1g}$	28.2 (29.5)	—	CC (2s)

[a] The numbers in the third column are scaled orbital energies (-0.92ε) from a 4-31G calculation based on the geometry given in footnote c of Table 4.
[b] Numbers in parentheses are the negative orbital energies $(-\varepsilon)$ of Ref. 43, based on D_{4h} geometry.
[c] Data cited in Ref. 43.

splitting of the states arising from 3ε and 4ε ionization, and also with the assumption that the $3a_1$ and $3b_2$ ionizations are not experimentally resolved. Although the 4ε level involves primarily carbon $2p$ orbitals, there is appreciable CH as well as CC bonding. The other Walsh-type orbitals are roughly identifiable with the $1b_1$ and $2a_1$ levels. The degree of bending in the cyclobutane bonds is much smaller than that in cyclopropane; e.g., the hybrids of Table 11 imply ~11°, correspondingly, the HCH angle is much closer to tetrahedral form.

We have previously emphasized the large degree of ring puckering in cyclobutane (angle ω in the profile displayed in Fig. 4), an effect which *ab initio* calculations show as arising only if rocking of the methylene groups is allowed.[20,43] The view looking along the C_1C_4 bond in Fig. 5 clearly shows that *endo* rocking of the methyenes (i.e., in the sense of the *endo* hydrogens H_1' and H_2' moving towards the opposite carbon atoms, C_3 and C_4, respectively) relieves eclipsing of adjacent methylene groups. Calculated equilibrium rocking angles, $(\beta - \alpha)/2$ from Fig. 4, are shown in Table 15 to be in good agreement with experiment.

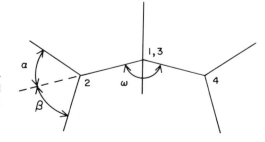

Fig. 4. Profile of the cylobutane (II), showing the framework dihedral angle (ω) and the orientation of the methylene groups. Dotted line denotes extension of framework plane.

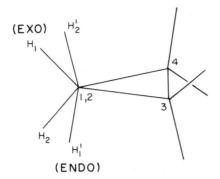

Fig. 5. View of cyclobutane (II) along C_1C_2 bond (C_2 in foreground) showing near-eclipsing of C_1 and C_2 methylene protons. The primed and unprimed protons are, respectively, *endo* and *exo* with respect to the puckered framework. The *endo*, *exo* labels refer to the C_1 hydrogens.

Table 15. Methylene Rocking Angles[a]

Molecule	Structure	Tilt angle, degrees	
		Calculated	Experimental
Cyclobutane	II	$+2.4,^b$ $+4.3,^c$ $+7.0^d$	$+4^e$
Bicyclo(1.1.0)butane	V	-2.2^f	$-1.4,^g$ $+0.7^h$
Bicyclo(2.1.0)pentane (at C_5)	VI	-2.8^i	-2.7^j

[a] Defined as angle between bisector of HCH angle and the plane of the methylene carbon atom and its two neighboring carbon atoms. In Figs. 4 and 9a the tilt angle is $(\beta - \alpha)/2$; i.e., a positive angle implies tilting in the *endo* sense.
[b] STO basis (Ref. 43).
[c] 4-31G basis (unpublished). Other geometrical parameters are given in footnote c of Table 4.
[d] FSGO basis (Ref. 20).
[e] Reference 103.   [f] Reference 19.   [g] Reference 113.   [h] Reference 114.
[i] Obtained with STO-3G basis, using framework geometry referred to in footnote p, Table 4.
[j] S. N. Mathur, M. D. Harmony, and R. D. Suenram, Microwave spectra of deuterated forms of bicyclo[2.1.0]pentane and the complete molecular structure, *J. Chem. Phys.* **64**, 4340–4344 (1976).

4.3. Fused 3- and 4-Membered Ring Systems and the Nature of Bonding between Bridgehead Carbon Atoms

4.3.1. Bicyclo[1.1.0]butane

Experimental properties of bicyclo[1.1.0]butane (V), such as the relative acidity of its bridgehead protons and the apparent ability of the C_1C_3 bond to allow conjugation between bridgehead substituents such as phenyl groups,[109] have suggested ethylenic or acetylenic character in the central bond (C_1C_3).[110] *Ab initio* calculations provide a simpler model in terms of an essentially pure $2p$–$2p$ bonding between C_1 and C_3, with an appreciable π component, and correspondingly high $2s$ character in the bridgehead CH bonds.[25,26] Schulman and Fisanick suggested this model on the basis of the overlap population obtained from the total density.[25] It is also of interest to note that the highest filled delocalized MO ($7a_1$ in Table 8) is in fact primarily confined to C_1 and C_3,

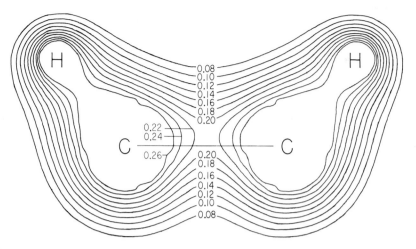

Fig. 6. Total calculated (4-31G) electronic density for bicyclo[1.1.0]butane (V) in the plane of the bridgehead CH groups (e/a.u.^{3}).

and has very little $2s$ character. The 4-31G total electronic density in the $H_1C_1C_3H_3$ plane is given in Fig. 6, showing a degree of bent character in the C_1C_3 bond similar to that for cyclopropane (Fig. 2). Other types of density maps give a more striking picture. Figure 7 displays the density of two electrons in the LMO, obtained from a minimal basis set calculation.[26] The $2p$ nature of

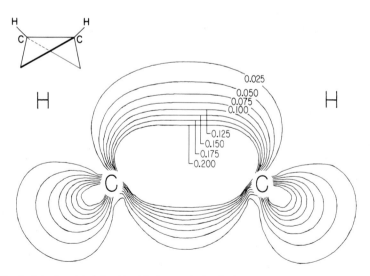

Fig. 7. Calculated (minimal basis) electronic density for two electrons in the localized molecular orbital for the C_1C_3 bond in bicyclo[1.1.0]butane (V) (e/a.u.^{3}). (Same plane as Fig. 6.) [Reproduced by permission from *J. Amer. Chem. Soc.* **94**, 767 (1972).]

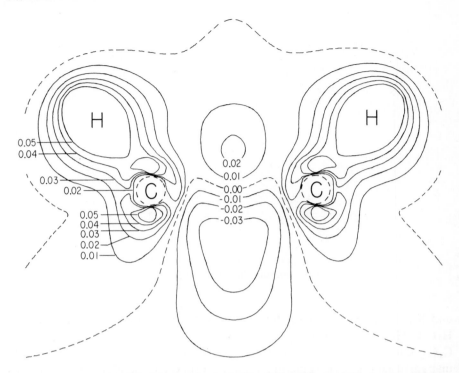

Fig. 8. Calculated (4-31G) difference density for bicyclo[1.1.0]butane (V) in the plane of the bridgehead CH groups (e/a.u.^{3}). (Density defined as in Fig. 3.)

the bonding hybrids is clear and corresponds to ~26% $2p\pi$ character, as obtained by squaring the C_π coefficient in the following normalized hybrid:

$$h_{C_1} = C_\pi(2p\pi) + C_\sigma(2p\sigma) \tag{3}$$

The STO-3G C_1 hybrid given in Table 11 yields a similar value (~29%). The $2p$ character of the bonding implies the existence of significant underside lobes. Although not prominent in the total density map (Fig. 6), diffuse underside lobes are clearly revealed in terms of the difference density (4-31G) of Fig. 8. Furthermore, the angle made by these "difference" lobes with the C_1C_3 bond (~130°) is the same as the optimal approach angle for a proton, as determined by other *ab initio* calculations.[23] The existence of appreciable underside electron density is consistent with the observed molecular Zeeman effect[111]* and also with underside attack of reactants, which has been inferred from product stereochemistry in the case of reaction with *o*-benzyne.[112] The accessibility of the underside region of the bridgehead–bridgehead bond is

---

*Reference 111 also lists second moments of the charge density, which are reproduced by a double-zeta level wave function to within 1%.[11]

groups (i.e., $\alpha > \beta$). This prediction is in agreement with conclusions based on NMR studies[113] and differs from the microwave result.[114] In both experiments, however, the uncertainty is of the order of the effect itself. Further support for rocking in the *exo* direction is given by the fact that atomic hybridizations (for an STO-3G basis) based on the microwave rocking angle (*endo*)[114] are reversed from the order given in Table 11 and, hence, do not correlate as expected with the known magnitudes of CH coupling constants for CH_{endo} and CH_{exo} (cf. the discussion with regard to Table 11). As in the case of cyclobutane, the predicted methylene rocking can be rationalized in terms of relief of eclipsing between vicinal CH bonds.

Calcuations at the 4-31G level corresponding to the C_1C_3 (D_{2h} planar framework) and C_1C_2 diradical triplet states of V suggest that such states may be quite low lying, with energies $\sim 1-2$ eV above the C_{2v} $^1A_1$ ground state.[26] The activation energies for their formation are not known. These diradicals do not appear to be involved in the thermal rearrangement of V to butadiene, which has an activation energy of 41 kcal/mole.[115] However, *endo-exo* isomerization, with a much smaller activation energy (26 ± 2 kcal/mole), has been observed for a 1,3-diphenyl derivative, thus implicating breakage of the C_1C_3 bond.[116] Even if a diradical were not involved, the energetic cost of the expected increase in bridgehead angle γ (Fig. 9b) in the transition state would be at least partially offset by conjugation effects, as indicated by model calculations on V which show, for example, that an increase of γ (at both C_1 and C_3) by 15° requires only ~ 6 kcal/mole and is attended by a 38% increase in the $2p\pi-2p\pi$ overlap population in the C_1C_3 bond.[26]

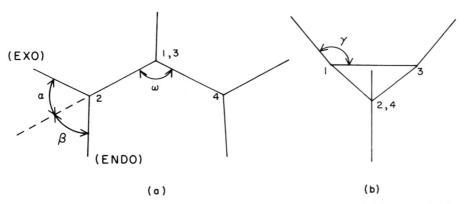

(a)   (b)

Fig. 9(a) Profile of bicyclo[1.1.0]butane (V), showing the dihedral angle (ω) between the 3-membered rings, and the orientation of the methylene group.* Dotted line denotes extension of framework plane. (b) View of the plane containing the bridgehead CH groups in bicyclo[1.1.0]butane (V), showing large HC_1C_3 angles (γ).

*See Table 15.

4.3.2. Other Related Systems

The features of the bridgehead CH groups just outlined for bicyclo[1.1.0]butane (V) also show up in varying degrees in a number of related systems, including (1) bicyclo[2.1.0]pentane (VI), which has a somewhat less bent and less $2p$-rich bridgehead–bridgehead bond (Table 11) but is also thought to have $2p$ lobes facilitating underside attack by electrophilic reactants[23,112c]; and (2) benzvalene (VII), which contains the bicyclobutane moiety, and hence some common properties, especially the predicted small (negative) spin–spin coupling constant for the C_1C_6 bond.[32,79] The *ab initio* electronic structure of benzvalene and the other familiar benzene valence isomers, prismane (VIII) and Dewar benzene (IX), has been analyzed in some detail.[24a,30–32] Examination of the sequence

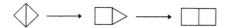

in terms of the *ab initio* hybrid orbital analysis given by Wipff[24a] reveals that it is the presence of the 3-membered ring which is crucial for the very high $2p$ character of the bridgehead–bridgehead bond. Thus, in this respect the central bond in bicyclo[2.2.0]hexane (⊓), and also in cubane, is not very different from the CC bond in cyclobutane itself. Wiberg and Ellison have noted another striking property of the above sequence of three molecules, which once again points to the unique nature of the cyclopropane ring: namely, that the relative reactivities toward electrophilic reagents drop by factors of $\sim 10^6$ as one proceeds to the right.[106]

Turning now from the localized hybrid orbital model of bridgehead bonding, we note that a rough correlation exists between the ionization energies for certain high-lying occupied molecular orbitals which principally involve the interaction of $2p$ orbitals on adjacent singly bonded carbon atoms, and the degree of σ and π character in this interaction. Thus, in Table 16, one proceeds from the rather low predicted IPs for V, VI, and VII, where the appreciable $2p\pi$–$2p\pi$ component in the MOs of interest would lead to relatively low $2p$–$2p$ overlap, to the intermediate case of Dewar benzene (IX), with primarily $2p\sigma$–$2p\sigma$ overlap, and finally to the case of ethane, where, by symmetry, the $2p$–$2p$ interaction in the $3a_{1g}$ orbital must be pure sigma. Small variations in CC bond lengths for the molecules shown would also have some effect on the degree of overlap.

4.3.3. Tetrahedrane

Tetrahedrane is one of the most fascinating of the small fused systems because of its high-symmetry, high expected strain energy (Table 1), and its elusive nature which has so far prevented its isolation in spite of indications of a

Table 16. *Trends in Ionization Energies Associated with Bridgehead–Bridgehead and Other Carbon–Carbon Bonds*

Molecule	Structure	Ionization energy (eV)[a]		Orbital	Bond	Nature of 2p–2p interactions
		Calculated	Experimental			
Bicyclo[1.1.0]butane	V	8.7	9.1	$7a_1\,(C_{2v})$	C_1C_3	Appreciable π character
Bicyclo[2.1.0]pentane	$12a'\,(C_s)$	9.2	—	$12a'\,(C_s)$	C_1C_4	
Benzvalene	VII	9.4	—	$10a_1\,(C_{2v})$	C_1C_6	
Dewar benzene	IX	10.7	—	$7a_1\,(C_{2v})$	C_1C_4	Little π character
Ethane		12.2	13	$3a_{1g}\,(D_{2d})$	CC	Pure σ

[a] All pertinent data are cited in Ref. 32. The calculated values are obtained as -0.92ε. In Ref. 32 the orbital energies ε were scaled by a slightly different factor (-0.9), which gave somewhat better agreement with experiment for ionization from the lowest-lying valence MOs (i.e., largest IPs).
[b] A. D. Baker, C. Baker, C. R. Brundle, and D. W. Turner, The electronic structures of methane, ethane, ethylene and formaldehyde studied by high-resolution molecular photoelectron spectroscopy, *Int. J. Mass Spectrosc. Ion Phys.* **1**, 285–295 (1968).

possible transitory existence.[58,59] Accurate *ab initio* calculations have been carried out by Buenker and Peyerimhoff[34] and by Schulman and Venanzi[35] to determine both the electronic structure of tetrahedrane and its stability with respect to various distortions. Schulman and Venanzi showed that the tetrahedral structure does indeed correspond to a local minimum in the 18-dimensional space for vibrational motion, and the equilibrium geometry ($R_{CC} = 1.482$ Å, $R_{CH} = 1.054$ Å), force constants (see Table 6), and vibrational frequencies appear to have physically reasonable magnitudes. Dipole moment derivatives with respect to the ir-active T_{2a} vibrational modes have magnitudes of ~0.1–1.0/Å, which suggest that the ir intensity should be observable if the molecule could be isolated.

The *ab initio* energy for the last reaction in Table 3 in conjunction with known thermochemical data leads to an estimate of ~124 kcal/mole for $\Delta H_f(0°)$, which turns out to be about the same as the strain energy (cf. Table 1).[35] The MINDO/3 value of $\Delta H_f(0°)$ is 128 kcal/mole.[101]

Since the concerted thermal conversion of tetrahedrane to the more stable cyclobutadiene or two acetylene molecules is symmetry forbidden,[117] most attention with respect to the stability of tetrahedrane has focused on the energy needed to break a single bond, leading to the bicyclobutadienyl biradical or biradicaloid (·◇·).[35,101] Limited *ab initio* CI calculations at an early point on the assumed reaction coordinate suggest that the barrier for this process would be >18 kcal/mole,[35] while MINDO/3 calculations lead to an estimate of ~11 kcal/mole for the barrier.[101]

The highest filled orbital of tetrahedrane (1ε) has the interesting (symmetry-imposed) property of containing only carbon $2p$ orbitals (those directed perpendicular to the radial direction from the center of the tetrahedron).[35]

4.4. Propellanes

The discussion so far has involved the nature of bridgehead methine (CH) groups. We now consider a geometrically more constrained situation—namely, a formally bonded pair of quaternary bridgehead carbon atoms whose other CC bonds are all involved in rings—i.e., the propellanes,[60] exemplified by XI, XIV, and XV.

4.4.1. [1.1.1]Propellane

Perhaps not surprisingly, the prototype propellane (XI), with its three fused cyclopropanes has never been experimentally detected. Nevertheless, theoretical studies of the interaction between its bridgehead atoms have been instructive in showing how one may attempt to assess the degree of bonding between a given pair of atoms.[27] Within the constraints of D_{3h} symmetry, the equilibrium geometry of XI is characterized by a normal side-bond length ($R_{12} = 1.534$ Å) and a rather long central internuclear distance ($R_{13} = 1.600$ Å), which nevertheless suggests the possibility of appreciable bridgehead–bridgehead bonding. Furthermore, the equilibrium D_{3h} triplet state ($^3A_2''$, $R_{13} = 1.80$ Å) is found to be >3 eV above the $^1A_1'$ state, indicating that the C_1C_3 diradical is not the D_{3h} ground state. As indicated in Table 1, the calculated strain energy is large, but still only ~20% of a normal CC bond energy on a per bond basis. Examination of the wave function gives the first strong indication that the C_1C_3 atoms are involved in an antibonding or nonbonded interaction. The total density (Fig. 10) along the C_1C_3 vector shows a *minimum* at the midpoint. More striking is the fact that the hybrids in the C_1C_3 LMO (Fig. 11) are directed *away* from each other. An interesting comparison is also provided by the corresponding *total* electron density for the related D_{3h} molecule bicyclo[1.1.1]pentane (XII), which shows a very similar minimum at the C_1C_3 midpoint (Fig. 12). Overlap populations for the two

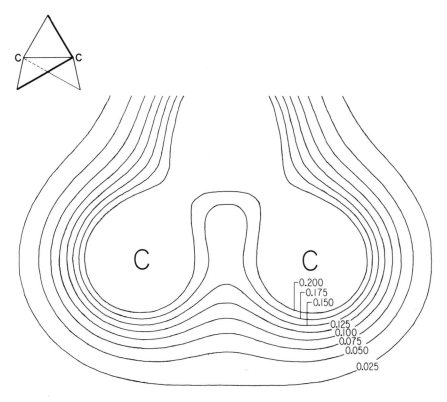

Fig. 10. Total calculated (4-31G) electronic density for [1.1.1]propellane (XI) in the plane of one of the 3-membered rings (e/a.u.³). [Reproduced by permission from *J. Amer. Chem. Soc.* **94**, 773 (1972).]

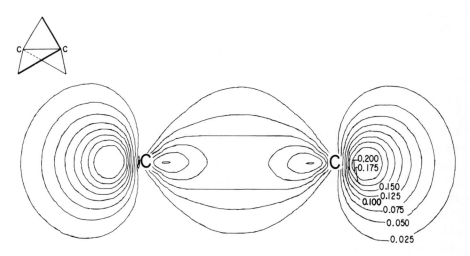

Fig. 11. Calculated (minimal basis) electronic density for two electrons in the localized molecular orbital for the C_1C_3 "bond" in [1.1.1]propellane (XI). (Same plane as Fig. 10.)

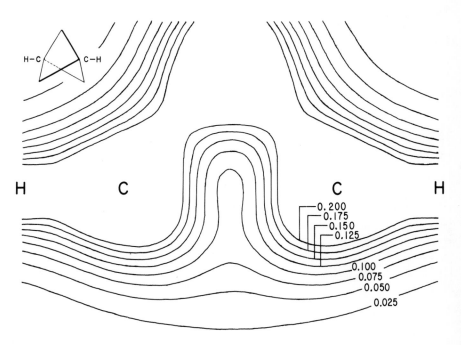

Fig. 12. Total calculated (4-31G) electronic density for bicyclo[1.1.1]pentane (XII) in the $C_1C_2C_3$ plane (e/a.u.^{3}). [Reproduced by permission from *J. Amer. Chem. Soc.* **94**, 773 (1972).]

Table 17. Comparison of Overlap Populations in Bicyclo[1.1.1]pentane and [1.1.1]Propellane[a]

		Overlap population	
Atom pair	Type of interaction	Bicyclo[1.1.1]pentane (XII)	[1.1.1]Propellane (XI)
C_1C_2	Side bond	0.494	0.346
C_1C_3	Nonbonded bridgehead pair	−0.908	−0.252
C_2C_4	Nonbonded methylene pair	−0.514	−0.285

[a] Reproduced by permission from *J. Am. Chem. Soc.* **94**, 773 (1972), Table II.

molecules are compared in Table 17. Both molecules have strongly negative C_1C_3 overlap populations, although the magnitude is much greater in bicyclo[1.1.1]pentane (XII), due primarily to an antisymmetric combination of bridgehead CH bonds in the highest filled MO.[21,27,28] The corresponding antibonding a_2'' MO in [1.1.1]propellane is unoccupied, thus helping to explain why the C_1C_3 distance is shorter [cf. 1.600 Å with the calculated value of 1.885 Å for $R_{C_1C_3}$ in (XII), which is in good agreement with a recent experimental value of 1.877 Å].[118] Table 17 also reveals significant nonbonded repulsion effects between methylene groups. It is clear that relief of this strong repulsion (in D_{3h} symmetry) would be at the expense of increasing the equally significant C_1C_3 repulsion. Given these geometrical constraints, it is not surprising that the $C_1C_2C_3$ bending mode in bicyclo[1.1.1]pentane is characterized by an unusually large force constant (Table 6).

4.4.2. [2.2.2]Propellane

Inspection of the trigonal geometry at the bridgehead carbons of XIII suggests that [2.2.2]propellane may be another polycyclic molecule characterized by $2p$–$2p$ bonding (purely σ-type in this case due to the D_{3h} symmetry) between the bridgeheads, in contrast to the case of only two fused 4-membered rings, where no unusual hybridization was found (*vide supra*).[24a] In addition, the antisymmetric combination of bridgehead $2p$ orbitals is expected to be stabilized relative to the symmetric combination due to through-bond coupling,[119] as noted by Stohrer and Hoffmann,[61] thus implying the occurrence of an orbital energy crossing for sufficiently large C_1C_4 separation. This in turn suggests the possibility of a double-well potential for stretching of the C_1C_4 bond, with an inner minimum (XIII), corresponding to a single-configuration wave function (denoted ψ_{S^2}) with a highest-filled orbital (ϕ_S) *symmetric* in the axial $2p$ orbitals on C_1 and C_4, and an outer minimum (XIV) with the analogous antisymmetric orbital (ϕ_A) occupied, leading to wave function ψ_{A^2}.[61] Note that both wave functions are totally symmetric $^1A_1'$

states. This possibility of bond-stretch isomerism, as the phenomenon was described by Stohrer and Hoffmann,[61] stimulated *ab initio* calculations by Newton and Schulman at the STO-3G level.[29] One objective of these calculations was to obtain an estimate of the magnitude of the intervening barrier, which is clearly of importance in estimating the stability of XIII. Stohrer and Hoffmann noted that if [2.2.2]propellane did attain the isomeric form (XIV), it would be expected to rearrange to dimethylene cyclohexane in a thermally allowed reaction.[61] Thus [2.2.2]propellane could, most likely, only be isolated in the form of XIII, and then only if it were separated from XIV by an appreciable barrier. The transformation of XIII → XIV, as represented by $\psi_{S^2} \rightarrow \psi_{A^2}$, is formally symmetry forbidden.[61,117] However, since the outer minimum isomer (XIV) is expected to be a diradical, it can be anticipated that the true wave function for XIV would not be very well represented by ψ_{A^2} alone, but rather by an appropriate linear combination of ψ_{A^2} and ψ_{S^2}. This situation is analogous to the case of the simple two-electron hydrogen molecule in its ground state, where σ_g^2 and σ_u^2 both make appreciable contributions for large distances. Accordingly, *ab initio* 2-configuration wave functions of the form

$$\Psi = C_{S^2}\psi_{S^2} + C_{A^2}\psi_{A^2} \tag{4}$$

were used to obtain the potential energy profile for the C_1C_4 bond, with all other parameters kept constant. The MOs for the CI calculations were obtained from the SCF wave functions at each value of $R_{C_1C_4}$. The functions ψ_{S^2} and ψ_{A^2} appearing in Eq. (4) differ only in the occupation of ϕ_S and ϕ_A and have a common core for the remaining electrons. The resulting CI energies are displayed in Fig. 13, along with the SCF energies (for ψ_{S^2} and ψ_{A^2}) and the 1-electron energies (for ϕ_S and ϕ_A), both of which cross at ~2.35 Å. The final prediction is a normal inner equilibrium geometry ($R_{C_1C_4} = 1.54$ Å) separated by a barrier of ~29 kcal/mole from an outer minimum at $R_{C_1C_4} \sim 2.51$ Å. To the extent that ϕ_S and ϕ_A differ only in the phases of identical hybrids (on C_1 and C_4) with negligible overlap, a connection may be established with simple valence bond theory in terms of the ratio of the covalent and ionic character in the CI wave function

$$\psi = C_{\text{cov}}\psi_{\text{cov}} + C_{\text{ion}}\psi_{\text{ion}} \tag{5}$$

$$\frac{C_{\text{cov}}}{C_{\text{ion}}} = \frac{C_{S^2} - C_{A^2}}{C_{S^2} + C_{A^2}} \tag{6}$$

In this sense 97% covalent character can be attributed to the outer minimum (XIV). INDO calculations confirm the expected $2p$ character of the C_1C_4 bond hybrids ($\sim sp^{9.1}$).[29]

Subsequent to the above predictions, a derivative of XIII was synthesized and an activation energy of ~22 kcal/mole was inferred from its unimolecular

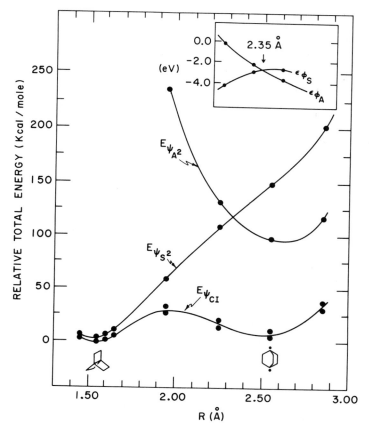

Fig. 13. Total SCF energies ($E_{\psi_{S_2}}$ and and $E_{\psi_{A_2}}$) and CI energies (E_{CI}) as a function of the central CC bond length in [2.2.2]propellane (XIII, XIV). Energies on the ordinate are in kcal/mole relative to the minimum CI energy. The inset shows the crossing of the orbital energies for ϕ_S and ϕ_A, obtained from the ψ_{S^2} and ψ_{A^2} solutions, respectively (errors in the original figure (38) have been corrected). The double entries for $E_{\psi_{CI}}$ ($R_{CC} > 1.9$ Å) show the similar results obtained by taking CI orbitals from either the ψ_{S^2} or ψ_{A^2} SCF wave functions. [Reproduced by permission from *J. Amer. Chem. Soc.* **94**, 4391 (1972).]

decay rate,[62] in satisfactory accord with the theoretical estimate of ~29 kcal/mole. The parent (XIII) has also been detected.[120,121] It is indicative of the current development of *ab initio* techniques and capabilities that a system as large as XIII (58 electrons) could be treated at a CI level (albeit only two configurations) and yield apparently accurate predictions. It is, of course, fortunate that the symmetry of XIII, as well as the localization of the effect in a single CC bond, greatly helped in reducing the scope of the computational problem. We have already alluded to other detailed CI studies of bond-breaking processes in strained rings,[35,39–45] and studies of this type can be expected to have a major impact on our knowledge of organic reaction mechanisms in the future.

4.5. Strained Conjugated Organic Molecules

In this final section of our discussion we mention briefly some properties of two important strained organic molecules: *o*-benzyne (XVIII) and cyclobutadiene (XIX and XX).

4.5.1. *o*-Benzyne

Most theoretical studies of *o*-benzyne have been based on the benzene hexagonal framework geometry (XVI). Even with this assumption it was possible to show convincingly that the ground state of *o*-benzyne is a singlet.[37] This was demonstrated through the use of both simple 2-configuration and more extensive CI wave functions. With regard to the electronic structure of *o*-benzyne, an important question is the relative importance of the resonance structures (XVIIIa) and (XVIIIb). Calculations employing the benzene geometry suggest an appreciable contribution from structure XVIIIa: i.e., the total overlap population for the C_4C_5 bond is larger than that for the adjacent C_3C_4 and C_5C_6 bonds (see Table 18 and Refs. 37 and 119), and the π-electron overlap populations alternate around the ring in the sense of XVIIIa (Table XX). A very different picture is provided by the equilibrium geometry (Fig. 14), which is characterized by a very short C_1C_2 "acetylenic" bond length (SCF, 1.22 Å; 2-configuration CI, 1.24 Å), a slight *extension* of the opposite bond ($R_{C_4C_5} = 1.42$ Å), and negligible departure of the other bond lengths from the benzene value. Recent MINDO/3 results[65] are in general accord with the *ab initio* equilibrium geometry. In view of the lack of any pronounced alternation in the equilibrium bond lengths corresponding to XVIIIb, it is perhaps surprising that the total overlap populations *do* alternate in this sense (Table 18), an effect due primarily to the π-electron contribution. The absence of lengthening in the bonds adjacent to the "acetylenic" bond (C_1C_2) can be explained in terms of an *ab initio* hybrid orbital analysis (Fig. 14)[38] which shows that the strong *s* character in the C_1C_6 and C_2C_3 bond hybrids roughly cancels the weakening due to a reduced π-electron population. The relationship between

Table 18. o-Benzyne Overlap Population (STO-3G)[a]

Bond	Benzene geometry[b]	Equilibrium geometry[b]
1–2	1.294 (0.218)	1.506 (0.324)
2–3	1.002 (0.222)	0.970 (0.178)
3–4	1.006 (0.214)	1.050 (0.250)
4–5	1.060 (0.228)	0.996 (0.184)
1–2 (benzene)	1.020 (0.220)	—

[a] π-Electron contribution given in parentheses.
[b] Calculated equilibrium geometry at the STO-3G level (see Refs. 38 and 95, and Fig. 14).

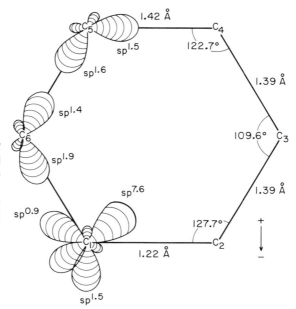

Fig. 14. Equilibrium structure of *o*-benzyne (XVIII): CC bond lengths and CCC bond angles (right-hand side) and CC σ-framework hybrid atomic orbitals (left-hand side). Dipole moment polarity is indicated. Geometry and hybrid directions are drawn approximately to scale. The notation $sp^{7.6}$ (equivalently, $s^{0.12}p^{0.88}$) implies 88% 2p character. [Reproduced by permission from *Chem. Phys. Lett.* **18**, 244 (1973).]

hybridization, bond lengths, and bond strengths has been discussed in detail by Bent.[122]

Since *o*-benzyne chooses to form a strong short C_1C_2 bond, a price in angle strain must be paid (see Fig. 14). Table 1 indicates that the resultant of σ-electron strain energy and π-electron resonance energy is +27 kcal/mole. This value actually implies considerable strain energy, since it undoubtedly contains a large *negative* resonance contribution, as inferred by inspection of the π-electron overlap populations in Table 18. The resonance energy can be expected to be an appreciable fraction of the value found for Kekulé benzene (37 kcal/mole,* Table 1).

Relative energies for various singlet and triplet states of *o*-benzyne are collected in Table 19. They are consistent with earlier results based on the benzene geometry[37] and show that at the SCF level the benzene-geometry triplet (XVII) lies slightly below the equilibrium singlet (XVIII). This effect disappears at the 2-configuration CI level, which yields a singlet ground state.

Since the matrix isolation of *o*-benzyne has recently been reported,[66] it should be noted that the vibrational frequency associated with the short C_1C_2 bond is expected (cf. Table 7) to be well above the frequency range examined in the experimental study,[66] and should be somewhere between ethylene and acetylene values, but closer to the latter. Furthermore, due to the bent nature of C_1C_2 bond (Fig. 14), there is an appreciable molecular dipole moment

*The so-called vertical resonance energy of Kekulé benzene is larger.[123]

Table 19. o-Benzyne Energies (STO-3G)

Geometry	State	Relative energy, eV
Equilibrium^a	$^1A_1$ (SCF)	0.00
	$^3B_2$ (SCF)^{b,c}	+0.56
	$^1A_1$ (2×2 CI)	−1.27
Benzene^a	$^1A_1$ (SCF)	+1.23
	$^3B_2$ (SCF)^{b,c}	−0.10
	$^1A_1$ (2×2 CI)	−0.49

^{a}See footnote b in Table 18.
^{b}UHF calculation.
^{c}The B_2 representation (C_{2v}) is antisymmetric in the plane perpendicular to the C_1C_2 bond.

($\sim$1 D) and also a sizable dipole moment derivative ($\sim$1 D/Å) for C_1C_2 stretching. Thus, this vibration should have observable ir intensity.*

4.5.2. Cyclobutadiene

Like the other strained $(CH)_4$ isomer, tetrahedrane, cyclobutadiene (IX, X) has been the focus of intensive experimental and theoretical effort.[67,68]† Although, in contrast to the case of tetrahedrane, much data are available for this species (both experimental[67,72] and theoretical[19,33,34,52,87,72,101,124] there is still little consensus as to the detailed nature of its ground state. A major cause for confusion is the generally agreed upon fact that the lowest singlet and triplet states lie quite close together, and one or the other may be the ground state, depending perhaps on the nature of the substituents. The parent C_4H_4 and three deuterated variants have been isolated in low-temperature matrices,[69–72] and the ir evidence‡ (including a 1236 cm^{-1} framework stretching frequency) and isotope distribution in the acetylene decomposition products[71] appear to be compatible only with a *square* equilibrium geometry. Since a square geometry is expected to imply a triplet state (*vide infra*), it is important to note that *no* ESR signal has been detected so far for the matrix-isolated species. Furthermore, magnetic and structural data for various derivatives with bulky alkyl groups are consistent with a closed-shell antiaromatic[125] $4n$ π-electron system,[126] although steric repulsion in these derivatives may cause appreciable distortion of the parent framework geometry. A very recent review of the available data for cyclobutadiene has been given by Maier.[68]

Before commenting on the current status of *ab initio* calculations for cyclobutadiene we summarize, in Table 20, the π-electron model for a square geometry. The π MOs are constructed from the four $2p\pi$ orbitals (χ_i) located

*See note 2 of Notes Added in Proof on page 275.
†Original suggestions that the molecule was too strained to exist were convincingly disputed by subsequent calculations.[52,67]
‡See note 3 of Notes Added in Proof on page 275.

Table 20. π-Electron MO Picture of Cyclobutadiene $(D_{4h})^a$

π-electron MOs^b	Reflection symmetry
$\phi_{a_{2u}} = N(\chi_1 + \chi_2 + \chi_3 + \chi_4)$	symmetric (S)
$\phi_{\varepsilon_g}^S = N(\chi_1 - \chi_2 - \chi_3 + \chi_4)$	symmetric (S)
$\phi_{\varepsilon_g}^A = N(\chi_1 + \chi_2 - \chi_3 - \chi_4)$	antisymmetric (A)
$\phi_{b_{2u}} = N(\chi_1 - \chi_2 + \chi_3 - \chi_4)$	antisymmetric (A)

Lowest electronic states^c

$$\psi^3 A_{2g} = |(\phi_{a_{2u}})(\bar{\phi}_{a_{2u}})(\phi_{\varepsilon_g}^S)(\phi_{\varepsilon_g}^A)|$$

$$\psi^1 B_{1g} = |(\phi_{a_{2u}})(\bar{\phi}_{a_{2u}})(\delta_{\varepsilon_g}^S)(\bar{\phi}_{\varepsilon_g}^S)| - |(\phi_{a_{2u}})(\bar{\phi}_{a_{2u}})(\phi_{\varepsilon_g}^A)(\bar{\phi}_{\varepsilon_g}^A)|$$

^{a}The dotted line represents the reflection plane bisecting the 1–4 and 2–3 bonds.
^{b}The χ are $2p\pi$ atomic orbitals; N is a normalization constant.
^{c}The above choice for ϕ_{ε_g} insures that $\psi^3 A_{2g}$ and $\psi^1 B_{1g}$ are properly converted into $\psi^3 B_{1g}$ and $\psi^1 A_{1g}$ under a rectangular (D_{2h}) distortion.

on the carbon atoms. The degenerate ε_g pair has been grouped into components either symmetric or antisymmetric in the reflection plane bisecting the C_1C_4 and C_2C_3 bonds. The $^3A_{2g}$ state is seen to involve a single configuration, while the $^1B_{1g}$ state requires an equal mixture of the $(\varepsilon_g^S)^2$ and $(\varepsilon_g^A)^2$ configurations.* We must now consider the likelihood that the square would be the equilibrium geometry for either the triplet or the singlet. The first excited triplet is the ^3E_u state, which requires that an electron be promoted from ε_g to b_{2u}, or from a_{2u} to ε_g. Perturbation theory tells us that a non-totally symmetric vibrational distortion may lead to a lower energy, provided this vibration can couple the ground state and a sufficiently low-lying excited state; i.e., the direct product of the irreducible representation of the two electronic states and the vibrational (distortion) mode must have a totally symmetric component.[128] In this sense the E_u vibrational mode (which in the following discussion can be taken as the trapezoidal distortion of the square) couples the $^3A_{2g}$ and ^3E_u states. However, it is not immediately obvious whether the ^3E_u state lies low enough (calculations[33] place it ~4 eV above the $^3A_{2g}$ state) to cause a lower symmetry equilibrium geometry (i.e., a C_{2v} trapezoid) or perhaps merely lead to an unusually low force constant (and hence frequency) for the E_u vibration of the square. In the case of the singlet ($^1B_{1g}$), the first excited state ($^1A_{1g}$) is very low lying since it arises from the same MO configuration, and the B_{1g}

*If rectangular distortions are introduced, the CI coefficients must be variationally determined. This is done here simultaneously with the self-consistent determination of the LCAO coefficients of the occupied MOs, using the procedure outlined in Ref. 127.

(rectangular) vibration should cause the $^1B_{1g}$ electronic state to distort to an equilibrium rectangular geometry for the ground state singlet. Bearing these expectations in mind, we now turn to the calculated results.

The first extensive *ab initio* calculations for cyclobutadiene were carried out by Buenker and Peyerimhoff,[33,34] using a minimal set of $2p$ atomic orbitals and different levels of CI for a small set of rectangular geometries. Their results suggested a nearly square equilibrium geometry for the triplet and a strongly distorted rectangular singlet. Their most elaborate CI wave function indicated that the singlet lies below the triplet for all values of R_{12} and R_{23}, with ~0.6 eV separating the square singlet and triplet. More recent calculations using a 4-31G basis and one- and two-configuration SCF wave functions (see Table 20) lead to the data given in Table 21.[72] As in earlier calculations,[33] they show the absolute ground state to be the $^1A_{1g}(D_{2h})$ singlet (at the 2-configuration level) with energy slightly less than that for the $^3A_{2g}$ state. However, the latter state is the ground state for square geometries. Although this $^3A_{2g}$ state in fact turns out to be stable under all possible distortions of the square, it is interesting (in light of the above discussion) to note that the calculated force constant for the E_u deformation of the carbon framework is indeed unusually small (<1 mdyn/Å). Even more striking is the fact that with the less flexible STO-3G basis, C_{2v} trapezoidal triplet geometries have energies below that for the equilibrium square. These triplet calculations employed the unrestricted Hartree–Fock formalism (UHF).[129] The safest inference from the calculated results for the E_u distortion is simply that the qualitative picture obtained at the 4-31G level is correct, i.e., the E_u force constant is probably rather small relative to those for the other framework stretching modes (cf. Table 7). The actual calculated magnitude and the qualitatively different picture obtained with the smaller STO-3G basis should perhaps serve as a warning that nonsymmetric stretching force constants in highly strained systems may be especially unreliable, particularly since the degree of strain changes in these nonsymmetric distortions. In addition, the quantitative results may to some extent be an artifact of the UHF framework, and a reexamination of the E_u distortions with the restricted Hartree–Fock (RHF) formalism[130] would be desirable.

Table 21. Calculated Energies of Cyclobutadiene (4-31G)

| | CC bond lengths, Å | | Relative energy, kcal/mole | |
| | | | $^3A_{2g}$ ($^3B_{1g}$)[a] (RHF SCF)[b] | $^1B_{1g}$ ($^1A_{1g}$)[a] (CI SCF)[b] |
Geometry	R_{12}	R_{23}		
$^3A_{2g}$ equilibrium (D_{4h})	1.433[c]	1.433[c]	0.0	4.8
$^1A_{1g}$–$^3B_{1g}$ crossing (D_{2h})	1.400	1.466	2.3	2.3
$^1A_{1g}$ equilibrium (D_{2h})	1.334	1.564	10.2	−2.9

[a] The irreducible representations refer to D_{4h} (D_{2h}) symmetry.
[b] See footnote h of Table 4.
[c] A similar equilibrium R value (1.463 Å) is found for the D_{4h} singlet ($^1B_{1g}$).

The rectangular singlet (at the 4-31G level) corresponds to reasonably normal single and double bonds in terms of bond lengths and force constants (see Tables 7 and 21) and is stable with respect to in-plane framework distortions. Rapid (on a vibrational time scale) interconversion of the rectangular singlets (XXa–XXb) seems unlikely in view of the 8-kcal/mole barrier implied by the energy of the square singlet.*

The energies (4-31G) of the triplet (RHF)[130] and singlet (2-configuration SCF)[127] are calculated to cross at an energy of 2.3 kcal/mole above the triplet minimum (Table 21). Since the zero-point energy for the rectangular ($^1B_{1g}$) vibration of the triplet would have a similar magnitude, there is no evidence at this level of calculation that the triplet could be isolated from the singlet (i.e., even if a crossing of the $^1A_{1g}$ and $^3B_{1g}$ states occurs in D_{2h} symmetry, it might lie below the zero-point level of the $^3A_{2g}$ state; cf. Ref. 101a).

Returning to the experimental ir data, we note that the magnitude of the observed frequency for the framework deformation (1236 cm^{-1}) would be hard to rationalize in terms of the "normal" single and double bonds calculated for the singlet. Typical single and double bond frequencies are 1000 and 1650 cm^{-1}, respectively.[131] The 1236 cm^{-1} cyclobutadiene frequency is most easily attributed to the E_u stretching mode of the square triplet. The rather low frequency (1482 cm^{-1} for the benzene E_{1u} framework mode[131]) is not surprising in view of the calculations for the E_u distortion (*vide supra*). Unfortunately there is no other experimental evidence supporting the assignment of the triplet electronic state. Furthermore, available CI wave functions make it appear unlikely that the cyclobutadiene triplet could be trapped for the long times (many hours) characteristic of the matrix-isolated species. Clearly, more elaborate CI calculations would be invaluable in helping to identify the source of the matrix ir spectrum.

5. Summary

The data and examples discussed have demonstrated that the same computationally feasible techniques which have previously given good accounts of small, generally unstrained systems, also lead to accurate, systematic results for strained polycyclic hydrocarbons. Of particular note is the ability to reproduce the unusual range of geometrical parameters associated with bridgehead carbon atoms and the corresponding simple analysis in terms of hybrid orbitals, which gives excellent correlation with both the geometric and the spin–spin coupling data. The ability to calculate consistently accurate dipole moments in strained hydrocarbons helps to justify the analysis of strained bonds in terms of calculated electronic densities, especially difference densities, for which comparisons with experiment will become more, available in the future. Polarization basis functions do not appear to be crucial in

*See note 4 of Notes Added in Proof on page 275.

obtaining meaningful electron densities for the strained hydrocarbons we have considered, although they would undoubtedly be more important for heterocyclic variants, where lone pairs are present, and in cases involving second row atoms.[18]

While *ab initio* levels short of the Hartree–Fock limit (e.g., the double-zeta level) are not useful in direct calculation of strain energies, a wealth of reliable energetic data is provided at this level including energies of isodesmic reactions, ionization energies, and assignment of photoelectron spectra, force constants, and the energetics of bond-breaking processes. An example of the latter was the successful model of the central bond in [2.2.2]propellane (XIII), which yielded a reasonable estimate of its lifetime before the molecule had been experimentally detected. Among the outstanding challenges to *ab initio* theory which remain in the realm of strained molecules we have noted the unresolved aspects of the cyclobutadiene problem and, in general, the difficulties in accounting for the nonadditive aspects of strain energies in fused ring systems.

By providing details of the energetics in strained systems which are not directly observable by experiment, *ab initio* calculations are not only indispensible in the elucidation of microscopic details of reaction mechanisms involving strained molecules, but can perhaps also be of utility in helping to refine powerful semiempirical techniques, such as MINDO/3, which fill a role complementary to that of *ab initio* methods.

ACKNOWLEDGMENTS

We are indebted to numerous people for supplying preprints and results prior to publication: Professor P. Coppens, Professor M. J. S. Dewar, Professor H. J. Hehre, Professor J. M. Lehn, Professor J. A. Pople, and Professor K. B. Wiberg; Dr. B. G. Ellison, Dr. B. Roos, and Dr. I. Shavitt; and Mr. J. S. Binkley. We also gratefully acknowledge many helpful conversations with the above colleagues, and also with Professor J. M. Schulman and Mr. K. Peters. Dr. H. J. Bernstein provided assistance with the computer-generated figures, which were produced with the CRYSNET[132] facilities. This work was performed under the auspices of the U.S. Energy Research and Development Administration.

References

1. L. C. Allen, Quantum theory of structure and dynamics, *Ann. Rev. Phys. Chem.* **20**, 315–356 (1969).
2. M. J. S. Dewar, *The Molecular Orbital Theory of Organic Chemistry*, McGraw-Hill, New York (1969).

3. (a) J. A. Pople and D. L. Beveridge, *Approximate Molecular Orbital Theory*, McGraw-Hill, New York (1970); (b) L. Radom and J. A. Pople, *ab initio* molecular orbital theory of organic molecules, *MTP Int. Rev. Sci.* **1**, 71–112 (1972).

4. H. F. Schaefer III, *The Electronic Structure of Atoms and Molecules*, Addison-Wesley, Reading, Massachusetts (1972).

5. J. Hinze, An overview of computational methods for large molecules, *Adv. Chem. Phys.* **XXVI**, 213–263 (1974).

6. Henry F. Schaefer III, Molecular electronic structure theory: 1972–1975, *Ann. Rev. Phys. Chem.*, to be published.

7. (a) D. T. Clark, Physical methods. Part (iii) Theoretical organic chemistry and ESCA, *Ann. Rep. Chem. Soc.* **B69**, 40–65 (1972); (b) J. M. Mellor, Alicyclic compounds, *Ann. Rep. Chem. Soc.* **B69**, 403–424 (1972).

8. W. A. Lathan, L. Radom, P. C. Hariharan, W. J. Hehre and J. A. Pople, Structures and stabilities of three-membered rings from *ab initio* molecular orbital theory, *Fortsch. Chem. Forsch.* **40**, 1–45 (1973).

9. M. Randić and Z. B. Maksić, Hybridization by the maximum overlap method, *Chem Rev.* **72**, 43–53 (1972).

10. M. J. S. Dewar, Quantum organic chemistry, *Science* **187**, 1037–1044 (1975).

11. L. C. Snyder and H. Basch, *Molecular Wavefunctions and Properties*, John Wiley and Sons, New York (1972).

12. J. D. Petke and J. L. Whitten, Self-consistent-field calculation of the geometry of protonated cyclopropane, *J. Am. Chem. Soc.* **90**, 3338–3343 (1968).

13. R. J. Buenker and S. D. Peyerimhoff, Theoretical study of the geometry, reactivity, and spectrum of cyclopropane, *J. Phys. Chem.* **73**, 1299–1313 (1969).

14. E. Kochanski and J. M. Lehn, The electronic structure of cyclopropane, cyclopropene and diazirine. An *ab initio* SCF–LCAO–MO study, *Theor. Chim. Acta* **14**, 281–304 (1969).

15. (a) H. Basch, M. B. Robin, N. A. Kuebler, C. Baker, and D. W. Turner, Optical and photoelectron spectra of small rings. III The saturated three-membered rings, *J. Chem. Phys.* **51**, 52–66 (1969); (b) M. B. Robin, C. R. Brundle, N. A. Kuebler, G. B. Ellison, and K. B. Wiberg, Photoelectron spectra of small rings. IV. The unsaturated three-membered rings, *J. Chem. Phys.* **57**, 1758–1763 (1972).

16. R. Bonaccorsi, E. Scrocco, and J. Tomasi, Molecular SCF calculations for the ground state of some three-membered ring molecules: $(CH_2)_3$, $(CH_2)_2NH$, $(CH_2)_2NH_2^+$, $(CH_2)_2O$, $(CH_2)_2S$, $(CH)_2CH_2$, and N_2CH_2, *J. Chem. Phys.* **52**, 5270–5284 (1970).

17. (a) M. D. Newton, E. Switkes, and W. N. Lipscomb, Localized bonds in SCF wavefunctions for polyatomic molecules. III. C—H and C—C bonds, *J. Chem. Phys.* **53**, 2645–2657 (1970); (b) R. M. Stevens, E. Switkes, E. A. Laws, and W. N. Lipscomb, Self-consistent-field studies of the electronic structures of cyclopropane and benzene, *J. Am. Chem. Soc.* **93**, 2603–2609 (1971).

18. M.-M. Rohmer and D. Roos, An *ab initio* study of ethylene episulfide, episulphoxide, and episulphone, *J. Am. Chem. Soc.* **97**, 2025 (1975).

19. (a) W. J. Hehre and J. A. Pople, Molecular orbital theory of the electronic structure of organic compounds. XXVI. Geometries, energies and polarities of C_4 hydrocarbons, *J. Am. Chem. Soc.* **97**, 6941 (1975); (b) J. S. Binkley, J. A. Pople, and W. J. Hehre, extended basis set studies of hydrocarbon molecular orbital energies, *Chem. Phys. Lett.* **36**, 1 (1975).

20. J. L. Nelson and A. A. Frost, A floating spherical Gaussian orbital model of molecular structure. X. C_3 and C_4 saturated hydrocarbons and cyclobutane, *J. Am. Chem. Soc.* **94**, 3727–3731 (1972).

21. J.-M. Lehn and G. Wipff, The electronic structure of bicyclo[1.1.1]pentane systems, *Chem. Phys. Lett.* **15**, 450–454 (1972).

22. J.-M. Lehn and G. Wipff, The electronic structure of bicyclo[2.1.1]hexane, *Theor. Chim. Acta* **28**, 223–233 (1973).

23. J.-M. Lehn and G. Wipff, Theoretical study of proton approach towards strained hydrocarbon molecules, *J. Chem. Soc. D* 747–748 (1973).

24. (a) G. Wipff, Electronic Structure of Hydrocarbons and Stereoelectronic Effects in the Reactivity of Carbonyl Groups, Ph.D. thesis, Université Louis Pasteur, Strasbourg, 1974; (b) G. Wipff and J.-M. Lehn, unpublished.

25. J. M. Schulman and G. J. Fisanick, A new model for the bonding in bicyclobutanes, *J. Am. Chem. Soc.* **92**, 6653–6654 (1970).
26. M. D. Newton and J. M. Schulman, Theoretical studies of bicyclobutane, *J. Am. Chem. Soc.* **94**, 767 (1972).
27. M. D. Newton and J. M. Schulman, Theoretical studies of tricylo[1.1.1.0^{1,3}]pentane and bicyclo[1.1.1.]pentane, *J. Am. Chem. Soc.* **94**, 773 (1972).
28. D. R. Whitman and J. F. Chiang, Electronic structures of bicyclo[1.1.0]butane and bicyclo[1.1.1]pentane, *J. Am. Chem. Soc.* **94**, 1126–1129 (1972).
29. M. D. Newton and J. M. Schulman, Electronic structure of [2.2.2]propellane, *J. Am. Chem. Soc.* **94**, 4391–4392 (1972).
30. G. Berthier, A. Y. Meyer, and L. Praud, All-electron calculations on benzene isomers, *Jerusalem Symp. Quantum Chem. Biochem.* **3**, 174–188 (1971).
31. R. E. Christoffersen, *Ab initio* calculations on large molecules using molecular fragments. Benzene and naphthalene isomer characterizations and Aromaticity considerations, *J. Am. Chem. Soc.* **93**, 4104–4111 (1971).
32. M. D. Newton, J. M. Schulman, and M. M. Manus, Theoretical studies of benzene and its valence isomers, *J. Am. Chem. Soc.* **96**, 17–23 (1974).
33. R. J. Buenker and S. D. Peyerimhoff, *Ab initio* study on the stability and geometry of cyclobutadiene, *J. Chem. Phys.* **48**, 354–373 (1968).
34. R. J. Buenker and S. D. Peyerimhoff, Theoretical comparison of tetrahedrane and cyclobutadiene by *ab initio* techniques, *J. Am. Chem. Soc.* **91**, 4342–4346 (1969).
35. J. M. Schulman and T. J. Venanzi, A theoretical study of the tetrahedrane molecule, *J. Am. Chem. Soc.* **96**, 4739–4746 (1974).
36. P. Millie, L. Praud, and J. Serre, The nature of the "triple bond" of 1,2-benzyne, *Int. J. Quantum Chem.* **4**, 187–193 (1971).
37. D. L. Wilhite and J. L. Whitten, Theoretical studies of the ground and excited electronic states of the benzynes by *ab initio* self-consistent-field and configuration-interaction methods, *J. Am. Chem. Soc.* **93**, 2858–2864 (1971).
38. M. D. Newton and H. A. Fraenkel, The equilibrium geometry, electronic structure, and heat of formation of ortho-benzyne, *Chem. Phys. Lett.* **18**, 244–246 (1973).
39. A. K. Q. Siu, W. M. St. John III, and E. F. Hayes, Theoretical studies of trimethylene, *J. Am. Chem. Soc.* **92**, 7249–7251 (1970).
40. P. J. Hay, W. J. Hunt, and W. A. Goddard III, Theoretical investigations of the trimethylene biradical, *J. Am. Chem. Soc.* **94**, 638–640 (1972).
41. J. A. Horsley, Y. Jean, C. Moser, L. Salem, R. M. Stevens, and J. S. Wright, An organic transition state, *J. Am. Chem. Soc.* **94**, 279–282 (1972).
42. W. J. Hehre, M. R. Willcott, and L. Salem, Organic transition states. II. The methylenecyclopropane rearrangement. A two-step diradical pathway with a secondary minimum, *J. Am. Chem. Soc.* **96**, 4328–4330 (1974).
43. (a) J. S. Wright and L. Salem, *Ab initio* calculation of the equilibrium structure and coplanar decomposition of cyclobutane, *J. Am. Chem. Soc.* **94**, 322–329 (1972); (b) G. A. Segal, Organic transition states. III. An *ab initio* study of the pyrolysis of cyclobutane *via* the tetramethylene diradical, *J. Am. Chem. Soc.* **96**, 7892–7898 (1974).
44. (a) K. Hsu, R. J. Buenker, and S. D. Peyerimhoff, Theoretical determination of the reaction path in the prototype electrocyclic transformation between cyclobutene and *cis*-butadiene. Thermochemical process, *J. Am. Chem. Soc.* **93**, 2117–2127 (1971); (b) K. Hsu, R. J. Buenker, and S. D. Peyerimhoff, The role of ring torsion in the electrocyclic transformation between cyclobutene and butadiene. A theoretical study, *J. Am. Chem. Soc.* **94**, 5639–5644 (1972).
45. R. J. Buenker, S. D. Peyerimhoff, and K. Hsu, Analysis of qualitative theories for electrocyclic transformations based on the results of *ab initio* self-consistent-field and configuration-interaction calculations, *J. Am. Chem. Soc.* **93**, 5005–5013 (1971).
46. G. B. Ellison, K. Peters, and K. B. Wiberg, private communication; G. B. Ellison, Electronic States of Olefins, Ph.D. thesis, Yale University, 1974.
47. A. Baeyer, Über Polyacetyenverbindung, *Chem. Ber.* **18**, 2269–2281 (1885).

48. H. B. Burgi and L. S. Bartell, Structure of tri-*tert*-butylmethane. I. An electron diffraction study, *J. Am. Chem. Soc.* **94**, 5236–5238 (1972).

49. (a) J. L. Franklin, Prediction of heat and free energies of organic compounds, *Ind. Eng. Chem.* **41**, 1070–1076 (1949); (b) S. W. Benson, *Thermochemical Kinetics*, John Wiley and Sons, New York (1968).

50. P. v. R. Schleyer, J. E. Williams, and K. R. Blanchard, The evaluation of strain in hydrocarbons. The strain in adamantane and its origin, *J. Am. Chem. Soc.* **92**, 2377–2386 (1970).

51. P. Coppens, Measurement of electron densities in solids by x-ray diffraction, *MTP Int. Rev. Sci.* **11**, 21 (1975).

52. C. A. Coulson and W. E. Moffitt, The properties of certain strained hydrobarbons, *Philos. Mag.* **40**, 1–35 (1949).

53. A. D. Walsh, The structures of ethylene oxide, cyclopropane, and related molecules, *Trans. Faraday Soc.* **45**, 179–190 (1949).

54. C. F. Fritchie, Jr., The crystal and molecular structure of 2,5-dimethyl-7,7-dicyanonorcaradiene, *Acta crystallogr.* **20**, 27–36 (1966).

55. A. Hartman and F. L. Hirschfeld, Structure of cis-1,2,3-tricyano-cylopropane, *Acta Crystallogr.* **20**, 80–81 (1966).

56. D. A. Matthews and G. D. Stuckey, Bonding and Valence Electron distributions in molecules. Experimental determination of aspherical electron charge density in tetracyanoethylene oxide, *J. Am. Chem. Soc.* **93**, 5954–5959 (1971).

57. T. Ito and T. Sakurai, The structure and electron density of ethyleneimine quinone, *Acta Crystallogr.* **B29**, 1594–1603 (1973).

58. R. F. Peterson, Jr., R. T. K. Baker, and R. L. Wolfgang, Reactions of atomic carbon and carbonylcarbene with cyclopropene: another search for tetrahedrane, *Tetrahedron Lett.* 4749–4752 (1969).

59. P. B. Shevlin and A. P. Wolf, On the probable intermediacy of tetrahedrane, *J. Am. Chem. Soc.* **92**, 406–408 (1970).

60. D. Ginsburg, Carbocyclic and Heterocyclic Propellanes, *Acc. Chem. Res.* **2**, 121–128 (1969).

61. W.-D. Stohrer and R. Hoffmann, The electronic structure and reactivity of strained tricyclic hydrocarbons, *J. Am. Chem. Soc.* **94**, 779–786 (1972).

62. P. E. Eaton and G. H. Temme III, The [2.2.2]propellane system, *J. Am. Chem. Soc.* **95**, 7508–7510 (1973).

63. K. B. Wiberg, G. J. Burgmaier, K.-w. Shen, S. J. La Placa, W. C. Hamilton, and M. D. Newton, "Inverted" tetrahedral geometry at a bridgehead carbon. The x-ray crystal, molecular, and electronic structure of 8,8-dichlorotricyclo[3.2.1.0$^{1.5}$]octane ($C_8H_{10}Cl_2$) at $-40°$, *J. Am. Chem. Soc.* **94**, 7402–7406 (1972).

64. E. Haselbach, Theoretical study of the structure and the physicochemical properties of 1,2-benzyne, *Hel. Chim. Acta* **54**, 1981–1988 (1971).

65. M. J. S. Dewar and W.-K. Li, MINDO/3 study of the bisdehydrobenzenes, *J. Am. Chem. Soc.* **96**, 5569–5571 (1974).

66. O. L. Chapman, K. Mattes, C. L. McIntosh, J. Pacansky, G. V. Calder, and G. Orr, Benzyne, *J. Am. Chem. Soc.* **95**, 6134–6135 (1973).

67. M. P. Cava and M. J. Mitchell, *Cyclobutadiene and Related Compounds*, Academic Press, New York (1967).

68. G. Maier, The cyclobutadiene problem, *Angew. Chem., Int. Ed. Engl.* **13**, 425–438 (1974).

69. C. Y. Lin and A. Krantz, Matrix preparation of cyclobutadiene, *Chem. Commun.* 1111 (1972).

70. O. L. Chapman, C. L. McIntosh, and J. Pacansky, Cyclobutadiene, *J. Am. Chem. Soc.* **95**, 614–617 (1973).

71. O. L. Chapman, D. De La Cruz, R. Roth, and J. Pacansky, Mono- and dideuteriocyclobutadienes, *J. Am. Chem. Soc.* **95**, 1337–1338 (1973).

72. A. Krantz, C. Y. Lin, and M. D. Newton, Cyclobutadiene. II. On the geometry of the matrix-isolated species, *J. Am. Chem. Soc.* **95**, 2744–2746 (1973).

73. E. M. Engler, J. D. Andose, and P. v. R. Schleyer, Critical evaluation of molecular mechanics, *J. Am. Chem. Soc.* **95**, 8003–8025 (1973).

74. R. Hoffmann, An extended Hückel theory I. Hydrocarbons, *J. Chem. Phys.* **39**, 1397–1412 (1963).

75. J. A. Pople and G. A. Segal, Approximate self-consistent molecular orbital theory. III. CNDO results for AB_2 and AB_3 systems, *J. Chem. Phys.* **44**, 3289–3296 (1966).

76. J. A. Pople, D. L. Beveridge, and P. A. Dobosh, Approximate self-consistent molecular-orbital theory. V. Intermediate neglect of differential overlap, *J. Chem. Phys.* **47**, 2026–2033 (1967).

77. Z. B. Maksić and M. Randić, Comparative study of hybridization in hydrocarbons, *J. Am. Chem. Soc.* **95**, 6522–6530 (1973).

78. R. Hoffmann and R. B. Davidson, The valence orbitals of cyclobutane, *J. Am. Chem. Soc.* **93**, 5699–5705 (1971).

79. J. M. Schulman and M. D. Newton, Contributions to the nuclear spin–spin coupling constants of directly bonded carbons, *J. Am. Chem. Soc.* **96**, 6295–6297 (1974).

80. R. C. Bingham, M. J. S. Dewar, and D. H. Lo, Ground states of molecules. XXV. MINDO/3. An improved version of the MINDO semiempirical SCF-MO method, *J. Am. Chem. Soc.* **97**, 1285–1293 (1975); R. C. Bingham, M. J. S. Dewar, and D. H. Lo, Ground states of molecules. XXVI. MINDO/3 calculations for hydrocarbons, *J. Am. Chem. Soc.* **97**, 1294–1301 (1975).

81. (a) H. Preuss and G. Diercksen, Wave mechanical absolute calculations of molecular and atomic systems in the S.C.F.–M.O.–L.C.(L.C.G.O.) method. I. Cyclopentadienyl anion, *Int. J. Quantum Chem.* **1**, 349–355 (1967); (b) A. A. Frost, Floating spherical Gaussian orbital model of molecular structure. I Computational procedure. LiH as an example, *J. Chem. Phys.* **47**, 3707–3713 (1967); (c) R. E. Christoffersen, *Ab initio* calculations on large molecules, *Adv. Quantum Chem.* **6**, 333–393 (1972); (d) A. A. Frost, in: *Modern Theoretical Chemistry, Vol. 3: Methods of Electronic Structure Theory* (H. Schaefer, ed.), Chap. 2, Plenum Publishing, New York (1977).

82. (a) J. L. Whitten, Gaussian lobe function expansions of Hartree–Fock solutions for the first-row atoms and ethylene, *J. Chem. Phys.* **44**, 359–364 (1966); (b) E. Clementi and D. R. Davis, Barrier to internal rotation in ethane, *J. Chem. Phys.* **45**, 2593–2599 (1966); (c) J. M. Schulman, J. W. Moskowitz, and C. Hollister, Ethylene molecule in a Gaussian basis. II Contracted bases, *J. Chem. Phys.* **46**, 2759–2764 (1967); (d) T. H. Dunning, Gaussian basis functions for use in molecular calculations. I. Contraction of $(9s5p)$ atomic basis sets for the first-row atoms, *J. Chem. Phys.* 2823–2833 (1970); (e) T. H. Dunning and P. J. Hay, in: *Modern Theoretical Chemistry, Vol. 3: Methods of Electronic Structure Theory* (H. Schaefer, ed.), Chap. 1, Plenum Publishing, New York (1977).

83. H. Basch, M. B. Robin, and N. A. Kuebler, Electronic states of the amide group, *J. Chem. Phys.* **47**, 1201–1210 (1967).

84. W. J. Hehre, R. F. Stewart, and J. A. Pople, Self-consistent molecular-orbital methods. I. Use of Gaussian expansions of Slater-type atomic orbitals, *J. Chem. Phys.* **51**, 2657–2664 (1969).

85. R. Ditchfield, W. J. Hehre, and J. A. Pople, Self-consistent molecular-orbital methods. IX. An extended Gaussian-type basis for molecular-orbital studies of organic molecules, *J. Chem. Phys.* **54**, 724–728 (1971).

86. (a) P. C. Hariharan, W. A. Lathan, and J. A. Pople, Molecular orbital theory of simple carbonium ions, *Chem. Phys. Lett.* **14**, 385–388 (1972); (b) P. C. Hariharan and J. A. Pople, The effect of d-functions on molecular orbital energies for hydrocarbons, *Chem. Phys. Lett.* **16**, 217–219 (1972); (c) P. C. Hariharan and J. A. Pople, The influence of polarization functions on molecular orbital hydrogenation energies, *Theor. Chim. Acta* **28**, 213–222 (1973).

87. R. Ditchfield, W. J. Hehre, and J. A. Pople, Self-consistent molecular orbital methods. VI. Energy optimized Gaussian atomic orbitals, *J. Chem. Phys.* **52**, 5001–5007 (1970).

88. W. J. Hehre, R. Ditchfield, and J. A. Pople, Self-consistent molecular-orbital methods. VIII. Molecular studies with least energy minimal atomic orbitals, *J. Chem. Phys.* **53**, 932–935 (1970).

89. J. M. Schulman, C. J. Hornback, and J. W. Moskowitz, An optimized minimal contracted-Gaussian basis Set for organic PI-systems, *Chem. Phys. Lett.* **8**, 361–365 (1971).

90. W. J. Hehre, W. A. Lathan, R. Ditchfield, M. D. Newton, and J. A. Pople, FORTRAN IV. GAUSSIAN 70: *Ab initio* SCF–MO Calculations on Organic Molecules. QCPE 236, Quantum Chemistry Program Exchange, Indiana University, Newsletter 43, Bloomington, Indiana 47401, November, 1973.

91. C. Edmiston and K. Ruedenberg, Localized atomic and molecular orbitals. II, *J. Chem. Phys.* **43**, S97–S115 (1965).

92. J. E. Lennard-Jones and J. A. Pople, The molecular orbital theory of chemical valency IV. The significance of equivalent orbitals, *Proc. R. Soc. London, Ser. A* **202**, 166–180 (1951).

93. J. M. Foster and S. F. Boys, Canonical configurational interaction procedure, *Rev. Mod. Phys.* **32**, 300–302 (1960).

94. D. A. Kleier, T. A. Halgren, J. H. Hall, Jr., and W. N. Lipscomb, Localized molecular orbitals for polyatomic molecules. I. A comparison of the Edmiston–Ruedenberg and Boys localization methods, *J. Chem. Phys.* **61**, 3905–3919 (1974).

95. M. D. Newton, W. A. Lathan, W. J. Hehre, and J. A. Pople, Self-consistent molecular orbital methods. V. *Ab initio* calculation of equilibrium geometries and quadratic force constants, *J. Chem. Phys.* **52**, 4064–4072 (1970).

96. (a) W. A. Lathan, W. J. Hehre, and J. A. Pople, Molecular orbital theory of the electronic structure of organic compounds. VI. Geometries and energies of small hydrocarbons, *J. Am. Chem. Soc.* **93**, 808–815 (1971); (b) L. Radom, W. A. Lathan, W. J. Hehre, and J. A. Pople, Molecular orbital theory of the electronic structure of organic compounds. VIII. Geometries, energies, and polarities of C_3 hydrocarbons, *J. Am. Chem. Soc.* **93**, 5339–5342 (1971).

97. J. A. Pople, *A priori* geometry predictions, this volume, Chapter 1.

98. W. J. Hehre, R. Ditchfield, L. Radom, and J. A. Pople, Molecular orbital theory of the electronic structure of organic compounds. V. Molecular theory of bond separation, *J. Am. Chem. Soc.* **92**, 4796–4801 (1970).

99. N. C. Baird and M. J. S. Dewar, Ground states of σ-bonded molecules. IV. The MINDO method and its application to hydrocarbons, *J. Chem. Phys.* **50**, 1262–1274 (1969).

100. N. C. Baird, The calculation of strain energy by molecular orbital theories, *Tetrahedron* **26**, 2185–2190 (1970).

101. (a) M. J. S. Dewar and H. W. Kollmar, MINDO/3 study of cyclobutadiene, *J. Am. Chem. Soc.* **97**, 2933–2934 (1975); (b) M. J. S. Dewar, MO studies of some nonbenzenoid aromatic systems, *Pure Appl. Chem.* **44**, 767–782 (1975).

102. J. M. R. Stone and I. M. Mills, Puckering structure in the intrafred spectrum of cyclobutane, *Mol. Phys.* **18**, 631–652 (1970).

103. S. Meiboom and L. C. Snyder, Molecular structure of cyclobutane from its proton NMR in a nematic solvent, *J. Chem. Phys.* **52**, 3857–3863 (1970).

104. G. B. Ellison, private communication; K. B. Wiberg, G. B. Ellison, M. B. Robin, and C. R. Brundle, unpublished.

105. R. C. Benson and W. H. Flygare, Molecular g values, magnetic susceptibilities, molecular quadrupole moments, and sign of the electric dipole moment in cyclopropene, *J. Chem. Phys.* **51**, 3087–3096 (1969).

106. (a) K. B. Wiberg and G. B. Ellison, Distorted geometries at carbon, *Tetrahedron* **30**, 1573–1578 (1974); (b) S. Palalikit, P. C. Hariharan, and I. Shavitt, unpublished; S. Palalikit, M.Sc. thesis, The Ohio State University, Columbia (1974).

107. L. S. Bartell, On the effects of intramolecular van der Waals forces, *J. Chem. Phys.* **32**, 827–831 (1960).

108. G. D. With and D. Feil, Molecular charge distribution of CO, *Chem Phys. Lett.* **30**, 279–283 (1975).

109. K. B. Wiberg, G. M. Lampman, R. P. Ciula, D. S. Connor, P. Schertler, and J. Lavanish, Bicyclo[1.1.0]butane, *Tetrahedron* **21**, 2749–2769 (1965).

110. M. Pomerantz and E. W. Abrahamson, The electronic structure and reactivity of small ring compounds. I. Bicyclobutane, *J. Am. Chem. Soc.* **88**, 3970–3972 (1966).

111. T. D. Giercke, R. C. Benson, and W. H. Flygare, Calculation of molecular electric dipole and quadrupole moments, *J. Am. Chem. Soc.* **94**, 330–338 (1972).
112. (a) M. Pomerantz, G. W. Gruber, and R. N. Wilke, The electronic structure and reactivity of small-ring compounds. III. Mechanistic studies of the bicyclobutane–benzyne reaction, *J. Am. Chem. Soc.* **90**, 5040–5041 (1968); (b) P. G. Gassman and G. D. Richmond, The reaction of highly strained polycyclic molecules with carbon–carbon multiple bonds, *J. Am. Chem. Soc.* **92**, 2090–2096 (1970); (c) P. Gassman, The thermal addition of carbon–carbon multiple bonds to strained carbocyclics, *Acc. Chem. Res.* **4**, 128–136 (1971).
113. S. Meiboom and L. C. Snyder, Nuclear magnetic resonance spectra in liquid crystals and molecular structures, *Acc. Chem. Res.* **4**, 81–87 (1971).
114. K. W. Cox, M. D. Harmony, G. Nelson, and K. B. Wiberg, Microwave spectrum of bicyclo[1.1.0]butane, *J. Chem. Phys.* **50**, 1976–1980 (1969); (Erratum) *J. Chem. Phys.* **53**, 858 (1970).
115. (a) H. M. Frey and I. D. R. Stevens, Thermal unimolecular isomerization of bicyclobutane, *Trans. Faraday Soc.* **61**, 90–94 (1965); (b) K. B. Wiberg and J. M. Lavanish, Formation and thermal decomposition of bicyclo[1.1.0]butane-2-*exo-d₁*, *J. Am. Chem. Soc.* **88**, 5272 (1966); (c) K. B. Wiberg, Application of the Pople–Santry–Segal CNDO method to the cyclopropylcarbinyl and cyclobutyl cation and to bicyclobutane, *Tetrahedron* **24**, 1083–1096 (1968); (d) G. L. Closs and P. E. Pfeffer, The steric course of the thermal rearrangements of methylbicyclobutanes, *J. Am. Chem. Soc.* **90**, 2452–2453 (1968).
116. R. B. Woodward and D. L. Dalrymple, Dimethyl 1,3-diphenylbicyclobutane-2,4-dicarboxylates, *J. Am. Chem. Soc.* **91**, 4612–4613 (1969).
117. R. B. Woodward and R. Hoffmann, *The Conservation of Orbital Symmetry*, Academic Press, New York (1970).
118. A. Almenningen, B. Andersen, and B. A. Nyhus, On the molecular structure of bicyclo(1.1.1)pentane in the vapour phase determined by electron diffraction, *Acta Chem. Scand.* **25**, 1217–1223 (1971).
119. R. Hoffmann, A. Imamura, and W. J. Hehre, Benzynes, Dehydroconjugated molecules, and the interaction of orbitals separated by a number of intervening σ-bonds, *J. Am. Chem. Soc.* **90**, 1499–1509 (1967).
120. K. B. Wiberg, G. A. Epling, and M. Jason, Electrochemical reduction of 1,4-dibromobicyclo[2.2.2]octane. Formation of the [2.2.2]propellane, *J. Am. Chem. Soc.* **96**, 912–913 (1974).
121. J. J. Dannenberg, T. M. Prociv, and C. Hutt, Preparation and trapping of [2.2.2]propellane, *J. Am. Chem. Soc.* **96**, 913–914 (1974).
122. H. A. Bent, An appraisal of valence-bond structures and hybridization in compounds of the first-row elements, *Chem. Rev.* **61**, 275–311 (1961).
123. C. A. Coulson, *Valence*, p. 252, Oxford Univ. Press, London (1961).
124. M. J. S. Dewar, M. C. Kohn, and N. Trinajstić, Cyclobutadiene and diphenylcyclobutadiene, Quantitative studies on aromaticity and antiaromaticity, *Pure appl. Chem.* **28**, 111–130 (1971).
126. (a) S. Masamune, N. Nakamura, M. Suda, and H. Ona, Properties of the [4]annulene system. Induced paramagnetic ring current, *J. Am. Chem. Soc.* **95**, 8481–8483 (1973); (b) L. T. J. Delbaere, M. N. G. James, N. Nakamura, and S. Masamune, The [4]annulene system. Direct proof for its rectangular geometry, *J. Am. Chem. Soc.* **97**, 1973–1974 (1975); (c) H. Irngartinger and H. Rodewald, The structure of a rectangular cyclobutadiene, *Angew. Chem. Int. Ed. Engl.* **13**, 740–741 (1974); (d) G. Maier and A. Alzerreca, Tri-*tert*-butylcyclobutadiene, *Angew. Chem. Int. Ed.* **12**, 1015–1016 (1973).
127. W. J. Hunt, P. J. Hay, and W. A. Goddard III, Self-consistent procedures for generalized valence bond wavefunctions. Applications to H₃, BH, H₂O, C₂H₆ and O₂, *J. Chem Phys.* **57**, 738–748 (1972).
128. R. G. Pearson, Symmetry rules for chemical reactions, *Acc. Chem. Res.* **4**, 152–160 (1971).
129. J. A. Pople and R. K. Nesbet, Self-consistent orbitals for radicals, *J. Chem. Phys.* **22**, 571–572 (1954).

130. (a) C. C. J. Roothaan, Self-consistent field theory for open shells of electronic systems, *Rev. Mod. Phys.* **32**, 179–185 (1960); (b) W. J. Hunt, W. A. Goddard, and T. H. Dunning, Jr., The incorporation of quadratic convergence into open-shell self-consistent field equations, *Chem. Phys. Lett.* **6**, 147–151 (1970); (c) J. S. Binkley, J. A. Pople, and P. A. Dobosh, The calculation of spin-restricted single-determinant wavefunctions, *Mol. Phys.* **28**, 1423–1429 (1974).

131. G. Herzberg, *Molecular Spectra and Molecular Structure, II. Infrared and Raman Spectra of Polyatomic Molecules*, p. 92, D. Van Nostrand Co., Inc. Princeton, New Jersey (1945).

132. H. J. Bernstein, L. C. Andrews, H. M. Berman, F. C. Bernstein, G. H. Campbell, H. L. Carrell, H. B. Chiang, W. C. Hamilton, D. D. Jones, D. Klunk, T. F. Koetzle, E. F. Mayer, C. N. Morimoto, S. S. Sevian, R. K. Stodola, M. M. Strongson, and T. V. Willoughby, CRYSNET–A Network of Intelligent Remote Graphics Terminals, presented at the Second Annual AEC Scientific Computer Information Exchange Meeting, Proceedings of the Technical Program, 1974, Report BNL 18803, pp. 148–158, Brookhaven National Laboratory.

133. K. R. Ramaprasad, V. W. Laurie, and N. C. Craig, Microwave spectrum, structure, and dipole moment of 3,3-difluorocyclopropene, *J. Chem. Phys.* **64**, 4832–4836 (1976).

134. O. L. Chapman, C.-C. Chang, J. Kolc, N. R. Rosenquist, and H. Tomioka, A photochemical method for the introduction of strained multiple bonds: Benzyne C≡C stretch, *J. Am. Chem. Soc.* **97**, 6586–6588 (1975).

135. J. W. Laing and R. S. Berry, Normal coordinates, structure, and bonding of benzyne, *J. Am. Chem. Soc.* **98**, 660–664 (1976).

136. S. Masamune, Y. Sugihara, K. Morio, and J. E. Bertie, [4]Annulene, comments on its infrared spectrum, *Canad. J. Chem.* **54**, 2679–2680 (1976).

137. M. D. Newton, unpublished work.

138. M. D. Newton and J. Jafri, unpublished work.

Notes Added in Proof

1. A recent microwave study of the 3,3-difluoro derivative of cyclopropene appears to confirm the dipole moment direction predicted by theory.[133]

2. The acetylenic CC stretching vibration for the matrix-isolated species has subsequently been observed in the expected region (2085 cm^{-1}).[134] A molecular structure fairly similar to the *ab initio* predicted structure has been obtained [135] by fitting a molecular force field to the earlier vibrational data.[66]

3. Recent experimental work has implied that the band observed at 650 cm^{-1}[70] or 653 cm^{-1}[72] and previously assigned to the in-plane CH bending mode may derive most of its intensity from a perturbed bending mode of CO_2.[136] In this connection, it is worth noting that calculated ir intensities[137] based on the INDO method and the normal coordinates derived in Ref. 72 indicate that the in-plane CH bending mode would be at least an order of magnitude weaker than the out-of-plane mode observed at 570 cm^{-1}[70] or 573 cm^{-1}[72]. Thus a definitive location of the in-plane bending mode will clearly require further detailed spectroscopic investigation.

4. A similar barrier is obtained using the more flexible 6-31G* basis.[138] Furthermore, the relative energies of the square singlet and triplet at $R = 1.433$ Å are found to be similar at the 4-31G and 6-31G* levels.[138]

7

Carbonium Ions: Structural and Energetic Investigations

Warren J. Hehre

1. Introduction

Of the many inroads made by *ab initio* molecular orbital theory in recent years, perhaps the most impressive have been efforts directed at questions of structure and stability of reactive chemical intermediates. Such species are generally too short-lived to be amenable to direct spectroscopic observation and characterization. Hence, even the most rudimentary knowledge about them—information that might normally be taken for granted—must be gained from theory and not by experimental probes. For example, a half-century of chemical intuition, resting primarily upon experiment, leads us to suggest a tetrahedral geometry for methane, but this chemical intuition has little to offer regarding possible structures which the molecule might adopt after being protonated. That is to say, there is no precedent in our accumulated experience for such a species, nor do we have the tools ready in our experimental arsenal to explore the issue.

Carbocations occupy a special place in the realm of reactive chemical intermediates, for it is now apparent that in order to account properly for their properties, some rethinking of our familiar concepts of chemical bonding is necessary.[1] For instance, the notion of charge delocalization in π-electron systems must certainly be generalized to include the σ framework (other examples are numerous). *Ab initio* molecular orbital theory provides an ideal

Warren J. Hehre • Department of Chemistry, University of California, Irvine, California

vessel for reestablishing our thoughts. We can ask direct questions of it about structure and stability, and expect to receive unbiased answers. We can easily perform "numerical experiments" on molecules far too unstable ever to be physically observed. Finally, we can look directly into the properties not only of molecules which correspond to minima on a potential energy surface, but also those of transition states for chemical reactions (species which have no lifetime in the normal sense of the word).

It is, of course, absolutely essential that any molecular orbital method which we choose to apply to problems of this sort (where comparison with experimental data is not possible) be subject to thorough scrutiny in advance. Thus, a theory's prediction, say, of the structure of some reactive species yet to be observed should carry very little weight, indeed, unless the model itself had previously shown its ability actually to reproduce the geometries of known systems. If, on the other hand, our theory were capable of meeting such a stringent criterion, then we might be inclined to look with favor upon its predictions in those instances where experimental data were lacking.

The methods which we shall consider in the present review have been intensively scrutinized with regard to reproducing geometrical structures and relative energies of simple organic molecules. Hence, there is good reason to believe that, if carefully applied, they will lead to reasonable predictions of these quantities where they have not as yet been experimentally determined. Our arsenal is three-tiered. At the foundation is a Gaussian simulation of a minimal Slater basis set. Here, each Slater-type orbital (STO) has been replaced with a least-squares-fitted linear combination of three Gaussian functions. The resulting set—termed STO-3G[2]—has been repeatedly shown to be capable of reproducing the geometrical structures of a wide variety of simple organic molecules,[3-6] and it is for this purpose that we employ it here. A detailed discussion has been presented elsewhere in this monograph.[7] The minimal STO-3G basis, despite all its virtues, is unreliable for accurately assessing relative molecular energies, particularly where comparisons between small ring compounds and acyclics are involved. One way to correct for this deficiency is to supplement the STO-3G results by a limited number of calculations using a basis-set representation extended somewhat beyond the minimal level. We have found that the split-valence-shell 4-31G basis[8] is generally adequate for such a purpose, even though it is still small enough to be easily and economically applied to molecules of moderate complexity. This increase in size of the 4-31G set over the minimal expansion does, however, severely restrict our ability to explore thoroughly geometric potential surfaces (except for the smallest systems), and normally we must be satisfied with but a single calculation at the optimum geometry predicted by the STO–3G level. Using the split-valence-shell 4-31G basis, energy differences, especially those between cyclic and acyclic molecules, may still not be entirely satisfactory, although the errors here are markedly smaller than those resulting from comparison at the STO-3G level.[5,6,9] Most of this residual error is removed if polarization-type functions (mainly d orbitals on the heavy atoms) are added to the basis description. The basis which we have come to employ—termed 6-31G*[10]—is among the most simple of this type, formed by appending a single Gaussian d function to the split-valence-shell 6-31G basis.[11] It must be realized, however, that at this level (split-valence-shell plus polarization) individual calculations are quite costly and even prohibitive for molecules containing more than three or four heavy (nonhydrogen) atoms. Whereas we have found, in our studies on neutral hydrocarbons, that corrections for the effects of polarization functions may be reasonably well approximated by additivity schemes,[6] we seriously doubt whether such an approach will be fruitful when dealing with carbocations, where even gross structural features are not easily transferable from one system to another.

For the most part we have limited ourselves in the following discussion to results obtained from one or more of these three methods. Some exceptions have been made for the smaller—one and two heavy atom—systems, where *ab initio* results equal to or better than those obtained by the best of our schemes are available. With this limitation in mind, we shall attempt to survey what is known from *ab initio* theory regarding the geometries and relative energies of singly charged carbocations. In addition, we have chosen to restrict our discussion to molecules which may be represented in terms of a closed-shell electronic configuration, but have considered a small number of particularly interesting ions believed to possess triplet ground states as well (e.g., cyclopentadienyl cation). Calculated equilibrium geometries for all carbocations which we shall discuss are found in Table 1 and in Figs. 1–13. Theoretical energies (STO-3G, 4-31G, and 6-31G*) corresponding to each of these structures are given in Table 2. Although we will not refer further to these data as such, we shall use them to construct the energy comparisons given in the remaining tables.

Table 1. Theoretical Equilibrium Geometries

Molecule	Figure	Geometrical parameter[a]	Value[b]
CH^+	1	$r(CH)$	1.185 (1.108) [1.131]
CH_3^+	1	$r(CH)$	1.120 (1.076)
CH_5^+			
C_s (I)	1	$r(CH_1)$	1.098 (1.077)
		$r(CH_3)$	1.106 (1.086)
		$r(CH_4)$	1.370 (1.242)
		$r(CH_5)$	1.367 (1.241)
		$\angle(H_1CH_2)$	117.7 (116.2)
		$\angle(H_{12}CH_3)$	140.0 (131.6)
		$\angle(H_3CH_4)$	83.8 (84.8)
		$\angle(H_4CH_5)$	37.2 (40.1)

—continued

Table 1—continued

Molecule	Figure	Geometrical parameter[a]	Value[b]
C_s (II)	1	$r(CH_1)$	1.095 (1.074)
		$r(CH_3)$	1.103 (1.083)
		$r(CH_4)$	1.364 (1.238)
		$\angle(H_1CH_2)$	111.6 (108.1)
		$\angle(H_{12}CH_3)$	141.8 (134.4)
		$\angle(H_4CH_5)$	37.5 (40.4)
C_{2v} (III)	1	$r(CH_1)$	1.094
		$r(CH_3)$	1.118
		$r(CH_4)$	1.154
		$\angle(H_1CH_2)$	117.7
		$\angle(H_3CH_4)$	63.9
C_{4v} (IV)	1	$r(CH_1)$	1.139 (1.117)
		$r(CH_3)$	1.083 (1.071)
		$\angle(H_1CH_2)$	81.2 (81.5)
D_{3h} (V)	1	$r(CH_1)$	1.114 (1.100)
		$r(CH_3)$	1.136 (1.118)
C_2H^+			
$^3\pi$	2	$r(CC)$	1.302
		$r(CH)$	1.110
$^1\Delta$	2	$r(CC)$	1.406
		$r(CH)$	1.112
$C_2H_3^+$			
Open (I)	2	$r(CC)$	1.281
		$r(C_1H_1)$	1.106
		$r(C_3H_3)$	1.106
		$\angle(H_1C_1H_2)$	118.3
Bridged (II)	2	$r(CC)$	1.227
		$r(CH_1)$	1.097
		$r(CH_2)$	1.334
		$\angle(H_1CC)$	177.5
$C_2H_5^+$			
Open, eclipsed (I)	2	$r(CC)$	1.488
		$r(C_1H_1)$	1.101
		$r(C_1H_3)$	1.088
		$r(C_2H_4)$	1.115
		$r(C_2H_5)$	1.115
		$\angle(H_1C_1H_2)$	105.9
		$\angle(H_{12}C_1H_3)$	128.0
		$\angle(H_3C_1C_2)$	112.9
		$\angle(H_4C_2C_1)$	122.7
		$\angle(H_5C_2C_1)$	120.6
Open, bisected (II)	2	$r(CC)$	1.484
		$r(C_1H_1)$	1.091
		$r(C_1H_3)$	1.110
		$r(C_2H_4)$	1.115
		$\angle(H_1C_1H_2)$	113.6
		$\angle(H_{12}C_1H_3)$	124.4
		$\angle(H_3C_1C_2)$	102.2
		$\angle(H_4C_2H_5)$	116.7
		$\angle(H_{45}C_2C_1)$	177.1

Table 1—*continued*

Molecule	Figure	Geometrical parameter[a]	Value[b]
Bridged (III)	2	$r(CC)$	1.403
		$r(CH_1)$	1.099
		$r(CH_3)$	1.348
		$\angle(H_1CH_2)$	118.8
		$\angle(H_{12}CC)$	177.5
$C_2H_7^+$			
D_{3d} (I)	2	$r(CC)$	2.486
		$r(CH_1)$	1.097
		$\angle(H_1CH_2)$	115.8
D_{3h} (II)	2	$r(CC)$	2.488
		$r(CH_1)$	1.097
		$\angle(H_1CH_2)$	115.8
C_s (III)	2	$r(CC)$	2.271
		$r(C_1H_1)$	1.093
		$r(C_1H_2)$	1.097
		$r(C_1H_4)$	1.253
		$r(C_2H_5)$	1.094
		$r(C_2H_6)$	1.099
		$\angle(H_1C_1C_2)$	86.2
		$\angle(H_1C_1H_{23})$	142.1
		$\angle(H_4C_1C_2)$	24.9
		$\angle(H_5C_2C_1)$	120.1
		$\angle(H_5C_2H_{67})$	142.0
C_{2v} (IV)	2	$r(CC)$	2.362
		$r(CH_1)$	1.097
		$r(CH_3)$	1.094
		$r(CH_4)$	1.251
		$\angle(H_1CH_2)$	143.2
		$\angle(H_{12}CH_3)$	115.6
		$\angle(H_3CC)$	89.5
C_s (V)	2	$r(CC)$	1.487
		$r(C_1H_1)$	1.091
		$r(C_1H_3)$	1.110
		$r(C_2H_4)$	1.115
		$r(C_2H_5)$	2.746
		$\angle(H_1C_1H_2)$	113.6
		$\angle(H_{12}C_1H_3)$	124.6
		$\angle(H_3C_1C_2)$	102.3
		$\angle(H_4C_2H_5)$	116.7
		$\angle(H_{45}C_2C_1)$	176.3
		$\angle(H_6C_2H_7)$	15.0
		$\angle(H_{67}C_2C_1)$	102.7
C_3H^+			
propa-1,2-dien-1-yl-3-ylidene (I)	3	$r(C_1C_2)$	1.380
		$r(C_2C_3)$	1.213
		$r(C_3H)$	1.102
Prop-2-en-1-yl-1,3-diylidene (II)	3	$r(C_1C_2)$	1.387
		$r(C_2H)$	1.101
		$\angle(C_1C_2C_3)$	123.9

—*continued*

Table 1—continued

Molecule	Figure	Geometrical parameter[a]	Value[b]
$C_3H_3^+$			
Cyclopropenyl (I)	4	$r(CC)$	1.377
		$r(CH)$	1.095
Propargyl (II)	4	$r(C_1C_2)$	1.360
		$r(C_2C_3)$	1.214
		$r(C_1H_1)$	1.109
		$r(C_3H_3)$	1.091
		$\angle(H_1C_1H_1)$	117.8
Prop-2-en-1-yl-3-ylidene (III)	4	$r(C_1C_2)$	1.396
		$r(C_2C_3)$	1.377
		$r(C_2H_1)$	1.090
		$r(C_3H_2)$	1.102
		$r(C_3H_3)$	1.102
		$\angle(C_1C_2C_3)$	115.8
		$\angle(H_1C_2C_1)$	123.6
		$\angle(H_2C_3C_2)$	121.2
		$\angle(H_3C_3C_2)$	121.9
Prop-2-en-1-yl-3-ylidene, perpendicular (IV)	4	$r(C_1C_2)$	1.333
		$r(C_2C_3)$	1.458
		$r(C_2H_1)$	1.100
		$r(C_3H_2)$	1.120
		$\angle(C_1C_2C_3)$	131.0
		$\angle(H_1C_2C_1)$	113.6
		$\angle(H_2C_3H_3)$	177.7
		$\angle(H_{23}C_3C_2)$	116.3
1-Propynyl (V)	4	$r(C_1C_2)$	1.394
		$r(C_2C_3)$	1.481
		$r(C_3H)$	1.098
		$\angle(HC_3C_2)$	109.0
Cycloprop-1-yl-2-ylidene (VI)	4	$r(C_1C_2)$	1.641
		$r(C_2C_3)$	1.456
		$r(C_1C_3)$	1.530
		$r(C_2H_1)$	1.109
		$r(C_3\text{-}H_2)$	1.104
		$\angle(H_1C_2C_3)$	145.1
		$\angle(H_{23}C_3C_2)$	144.7
		$\angle(H_2C_3H_3)$	110.7
Propan-1-yl-1, 3-diylidene (VII)	4	$r(C_1C_2)$	1.623
		$r(C_2C_3)$	1.457
		$r(C_1H_1)$	1.128
		$r(C_2H_2)$	1.104
		$\angle(C_1C_2C_3)$	106.0
		$\angle(H_1C_1C_2)$	99.1
		$\angle(H_{23}C_2C_3)$	124.8
		$\angle(H_2C_2H_3)$	109.6
Corner protonated cyclopropyne (VIII)	4	$r(C_1C_2)$	1.819
		$r(C_1C_3)$	1.461
		$r(C_2H_1)$	1.099
		$r(C_2H_2)$	1.097
		$\angle(H_{13}C_2C_{13})$	127.3
		$\angle(H_2C_2C_{13})$	106.6
		$\angle(H_1C_2H_3)$	107.6

Table 1—continued

Molecule	Figure	Geometrical parameter[a]	Value[b]
$C_3H_5^+$			
Allyl (I)	5	$r(C_1C_2)$	1.385
		$r(C_1H_1)$	1.100
		$r(C_1H_2)$	1.101
		$r(C_2H_3)$	1.084
		$\angle(C_1C_2C_3)$	118.9
		$\angle(H_1C_1C_2)$	122.0
		$\angle(H_2C_1C_2)$	121.4
2-Propenyl,	5	$r(C_1C_2)$	1.480
eclipsed (II)		$r(C_2C_3)$	1.282
		$r(C_1H_1)$	1.094
		$r(C_1H_3)$	1.102
		$r(C_3H_4)$	1.099
		$r(C_3H_5)$	1.100
		$\angle(C_1C_2C_3)$	179.0
		$\angle(H_{12}C_1C_2)$	129.0
		$\angle(H_1C_1H_2)$	112.3
		$\angle(H_3C_1C_2)$	105.0
		$\angle(H_4C_3C_2)$	121.1
		$\angle(H_5C_3C_2)$	120.3
2-Propenyl,	5	$r(C_1C_2)$	1.481
staggered (III)		$r(C_2C_3)$	1.282
		$r(C_1H_1)$	1.098
		$r(C_1H_3)$	1.092
		$r(C_3H_4)$	1.100
		$\angle(C_1C_2C_3)$	180.0
		$\angle(H_{12}C_1C_2)$	120.5
		$\angle(H_1C_1H_2)$	107.7
		$\angle(H_3C_1C_2)$	111.7
		$\angle(H_{45}C_3C_2)$	180.0
		$\angle(H_4C_3H_5)$	118.6
H-bridged	5	$r(C_1C_2)$	1.230
propenyl (IV)		$r(C_2C_3)$	1.506
		$r(C_1H_1)$	1.092
		$r(C_1H_2) \equiv r(C_2H_2)$	1.328
		$r(C_3H_3)$	1.092
		$r(C_3H_5)$	1.093
		$\angle(C_1C_2C_3)$	178.8
		$\angle(H_1C_1C_2)$	177.2
		$\angle(H_{34}C_3C_2)$	109.7
		$\angle(H_3C_3H_4)$	110.1
		$\angle(H_5C_3C_2)$	106.5
1-Propenyl,	5	$r(C_1C_2)$	1.283
eclipsed (V)		$r(C_2C_3)$	1.550
		$r(C_1H_1)$	1.091
		$r(C_1H_3)$	1.088
		$r(C_2H_4)$	1.106
		$r(C_3H_5)$	1.101
		$\angle(C_1C_2C_3)$	125.4
		$\angle(H_{12}C_1C_2)$	121.2
		$\angle(H_1C_1C_2)$	109.8
		$\angle(H_3C_1C_2)$	111.1

continued

Table 1—continued

Molecule	Figure	Geometrical parameter[a]	Value[b]
1-Propenyl, eclipsed (V)		$\angle(H_4C_2C_3)$	116.3
		$\angle(H_5C_3C_2)$	180.0
1-Propenyl, staggered (VI)	5	$r(C_1C_2)$	1.553
		$r(C_2C_3)$	1.283
		$r(C_1H_1)$	1.088
		$r(C_1H_2)$	1.091
		$r(C_2H_4)$	1.108
		$r(C_3H_5)$	1.101
		$\angle(C_1C_2C_3)$	126.8
		$\angle(H_1C_1C_2)$	108.3
		$\angle(H_{23}C_1C_2)$	124.5
		$\angle(H_2C_1H_3)$	110.2
		$\angle(H_4C_2C_3)$	114.2
		$\angle(H_5C_3C_2)$	178.9
Bridged, protonated allene (VII)	5	$r(C_1C_2)$	1.378
		$r(C_2C_3)$	1.295
		$r(C_1H_1)$	1.100
		$r(C_1H_3) \equiv r(C_2H_3)$	1.346
		$r(C_3H_4)$	1.093
		$r(C_3H_5)$	1.095
		$\angle(C_1C_2C_3)$	177.0
		$\angle(H_{12}C_1C_2)$	176.4
		$\angle(H_1C_1H_2)$	119.1
		$\angle(H_4C_3C_2)$	123.0
		$\angle(H_5C_3C_2)$	118.6
Allyl, perpendicular (VIII)	5	$r(C_1C_2)$	1.313
		$r(C_2C_3)$	1.489
		$r(C_1H_1)$	1.088
		$r(C_1H_2)$	1.088
		$r(C_2H_3)$	1.095
		$r(C_3H_4)$	1.119
		$\angle(C_1C_2C_3)$	126.4
		$\angle(H_1C_1C_2)$	123.5
		$\angle(H_2C_1C_2)$	119.7
		$\angle(H_3C_2C_1)$	124.4
		$\angle(H_{45}C_3C_2)$	176.5
		$\angle(H_4C_3H_5)$	115.8
Cyclopropyl (IX)	5	$r(C_1C_2)$	1.485
		$r(C_1C_3)$	1.518
		$r(C_1H_1)$	1.094
		$r(C_2H_3)$	1.110
		$\angle(H_{12}C_1C_3)$	156.9
		$\angle(H_1C_1H_2)$	114.2
Corner protonated cyclopropene, eclipsed (X)	5	$r(C_1C_2) \equiv r(C_2C_3)$	1.833
		$r(C_1C_3)$	1.219
		$r(C_1H_1)$	1.089
		$r(C_2H_2)$	1.095
		$r(C_2H_4)$	1.088
		$r(C_3H_5)$	1.087

Table 1—continued

Molecule	Figure	Geometrical parameter[a]	Value[b]
Corner protonated cyclopropene, eclipsed (X)		$\angle(H_1C_1C_3)$	114.2
		$\angle(H_{23}C_2C_{13})$	125.5
		$\angle(H_2C_2H_3)$	110.2
		$\angle(H_4C_2C_{13})$	106.6
		$\angle(H_5C_3C_1)$	170.2
Corner protonated cyclopropene, staggered (XI)	5	$r(C_1C_2) \equiv r(C_2C_3)$	1.834
		$r(C_1C_3)$	1.219
		$r(C_1H_1)$	1.088
		$r(C_2H_2)$	1.095
		$r(C_2H_3)$	1.090
		$\angle(XC_1C_3)^d$	171.3
		$\angle(H_1C_1X)^d$	0.7
		$\angle(H_{24}C_2C_{13})$	120.0
		$\angle(H_2C_2H_4)$	107.6
Propylidyne, staggered (XII)	5	$r(C_1C_2)$	1.584
		$r(C_2C_3)$	1.453
		$r(C_1H_1)$	1.089
		$r(C_1H_2)$	1.088
		$r(C_2H_4)$	1.104
		$\angle(C_1C_2C_3)$	109.7
		$\angle(H_1C_1C_2)$	106.3
		$\angle(H_{23}C_1C_2)$	127.6
		$\angle(H_{45}C_2C_3)$	120.4
Propylidyne, eclipsed (XIII)	5	$r(C_1C_2)$	1.591
		$r(C_2C_3)$	1.451
		$r(C_1H_1)$	1.084
		$r(C_1H_3)$	1.090
		$r(C_2H_4)$	1.104
		$\angle(C_1C_2C_3)$	107.9
		$\angle(H_{12}C_1C_2)$	122.6
		$\angle(H_3C_1C_2)$	112.2
		$\angle(H_{45}C_2C_3)$	120.8
Propan-1-yl-3-ylidene (XIV)	5	$r(C_1C_2)$	1.553
		$r(C_2C_3)$	1.508
		$r(C_1H_1)$	1.120
		$r(C_2H_2)$	1.103
		$r(C_3H_4)$	1.117
		$r(C_3H_5)$	1.116
		$\angle(C_1C_2C_3)$	112.3
		$\angle(H_1C_1C_2)$	102.9
		$\angle(H_{23}C_2C_3)$	117.0
		$\angle(H_4C_3C_2)$	120.7
		$\angle(H_5C_3C_2)$	121.9
$C_3H_7^+$			
2-Propyl (I)	6	$r(C_1C_2)$	1.500
		$r(C_1H_1)$	1.087
		$r(C_1H_2)$	1.097
		$r(C_2H_4)$	1.113

—*continued*

Table 1—continued

Molecule	Figure	Geometrical parameter[a]	Value[b]
$C_3H_7^+$			
2-Propyl (I)		$\angle(C_1C_2C_3)$	126.0
		$\angle(H_1C_1C_2)$	112.6
		$\angle(H_{23}C_1C_2)$	119.8
		$\angle(H_2C_1H_3)$	106.6
Corner protonated	6	$r(C_1C_2) = r(C_2C_3)$	1.803
cyclopropane,		$r(C_1C_3)$	1.399
staggered (II)		$r(C_1H_1)$	1.093
		$r(C_1H_2)$	1.093
		$r(C_2H_3)$	1.094
		$r(C_2H_4)$	1.085
		$\angle(XC_1C_3)^e$	58.2
		$\angle(H_1C_1X)^e$	15.3
		$\angle(X'C_1C_3)^e$	58.9
		$\angle(H_2C_1X')^e$	6.4
		$\angle(H_{35}C_2C_{13})$	123.0
		$\angle(H_3C_2H_5)$	105.4
		$\angle(H_4C_2C_{13})$	109.5
Corner protonated	6	$r(C_1C_2) \equiv r(C_2C_3)$	1.803
cyclopropane,		$r(C_1C_3)$	1.399
eclipsed (III)		$r(C_1H_1)$	1.092
		$r(C_2H_3)$	1.088
		$r(C_2H_5)$	1.098
		$r(C_3H_6)$	1.094
		$\angle(H_{12}C_1C_3)$	178.6
		$\angle(H_1C_1H_2)$	117.3
		$\angle(H_{34}C_2C_{13})$	129.3
		$\angle(H_3C_2H_4)$	113.9
		$\angle(H_5C_2C_{13})$	107.7
		$\angle(H_{67}C_3C_1)$	179.0
		$\angle(H_6C_3H_7)$	117.5
H-bridged	6	$r(C_{Me}C_1)$	1.533
propyl (IV)		$\angle(H_{Me}C_{Me}C_1C_2)^f$	12.0
		Structure based on geometry for bridged ethyl cation substituted by a standard model methyl group[g]	
Edge protonated	6	$r(C_1C_2) \equiv r(C_2C_3)$	1.516
cyclopropane (V)		$r(C_1C_3)$	1.849
		$r(C_1H_1)$	1.095
		$r(C_1H_7) \equiv r(C_3H_7)$	1.315
		$r(C_2H_3)$	1.087
		$\angle(H_{12}C_1C_3)$	151.4
		$\angle(H_1C_1H_2)$	117.8
		$\angle(H_3C_2H_4)$	114.0
Face protonated	6	$r(C_1C_2)$	1.544
cyclopropane (VI)		$r(C_1H_1)$	1.089
		$r(C_1H_2)$	1.106
		$r(OH_7)^h$	1.117
		$\angle(H_1C_1X)^i$	57.5
		$\angle(H_2C_1X')^i$	51.9

Table 1—continued

Molecule	Figure	Geometrical parameter[a]	Value[b]
$C_4H_5^+$			
Methylcyclopropenyl (I)	7	$r(C_{Me}C_{cyclo})$	1.505
		Structure based on geometry for cyclopropenyl cation substituted by a standard methyl group	
Cyclobutenyl (II)	7	$r(C_1C_2)$	1.392
		$r(C_2C_3)$	1.545
		$\angle(C_1C_{24}C_3)$	170.0
		$\angle(H_1C_1C_{24})$	174.8°
		$\angle(H_2C_2C_4)$	178.4°
		$\angle(H_3C_3C_{24})$	123.2°
		$\angle(H_4C_3C_{24})$	126.1°
		All $r(CH)$ bond lengths set at 1.10 Å; C_2H_2 bond constrained to bisect the angle $C_1C_2C_3$	
Bisected cyclopropenylcarbinyl (III)	7	$r(C_1C_2)$	1.250
		$r(C_1C_4)$	1.592
		$r(C_3C_4)$	1.384
		$\angle(C_{12}C_4C_3)$	119.6
		$\angle(H_1C_1C_4)$	132.1
		$\angle(H_3C_3C_4)$	128.1
		$\angle(H_4C_3C_4)$	121.6
		$\angle(H_5C_4C_{12})$	121.4
		All $r(CH)$ set at 1.10 Å; C_1H_1 bond constrained to lie in the plane formed by $C_1C_2C_4$	
Open eclipsed cyclopropenyl-carbinyl (IV)	7	$r(C_1C_2)$	1.287
		$r(C_1C_4)$	1.477
		$r(C_3C_4)$	1.524
		$\angle(C_{12}C_4C_3)$	119.6
		$\angle(H_1C_1C_4)$	143.7
		$\angle(H_{34}C_3C_4)$	173.3
		$\angle(H_3C_3H_4)$	112.0
		$\angle(H_5C_4C_{12})$	128.4
		All $r(CH)$ set at 1.10 Å; C_1H_1 bond constrained to lie in the plane formed by $C_1C_2C_4$	
Closed eclipsed cyclopropenyl-carbinyl (V)	7	$r(C_1C_2)$	1.349
		$r(C_1C_4)$	1.505
		$r(C_3C_4)$	1.467
		$\angle(C_{12}C_4C_3)$	75.6
		$\angle(H_1C_1C_4)$	148.1
		$\angle(H_{34}C_3C_4)$	171.2
		$\angle(H_3C_3H_4)$	115.5
		$\angle(H_5C_4C_{12})$	143.3
$C_4H_7^+$			
1-Methylallyl, *anti* (I)	8	$r(C_{Me}C_{allyl})$	1.514
		Structure based on geometry of allyl cation substituted by a standard methyl group	

—continued

Table 1—continued

Molecule	Figure	Geometrical parameter[a]	Value[b]
1-Methylallyl, syn (II)	8	$r(C_{Me}C_{allyl})$	1.526
		Structure based on geometry of allyl cation substituted by a standard methyl group	
2-Methylallyl (III)	8	$r(C_{Me}C_{allyl})$	1.536
		Structure based on geometry of allyl cation substituted by a standard methyl group	
Cyclopropylcarbinyl, bisected (IV)	8	$r(C_1C_2)$	1.609
		$r(C_1C_3)$	1.454
		$r(C_3C_4)$	1.384
		$r(C_1H_1)$	1.088
		$r(C_1H_2)$	1.087
		$r(C_3H_5)$	1.083
		$r(C_4H_6)$	1.104
		$r(C_4H_7)$	1.104
		$\angle(C_{12}C_3C_4)$	120.4
		$\angle(H_1C_1C_2)$	119.5
		$\angle(H_1C_1C_3)$	110.9
		$\angle(H_2C_1C_2)$	120.0
		$\angle(H_2C_1C_3)$	115.3
		$\angle(H_5C_3C_{12})$	121.0
		$\angle(H_6C_4C_3)$	121.6
		$\angle(H_7C_4C_3)$	122.0
Cyclobutyl, puckered (V)	8	$r(C_1C_2)$	1.457
		$r(C_2C_3)$	1.622
		$r(C_1C_3)$	1.756
		$r(C_1H_1)$	1.094
		$r(C_2H_2)$	1.089
		$r(C_2H_3)$	1.089
		$r(C_3H_4)$	1.086
		$r(C_3H_5)$	1.088
		$\angle(C_1C_{24}C_3)$	122.4
		$\angle(H_1C_1C_{24})$	148.9
		$\angle(H_2C_2C_1)$	117.4
		$\angle(H_2C_2C_3)$	119.9
		$\angle(H_3C_2C_1)$	118.7
		$\angle(H_3C_2C_3)$	108.6
		$\angle(H_4C_3C_2)$	118.7
		$\angle(H_5C_3C_2)$	106.5
Cyclobutyl, planar (VI)	8	$r(C_1C_2)$	1.516
		$r(C_2C_3)$	1.562
		$r(C_1C_3)$	2.156
		$r(C_1H_1)$	1.114
		$r(C_2H_2)$	1.097
		$r(C_3H_4)$	1.088
		$\angle(H_2C_2C_1)$	111.6
		$\angle(H_2C_2C_3)$	117.5
		$\angle(H_4C_3H_5)$	

Table 1—continued

Molecule	Figure	Geometrical parameter[a]	Value[b]
Methylcyclopropyl (VII)	8	$r(C_{Me}C_{cyclo})$	1.495

Structure based on geometry of cyclopropyl cation substituted by a standard methyl group

Molecule	Figure	Geometrical parameter[a]	Value[b]
Allylcarbinyl *trans* eclipsed (VII)	8	$r(C_1C_2)$	1.312
		$r(C_2C_3)$	1.572
		$r(C_3C_4)$	1.483
		$\angle(C_1C_2C_3)$	120.0
		$\angle(C_2C_3C_4)$	105.6
		$\angle(H_{67}C_4C_3)$	180.0

All $r(CH) = 1.10$ Å; $\angle(H_1C_1C_2) = \angle(H_2C_1C_2) = \angle(H_3C_2C_1) = \angle(H_6C_4C_3) = 120°$; $\angle(H_4C_3C_2) = \angle(H_4C_3C_4) = 109.5°$

Molecule	Figure	Geometrical parameter[a]	Value[b]
Allylcarbinyl, *trans* bisected (IX)	8	$r(C_1C_2)$	1.312
		$r(C_2C_3)$	1.530
		$r(C_3C_4)$	1.481
		$\angle(C_1C_2C_3)$	122.3
		$\angle(C_2C_3C_4)$	115.6

All $r(CH) = 1.10$ Å; $\angle(H_1C_1C_2) = \angle(H_2C_1C_2) = \angle(H_3C_2C_1) = \angle(H_6C_4C_3) = \angle(H_7C_4C_3) = 120°$; $\angle(H_4C_3C_2) = \angle(H_4C_3C_4) = 109.5°$

Molecule	Figure	Geometrical parameter[a]	Value[b]
Allylcarbinyl, *cis* eclipsed (X)	8	$r(C_1C_2)$	1.314
		$r(C_2C_3)$	1.553
		$r(C_3C_4)$	1.492
		$\angle(C_1C_2C_3)$	127.3
		$\angle(C_2C_3C_4)$	108.2
		$\angle(H_{67}C_4C_3)$	180.0

All $r(CH) = 1.10$ Å; $\angle(H_1C_1C_2) = \angle(H_2C_1C_2) = \angle(H_3C_2C_1) = \angle(H_6,C_4C_3) = 120°$; $\angle(H_4C_3C_2) = \angle(H_4C_3C_4) = 109.5°$

Molecule	Figure	Geometrical parameter[a]	Value[b]
Allylcarbinyl, *cis* bisected	8	$r(C_1C_2)$	1.313
		$r(C_2C_3)$	1.526
		$r(C_3C_4)$	1.481
		$\angle(C_1C_2C_3)$	128.1
		$\angle(C_2C_3C_4)$	117.3

All $r(CH) = 1.10$ Å; $\angle(H_1C_1C_2) = \angle(H_2C_1C_2) = \angle(H_3C_2C_1) = \angle(H_6C_4C_3) = \angle(H_7C_4C_3) = 120°$; $\angle(H_4C_3C_2) = \angle(H_4C_3C_4) = 109.5°$

Molecule	Figure	Geometrical parameter[a]	Value[b]
Cyclopropylcarbinyl, eclipsed	8	$r(C_1C_2)$	1.512
		$r(C_1C_3)$	1.513
		$r(C_3C_4)$	1.480
		$r(C_1H_1)$	1.085
		$r(C_1H_2)$	1.084
		$r(C_3H_5)$	1.098
		$r(C_4C_6)$	1.116
		$\angle(C_{12}C_3C_4)$	130.1

—continued

Table 1—continued

Molecule	Figure	Geometrical parameter[a]	Value[b]
Cyclopropylcarbinyl eclipsed		$\angle(H_1C_1C_2)$	118.8
		$\angle(H_1C_1C_3)$	116.6
		$\angle(H_2C_1C_2)$	118.3
		$\angle(H_2C_1C_3)$	118.2
		$\angle(H_5C_3C_{12})$	123.8
		$\angle(H_{67}C_4C_3)$	177.8
		$\angle(H_6C_4H_7)$	116.2
C_4H_9			
t-Butyl	9	$r(C_1C_2)$	1.513
		$r(C_2H_1)$	1.093
		$r(C_2H_3)$	1.086
		$\angle(H_{12}C_2C_1)$	121.2
		$\angle(H_1C_2H_2)$	107.3
		$\angle(H_3C_2C_1)$	111.8
$C_5H_5^+$			
Cyclopentadienyl (I)	10	$r(CC)$	1.428
			$r(CH) = 1.10$ Å
Cyclopentadienyl (II)	10	$r(C_1C_2)$	1.310
		$r(C_2C_3)$	1.538
		$r(C_3C_4)$	1.386
		$\angle(C_1C_2C_3)$	107.5
		$\angle(C_2C_3C_4)$	108.8
		$\angle(C_3C_4C_5)$	107.4
		$\angle(H_1C_1C_2)$	129.7
		$\angle(H_3C_3C_2)$	124.7
			$r(CH) = 1.10$ Å
Cyclopentadienyl (III)	10	$r(C_1C_2)$	1.604
		$r(C_2C_3)$	1.348
		$r(C_3C_4)$	1.464
		$\angle(C_1C_2C_3)$	107.5
		$\angle(C_2C_3C_4)$	107.0
		$\angle(C_3C_4C_5)$	111.0
		$\angle(H_1C_1C_2)$	123.5
		$\angle(H_3C_3C_2)$	127.7
			$r(Ch) = 1.10$ Å
Square-based pyramid (IV)	10	$r(C_1C_2)$	1.477
		$r(OC_3)$	1.195
		$\angle(H_1C_1O)$	174.6
			$r(CH) = 1.10$ Å
$C_6H_7^+$			
Protonated benzene, open (I)	11	$r(C_1C_6)$	1.472
		$r(C_1H_1)$	1.094
		$r(C_6H_6)$	1.106
		$\angle(C_1C_6C_5)$	110.9
		$\angle(H_6C_6H_7)$	118.3
		$\angle(H_6C_6H_7)$	105.3
		Other $r(CH) = 1.08$ Å; other $\angle(CCC)$ and $\angle(HCC) = 120°$	

Table 1—continued

Molecule	Figure	Geometrical parameter[a]	Value[b]
Protonated benzene, H-bridged (II)	11	$r(C_1C_2)$	1.414
		$r(C_1C_6)$	1.451
		$r(C_1H_1)$	1.090
		$r(C_1H_7)$	1.333
		$\angle(C_2C_1H_1)$	121.8°
		$\angle(H_7C_{16}C_{34})$	93.9
		$\angle(H_{16}C_{16}H_7)$	93.4

Other $r(CH) = 1.08$ Å; other $\angle(CCC)$ and $\angle(HCC) = 120°$

Molecule	Figure	Geometrical parameter[a]	Value[b]
Bicyclo[3.1.0] hexenyl (III)	11	$r(C_1C_2)$	1.463
		$r(C_1C_5)$	1.483
		$r(C_1C_6)$	1.617
		$r(C_2C_3)$	1.391
		$\angle(C_1C_2C_3)$	110.2
		$\angle(C_2C_3C_4)$	108.7
		$\angle(C_5C_1C_2)$	105.7
		$\angle(C_5C_6C_1)$	54.6
		$\angle(C_6C_{51}C_3)$	104.4
		$\angle(H_1C_1C_2)$	120.9
		$\angle(H_1C_1C_5)$	125.8
		$\angle(H_2C_2C_1)$	124.3
		$\angle(H_6C_6C_{15})$	120.7
		$\angle(H_7C_6C_{15})$	122.2

$r(CH) = 1.10$ Å; base 5-membered ring, except for
hydrogens 1 and 5, constrained to be planar

Molecule	Figure	Geometrical parameter[a]	Value[b]
1-Methylcyclo- pentadienyl (IV)	11	$r(C_{Me}C_1)$	1.511
		$\angle(C_{Me}C_1C_5)$	124.4

Structure based on geometry of cyclopentadienyl cation
substituted by a standard methyl group

Molecule	Figure	Geometrical parameter[a]	Value[b]
5-Methylcyclo- pentadienyl (V)	11	$r(C_{Me}C_5)$	1.516
		$\angle(C_{Me}C_5C_4)$	130.7

Structure based on geometry of cyclopentadienyl cation
substituted by a standard methyl group

Molecule	Figure	Geometrical parameter[a]	Value[b]
2-Methylcyclo- pentadienyl (VI)	11	$r(C_{Me}C_2)$	1.526

Structure based on geometry of cyclopentadienyl cation
substituted by a standard methyl group

Molecule	Figure	Geometrical parameter[a]	Value[b]
Cyclopentadienyl- carbinyl (VII)	11	$r(C_1C_2)$	1.519
		$r(C_2C_3)$	1.346
		$r(C_3C_4)$	1.470
		$r(C_1C_6)$	1.473
		$\angle(C_1C_2C_3)$	109.5
		$\angle(C_2C_3C_4)$	109.4
		$\angle(C_2C_1C_5)$	102.2
		$\angle(C_6C_1C_{34})$	103.5
		$\angle(H_1C_1C_{34})$	141.5

—*continued*

Table 1—continued

Molecule	Figure	Geometrical parameter[a]	Value[b]
Cyclopentadienyl-carbinyl (VII)		$\angle(H_2C_2C_1)$	125.3
		$\angle(H_3C_3C_2)$	126.9

$r(CH) = 1.10$ Å; $\angle(H_6C_6C_1) = \angle(H_7C_6C_1) = 120°$; except for H_1, 5-membered ring constrained to be planar

Molecule	Figure	Geometrical parameter[a]	Value[b]
Cyclopenta-dienylcarbinyl, eclipsed (VIII)	11	$r(C_1C_2)$	1.530
		$r(C_2C_3)$	1.319
		$r(C_3C_4)$	1.508
		$r(C_1C_6)$	1.490
		$\angle(C_1C_2C_3)$	109.5
		$\angle(C_2C_3C_4)$	109.4
		$\angle(C_2C_1C_5)$	102.3
		$\angle(C_6C_1C_{34})$	134.1
		$\angle(H_1C_1C_{34})$	126.5
		$\angle(H_2C_2C_1)$	125.3
		$\angle(H_3C_3C_2)$	126.9

$r(CH) = 1.10$ Å; $\angle(H_6C_6C_1) = 120°$; $\angle(H_{67}C_6C_1) = 180°$; except for H_1, 5-membered ring constrained to be planar

Molecule	Figure	Geometrical parameter[a]	Value[b]
Base methyl-cyclopentadienyl, square-based pyramid (IX)	11	$r(C_{Me}C_{cyclo})$	1.518
		$\angle(C_{Me}C_{cyclo}O)^j$	176.3

Structure based on geometry of square pyramid cyclopentadienyl cation substituted by a standard methyl group

Molecule	Figure	Geometrical parameter[a]	Value[b]
Apex methyl-cyclopentadienyl, square-based pyramid (X)	11	$r(C_{Me}C_{cyclo})$	1.520

Structure based on geometry of square pyramid cyclopentadienyl cation substituted by a standard methyl group

Molecule	Figure	Geometrical parameter[a]	Value[b]
$C_7H_7^+$ Tropylium (I)	12	$r(CC)$	1.398
		$r(CH) = 1.10$ Å	
Benzyl (III)	12	$r(C_1C_2)$	1.368
		$r(C_2C_3)$	1.441
		$r(C_3C_4)$	1.368
		$r(C_4C_5)$	1.417

$r(CH) = 1.08$ Å; all $\angle(CCC)$ and $\angle(CCH) = 120°$

Molecule	Figure	Geometrical parameter[a]	Value[b]
$C_8H_9^+$ Methyltropylium (I)	13		

Structure based on geometry of tropylium substituted by a standard methyl group; connecting bond length = 1.49 Å

Molecule	Figure	Geometrical parameter[a]	Value[b]
Para methylbenzyl (II); *meta* methyl-benzyl (IV)	13		

Structures based on geometry of benzyl cation substituted by standard methyl groups; connecting bond lengths = 1.49 Å

Table 1—*continued*

Molecule	Figure	Geometrical parameter[a]	Value[b]
Methylbenzyl (III)	13	$r(C_{Me}C_{benzyl})$	1.488
		Structures based on geometry of benzyl cation substituted by a standard methyl group	
Phenonium (V)	13	$r(c_1C_2)$	1.460
		$r(C_1C_3)$	1.598
		$r(C_3C_4)$	1.431
		$\angle(C_1C_3C_2)$	54.4
		$\angle(C_3C_4C_5)$	122.1
		$\angle(C_4C_3C_8)$	115.8
		$\angle(H_{12}C_1C_2)$	158.2
		$\angle(H_1C_1H_2)$	115.7
		$\angle(H_5C_4C_3)$	118.3
		$r(C_4C_5) = r(C_5C_6) = 1.40$ Å; all $r(CH) = 1.08$ Å; other $\angle(CCC)$ and $\angle(CCH) = 120°$	
Ring-protonated styrene, planar (VI); orthogonal (VII)	13		
		Structures based on geometry of protonated benzene substituted by standard vinyl group; connecting bond lengths = 1.52 Å	
Homotropylium (VIII)	13	$r(C_1C_2)$	1.471
		$r(C_1C_7)$	1.512
		$r(C_1C_8)$	1.510
		$r(C_2C_3)$	1.361
		$r(C_3C_4)$	1.438
		$\angle(C_7C_1C_2)$	127.6
		$\angle(C_1C_2C_3)$	128.5
		$\angle(C_2C_3C_4)$	126.7
		$\angle(C_3C_4C_5)$	134.5
		$\angle(C_1C_8C_7)$	60.1
		$\angle(H_1C_1C_7)$	116.3
		$\angle(H_1C_1C_2)$	112.8
		$\angle(H_2C_2C_1)$	114.6
		$\angle(H_3C_3C_2)$	115.8
		$\angle(H_8C_8C_{71})$	121.9
		$\angle(H_9C_8C_{71})$	121.4
		$r(CH) = 1.10$ Å; except for H_1, H_7, 7-membered ring constrained to be planar	
2-Phenylethyl, orthogonal bisected (IX); orthogonal eclipsed (X); planar bisected (XI); planar eclipsed (XII)	13		
		Structures based on geometry of ethyl cation substituted by a standard model benzene ring; connecting bond lengths = 1.49 Å	

continued

Table 1—continued

Molecule	Figure	Geometrical parameter[a]	Value[b]
Cyclohepta-trienylcarbinyl, eclipsed (XIII)	13	$r(C_1C_2)$	1.490
		$r(C_1C_8)$	1.465
		$r(C_2C_3)$	1.303
		$r(C_3C_4)$	1.493
		$r(C_4C_5)$	1.344
		$\angle(C_8C_1C_{45})$	153.0
		$\angle(H_1C_1C_{45})$	129.1
		$\angle(H_{89}C_8C_1)$	167.5
		$\angle(H_8C_8H_9)$	117.3

$r(CH) = 1.10$ Å; ring $\angle(CCC) = 128.6°$ $(5/7\pi)$; ring $\angle(HCC) = 115.7°$

Molecule	Figure	Geometrical parameter[a]	Value[b]
Cyclohepta-trienylcarbinyl, bisected (XIV)	13	$r(C_1C_2)$	1.491
		$r(C_1C_8)$	1.534
		$r(C_2C_3)$	1.305
		$r(C_3C_4)$	1.491
		$r(C_4C_5)$	1.348
		$\angle(C_8C_1C_{45})$	119.5
		$\angle(H_1C_1C_{45})$	131.9
		$\angle(H_8C_8C_1)$	119.3
		$\angle(H_9C_8C_1)$	124.4

$r(CH) = 1.10$ Å; ring $\angle(CCC) = 128.6°$ $(5/7\pi)$; ring $\angle(HCC) = 115.7°$

[a] Notations such as $\angle(H_{12}CH_3)$ indicate the angle between the bisector of the plane formed by the three atoms H_1, C, H_2, and the bond C–H_3.
[b] STO-3G except: ( )4-31G; [ ]experimental; or where otherwise mentioned.
[c] Reference 15.
[d] X is the projection of the C_1H_1 bond on the plane formed by the three atoms C_1, C_2, and C_3.
[e] X(X') is the projection of the C_1H_1 (C_1H_2) bond on the plane formed by the three atoms C_1, C_2, C_3.
[f] Dihedral angle.
[g] For definition of standard model geometries see J. A. Pople and M. S. Gordon, *J. Am. Chem. Soc.* **89**, 4253 (1967).
[h] O is a point lying in the plane at the center of the 3-membered ring.
[i] X(X') is the projection of the C_1H_1 (C_1H_2) bond on the plane formed by the three carbons.
[j] O is a point lying in the plane at the center of the square base.

Table 2. Theoretical Energy Data (hartrees)[a]

Cation	STO-3G	4-31G	6-31G*
CH^+	−37.45638	-37.83988^{b}	-37.89554^{c}
CH_3^+	−38.77948	-39.17512^{b}	−39.23064
CH_5^+			
$\quad C_s$ (I)	−39.91887	-40.32214^{d}	$-40.38822^{b,e}$
$\quad C_s$ (II)	−39.91887	−40.32207	
$\quad C_{2v}$ (III)	—	—	
$\quad C_{4v}$ (IV)	−39.91589	−40.31893	[g]
$\quad D_{3h}$ (V)	−39.90885	-40.31508^{h}	$-40.36763^{b,i}$

Table 2—*continued*

Cation	STO-3G	4-31G	6-31G*
C_2H^+			
$^3\pi$	−74.87832	−75.67440	—
$^1\Delta$	−74.80630	−75.60705[j]	—
$C_2H_3^+$			
Open (I)	−76.16540	−76.98983[b]	−77.08661[b,k]
Bridged (II)	−76.13652	−76.94913[b]	−77.07483[b,l]
$C_2H_5^+$			
Open, eclipsed (I)	−77.40806	−78.19852[b]	−78.30766[b,m]
Open, bisected (II)	−77.40770	−78.19788[b]	−78.30692[b]
Bridged (III)	−77.38986	−78.18680[b]	−78.30703[b,n]
$C_2H_7^+$			
D_{3d} (I)	−78.54388	−79.34085	—
D_{3h} (II)	−78.54314	−79.34051	—
C_s (III)	−78.54476	−79.33985	—
C_{2v} (IV)	−78.54363	−79.33910	—
C_s (V)	−78.52620	−79.32243	—
C_3H^+			
Propa-1,2-dien-1-yl-3-ylidene (I)	−112.27535	−113.49373	−113.65222
Prop-2-en-1-yl-1,3-diylidene (II)	−112.21407	−113.42066	—
Corner protonated cyclopropene, staggered (XI)	−114.74247	−115.94580	—
Propylidyne, staggered (XII)	−114.71272	−115.89515	—
Propylidyne, eclipsed (XIII)	−114.71007	−115.89288	—
Propan-1-yl-3-ylidene (XIV)	−114.65286	−115.85610	—
$C_3H_7^+$			
2-Propyl (I)	−116.02765	−117.20864	−117.37749
Corner protonated cyclopropane, staggered (II)	−115.99130	−117.18109	−117.35681
Corner protonated cyclopropane, eclipsed (III)	−115.99117	−117.18091	−117.35665
H-bridged propyl (IV)	−115.98959	−117.17957	—
Edge protonated cyclopropane (V)	−115.98450	−117.16541	−117.34713
Face protonated cyclopropane (VI)	−115.77102	−116.98612	−117.17013
$C_4H_5^+$			
Methylcyclopropenyl (I)	−152.23019	−153.81896	—
Cyclobutenyl (II)	−152.20726	−153.79574	—
Cyclopropenylcarbinyl, bisected (III)	−152.13722	−153.75315	—
Cyclopropenylcarbinyl, eclipsed (IV)	−152.09251	−153.70063	—
Cyclopropenylcarbinyl, eclipsed (V)	−152.10123	−153.69285	—

—*continued*

Table 2—continued

Cation	STO-3G	4-31G	6-31G*
$C_3H_3^+$			
Cyclopropenyl (I)	−113.62032	−114.81364	−115.00369
Propargyl (II)	−113.56391	−114.78923	−114.94894
Prop-2-en-1-yl-3-ylidene (III)	−113.51319	−114.72582	−114.89283
Prop-2-en-1-yl-3-ylidene, perpendicular (IV)	−113.45112	−114.67176	−114.83927
1-Propynyl (V)	−113.43715	−114.63828	—
Cycloprop-1-yl-2-ylidene (VI)	−113.42399	−114.60974	—
Propan-1-yl-1, 3-diylidene (VII)	−113.37232	−114.56726	—
Corner protonated cyclopropyne (VIII)	−113.34084	−114.52079	—
$C_3H_5^+$			
Allyl (I)	−114.80953	−116.02511	−116.19106
2-Propenyl, eclipsed (II)	−114.79296	−116.00048	−116.16451
2-Propenyl, staggered (III)	−114.79283	−116.00033	—
H-bridged propenyl (IV)	−114.75238	−115.95875	−116.13981
1-Propenyl, eclipsed (V)	−114.76859	−115.97503	−116.13925
1-Propenyl, staggered (VI)	−114.76668	−115.97306	—
Bridged protonated allene (VII)	−114.74340	−115.96211	−116.13687
Allyl, perpendicular (VIII)	−114.75464	−115.96939	−116.13561
Cyclopropyl (IX)	−114.76523	−115.95095	−116.12865
Corner protonated cyclopropene, eclipsed (X)	−114.74243	−115.94583	−116.12319
$C_4H_7^+$			
1-Methylallyl, *anti* (I)	−153.41560	−155.02684	—
1-Methylallyl, *syn* (II)	−153.40480	−155.01451	—
2-Methylallyl (III)	−153.39761	−155.00895	—
Cyclopropylcarbinyl, bisected (IV)	−153.39490	−154.99277	—
Cyclobutyl, puckered (V)	−153.40237	−154.97776	—
Cyclobutyl, planar (VI)	−153.40896	−154.97358	—
Methylcyclopropyl (VII)	−153.38578	−154.96626	—
Allylcarbinyl, *trans* eclipsed (VIII)	−153.34165	−145.96090	—
Allylcarbinyl, *trans* bisected (IX)	−153.33917	−154.95871	—
Allylcarbinyl, *cis* eclipsed (X)	−153.34274	−154.95719	—
Allylcarbinyl, *cis* bisected (XI)	−153.34000	−154.95576	—

Table 2—*continued*

Cation	STO-3G	4-31G	6-31G*
$C_4H_7^+$			
Cyclopropylcarbinyl, eclipsed (XII)	−153.35301	−154.94407	—
$C_4H_9^+$			
t-Butyl	−154.63918	−156.21460	—
$C_5H_5^+$			
Cyclopentadienyl,[3] (I)	−189.63283	−191.64648	—
Cyclopentadienyl (II)	−189.59995	−191.61514	—
Cyclopentadienyl (III)	−189.59868	−191.61377	—
Cyclopentadienyl, square-based pyramid structure (IV)	−189.56069	−191.52651	—
$C_6H_7^+$			
Protonated benzene, open (I)	−228.25309	−230.67543	—
protonated benzene, H-bridged (II)	−228.20891	−230.64254	—
Bicyclo [3.1.0]hexenyl (III)	−228.21903	−230.62194	—
1-Methylcyclopentadienyl (IV)	−228.20649	−230.61840	—
4-Methylcyclopentadienyl (V)	−228.19488	−230.60717	—
2-Methylcyclopentadienyl (VI)	−228.18873	−230.59982	—
Cyclopentadienylcarbinyl, bisected (VII)	−228.15960	−230.59156	—
Cyclopentadienylcarbinyl, eclipsed (VIII)	−228.14519	−230.57222	—
Base methylcyclopentadienyl, square-based pyramid structure (IX)	−228.15937	−230.52204	—
Apex methylcyclopentadienyl, square-based pyramid structure (X)	−228.15678	−230.51924	—
$C_7H_7^+$			
Tropylium (I)	−265.66771	−268.51634	—
Benzyl (II)	−265.65231	−268.50168	—
$C_8H_9^+$			
Methyltropylium (I)	−304.25431	—	—
Para methylbenzyl (II)	−304.24635	—	—
Methylbenzyl (III)	−304.24126	—	—
Meta methylbenzyl (IV)	−304.23891	—	—
Phenonium (V)	−304.23142	—	—
Ring protonated styrene, planar (VI)	−304.20813	—	—
Ring protonated styrene, orthogonal (VII)	−304.19394	—	—
Homotropylium (VIII)	−304.19104	—	—
2-Phenylethyl, orthogonal bisected (IX)	—	—	—
2-Phenylethyl, orthogonal eclipsed (X)	−304.16401	—	—
2-Phenylethyl, planar bisected (XI)	—	—	—

continued

Table 2—continued

Cation	STO-3G	4-31G	6-31G*
$C_8H_9^+$			
2-Phenylethyl, planar eclipsed (XII)	−304.15365	—	—
Cycloheptatrienylcarbinyl, eclipsed (XIII)	−304.09396	—	—
Cycloheptatrienylcarbinyl, bisected (XIV)	−304.08490	—	—

[a]Unless otherwise indicated, both 4-31G and 6-31G* energies calculated at optimum STO-3G geometry.
[b]Energy calculated using optimum 4-31G geometry.
[c]Energy calculated using optimum 6-31G* geometry.
[d]4-31G energy using 4-31G geometry, −40.32715. Value in table used for energy comparisons.
[e]6-31G** energy (4-31G geometry), −40.40571; from Ref. 15, best SCF energy, −40.4010; estimated total energy, −40.720.
[f]From Ref. 15, best SCF energy, −40.4012; estimated total energy, −40.720.
[g]From Ref. 15, best HF energy, −40.3957; estimated total energy, −40.710.
[h]4-31G energy using 4-31G geometry, −40.31571. Value in table used for energy comparisons.
[i]6-31G** energy (4-31G geometry), −40.38021; from Ref. 15, best SCF energy, −40.3837; estimated total energy, −40.703.
[j]The molecular orbitals were allowed to become complex.
[k]6-31G** energy (4-31G geometry), −77.09293; from Ref. 17, best SCF energy, −77.0330; estimated total energy, −77.3263.
[l]6-31G** energy (4-31G geometry), −77.08382; from Ref. 17, best SCF energy, −77.0246; estimated total energy, −77.3381.
[m]6-31G** energy (4-31G geometry), −78.31831: from Ref. 17, best SCF energy, −78.2647; estimated total energy, −78.5934.
[n]6-31G** energy (4-31G geometry), −78.31831: from Ref. 17, best SCF energy, −78.2547; estimated total energy, −78.5934.

2. CH⁺ (Fig. 1)

The electronic structure of CH^+, the simplest of all carbocations, indicates its potential to act either in the capacity of a nucleophile or an electrophile. Thus, the molecule's valence shell is essentially a loosely bound σ lone pair, while a low-lying degenerate set of π symmetry orbitals is ready to be filled.

3. CH₃⁺ (Fig. 1)

The methyl cation is predicted to be planar (D_{3h} symmetry) in accordance with the simple arguments set forth by Walsh some years ago.[12] As with CH^+, it is devoid of π electrons, although the lowest vacant molecular orbital is of this symmetry, being a carbon $2p$ function orthogonal to the molecular plane. This being the case we would expect CH_3^+ to participate head-on in donor–acceptor complexes with areas rich in electron density (e.g., heteroatom lone pairs, single, and multiple bonds). This, of course, closely parallels the behavior

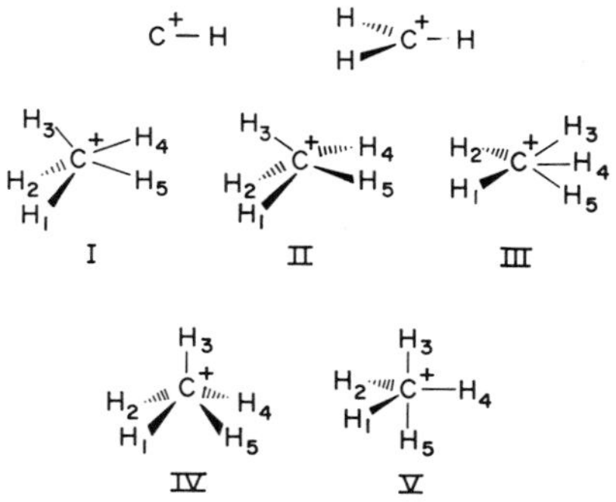

Fig. 1. Theoretical STO-3G (4-31G) geometries for CH^+, CH_3^+, and CH_5^+ cations. CH_5^+ structure III from Ref. 15.

of isoelectronic BH_3 is forming stable complexes with donor molecules such as carbon monoxide and ammonia.

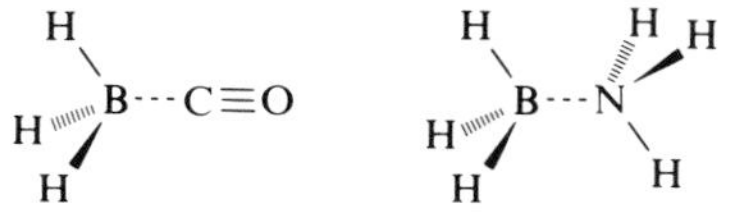

4. CH_5^+ (Fig. 1, Table 3, Refs. 3, 13–15)

The theory's assignment of the equilibrium structure of protonated methane (I) resembles to a great extent one such intermolecular complex involving CH_3^+, with the hydrogen molecule acting as an electron source. Here,

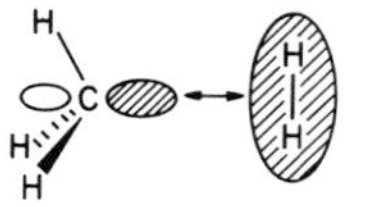

withdrawal of electrons from the σ bonding orbital of the hydrogen molecule leads to a lengthening of the H—H linkage (to 0.850 Å at the 4-31G level from a value of 0.730 Å in the free H_2 molecule).

Table 3. Relative Energies of CH_5^+ Structures[a]

Structure	Hartree–Fock					Ref. 12, SCF + valence shell correlation
	STO-3G	4-31G	6-31G*	6-31G**	Ref. 12, SCF	
C_s (I)	0.0	0.0	0.0	0.0	0.0	0.0
C_s (II)	0.0	0.0	—	—	—	—
C_{2v} (III)	—	—	—	—	5.6	0.2
C_{4v} (IV)	1.9	4.5	—	—	9.0	6.3
D_{3h} (V)	6.3	7.2	12.9	16.0	16.6	10.8

[a] Values given in kilocalories per mole.

A number of other possible structures for CH_5^+ have been considered (II–V) both by single-determinant methods closely approaching the Hartree–Fock limit,[14,15] and by computational schemes in which partial account is taken of the differential effects of electron correlation.[15] The energies of the two are comparable to that of the predicted ground state form (I). One of these (II) again has the geometry of an intermolecular complex involving the methyl cation and H_2, and differs from the lowest energy structure only in the relative orientation of the two groups. The energy corresponding to a geometrical structure of C_{2v} symmetry (III) is lowered so much upon consideration of correlation effects that it also cannot be ignored as a potential secondary form or even possibly the ion's ground state. A square-based pyramid structure (IV) is probably unstable with respect to distortion into lower symmetry, but no doubt represents a low-energy transition state to hydrogen scrambling in II.

Finally, the energies of both a trigonal bipyramid form (V)—the usual S_N2 transition state—and an ion of C_{3v} symmetry (corresponding to collinear approach of a hydrogen atom onto one of the CH bonds of methane) are far higher.

5. C_2H^+ (Fig. 2, Refs. 4, 13)

Both singlet and ground triplet states[16] of the ethynyl cation are predicted to be linear, with the carbon–carbon bond length in the former species being far longer than that found in acetylene. No structure incorporating a symmetric hydrogen bridge seems to have been investigated.

6. $C_2H_3^+$ (Fig. 2, Refs. 4, 13, 14, 17)

At the single-determinant level, the vinyl cation is seen to prefer an open "classical" geometry (I) rather than a structure in which one of the hydrogens bridges the carbon–carbon bond (II). The best estimate of the energy difference between the two (barrier to hydrogen migration) is 5.7 kcal/mol, as given by

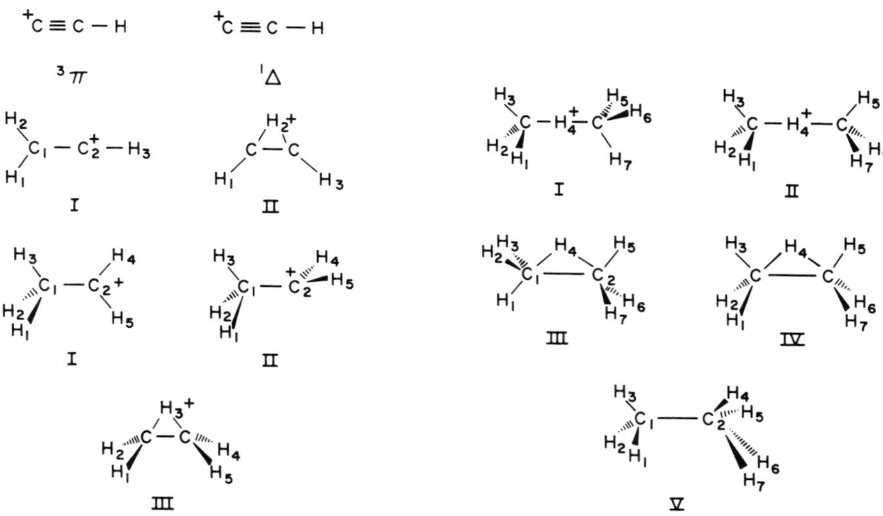

Fig. 2. Theoretical STO-3G (4-31G) geometries for C_2H^+, $C_2H_3^+$, $C_2H_5^+$, and $C_2H_7^+$ cations.

the 6–31G** full polarization basis set.† Kutzelnigg[17] has recently reported, however, that after due account is taken for the effects of electron correlation, the prediction of the single-determinant theory (that is a favoring of an open structure for $C_2H_3^+$) is overturned. Thus, it appears that the vinyl cation provides us with the simplest example of the possibility of a symmetrical hydrogen bridge between two carbons.

7. $C_2H_5^+$ (Fig. 2, Refs. 4, 13, 14, 17)

The polarization basis 6-31G** computations of Hariharan, Lathan, and Pople[14] suggest that the ethyl cation also adopts a hydrogen-bridged geometry (III) in its electronic ground state. Calculations extending beyond the Hartree–Fock limit further this notion,[17] not at all surprising in view of the effects of correlation energy on the relative stabilities of open and bridged isomers of the vinyl cation. In its open form, the ethyl cation exhibits no preference whatsoever for an eclipsed (I) or bisected (II) arrangement of the methyl group relative to the carbonium center.[18] This is in keeping with simple arguments based on the perturbation theoretic treatment of interacting orbitals.[19] The calculated carbon–carbon bond length in either of these two forms of the open ethyl cation (1.488 and 1.484 Å at the STO-3G level for I and II, respectively, and 1.440 and 1.443 Å using 4-31G basis set) is considerably shorter than that found in staggered ethane (1.538 and 1.529 Å, respectively, at the

---

†This is formed from the split-valence-shell 6-31G set[11] by addition of d-type polarization functions on first row atoms and p-type orbitals on hydrogen. See Ref. 10.

Table 4. Relative Energies of $C_2H_7^+$ Structures[a]

Structure	STO-3G	4-31G
D_{3d} (I)	0.0	0.0
D_{3h} (II)	0.5	0.2
C_s (III)	−0.6	0.6
C_{2v} (IV)	0.2	1.1
C_s (V)	11.1	11.6

[a]Values given in kilocalories per mole.

minimal and extended basis set levels). This shrinkage is presumably due to the formation of a new hyperconjugative linkage between the methyl group and the carbonium center.

8. $C_2H_7^+$ (Fig. 2, Table 4, Refs. 4, 13)

The marked resemblance of the ground state structure of CH_5^+ to that of a weak intermolecular complex between the methyl cation and H_2 provides a clue to one of the possible equilibrium geometries of protonated ethane (V). Indeed, here the notion of a complex is even more appropriate, for the geometries of the two partners (bisected open ethyl cation and the hydrogen molecule) differ only slightly from those of the free molecules. The $C_2H_7^+$ surface offers us yet another minimum-energy form, one in which the proton associates itself with the carbon–carbon bond of ethane rather than with one of the CH linkages. According to the split-valence-shell 4-31G calculations the structure which results (I) is of the same symmetry as ethane itself (i.e., the proton is inserted directly into the center of the C–C bond), and at this level represents the absolute minimum on the $C_2H_7^+$ potential surface. Polarization basis calculations would certainly be of use here in order to distinguish between these two candidates for the structure of protonated ethane (separated by 11.6 kcal/mol at the 4-31G level). The geometries of several other possible isomers of $C_2H_7^+$ have been investigated at the STO-3G level, and single-point calculations at each of the resulting structures have been carried out using the extended 4-31G basis set. Although each of the structures quoted no doubt corresponds to a local minimum at the STO-3G level (at least within the symmetry group to which optimization was restricted), we seriously doubt whether any structures other than I and V represent stable forms within the 4-31G approximation.

9. C_3H^+ (Fig. 3, Ref. 18)

Radom *et al.*[20] have recently reported geometries and relative energies for the two possible acyclic forms of C_3H^+, the propa-1,2-dien-1-yl-3-ylidene

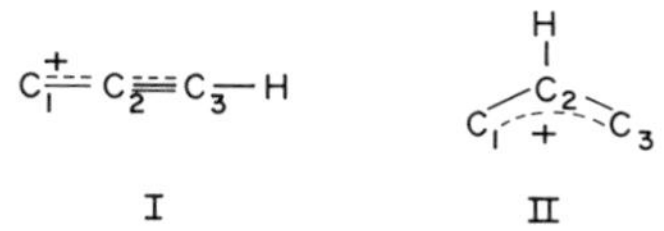

$$C_1^+ = C_2^+ = C_3 - H \qquad \overset{H}{\underset{C_1^+}{\overset{|}{C_2}}} \overset{C_3}{C_3}$$

Fig. 3. Theoretical STO-3G geometries for C_3H^+ cations.   I   II

(I) and prop-2-en-1-yl-1,3-diylidene (II) cations. At the 4-31G level the former was found to be 46 kcal/mol more stable. The possibility of cyclic isomers of C_3H^+ was also investigated, but all attempted structures collapsed to I without activation.

10. $C_3H_3^+$ *(Fig. 4, Table 5, Ref. 20)*

The cyclopropenyl cation (I) is the simplest example of a molecule stabilized by aromatic conjugation. The theoretical geometry of the molecule, an equilateral triangle with carbon–carbon bond lengths of 1.377 Å, is in essential agreement with a variety of x-ray crystal structures on a number of simple derivatives.[21] The exceptional stability of the cyclopropenyl cation

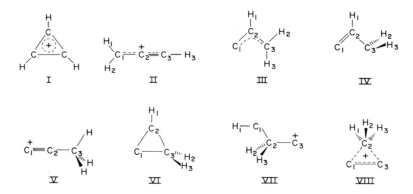

Fig. 4. Theoretical STO-3G geometries for $C_3H_3^+$ cations.

Table 5. *Relative Energies of* $C_3H_3^+$ *Structures*[a]

Structure	STO-3G	4-31G	6-31G*
Cyclopropenyl (I)	0.0	0.0	0.0
Propargyl (II)	35.4	15.3	34.4
Prop-2-en-1-yl-3-ylidene (III)	67.2	55.1	69.6
Perpendicular prop-2-en-1-yl-3-ylidene (IV)	106.2	89.0	103.2
1-Propynyl (V)	114.9	110.0	—
Cycloprop-1-yl-2-ylidene (VI)	123.2	128.0	—
Propan-1-yl-1,3-diylidene (VII)	155.6	154.6	—
Corner protonated cyclopropyne (VIII)	175.4	183.8	—

[a] Values given in kilocalories per mole.

may be seen by a comparison of its *isodesmic* bond separation energy [reaction (1)] to that of neutral cyclopropene [reaction (2)]. Thus, the strain energy inherent to incorporation of a double bond into a 3-membered ring is almost compensated for by the π electron stabilization which results.

$$\triangle + 2CH_4 + CH_3^+ \rightarrow 2CH_3-CH_2^+ + CH_2=CH_2 \tag{1}$$

$$\Delta E\ (6\text{-}31G^*) = -13.9\ \text{kcal/mol}$$

$$\triangle + 3CH_4 \rightarrow 2CH_3-CH_3 + CH_2=CH_2 \tag{2}$$

$$\Delta E\ (6\text{-}31G^*) = -50.4\ \text{kcal/mol}$$

The second lowest energy form on the $C_3H_3^+$ surface [34.4 kcal/mol above the ground-state structure (I) at the 6-31G* level] is the propargyl cation (II), the theoretical (STO-3G) geometry of which is intermediate between those implied by either of its classical resonance structures. Like the open form of the

vinyl cation,[4,13] one hydrogen is colinear to the carbon skeleton. All other forms of $C_3H_3^+$ which have been investigated (III–VIII) exhibit far higher energies; the best of these, prop-3-en-1-yl-3-ylidene (III), being 69.6 kcal/mol removed from the aromatic cyclopropenyl cation at the 6-31G* level. Of the remaining structures, we strongly suspect that the perpendicular form of prop-2-en-1-yl-3-ylidene (IV) freely rotates into a planar geometry (III), and that the 3-membered ring structure, cycloprop-1-yl-2-ylidene (VI), undergoes opening to III with little or no activation.

11. $C_3H_5^+$ *(Fig. 5, Table 6, Refs. 23, 24)*

The allyl cation (I) occupies the position of lowest energy on the $C_3H_5^+$ potential surface. In its planar arrangement, the carbon–carbon bond lengths are midway in between those of normal single and double linkages (1.385 Å for the allyl cation at the STO-3G level compared to 1.548 Å in ethane and 1.306 Å in ethylene),[4,13] an observation which might easily have been anticipated by examination of its principal resonance forms. Twisting one of the methylene groups by 90° is predicted to require some 35 kcal/mol (both at the

4-31G and 6-31G* levels) and corresponds to one possible transition for stereomutation in the parent system. This barrier is probably an upper bound

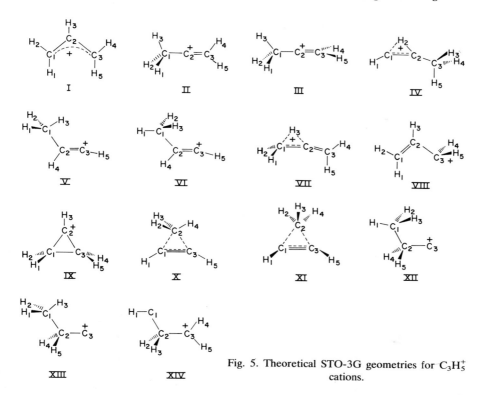

Fig. 5. Theoretical STO-3G geometries for $C_3H_5^+$ cations.

Table 6. *Relative Energies of $C_3H_5^+$ Structures*[a]

Structure	STO-3G	4-31G	6-31G*
Allyl (I)	0.0	0.0	0.0
2-Propenyl, eclipsed (II)	10.4	15.5	16.7
2-Propenyl, staggered (III)	10.5	15.6	—
H-Bridged propenyl (IV)	25.9	41.6	32.2
1-Propenyl, eclipsed (V)	25.7	31.4	32.5
1-Propenyl, staggered (VI)	26.9	32.7	—
Bridged protonated allene (VII)	41.5	39.5	34.0
Allyl, perpendicular (VIII)	34.4	35.0	34.8
Cyclopropyl (IX)	27.8	46.5	39.2
Corner protonated cyclopropene, eclipsed (X)	42.1	49.8	42.6
Corner protonated cyclopropene, staggered (XI)	42.1	49.8	—
Propylidyne, staggered (XII)	60.7	81.5	—
Propylidyne, eclipsed (XIII)	62.4	83.0	—
Propan-1-yl-3-ylidene (XIV)	98.3	106.1	—

[a] Values given in kilocalories per mole.

due to the inability of single-determinant theory to properly describe the diradical character of the orthogonal structure.[25] Radom, Pople, and Schleyer

have already noted that the activation energy for such a process is largely insensitive to substitution at the 2 position. An alternative mode by which stereomutation might occur is with the involvement of the cyclopropyl cation (IX). Although Radom *et al.* have shown that such a process is not energetically

competitive in the parent system, the stabilization afforded to the cyclopropyl species, relative to allyl, by a variety of substituents at the 2 position make it a viable pathway to stereomutation in these ions. Indeed, attachment of π-donor–σ-acceptor substituents such as NH_2 and OH actually lead to reversal in energies of the allyl and cyclopropyl cations. A summary is presented in Table 7. The parent cyclopropyl cation is itself unstable with respect to symmetry-allowed disrotatory opening into the planar allyl system. That is to say, it represents a maximum rather than a local minimum on the $C_3H_5^+$ energy surface. Opening of the 3-membered ring in a conrotatory manner requires an activation energy of at least 23 kcal/mol, a value which corresponds to the "cost" of violating orbital symmetry control.

Table 7. Effect of Substituents on the Relative Energies of Allyl and Cyclopropyl Cations

X	$E\left(\begin{smallmatrix}X\\ \triangle \end{smallmatrix}\right) - E\left(\begin{smallmatrix}X\\ \triangle \end{smallmatrix}\right)^a$
H	39.2
CH_3	18.8
NH_2	−33.5
OH	−9.8
F	14.9

[a] Except for X = H calculated from the STO-3G energy changes for:

and the 6-31G* relative energies for the parent ions. For X = H calculated directly from 6-31G* energies (kcal/mol).

The second lowest energy form on the $C_3H_5^+$ potential is the 2-propenyl cation, which shows a slight preference for an eclipsed arrangement of hydrogens (II), rather than one where they are staggered (III). The hydrogen-bridged structure (IV) linking this system to the 1-propenyl cation (V, VI) lies some 16 kcal/mol higher in energy, but slightly below that of 1-propenyl itself.

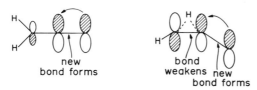

Comparison of this system with that of the vinyl cation which, recall, is predicted to adopt a bridged equilibrium structure is interesting. Thus, substitution by a methyl group on the carbonium center of the open vinyl cation is far more efficient energetically that that on the charge-delocalized hydrogen-bridged ion. Consider the following rationalization.[26] Charge transfer from the substituent methyl into the carbonium center of the open vinyl cation results primarily in the formation of a new hyperconjugative linkage between the donor and acceptor fragments. On the other hand, electrons donated into the empty orbital of the hydrogen-bridged vinyl cation go into a function which is not only delocalized over the carbon skeleton, but is also antibonding in character. In this instance, charge transfer leading to a new substituent–adduct linkage actually succeeds in weakening the carbon–carbon bond of the

new
bond forms

bond
weakens new
bond forms

bridged-vinyl system. We might then expect the overall stabilizing effect of the substituent methyl to be severely diminished. The theoretical (6-31G*) energies of reactions (3) and (4) indeed show this to be the case, stabilization by a methyl substituent imparted to the open vinyl cation being approximately one and a half times that given to the hydrogen-bridged system:

$$\Delta E \ (6\text{-}31G^*) = 27.9 \ \text{kcal/mol} \tag{3}$$

$$\Delta E \ (6\text{-}31G^*) = 19.8 \ \text{kcal/mol} \tag{4}$$

Interconversion of the allyl (I) and 2-propenyl (II) cations, via a hydrogen bridged form of protonated allene (VIII), is predicted to require an activation energy of 34 kcal/mol (18.3 kcal/mol in the reverse direction). Degenerate interconversion of equivalent 1-propenyl cations (V, VI) may occur by way of a corner protonated cyclopropene transition state (X, XI), with an activation of approximately 10 kcal/mol being necessary. All remaining forms on the $C_3H_5^+$ surface are far higher in energy.

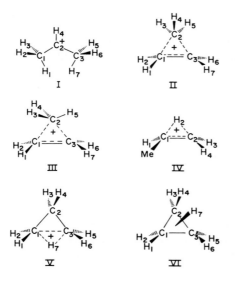

12. $C_3H_7^+$ (Fig. 6, Table 8, Ref. 26)

The isopropyl cation (I) is by far the most stable form on the $C_3H_7^+$ energy surface. It adopts an equilibrium conformation in which one hydrogen from

Fig. 6. Theoretical STO-3G geometries for $C_3H_7^+$ cations.

Table 8. Relative Energies of $C_3H_7^+$ Structures[a]

Structure	STO-3G	4-31G	6-31G*
2-Propyl (I)	0	9	0
Corner protonated cyclopropane, staggered (II)	22.8	17.3	13.0
Corner protonated cyclopropane, eclipsed (III)	22.9	17.4	13.1
H-Bridged propyl (IV)	23.9	18.2	—
Edge protonated cyclopropane (V)	27.1	27.1	19.1
Face protonated cyclopropane (VI)	161.0	139.6	130.1

[a] Values given in kilocalories per mole.

each methyl group eclipses the CH linkage at the carbonium center, quite the opposite tendency than that displayed in the structures of a wide variety of related double-rotor systems (e.g., propene and acetone). The rationale for this reversal in conformational preference has already been discussed at length.[29] In the ground-state conformations of molecules such as propene and acetone one might be tempted to visualize the possibility of overlap between the two pairs of out-of-plane methyl hydrogens, leading to the formation of a "π" electron cycle. Each methyl rotor supplies two electrons to the total, and the

overall character of the ring is determined by the contribution of the central grouping. If, as in the case of the neutral molecules noted above this approaches two electrons, then the cycle will contain 6π electrons in total and be "aromatic." In the isopropyl cation, however, the carbonium center makes no contribution, and hence we are left with the possibility of forming a 4π electron "antiaromatic" arrangement, sufficient cause for the methyl groups to rotate into a conformation where substantial overlap is no longer possible.

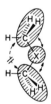

It is the nature of the second possible minimum energy form of $C_3H_7^+$ which has received the most attention in recent days.* Three structures are being considered. One form of protonated cyclopropane (II) is located 13 kcal/mol above the isopropyl cation (at the 6-31G* level). In view of the details of its geometrical structure, it is more adequately described, perhaps, in terms of a weak intermolecular complex between ethylene and the methyl cation. A similar structure for protonated cyclopropane (III) in which one of the methyl group hydrogens lies in the plane of the three carbons rather than being perpendicular to it, is found to be only slightly higher in energy. In their most recent work, Hariharan *et al.*[28] indicate a strong favoring of one of these two structures as the second stable form of $C_3H_7^+$. At the 4-31G level the H-bridged propyl cation (IV) is also of moderate energy, being within one kcal/mol of protonated cyclopropane. Unfortunately, calculations using the 6-31G* polarization basis set failed to give a converged solution for this

*See Ref. 24 for a review of the experimental and theoretical literature.

species, and hence did not provide a more accurate assessment of its energy disposition. It seems highly unlikely to this author, however, that such a structure would be stable with respect to the completion of hydrogen migration to the ground-state isopropyl cation. A third possibility (V) corresponds to protonation of cyclopropane along one of its edges. This is some 6 kcal/mol less stable than the corner protonated form and probably relaxes to such a structure with little or no activation. Finally, a form of face-protonated cyclopropane (VI) was investigated but found to be extremely high in energy.

$C_3H_7^+$ is perhaps the largest carbocation for which the potential surface has been examined in such detail as to carefully consider all reasonable structural possibilities. Beyond this, treatment has been far more sketchy and most attention has been focused on specific problems in which a carbocation is a likely intermediate.

13. $C_4H_5^+$ (Fig. 7, Table 9, Ref. 30)

Of the limited number of forms of $C_4H_5^+$ which have been investigated theoretically the methylcyclopropenyl cation is the most stable, not at all surprising in view of its incorporation of an aromatic 2π electron cycle. Of greater interest, though, is the possibility of a 2π electron homoatomic system, the homocyclopropenium cation. Here one might consider the formal

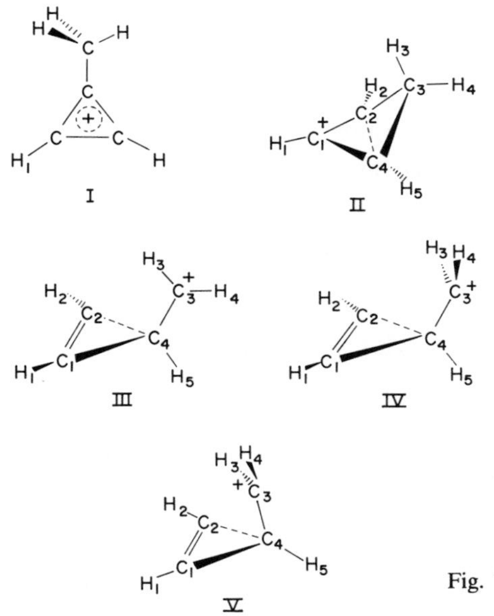

Fig. 7. Theoretical STO-3G geometries for $C_4H_5^+$ cations.

Table 9. *Relative Energies of* $C_4H_5^+$ *Structures*[a]

Structure	STO-3G	4-31G
Methylcyclopropenyl (I)	0.0	0.0
Cyclobutenyl (II)	14.4	14.6
Cyclopropenylcarbinyl, bisected (III)	58.3	41.3
Cyclopropenylcarbinyl, eclipsed (IV)	86.4	74.3
Cyclopropenylcarbinyl, eclipsed (V)	80.9	79.1

[a]Values given in kilocalories per mole.

double bond in the cyclopropenyl system to have been replaced by a cyclopropane moiety. The logic behind such a move is easily suggested by the overall similarity between the highest filled π and lowest empty π^* orbitals on an unsaturated linkage, and the occupied valence pair of (Walsh) orbitals on the

small ring.[32] Two different geometrical structures may result from such a substitution; one (the bicyclobutyl cation) in which the integrity of the cyclo-

propane ring has largely been maintained, and the other (a cyclobutenyl form) in which the ring-fused bond has undergone opening. The calculations suggest

 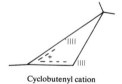

Bicyclobutyl cation   Cyclobutenyl cation

that of the two only the cyclobutenyl cation (II) is stable, and that the bicyclobutyl structure undergoes bond fission without activation. This is obviously akin to the instability of the cyclopropyl cation previously noted. In this instance, ring opening to an allyl cation also occurs without activation via a symmetry-allowed disrotatory pathway. A crystal structure of an $AlCl_3$ complex involving a substituted cyclobutenyl cation has recently been reported.[33] The details of the geometry are essentially the same as those given by the theory, except for the fact that the STO-3G method appears to have underestimated the degree to which the 4-membered ring is puckered. This is not entirely surprising in view of previous experience with cyclobutane,[6] where the theory's assignment of the deviation of the carbon skeleton from planarity (6.9°) is about one-fourth of what is observed experimentally. Parent cyclobutenyl cation has itself been observed by NMR under conditions of long life in super-acid media.[34,35] This ion has been established to have a puckered carbon skeleton and to undergo ring inversion—presumably through a planar arrangement—with a barrier of 8.4 kcal/mol. This is a far larger value than calculated by the theory (0.4 kcal/mol at the 4-31G level), but again is consistent with the poor description of ring puckering found by these methods for cyclobutane. Here the barrier to inversion using the 4-31G basis (at optimum STO-3G geometries) was a mere 0.1 kcal/mol, an order of magnitude below the experimental value (1.28 kcal/mol). It is interesting that although both the theoretical (STO-3G) and crystal structure geometries show the cyclobutenyl cation to incorporate a nearly "normal" allylic moiety (at least as far as bond lengths are concerned), the NMR spectra of the species are not easily interpretable as such. Further scrutiny is necessary.

1.392 Å   1.385 Å

Further attention on the $C_4H_5^+$ surface has been directed to the symmetry-allowed and symmetry-forbidden circumambulation processes open to the cyclobutenyl cation, via the eclipsed (IV or V) and bisected (III) forms of cyclopropenylcarbinyl, respectively. Normal orbital symmetry control would dictate that such a suprafacial [1,3]sigmatropic migration occur via an eclipsed cyclopropenylcarbinyl transition state or an intermediate (IV and V), rather

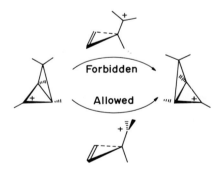

than a bisected structure (III). However, the calculations at the 4-31G level suggest just the opposite, the symmetry-forbidden circumambulation pathway,

$$II \rightarrow III \rightarrow II$$

requiring 33 kcal/mol less activation than the symmetry-allowed case. We have already suggested the motive behind this seeming violation of orbital symmetry control,[30] by indicating that the dominant interaction—at least as far as the stereochemistry of migration goes—is not that involving the π system of the polyene but rather the subjacent Walsh orbital on the 3-membered ring.

14. $C_4H_7^+$ (Fig. 8, Table 10, Ref. 34)

The most stable ion on the $C_4H_7^+$ potential surface is the 1-methylallyl cation (I), arranged such that the methyl group is *anti* to the allylic skeleton.

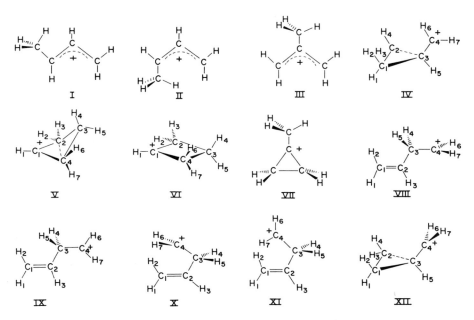

Fig. 8. Theoretical STO-3G geometries for $C_4H_7^+$ cations.

Table 10. *Relative Energies of $C_4H_7^+$ Structures*[a]

Structure	STO-3G	4-31G
1-Methylallyl, *anti* (I)	0.0	0.0
1-Methylallyl, *syn* (II)	6.8	7.7
2-Methylallyl (III)	11.3	11.2
Cyclopropylcarbinyl, bisected (IV)	13.0	21.4
Cyclobutyl, puckered (V)	8.3	30.8
Cyclobutyl, planar (VI)	4.2	33.4
Methylcyclopropyl (VII)	18.7	38.0
Allylcarbinyl, *trans* eclipsed (VIII)	46.4	41.4
Allylcarbinyl, *trans* bisected (IX)	48.0	42.8
Allylcarbinyl, *cis* eclipsed (X)	45.7	43.7
Allylcarbinyl, *cis* bisected (XI)	47.4	44.6
Cyclopropylcarbinyl, eclipsed (XII)	39.2	51.7

[a] Values given in kilocalories per mole.

The *syn* isomer (II) is some 7.7 kcal/mol higher in energy at the 4-31G level. The energy separation between the methylcyclopropyl cation (VII) and its ring-opening product, 2-methylallyl (III) (26.8 kcal/mol at the 4-31G level) is considerably less than that found in the parent systems (46 kcal/mol at 4-31G, 39 at 6-31G*). The cyclopropyl system is still the less stable of the two, although it is not apparent in this instance whether or not its conversion to the 2-methylallyl cation is in need of activation.

Perhaps the most interesting aspect of the $C_4H_7^+$ surface, and certainly the most controversial, is the nature of the intermediate involved both in the rapid interconversion of cyclopropylcarbinyl, cyclobutyl, and allylcarbinyl deriva-

tives in acid solution, and in the scrambling of carbon identities.[39] The tricyclobutonium ion was early suggested as a likely candidate, but more

careful scrutiny of the experimental data revealed that the carbon scrambling was incomplete, thus ruling out the possibility of a single intermediate of such

high (C_{3v}) symmetry. Perhaps one might now be inclined to think in terms of the stable existence of one of the cyclopropylcarbinyl, cyclobutyl, or allylcarbinyl cations, or of a rapid equilibration among two or more of them. Alternatively, the intermediate ions may be unrelated to any of its precursor molecules. Although there are clear examples in the literature both for substituted cyclopropylcarbinyl and cyclobutyl cations,[40] the proton and ^{13}C NMR

spectra of parent $C_4H_7^+$ formed upon super-acid ionization of either cyclopropylcarbinol or cyclobutanol are not easily interpretable as such.[40] Neither is it apparent that these data may be fit to a picture involving the dynamic equilibration of such species. Rather, it has been argued[40] that the NMR data are best represented in terms of a rapid equilibrium of σ delocalized ions with little or no structural resemblance to their precursors.

The bisected cyclopropylcarbinyl cation (IV) is this theory's choice for the structure of the $C_4H_7^+$ intermediate. Both puckered and planar cyclobutyl cations (V and VI) and all forms of allylcarbinyl (VIII–XI) are predicted to collapse to it without activation. Furthermore, the $C_4H_7^+$ potential surface has thus far revealed no other forms of comparable stability. The puckered cyclobutyl cation is itself one possible transition state for the skeletal scrambling in cyclopropylcarbinyl, an activation of 9.4 kcal/mol (at the 4-31G level)

being required for the process. The fact that the endo and exo hydrogens on the 3-membered ring and on the carbonium center maintain their integrity throughout all this is well accounted for by the calculations. Thus, isomerization about the ring–carbonium center bond—through an eclipsed cyclopropylcarbinyl cation (XII)—is predicted by the 4-31G basis to require 31 kcal/mol in the parent system, a value far greater than normally observed for single-

IV   XII   IV

bond torsions (1–5 kcal/mol) and only slightly below the range for twists about unsaturated linkages (40–60 kcal/mol). Experimentally the torsional barrier for the tertiary ion dimethylcyclopropylcarbinyl is found to be 13.7 kcal/mol,[41] in reasonable agreement with a theoretical (4-31G) value of 16.6 kcal/mol.[37]

The theoretical geometries of both bisected and eclipsed forms of cyclopropylcarbinyl are interesting in that they demonstrate the ability of simple orbital arguments to anticipate gross structural features.[26] In the bisected arrangement the carbonium center acts to remove electrons from the antisymmetric member of the degenerate set of highest occupied Walsh orbitals on cyclopropane. Thus, bonding density is removed from two of the small ring's linkages resulting in their weakening and, presumably, lengthening. At the

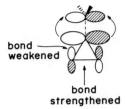

same time the antibonding character of the third linkage should be diminished, and in response it should contract. The STO-3G equilibrium structure of bisected cyclopropylcarbinyl clearly exhibits these geometrical distortions.

1.61 Å,   1.50 Å in cyclopropane

1.45 Å

Little change in the structure of the 3-membered ring incorporated into the eclipsed cyclopropylcarbinyl cation is to be expected, in this case because overlap between the empty orbital at the carbonium center and the symmetric

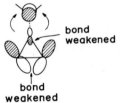

Walsh component is only slight. What little interaction there is should lead to lengthening of all three bonds in the small ring. The STO-3G calculations concur.

1.51 Å   1.50 Å in cyclopropane

1.51 Å,

15. $C_4H_9^+$ *(Fig. 9, Ref. 42)*

The *t*-butyl cation is the only form on the $C_4H_9^+$ potential surface whose structure has been investigated theoretically. This is predicted to adopt a

Fig. 9. Theoretical STO-3G geometries for $C_4H_9^+$ cations.

Table 11. *Relative Energies of Butyl Cations*[a]

Isomer	STO-3G	Experiment
t-Butyl	0	0
2-Butyl	18	16
i-Butyl	36	33
1-Butyl	39	35

[a] Values given in kilocalories per mole.

propellerlike geometry (C_{3h} symmetry) rather than an arrangement of C_{3v} symmetry, an observation consistent with the theory's findings on the isopropyl

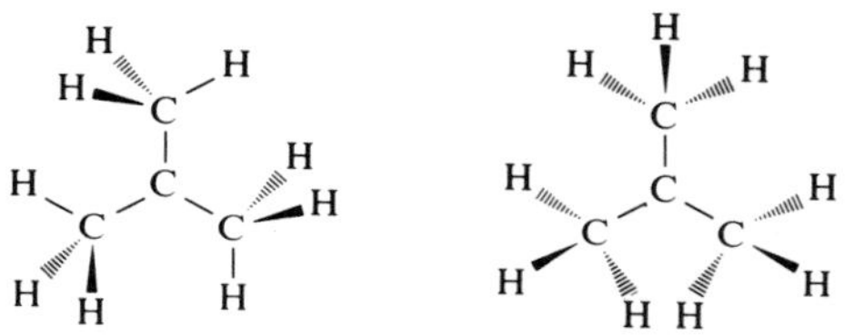

cation,[27,29] but opposite that reported by Olah and co-workers from investigation of the ir and Raman spectra of the species in super-acid media.[43] A study by Radom, Pople, and Schleyer,[44] at the STO-3G level and without the benefit of structure optimization, assigned the relative energies of the isomeric butyl cations in reasonable accord with available thermochemical data (Table 11).

16. $C_5H_5^+$ (Fig. 10, Table 12, Ref. 45)

The singlet cyclopentadienyl cation, if constrained to a planar D_{5h} symmetry, should be antiaromatic. Not only would we expect such a molecule to prefer a triplet ground-state electronic configuration, to which both experiment[46] and the quantitative *ab initio* calculations attest, but we would also

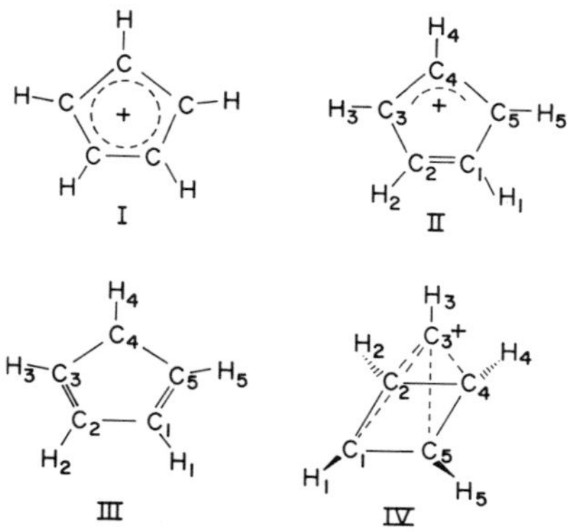

Fig. 10. Theoretical STO-3G geometries for $C_5H_5^+$ cations.

Table 12.   Relative Energies of $C_5H_5^+$ Structures[a]

Structure	STO-3G	4-31G
Cyclopentadienyl[3] (I)	0.0	0.0
Cyclopentadienyl (II)	20.6	17.2
Cyclopentadienyl (III)	21.4	18.1
Cyclopentadienyl, square-based pyramid structure (IV)	44.8	71.0

[a] Values given in kilocalories per mole.

anticipate that geometric distortions would occur to reduce the overall symmetry of the singlet species. Indeed, two such distorted forms of the singlet cyclopentadienyl cation appear on the $C_5H_5^+$ surface, one corresponding to a structure incorporating a solitary double bond (II), the other with two short linkages (III). It is certainly possible that if a further reduction of symmetry (below C_{2v}) had been allowed for in the calculations, both of these would have collapsed to a single form. The antiaromatic character of singlet cyclopentadienyl cation—even in a geometry distorted from fivefold symmetry—is evident from the energy of its bond separation reaction (5).

$$\text{(structure)} + 4CH_4 + CH_3^+ \rightarrow CH_3-CH_3 + 2CH_3-CH_2^+ + CH_2=CH_2 \qquad (5)$$

$$\Delta E \text{ (4-31G)} = -3 \text{ kcal/mol}$$

The possibility of a different type of geometrical structure for $C_5H_5^+$, one of a square-based pyramid (IV), was first proposed by Williams in a study of the isoelectronic carborane systems.[47] Stohrer and Hoffmann independently came to the same conclusion shortly thereafter, on the basis of orbital symmetry arguments.[48] Reports of experimental detection followed almost immediately.[49-51] Although the *ab initio* calculations do indicate that the square-based pyramid form of $C_5H_5^+$ is a stable entity, at least with respect to collapse to a number of other structural possibilities, its energy at the 4-31G level is more the 50 kcal/mol above that of planar singlet cyclopentadienyl. If this estimate is by any means reasonable, it seems unlikely that the species will ever be thoroughly characterized experimentally.

17. $C_6H_7^+$ (Fig. 11, Table 13, Refs. 52, 53)

Three of the structures of $C_6H_7^+$ which have been considered in detail correspond to forms of protonated benzene. The most stable is the open cyclohexadienyl cation (I), in line with numerous infrared, ultraviolet, and NMR studies on a variety of protonated aromatics (including benzene itself) in strong-acid media.[54] Also, the theoretical STO-3G geometry of (I) is consistent with an x-ray crystal structure of the tetrachloroaluminate salt of the

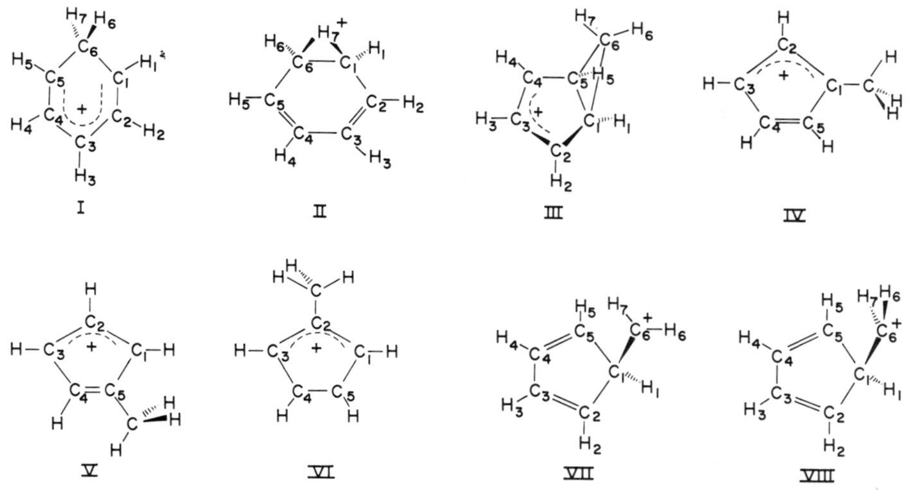

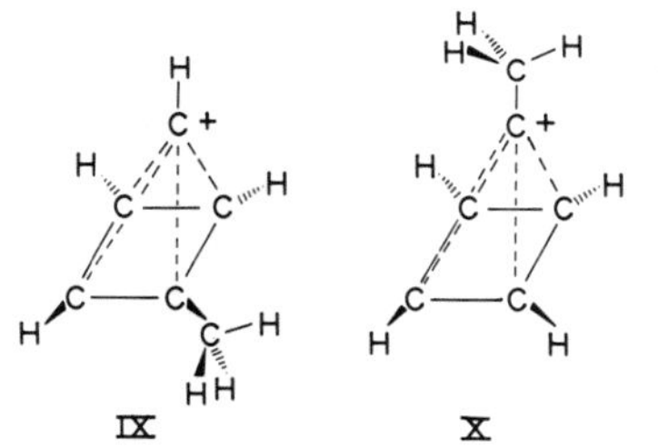

Fig. 11. Theoretical STO-3G geometries for $C_6H_7^+$ cations.

Table 13. *Relative Energies of $C_6H_7^+$ Structures*[a]

Structure	STO-3G	4-31G
Protonated benzene (open) (I),	0.0	0.0
Protonated benzene (H-bridged) (II)	27.7	20.6
Bicyclo[3.1.0]hexenyl (III)	21.4	33.6
1-Methylcyclopentadienyl (IV)	29.2	35.8
4-Methylcyclopentadienyl (V)	36.5	42.8
2-Methylcyclopentadienyl (VI)	40.4	47.4
Cyclopentadienylcarbinyl, bisected (VII)	58.7	52.6
Cyclopentadienylcarbinyl, eclipsed (VIII)	67.7	64.8
Base methylcyclopentadienyl, square-based pyramid structure (IX)	58.8	96.3
Apex methylcyclopentadienyl, square-based pyramid structure (X)	60.4	98.0

[a] Values given in kilocalories per mole.

heptamethylbenzene cation,[55] though a number of more recent investigations on other derivatives have suggested the possibility that the 6-membered ring is puckered.[56] The hydrogen-bridged structure (II), corresponding to protonation in the center of one of the carbon–carbon bonds in benzene, was found to be 20.6 kcal/mol less stable than the open form (I) at the 4-31G level. Although experience with protonated ethylene indicates that our estimate of the energy gap in the benzene system is probably too large, we seriously doubt that the theory's assignment of an open ground-state structure for the ion will be overturned by more extensive calculation. Experimentally, hydrogen scrambling in protonated benzene (presumably via such a bridged structure as II) has been estimated to require activation of around 10 kcal/mol.[57] Finally, a structure corresponding to face protonation of benzene was investigated, but found to be considerably less stable than the other two geometrical possibilities.

Ab initio calculations have also been carried out on the protonated forms of a number of simple alkyl substituted benzenes, and were paralleled by experimental proton affinity measurements using gas-phase ion cyclotron resonance techniques.[58] One of the concerns in this study was to determine whether it is the inductive or Baker–Nathan ordering of substituent effects which is followed in determining the effect of alkyl groups on the stability of protonated benzene. That is to say, would we expect a methyl group to be more effective than *t*-butyl, for example, in stabilizing a carbonium center (the Baker–Nathan ordering which is observed in solution for these species),[59] or would the larger substituent provide a better means of dispersing the positive charge in the absence of solvent. Both theory and gas-phase experiment favor the latter, (Table 14), putting to rest any electronic interpretations for the Baker–Nathan ordering. The data on the species in the gas-phase suggest that

Table 14. *Alkyl Substituent Effects on the Stability of Protonated Benzene*[a]

Alkyl group R	ΔE (STO-3G)	ΔG (icr)
Methyl	0.0	0.0
Ethyl	1.4	0.85
n-Propyl	2.3	1.7
i-Propyl	2.3	2.1
n-Butyl	2.7	2.1
i-Butyl	3.1	2.3 ± 0.5

[a] Energy of the reaction (kcal/mol):

this reverse inductive ordering is due to the bulkiness of the *t*-butyl group, making effective solvation of the carbonium center impossible.

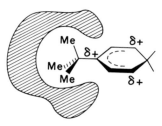

The bicyclo[3.1.0]hexenyl cation (III) probably owes its existence to the fact that thermal interconversion to the thermodynamically more stable isomer, protonated benzene (I), is a symmetry-forbidden process. Formally, the bicyclohexenyl cation is the homoaromatic—or more precisely, homoantiaromatic—derived from cyclopentadienyl by replacement of one of its double bonds by a cyclopropane moiety. Hence, it should exhibit some of the instability inherent in the parent polyene. The energy of the bond separation reaction (6) shows this to be the case. Although positive, it is far smaller than a value characteristic of an aromatic arrangement.

$$\text{(cyclopentadiene)} + 4CH_4 + CH_3^+ \longrightarrow \triangle + CH_3-CH_3 + 2CH_3-CH_2^+ + CH_2{=}CH_2 \qquad (6)$$

$$\Delta E \text{ (STO-3G)} = 27 \text{ kcal/mol}$$
$$\Delta E \text{ (4-31G)} = 26 \text{ kcal/mol}$$

The pathways open for degenerate circumambulation in the bicyclohexenyl cations have attracted considerable attention, and have in fact provided some of the earliest evidence in support of the Woodward–Hoffmann rules.[60] In keeping with orbital symmetry considerations for a suprafacial [1,4] sigmatropic shift, passage from one positional form of bicyclohexenyl to another requires progression through a bisected cyclopentadienylcarbinyl transition state or intermediate (VII) in order to maintain continuity of overlap. Interven-

tion of eclipsed cyclopentadienylcarbinyl (VIII) would, on the other hand, necessitate a break in this continuity. The *ab initio* calculations support such a contention (as do a variety of experimental studies on the stereochemistry of rearrangement) and suggest that the symmetry-allowed circumambulation pathway,

$$III \to VII \to III$$

is 12 kcal/mol more facile than the symmetry-forbidden process,

$$III \to VIII \to III$$

Methyl substituent effects on the most stable form of the singlet cyclopentadienyl cation support the notion that is best represented by the valence structures

Indeed, the ineffectiveness of methyl substitution at the 2 position compared to that

$$\Delta E \text{ (4-31G)} = 3.5 \text{ kcal/mol}$$

directly at a site of positive charge, largely mirrors the

$$\Delta E \text{ (4-31G)} = 15.2 \text{ kcal/mol}$$

behavior found in allyl cation itself:

$$\Delta E \text{ (4-31G)} = 4.9 \text{ kcal/mol}$$

$$\Delta E \text{ (4-31G)} = 16.1 \text{ kcal/mol}$$

Methyl substitution on the square pyramidal form of cyclopentadienyl cation is suggested by the *ab initio* calculations to be more effective on the base (IX) than at the apex (X). This appears to be at odds with the observation (in solution) that of the possible positional isomers of the dimethyl substituted ion, the one with a single alkyl group at each of the base and apex sites appears to be the most stable.[50] No explanation for the discrepancy is apparent at this time.

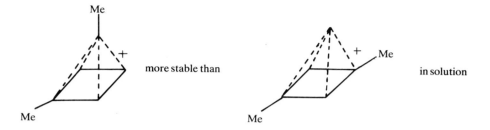

more stable than ... in solution

18. $C_7H_7^+$ (Fig. 12, Refs. 61, 62)

Only two forms of $C_7H_7^+$ have been investigated, the tropylium (I) and benzyl (II) cations. The former is found to be 9.2 kcal/mol more stable, not at all surprising once it is recognized that this ion is constructed around a 6π-electron aromatic arrangement. In fact, the bond-separation energy for tropylium, providing an estimate of the stabilization resulting from association of its component linkages, is actually greater than that for benzene itself [reactions (11) and (12), respectively].

$$+6CH_4+CH_3^+ \rightarrow 2CH_3-CH_3+2CH_3-CH_2^++3CH_2=CH_2 \qquad (11)$$
$$\Delta E\ (4\text{-}31G)=87\ \text{kcal/mol}$$

$$+6CH_4 \rightarrow 3CH_3-CH_3+3CH_2=CH_2 \qquad (12)$$
$$\Delta E\ (4\text{-}31G)=64\ \text{kcal/mol}$$
$$\Delta E\ (\text{expt.})=61\ \text{kcal/mol}$$

No doubt, the most contraversial aspect of the $C_7H_7^+$ potential surface is the energetic disposition of the benzyl cation. Is it a local minimum and hence, in principle, susceptible to experimental detection and characterization, or does it rest instead on the side or on top of a potential hill, thus making its collapse to some other form of $C_7H_7^+$, presumably tropylium, imminent? Although the *ab initio* calculations which have been done do not enable us to comment on this matter, the numerical data themselves (in particular the

Fig. 12. Theoretical STO-3G geometries for $C_7H_7^+$ cations.

I   II

difference in stabilities between the tropylium and benzyl cation) might provide a basis upon which such a judgment could be made. For example, if the $C_7H_7^+$ cation could be generated, say by removal of the halogen from benzyl bromide, and the heat of such an abstraction determined (by the use of trapped icr techniques) this value could be compared with the theoretical energies of the

separate processes leading to the formation of benzyl and tropylium species. Experiments of this nature are in progress.[62]

19. $C_8H_9^+$ (Fig. 13, Table 15, Refs. 53, 61, 63, 64)

$C_8H_9^+$ is the largest system explored to date in which a fair number of structural possibilities have been considered in detail. Systems of this size are just within the capability of our present generation of computer programs, and even individual calculations are rather costly. Thus, for the most part, geometric structures in this case have been only partially optimized, and for all cases consideration has been limited to the STO-3G minimal basis set. Because of this latter restriction it is necessary that we exhibit caution in making energetic comparisons between different $C_8H_9^+$ isomers, and where possible draw on our accumulated experience from smaller systems on the probable effects of improvement in basis description.

Methyl substitution is seen to do little to alter the relative stabilities of the benzyl and tropylium cations. Indeed, the STO-3G difference in stabilities of 9.7 kcal/mol between the two parent systems is reduced only slightly (to 8.2 kcal/mol) upon methyl substitution. It might be mentioned that the methyl benzyl cation has been observed by NMR under stable ion conditions in strong-acid media.[65] Its separate existence from the analogous tropylium species together with the predicted small differential effect of methyl substitution, perhaps, provide additional support for the stability of parent benzyl cation. It is interesting to note that the STO-3G calculations predict methyl substitution at the para position on the ring to be more effective in stabilizing the benzyl cation than substitution directly on the formal carbonium center.

Although part of the effect is no doubt due to steric interactions present in the latter, it does imply that a valence contribution of the form,

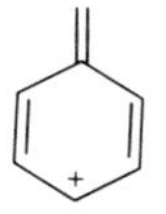

is of equal importance energetically with that corresponding to our normal

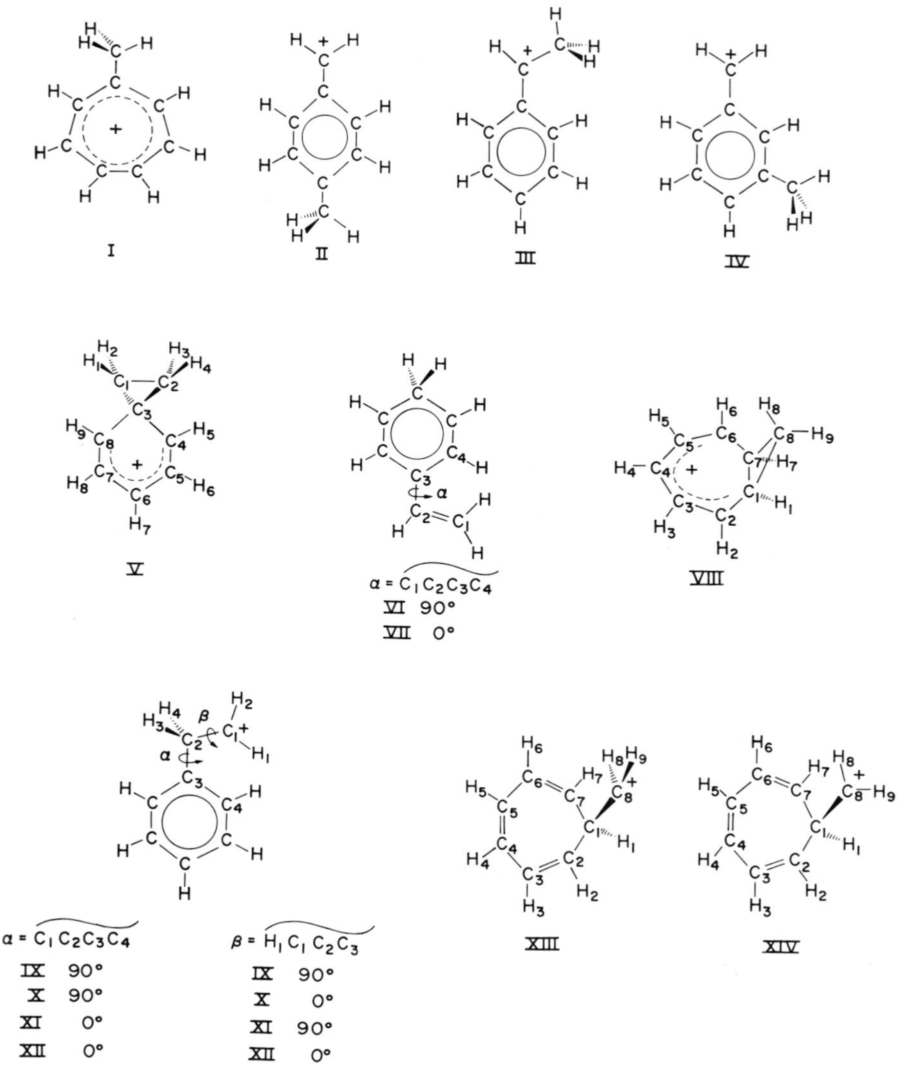

$$\alpha = C_1 C_2 C_3 C_4$$
VI 90°
VII 0°

$$\alpha = C_1 C_2 C_3 C_4 \qquad \beta = H_1 C_1 C_2 C_3$$

	α	β
IX	90°	90°
X	90°	0°
XI	0°	90°
XII	0°	0°

Fig. 13. Theoretical STO-3G geometries for $C_8H_9^+$ cations.

Table 15. *Relative Energies of* $C_8H_9^+$ *Structures*

Structure	STO-3G, kcal/mol
Methyltropylium (I)	0.0
Para methylbenzyl (II)	5.0
Methylbenzyl (III)	8.2
Meta methylbenzyl (IV)	9.7
Phenonium (V)	14.4
Ring protonated styrene, planar (VI)	29.0
Ring protonated styrene, orthogonal (VII)	37.9
Homotropylium (VIII)	39.7
2-Phenylethyl, orthogonal bisected (IX)	49.8
2-Phenylethyl, orthogonal eclipsed (X).	56.7
2-Phenylethyl, planar bisected (XI)	60.9
2-Phenylethyl, planar eclipsed (XII)	63.2
Cycloheptatrienylcarbinyl, eclipsed (XIII)	100.6
Cycloheptatrienylcarbinyl, bisected (XIV)	106.3

representation of the ion:

Meta methyl substitution also appears to be highly effective in stabilizing the benzyl cation, an observation which is not expected from normal resonance arguments.

For many years the phenonium ion (V) was center stage in an important controversy regarding the possible intermediacy of bridged carbonium ions in solvolytic reactions.[66] Thus, the observed rapid equilibration of 2-phenylethyl derivatives in acid solution could be explained either on the basis of a stable

bridged intermediate or in terms of rapid equilibration of open ions through such a form. The *ab initio* calculations suggest that bridged phenonium (V) is indeed a stable entity, and that any of the open 2-phenylethyl cations (IX–XII) collapse to it without activation. This is consistent with the proton and ^{13}C NMR spectra of the species formed upon super-acid ionization of 2-phenylethylchloride.[67] It should be mentioned, that both the ^{13}C spectrum of phenonium and its theoretical (STO-3G) geometry are not at all consistent with description of the ion in terms of a weak complex involving ethylene.

Rather, the data better fit the picture of a spirocyclic arrangement between protonated benzene and cyclopropane, though the two cyclopropane linkages

connecting it to the ring are considerably longer than normal (1.60 vs 1.50 Å in cyclopropane), while the third bond is rather short (1.46 Å). All of this is highly reminiscent of the geometrical structure of bisected cyclopropylcarbinyl, and indeed, close examination of the two shows them to be little different.

Bisected cyclopropylcarbinyl   Phenonium

Styrene may protonate not only on the vinyl group [leading to either the methylbenzyl (III) or 2-phenylethyl cations (IX–XII)] but also on the ring itself. The STO-3G calculations indicate the unlikelihood of such a process, suggesting that even in its planar arrangement, where extensive delocalization of the charge is possible, ring protonated styrene (VII) is more than 20 kcal/mol less stable than the methylbenzyl cation (III). The orthogonal form (VII) is still higher in energy.

Perhaps the most curious isomer of $C_8H_9^+$ is homotropylium (VIII), the homoaromatic formed by the replacement of one of the double bonds in tropylium by a cyclopropane ring. That the aromatic character of precursor tropylium largely remains is evidenced by the energy of the bond-separation reaction (13). (The corresponding STO-3G bond separation energy for

$$+ \ 6CH_4 + CH_3^+ \longrightarrow \ + 2CH_3-CH_3 + 2CH_3-CH_2^+ + 2CH_2{=}CH_2 \quad (13)$$

$$\Delta E \ (STO\text{-}3G) = 56 \ kcal/mol$$

tropylium itself is 99 kcal/mol.) One might anticipate that the "cyclopropane ring" in homotropylium, like that in the bicyclo[3.1.0]hexenyl cation, would undergo facile circumambulation about the cyclic polyene. In this instance symmetry conservation for the suprafacial [1.6] sigmatropic migration dictates the transition state or intermediate to be an eclipsed cycloheptatrienylcarbinyl cation (XIII), rather than a bisected arrangement (XII). The *ab initio* calculations concur; suggesting that the difference in energy between the two possibilities (the "cost" of violating orbital symmetry control) is some 6 kcal/mol. Degenerate circumambulation in homotropylium has not as yet actually been observed experimentally, although several attempts have been made.[68] At first glance this appears to be strange, particularly in view of the facility of the corresponding rearrangement process in the bicyclo[3.1.0]hexenyl cation. Indeed, the *ab initio* calculations (at the STO-3G level) show a 23 kcal/mol difference in activation required for circumambulation in the two systems. Closer scrutiny shows why. Thus, if we ignore any differences in the stabilities of the (symmetry allowed) transition-state structures, we see that the

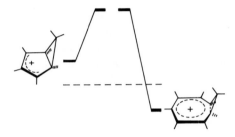

bicyclo[3.1.0]hexenyl cation, which is destabilized by antiaromatic conjugation, has only a small barrier to surmount in order to reach its rearrangement transition state, while homotropylium (an aromatic) has much further to climb.

20. Conclusion

The application of *ab initio* molecular orbital theory to problems of structure and stability of carbocation intermediates is still in its infancy. The progress which has been made certainly indicates the viability of the general approach, and suggests that its pursual would, without doubt, uncover further fruitful areas of research. It is the opinion of this author that in view of the emergence of ion cyclotron resonance techniques, with which the stabilities and reactivities of ions in their dilute gas phase may be directly studied, many of the predictions of the quantitative theory will shortly be brought to test. It is this parallel development of experiment and theory which should be watched closely.

References

1. G. A. Olah, Carbocations and electrophilic reactions *Angew. Chem., Int. Ed. Engl.* **12**, 173–212 (1973).
2. W. J. Hehre, R. F. Stewart, and J. A. Pople, Self-consistent molecular orbital methods. I. Use of Gaussian expansions of Slater-type atomic orbitals, *J. Chem. Phys.* **51**, 2657–2664 (1969).
3. W. A. Lathan, W. J. Hehre, L. A. Curtiss, and J. A. Pople, Molecular orbital theory of the electronic structure of organic compounds. X. A systematic study of geometries and energies of AH_n molecules and cations, *J. Am. Chem. Soc.* **93**, 6377–6387 (1971).
4. W. A. Lathan, L. A. Curtiss, W. J. Hehre, J. B. Lisle, and J. A. Pople, Molecular orbital structures for small organic molecules and cations, *in*: *Progress in Physical Organic Chemistry* (A. Streitwieser, Jr., and R. W. Taft, eds.), Vol. 11, pp. 175–261, Wiley–Interscience, New York (1974).
5. L. Radom, W. A. Lathan, W. J. Hehre, and J. A. Pople, Molecular orbital theory of the electronic structure of organic componds. VIII. Geometries, energies and polarities of C_3 hydrocarbons, *J. Am. Chem. Soc.* **93**, 5339–5342 (1971).
6. W. J. Hehre and J. A. Pople, Molecular orbital theory of the electronic structure of organic compounds. XXVI. Geometries, energies and polarities of C_4 hydrocarbons, *J. Am. Chem. Soc.* **97**, 6941–6955 (1975).
7. J. A. Pople, *A priori* geometry predictions, this volume, Chapter 1.
8. R. Ditchfield, W. J. Hehre, and J. A. Pople, Self-consistent molecular orbital methods. IX. An extended Gaussian-type basis for molecular orbital studies of organic molecules, *J. Chem. Phys.* **54**, 724–728 (1971).
9. W. A. Lathan, L. Radom, P. C. Hariharan, W. J. Hehre, and J. A. Pople, Structures and stabilities of three-membered rings from *ab initio* molecular orbital theory, *Fortsch. Chem. Forsch.* **40**, 1–45 (1973).
10. P. C. Hariharan and J. A. Pople, The influence of polarization functions on molecular orbital hydrogenation energies, *Theor. Chim. Acta* **28**, 213–222 (1973).
11. W. J. Hehre, R. Ditchfield, and J. A. Pople, Self-consistent molecular orbital methods. XII. Further extensions of Gaussian-type basis sets for use in molecular orbital studies of organic molecules, *J. Chem. Phys.* **56**, 2257–2261 (1972).
12. A. D. Walsh, The electronic orbital, shapes and spectra of polyatomic molecules, parts I–X, *J. Chem. Soc.* 2260–2331 (1953).
13. W. A. Lathan, W. J. Hehre, and J. A. Pople, Molecular orbital theory of the electronic structure of organic compounds. VI. Geometries and energies of small hydrocarbons, *J. Am. Chem. Soc.* **93**, 808–815 (1972).
14. P. C. Hariharan, W. A. Lathan, and J. A. Pople, Molecular orbital theory of simple carbonium ions, *Chem. Phys. Lett.* **14**, 385–388 (1972).
15. V. Dyczmons and W. Kutzelnigg, *Ab initio* calculations of small hydrides including electron correlation. XII. The ions CH_5^+ and CH_5^-, *Theor. Chim. Acta* **33**, 239–247 (1974).
16. J. A. Pople and R. K. Nesbet, Self-consistent orbitals for radicals, *J. Chem. Phys.* **22**, 571–672 (1954).
17. B. Zurawski, R. Ahlrichs, and W. Kutzelnigg, Have ions $C_2H_3^+$ and $C_3H_5^+$ classical or non-classical structure? *Chem. Phys. Lett.* **21**, 309–313 (1973).
18. J. E. Williams. J. V. Buss, L. C. Allen, P. v. R. Schleyer, W. A. Lathan, W. J. Hehre, and J. A. Pople, Barriers in ethyl cations, *J. Am. Chem. Soc.* **92**, 2141–2143 (1970).
19. R. Hoffmann, L. Radom, J. A. Pople, P. v. R. Schleyer, W. J. Hehre, and L. Salem, Strong conformational consequences of hyperconjugation, *J. Am. Chem. Soc.* **94**, 6221–6223 (1972).
20. L. Radom, P. C. Hariharan, J. A. Pople; and P. v. R. Schleyer, Molecular orbital of the electronic structure of organic compounds. XXII. Structures and stabilities of $C_3H_3^+$ and C_3H^{+-} cations, *J. Am. Chem. Soc.* **98**, 10–14 (1976).
21. M. Sundaralingam and L. H. Jensen, The structure of a carbonium ion. Refinement of the crystal and molecular structure of sym-triphenylcyclopropenium perchlorate, *J. Am. Chem. Soc.* **88**, 198–204 (1966).
22. A. T. Ku and M. Sundaralingam, X-ray studies on cyclopropenyl cations. II. Crystal and molecular structure of 1,2,3-trisdimethylaminocyclopropenium perchlorate, *J. Am. Chem. Soc.* **94**, 1688–1692 (1972).

23. L. Radom, P. C. Hariharan, J. A. Pople, and P. v. R. Schleyer, Molecular orbital theory of the electronic structure of organic compounds. XIX. Geometries and energies of $C_3H_5^+$ cations. Energy relationships among allyl, vinyl and cyclopropyl cations, *J. Am. Chem. Soc.* **95**, 6531–6544 (1973).

24. L. Radom, J. A. Pople, and P. v. R. Schleyer, Effects of substituents on the mechanism of stereomutation of allyl cations, *J. Am. Chem. Soc.* **95**, 8193–8195 (1973).

25. L. Salem and C. Rowland, the electronic properties of diradicals, *Angew. Chem., Int. Ed. Engl.* **11**, 92–111 (1972).

26. W. J. Hehre, Theoretical approaches to the structure of carbocations, *Acc. Chem. Res.* **8**, 369–376 (1975).

27. L. Radom, J. A. Pople, V. Buss, and P. v. R. Schleyer, Molecular orbital theory of the electronic structure of organic compounds. XI. Geometries and energies of $C_3H_7^+$ cations, *J. Am. Chem. Soc.* **94**, 311–321 (1972).

28. P. C. Hariharan, L. Radom, J. A. Pople, and P. v. R. Schleyer, Molecular orbital theory of the electronic structure of organic compounds. XX. $C_3H_7^+$ cations with a polarized basis set, *J. Am. Chem. Soc.* **96**, 599–601 (1974).

29. D. Cremer, J. S. Binkley, J. A. Pople, and W. J. Hehre, Molecular orbital theory of the electronic structure of organic compounds. XXI. Rotational potentials for geminal methyl groups, *J. Am. Chem. Soc.* **96**, 6900–6903 (1974).

30. A. J. P. Devaquet and W. J. Hehre, Degenerate rearrangement in homocyclopropenyl cation. Violation of orbital symmetry control for a sigmatropic migration, *J. Am. Chem. Soc.* **96**, 3644–3645 (1974).

31. W. J. Hehre and A. J. P. Devaquet, Theoretical approaches to rearrangements in carbocations. III. The $C_4H_5^+$ system, *J. Am. Chem. Soc.* **98**, 4370–4377 (1976).

32. R. Hoffmann and R. B. Davidson, The valence orbitals of cyclobutane, *J. Am. Chem. Soc.* **93**, 5699–5705 (1971).

33. C. Krüger, P. J. Roberts, Y. H. Tsay, and J. B. Koster, The molecular structure of a complex of tetramethylcyclobutadiene and aluminium chloride, containing a σ A1–C bond, *J. Organometal. Chem.* **78**, 69–74 (1974).

34. G. A. Olah, J. S. Staral, and G. Liang, Novel aromatic systems. I. The homocyclopropenyl cation, *J. Am. Chem. Soc.* **96**, 6233–6235 (1974).

35. G. A. Olah, J. S. Staral, R. J. Spear, and G. Liang, Novel aromatic systems. II. Preparation and study of the homocyclopropenium ion, the simplest homoaromatic system, *J. Am. Chem. Soc.* **97**, 5489–5497 (1975).

36. W. J. Hehre and P. C. Hiberty, The homoallyl cation, *J. Am. Chem. Soc.* **94**, 5917–5918 (1972).

37. W. J. Hehre and P. C. Hiberty, Interconverting cyclopropylcarbinyl cations, *J. Am. Chem. Soc.* **96**, 302–304 (1974).

38. W. J. Hehre, P. C. Hiberty, and J. S. Binkley, Theoretical approaches to rearrangements in carbocations. IV. The $C_4H_7^+$ system, *J. Am. Chem. Soc.*, to be published.

39. K. B. Wiberg, B. A. Hess, Jr., and A. J. Ashe III. Cyclopropylcarbinyl and cyclobutyl cations, in: *Carbonium ions* (G. A. Olah and P. v. R. Schleyer, eds.), Vol. 3, pp. 1295–1345, Wiley, New York (1972).

40. G. A. Olah, C. L. Jeuell, D. P. Kelly, and R. D. Porter, Stable carbocations. CXIV. The structure of cyclopropylcarbinyl and cyclobutyl cations, *J. Am. Chem. Soc.* **94**, 146–156 (1972).

41. D. S. Kabakoff and E. Namanworth, Nuclear magnetic double resonance studies of the dimethylcyclopropylcarbinyl cation. Measurement of the rotation barrier, *J. Am. Chem. Soc.* **92**, 3234–3235 (1970).

42. W. J. Hehre, unpublished.

43. G. A. Olah, J. R. DeMember, A. Commeyras, and J. L. Bribes, Stables carbonium ions. LXXXV. Lasar raman and infrared spectroscopic study of alkylcarbonium ions, *J. Am. Chem. Soc.* **93**, 459–463 (1971).

44. L. Radom, J. A. Pople, and P. v. R. Schleyer, Molecular orbital theory of the electronic structure of organic compounds. XVI. Conformations and stabilities of substituted ethyl, propyl and butyl cations, *J. Am. Chem. Soc.* **94**, 5935–5945 (1972).

45. W. J. Hehre and P. v. R. Schleyer, Cyclopentadienyl and related $(CH)_5^+$ cations, *J. Am. Chem. Soc.* **95**, 5837–5839 (1973).

46. M. Saunders *et al.*, Unsubstituted cyclopentadienyl cation, a ground state triplet, *J. Am. Chem. Soc.* **95**, 3017–3018 (1973).

47. R. E. Williams, Carboranes and boranes; polyhedra and polyhedral fragments, *Inorg. Chem.* **10**, 210–214 (1971).

48. W. D. Stohrer and R. Hoffmann, Bond-stretch isomerism and polytropal rearrangements in $(CH)_5^+$ $(CH)_5^-$ and $(CH)_4CO$, *J. Am. Chem. Soc.* **94**, 1661–1668 (1972).

49. S. Masamune, M. Sakai, and H. Ona, Nature of the $(CH)_5^+$ species. I. Solvolysis of 1,5-dimethyltricyclo[2.1.0.0]pent-3-yl benzoate, *J. Am. Chem. Soc.* **94**, 8955–8956 (1972).

50. S. Masamune, M. Sakai, H. Ona, and A. J. Jones, Nature of the $(CH)_5^+$ species. II. Direct observation of the carbonium ion of 3-hydroxyhomotetrahedrane derivatives, *J. Am. Chem. Soc.* **94**, 8956–8958 (1972).

51. H. Hart and M. Kuzuya, Evidence concerning the structure of $(CH)_5^+$ type carbonium ions, *J. Am. Chem. Soc.* **94**, 8958–8960 (1972).

52. W. J. Hehre and J. A. Pople, Molecular orbital theory of the electronic structure of organic compounds. XV. The protonation of benzene, *J. Am. Chem. Soc.* **94**, 6901–6904 (1972).

53. W. J. Hehre, Theoretical approaches to rearrangements in carbocations. II. Degenerate rearrangements in bicyclo[3.1.0]hexenyl and homotropylium cations. On the stability of homoaromatic molecules, *J. Am. Chem. Soc.* **96**, 5207–5217 (1974).

54. D. M. Brouwer, E. L. Mackor, and C. MacLean, Arenonium ions, *in*: *Carbonium Ions* (G. A. Olah and P. v. R. Schleyer, eds.), Vol. 2, pp. 837–897, Wiley, New York (1970).

55. N. C. Baenziger and A. D. Nelson, The crystal structure of the tetrachloroaluminate salt of the heptamethylbenzene cation, *J. Am. Chem. Soc.* **90**, 6602–6607 (1968).

56. P. Menzel and F. Effenberger, σ-complex intermediates in acylation and sulfonylation of 1,3,5-tripyrrolidinobenzene, preparation, reactions and structure, *Angew. Chem., Int. Ed. Engl.* **14**, 62–63 (1975).

57. G. A. Olah, R. H. Schlosberg, D. P. Kelly, and G. D. Mateesu, Stable carbonium ions. IC. The benzenonium ion $(C_6H_7^+)$ and its degenerate rearrangement, *J. Am. Chem. Soc.* **92**, 2546–2548 (1972).

58. W. J. Hehre, R. T. McIver, Jr., J. A. Pople, and P. v. R. Schleyer, Alkyl substituent effects on the stability of protonated benzene, *J. Am. Chem. Soc.* **96**, 7162–7163 (1974).

59. E. M. Arnett and J. W. Larsen, Stabilities of carbocations in solution. IV. A large Baker–Nathan effect for alkylbenzenonium ions. *J. Am. Chem. Soc.* **91**, 1438–1442 (1968).

60. R. B. Woodward and R. Hoffmann, *The Conservation of Orbital Symmetry*, Academic Press, New York (1970).

61. J. F. Wolf, P. G. Harch, R. W. Taft, and W. J. Hehre, Substituent effects on the stability of carbocations. The anomalous case of phenyl vs. Cyclopropyl substitution, *J. Am. Chem. Soc.* **97**, 2902–2904 (1975).

62. J. L. M. Abboud, R. W. Taft, and W. J. Hehre, research in progress; and J. L. M. Abboud, W. J. Hehre, and R. W. Taft, Benzyl cation. A long lived species in the gas phase, *J. Am. Chem. Soc.* **98**, 6072–6073 (1976).

63. W. J. Hehre, The ethylenebenzenium cation, *J. Am. Chem. Soc.* **94**, 5919–5920 (1972).

64. W. J. Hehre, unpublished.

65. G. A. Olah, R. D. Porter, C. L. Jeuell, and A. M. White, Stable carbocations. CXXV. Proton and carbon 13 magnetic resonance studies of phenylcarbenium ions (benzyl cations). The effects of substituents on the stability of carbocations, *J. Am. Chem. Soc.* **94**, 2044–2052 (1972).

66. C. J. Lancelot, D. J. Cram, and P. v. R. Schleyer, Phenonium ions. The solvolysis of β-arylalkyl systems, *in*: *Carbonium Ions* (G. A. Olah and P. v. R. Schleyer, eds.), Vol. 3, pp. 1347–1483, Wiley, New York (1972).

67. G. A. Olah and R. D. Porter, Stable carbocations. CXXI. Carbon-13 magnetic resonance spectroscopy study of ethylenarenium ions (spiro-[2.5]-octadienyl cations, *J. Am. Chem. Soc.* **93**, 6877–6887 (1971).

68. J. A. Berson and J. A. Jenkins, Homotropylium-4-d ion. On the energy barrier to a non-least motion circumambulatory rearrangement, *J. Am. Chem. Soc.* **94**, 8907–8908 (1972).

8

Molecular Anions

Leo Radom

1. Introduction

Anions present a tempting target for theoretical study because their examination by experimental means is not straightforward. The experimental difficulties arise because isolated anions are extremely fragile. Electron affinities are normally less than a few electron volts and, hence, the extra electron is only loosely bound. It is therefore difficult to study many anionic species by conventional procedures. However, recently developed techniques such as matrix isolation[1] (for structural information) and ion cyclotron resonance[2] (for energetic information) provide promising new sources of experimental data.

Unfortunately, theoretical study of anions is hindered by the same factor which makes experimental study difficult, i.e., the low electron affinity. It is found that single-determinant calculations on anions often produce energies higher than those of their neutral precursors. Under these circumstances, one might argue that given sufficient flexibility in the molecular orbital procedure (i.e., large basis set, unrestricted SCF), the best description of the anion will correspond to the neutral molecule and a separated electron. Such considerations have undoubtedly had an inhibiting effect on anion calculations, and these are much less common in the literature than are calculations on neutral species and positively charged ions.

In this review, we shall examine whether one should indeed be inhibited from carrying out molecular orbital calculations on anions or, whether there are circumstances under which anion calculations can provide useful chemical information. We shall largely be concerned with single-determinant

Leo Radom • Research School of Chemistry, Australian National University, Canberra, ACT 2600, Australia

calculations, although we shall occasionally turn to calculations beyond the Hartree–Fock limit to test our conclusions. In order that we may focus our attention totally on problems associated with *anion* calculations, we have omitted discussion of calculations on anionic species for which other complications might be present. Thus, we will not treat radical anions (e.g., C_2^-) for which results are influenced by the choice of SCF procedure, nor anions containing second row (e.g., SO_4^{2-}, PO_4^{3-}) or transition metal (e.g., $CuCl_4^{2-}$) atoms, for which the importance of d or higher angular functions, respectively, in the basis set is not yet fully resolved. We therefore restrict ourselves to anions containing first row elements. We shall concentrate on the performance of the theory with respect to those properties which are of greatest interest to chemists. These can be classified under the headings structures, stabilities, and reaction mechanisms.

2. Background

2.1. Calculations of Electron Affinities

Many of the early *ab initio* calculations on anions were directed towards the determination of electron affinities. This is not surprising since the electron affinity (EA) is perhaps the most important property associated with negative ion formation. It gives the energy of the anion (A^-) relative to its neutral precursor (A) and a separated electron, i.e., the enthalpy change in reaction (1)

$$A^- \rightarrow A + e \qquad (1)$$

A positive EA means that the anion is stable, while a negative EA means that the anion is unstable with respect to electron loss. In most cases where experimental EAs have been determined, the values are small and positive, indicating that anions are generally weakly bound.

A theoretical determination of the electron affinity of a molecule A requires the calculation of the energy of A and A^-. In order that the electron affinity be accessible within the Hartree–Fock framework, we would require that the electron affinity be much greater than the difference in correlation energy (E_{corr}) of A and A^-. In fact, detailed calculations on atoms by Clementi *et al.*[3,4] and on molecules by Cade[5] and others have shown that, generally,

$$EA \approx |E_{corr}^A - E_{corr}^{A^-}| \qquad (2)$$

Hence, single-determinant molecular orbital theory *alone* is unsuitable for the determination of electron affinities.

Some representative results for electron affinity calculations[3,5,6] for F, OH, and CH_3 are collected in Table 1. We can see that the near Hartree–Fock SCF values, EA (SCF) are very poor. In fact, for OH and CH_3, the neutral

Table 1. Estimates of Electron Affinities (eV) for F, OH, and CH₃

EA	F	OH	CH₃
SCF	1.36	−0.10	−1.38
Koopmans	4.92	2.90	—
Correlation	3.08[a]	1.40,[a] 1.91[b]	0.1–0.3[c]
Experiment	3.45[d]	1.83[e]	1.1[f,g]
Reference	3	5	6

[a] These results employ EA (SCF) plus a correction for the *change* in the correlation energy for an isoelectronic process (ionization of the isoelectronic neutral, i.e., Ne, HF).
[b] This result employs EA (SCF) plus a correction for the *change* in the correlation energy for the corresponding united atom process (i.e., F → F⁻).
[c] Correlation energy estimated using the IEPA–PNO procedure.[31]
[d] R. S. Berry and C. W. Reimann, *J. Chem. Phys.* **38**, 1540 (1963).
[e] L. M. Branscomb, *Phys. Rev.* **148**, 11 (1966).
[f] P. M. Page and P. C. Goode, *Negative Ions and the Magnetron*, Wiley Interscience, London (1969).
[g] Some doubt has been expressed concerning the validity of this estimate (see Ref. 6).

radicals have calculated energies lower than those for the anions, resulting in negative electron affinities. The Koopmans' theorem estimates, EA (Koopmans), obtained as the energy of the highest occupied orbital in the SCF calculation on the anion, are qualitatively in agreement with experiment but tend to be too large. Best agreement is found when correlation corrections are applied to the SCF results.

Thus, a reliable theoretical prediction of electron affinities requires some means of taking into account electron correlation. Moreover, because of the small values of molecular electron affinities the correlation energy *difference* between A and A⁻ has to be accurate to within a few tenths of an electron volt. This is often quite a difficult task. A promising alternative method of calculating molecular electron affinities which has been recently developed by Simons and co-workers[7] uses the equations-of-motion approach to determine the ion–molecule energy difference in a single calculation, rather than by separate variational calculations on the molecule and the ion.

2.2. A Less Ambitious Target

As mentioned earlier, the inability of calculations within the Hartree–Fock framework to accurately predict electron affinities has probably had a severe inhibiting effect on anion studies. However, before dismissing anion calculations out of hand, we should examine how seriously this shortcoming will affect the problems we wish to tackle. Although the electron affinity is of fundamental importance in telling us whether an anion is capable of existence or not, there are many chemical problems involving anion calculations which may not require accurate knowledge of electron affinities. For example, we may wish to know the structure of an experimentally observed anion or the energy change in a reaction involving anions. Is it possible to examine *these* kinds of problems within the Hartree–Fock framework?

The main difficulty associated with electron affinity calculations is that the correlation energy in the anion (A^-) with its additional electron is generally considerably greater than that of the neutral molecule A. Similar difficulties are encountered in examining energy changes for reactions involving neutral species with differing numbers of electron pairs, notably the dissociation of a diatomic molecule into separate atoms. For neutral molecules, good agreement of calculated and experimental heats of reaction may be obtained at the single-determinant level if we restrict ourselves to reactions involving closed-shell reactants and closed-shell products, in which case the correlation energy errors approximately cancel.[8,9] Furthermore, for *isodesmic* reactions, i.e., reactions in which the number of each type of bond is conserved, Pople and co-workers[10-12] have found that good results are obtained even with quite modest basis sets. We might expect that, for anions also, better results might be produced for these types of reaction than for the electron affinity calculations and these approaches are discussed in Section 4.

Another property which is likely to be more readily predicted than the electron affinity is molecular structure. Here, we would require that the correlation energy does not change significantly with small changes in bond length and bond angle in the vicinity of the potential minimum on the Hartree–Fock surface. It has been shown[13] that one-electron properties calculated from Hartree–Fock wave functions are correct to first order and that equilibrium geometry can be considered a one-electron property in this context.[14] Although "correct to first order" does not in itself imply any absolute accuracy, extensive calculations on neutral molecules and cations have shown[15,16] that single-determinant molecular orbital theory produces geometries in close agreement with experimental values. These considerations are sufficient to encourage the structural studies on anions reported in Section 3.

For molecules for which the electron affinity is so poorly predicted as to be the *wrong sign* (i.e., negative), one has to consider carefully the validity of the anion calculation. A negative electron affinity implies that the anion is less stable than the neutral species, from which is has been derived, together with an electron at infinity. Given sufficient flexibility in the basis set and the SCF procedure, the description of the anion that should emerge from the calculation would, therefore, be that of the neutral molecule plus a separated electron. In practice, this does not normally happen for two reasons. In the first place, calculations are usually performed with limited basis sets which would restrict the departing electron. The second and more important point which applies to closed-shell systems is that a closed-shell SCF procedure would not permit the dissociation of an anion into a radical and a separated electron.

3. Structural Studies

3.1. Experimental Comparison

3.1.1. The OH⁻ Ion

The OH⁻ ion has been studied by a number of workers[5,17-26] at various levels of sophistication and is a useful starting point in discussing the basis-set dependence of the geometry predictions for anions. The results for OH⁻ are summarized in Table 2.

The minimal STO-3G basis set (with standard exponents)[27] fares poorly, predicting a bond length which is too long. The result is even worse when the STO-3G exponents are optimized. The exponents, particularly for hydrogen ($\zeta = 0.932$ compared with the standard value of 1.24), become smaller, corresponding to more diffuse orbitals, and the bond lengths become much longer (1.265 Å). Similar observations of low-optimum hydrogen exponents have been reported[28,29] for BH_4^- and CH_3^-. In the latter case,[29] exponent optimization again yields bonds which are too long. Although it is not an ideal situation from a theoretical point of view, it seems that exponent optimization is not suitable for minimal basis-set determinations of the structure of anions.

All the other basis sets in Table 2 lead to reasonable bond lengths for OH⁻. The split-valence 4-31G basis[30] yields a bond slightly too long, while addition of polarization functions leads to bonds which are too short. Inclusion of correlation via the *independent electron-pair approximation* (IEPA)[31] increases the bond length to a value slightly greater than the experimental one. Very similar behavior is observed for bond lengths in neutral systems.[16,25,32]

3.1.2. Polyatomic Anions

Individual calculations[33-42] on the BH_4^-, HF_2^-, N_3^-, and NO_2^- anions have produced geometries in reasonable agreement with experiment. It would, however, be useful to have a *general* procedure which is both economical and reliable for determining the structures of anions. For neutral molecules[30,43,44] and cations,[44-46] the minimal STO-3G and split-valence 4-31G basis sets have

Table 2. *Calculated and Experimental Bond Lengths (Å) for OH⁻*

Level	STO-3G[a]	4-31G	Ext[b]	HF[c]	IEPA[b]	Experiment
Length	1.068	0.985	0.945	0.942	0.974	0.970[d]
Reference	26	23	25	5	25	

[a] Standard exponents.
[b] Extended basis set including polarization functions.
[c] Near-Hartree–Fock calculation.
[d] L. M. Branscomb, *Phys. Rev.* **148**, 11 (1966).

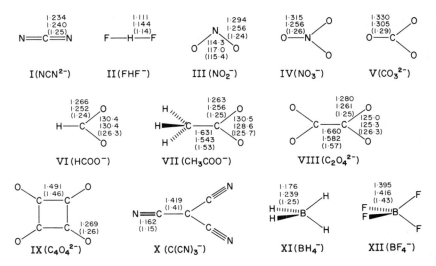

Fig. 1. Comparison[26] of theoretical and experimental structures for anions. Bond lengths are in Å, bond angles in degrees. Values given in the order STO-3G, 4-31G (when available), and experimental (in parentheses).

usually been found to be successful in this regard with exceptions that can normally be rationalized. The performance of these basis sets in predicting the geometries of anions for which experimental structures, in most cases from x-ray crystal studies, are available has recently been tested.[26] Some of the theoretical and experimental geometries are compared in Fig. 1.

The STO-3G bond lengths are often, but not always, too long. Large errors are noted for essentially pure single bonds such as the C—C lengths in the acetate and oxalate anions (too long) and the B—H lengths in BH_4^- (too short). The mean absolute difference between the STO-3G and experimental values for the 16 unique bond lengths in Fig. 1 is 0.04 Å, a figure similar to corresponding differences in studies on neutral molecules.[43,44] However, more data would be desirable before reaching a verdict concerning the relative performance of the STO-3G set for geometry predictions in anions and neutral molecules.

Much better geometries are obtained with the 4-31G basis set. The mean absolute error for the bond lengths is now 0.01 Å and no bond length differs from the experimental value by more than 0.02 Å. The C—C single bonds in acetate and oxalate anions are both close to the experimental values.

Sufficient data are not available to make too strong a statement at this stage on the prediction of bond angles. From the few results that are available, however, it seems likely that these may be predicted with an accuracy comparable to that for neutral molecules.

The results summarized in this section suggest quite strongly that accurate predictions of the geometries of anions may be made within the Hartree–Fock framework. Indeed, the split-valence 4-31G basis set is particularly recommended for this purpose.

3.2. The Methyl Anion

The methyl anion is undoubtedly the polyatomic anion which has received the greatest attention from theoreticians[6,20,22,29,47-58] and therefore warrants detailed discussion here. The targets of most of these theoretical studies have been the structure of the methyl anion and its inversion barrier. A summary of the more important results is presented in Table 3.

The minimal STO-4G basis set produces barriers which are very high except when the p exponents are anisotropically optimized.[29] Use of an extended sp basis gives barriers of about 3–4 kcal mol^{-1}. These are increased to 5–6 kcal mol^{-1} when polarization functions are added to the basis. Very diffuse functions are found to be more important in the planar than the nonplanar structures, and addition of such functions to the basis leads to a large decrease in barrier values. Duke,[58] and Driessler *et al.*[6] find similar barriers of 1.7 and 2.0 kcal mol^{-1}, respectively. Finally, Driessler *et al.*[6] have found that inclusion of electron correlation has little effect on the calculated barrier. They do not regard their estimated barrier lowering from 2.0 to 1.5 kcal mol^{-1} as significant.

In general, the inversion barrier calculations for CH_3^- show features similar to those noted[59-62] for inversion in the isoelectronic molecule ammonia and which are, therefore, not peculiar to anionic species. Thus the minimal basis-set calculations produce high barriers, extended sp basis sets low values, polarization functions increase the barrier, and correlation is unimportant for the inversion process. There may be some difference in the relative importance of diffuse functions in CH_3^- and NH_3. For NH_3, the most diffuse s and p functions of N included in the Gaussian basis set[59] which produces a good value of the inversion barrier have exponents of ~ 0.08 compared with the diffuse exponents on C in CH_3^- of ~ 0.02. However, until a careful and systematic comparison of the two species has been carried out, it would be dangerous to make too strong a conclusion on this point. Finally, however, we should note that for the isoelectronic H_3O^+ cation, preliminary results[63] suggest that the correlation contribution may lead to a significant increase in the barrier.

3.3. Conformational Studies

3.3.1. Inversion

Clark and Armstrong[64,65] have investigated the inversion process in the cyclopropenyl ($C_3H_3^-$) and cyclopropyl ($C_3H_5^-$) anions. A hypothetical planar

Table 3. Ab Initio Calculations on CH_3^-

Reference	Basis set[a]	Symmetry	r_{C-H},[e] Å	$\gamma^{o,e,f}$	E_{total}, hartrees	Barrier, kcal mol^{-1}
29	STO-4G[b]	C_{3v}	1.109	27.0	−39.11610	21.5
		D_{3h}	1.060	0	−39.08164	
	STO-4G[c]	C_{3v}	1.189	29.8	−39.21230	25.2
		D_{3h}	1.130	0	−39.17194	
	STO-4G[d]	C_{3v}	1.160	24.0	−39.24538	1.4
		D_{3h}	1.090	0	−39.24316	
	Ext cont	C_{3v}	1.103	21.4	−39.47762	3.7
		D_{3h}	1.080	0	−39.47166	
53–55	Ext uncont	C_{3v}	1.106	20.5	−39.48720	3.5
		D_{3h}	1.079	0	−39.50837	
	Ext uncont+pol	C_{3v}	(1.106)	23.5	−39.49720	7.0
		D_{3h}	(1.079)	0		
	Ext cont+pol	C_{3v}	(1.106)	(23.5)	−39.51292	5.5
		D_{3h}	(1.079)	0	−39.50422	
51	Ext cont+pol	C_{3v}	(1.079)	22	−39.5125	5.6
		D_{3h}	(1.079)	0	−39.50352	
58	Ext cont+pol	C_{3v}	(1.093)	(19.1)	−39.51388	5.2
		D_{3h}	(1.073)	0	−39.50561	
	Ext cont+pol+diff	C_{3v}	1.093	19.1	−39.52225	1.7
		D_{3h}	1.073	0	−39.51952	
6	Ext cont+pol+diff	C_{3v}	1.095	18.5	−39.51995	2.0
		D_{3h}	1.075	0	−39.51675	
	Correlation	C_{3v}	(1.095)	18.0	−39.7588	1.5

[a] Abbreviated descriptions are as follows: cont (contracted), uncont (uncontracted), ext (extended), pol (polarization functions), and diff (diffuse functions).
[b] Standard exponents.
[c] Isotropically ($s = p_x = p_y = p_z$) optimized exponents.
[d] Anisotropically ($s \neq p_x = p_y \neq p_z$) optimized exponents. The z direction coincides with the C_3 axis of the ion.
[e] Assumed values in parentheses.
[f] γ is the out-of-plane angle for the C—H bonds.

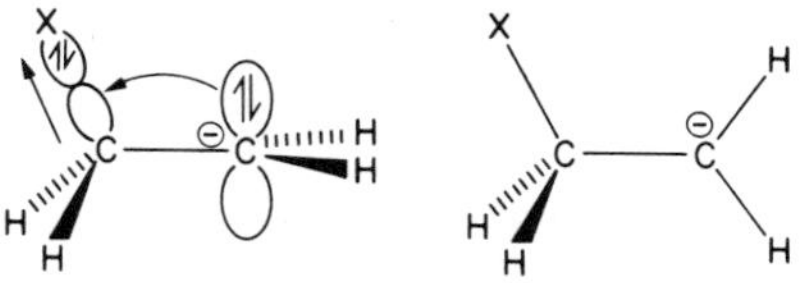

Fig. 2. Perpendicular (XIII) and eclipsed (XIV) conformations of substituted ethyl anions.

XIII   XIV

cyclopropenyl anion has four π electrons and is therefore antiaromatic, hence there is a strong driving force for a C—H bond to move out of plane. The calculated (*sp* basis set) barriers of 52.3 (cyclopropenyl) and 20.8 kcal mol^{-1} (cyclopropyl) are consistent with this line of reasoning.

The linear inversion barrier in the vinyl anion has been examined by Lehn, Munsch, and Millie.[66] They find a barrier of 38.9 kcal mol^{-1}, considerably larger than the barrier (26.2 kcal mol^{-1}) in the isoelectronic nitrogen analog CH_2NH, and an equilibrium bond angle corresponding to a 71.4° distortion from linearity.

3.3.2. Rotation

Factors influencing the conformations of β-substituted ethyl anions $XCH_2CH_2^-$ have been discussed by Hoffmann *et al.*[67] Using the STO-3G basis set and an approximate model with a planar carbanion center, these workers find that the conformational preference between extreme perpendicular (XIII, Fig. 2) and eclipsed (XIV, Fig. 2) conformations strongly depends on the electronegativity of X. When X is more electronegative than H, conformation XIII is favored, while XIV is favored (i.e., XIII is destabilized) if X is less electronegative than H. The calculated energy differences are contrasted with values for corresponding radicals and cations in Table 4. The results can be readily rationalized in terms of the hyperconjugative interaction shown in XIII or by means of simple perturbation molecular orbital arguments.

3.3.3. Rotation–Inversion

Wolfe, Csizmadia and co-workers[68,69] have investigated rotation *and* inversion in several anions of the type XCH_2^- (X = H, CH_3, CH_2^-, OH). The

Table 4. Calculated Energy Differences[67] [E(XIV)–E(XIII), kcal mol^{-1}] for Substituted Ethyl Cations, Radicals, and Anions

Species	Cation	Radical	Anion
BCH_2–CH_2	+10.4	+0.3	−6.2
FCH_2–CH_2	−8.4	0.0	+9.2

favored conformations for $HOCH_2^-$ and $[CH_2CH_2]^{2-}$ are similar to those of the isoelectronic nitrogen analogs hydroxylamine and hydrazine. The inversion barriers are in the order $X = CH_3 < H < CH_2^-$. The barrier to double inversion in $[CH_2CH_2]^{2-}$ is almost twice that of the single inversion.

4. Heats of Reaction

4.1. Proton Affinities

Since the protonation reaction

$$H^+ + A^- \rightarrow HA \tag{3}$$

does not lead to any change in the number of electron pairs, we might expect that the change in correlation energy should be small. Detailed studies on this point have been carried out by Hopkinson et al.[19] and by Bonaccorsi et al.[70] Some representative results are shown in Tables 5 and 6.

For heats of reactions involving neutral molecules alone, it is well known that minimal basis sets do not always perform well. It is not surprising then that Hopkinson et al.[19] obtain very poor proton affinities [heats of reaction (3)] with their near-minimal basis set of uncontracted Gaussians. Proton affinities for OH^- and CH_3O^-, calculated with the minimal STO-3G basis set, are likewise found[21,22,71] to be in error by about 150 kcal mol^{-1}.

For intermediate sized basis sets, i.e., roughly double-zeta quality, studies of the protonation of the acetylide and diacetylide anions[72] and of the hydroxide and methoxide anions[71,73] show that there are still errors of 30–40 kcal mol^{-1}. It is not until near-Hartree–Fock level calculations are performed that we find close agreement with experiment. This is the case for the results of Bonaccorsi et al.[70] in Table 6.

Table 5. Proton Affinities $(-\Delta H$, kcal mol$^{-1})$ with Different Basis Sets of Uncontracted Gaussians[19]

Ion	Basis^a				
	$(3s, 1p; 1s)$	$(5s, 2p; 2s)$	$(7s, 3p; 3s)$	$(10s, 6p; 4s)$	Experiment
H^-	434	448	433	422	406
N_3^-	1357	1182	988	813	820
NH^{2-}	1493	779	683	624	650
NH_2^-	627	478	438	427	380
O^{2-}	1445	882	735	636	615
OH^-	785	496	425	403	390
F^-	1022	504	403	387	370

^{a} The notation $(3s, 1p; 1s)$, for example, means three s functions and one p_x, p_y, and p_z function on each heavy atom and one s function on hydrogen.

Table 6. Proton Affinities $(-\Delta H, kcal\ mol^{-1})$ for Near-Hartree–Fock Wave Functions

Ion	Theory	Experiment
H⁻	405.2	405.8
Li⁻	350.8	357.4
B⁻	402.2	400.0
F⁻	383.5	370.4
CN⁻	371.9	368

Hopkinson *et al.* (Table 5) find that the smaller basis sets almost invariably give values of proton affinities whose magnitudes are too high. This means that the negative ion plus positive ion combination is not being described as well as the neutral molecules with the small basis sets. It seems, therefore, that near-Hartree–Fock basis sets may be required to provide a description of negative ions of comparable quality to the neutral molecules and hence to provide accurate absolute values of proton affinities.

4.2. Relative Acidities

4.2.1. Experimental Comparison

It would be an unsatisfactory state of affairs as far as large molecules are concerned if we always had to carry out calculations at the Hartree–Fock level to get reasonable results. It is, therefore, important to establish under which circumstances smaller basis sets such as STO-3G and 4-31G may be used for calculating the heats of reaction.

In a recently completed study[74] the series of isodesmic proton transfer reactions (4)–(6),

$$\text{(4)}$$

$$RO^- + MeOH \rightleftharpoons MeO^- + ROH \qquad (5)$$
$$(R = Me, Et, Pr^i, Bu^t)$$

$$RCC^- + HCCH \rightleftharpoons HCC^- + RCCH \qquad (6)$$
$$(R = Me, Et, Pr^i, Bu^t)$$

has been studied with the STO-3G basis set. These reactions provide a means of determining the relative acidities of similar molecules. The theoretical heats of reaction are compared with recent gas-phase experimental estimates in Table 7. It should be noted that the experimental values come from studying

Table 7. STO-3G and Experimental Heats (kcal mol^{-1}) of Proton Transfer Reactions[74]

Reaction	Substituent	Theoretical	Experimental[a]
(4)	*o*-F	+2.7	+2.8
	m-F	+5.0	+4.8
	p-F	+1.7	+2.1
	o-Me	+0.3	+0.3
	m-Me	−0.6	−0.5
	p-Me	−1.0	−1.2
(5)	Me	0	0
	Et	+1.9	+2.9
	Pr^i	+4.3	+4.8
	Bu^t	+6.7	+5.6
(6)	Me	−3.0	−3.1
	Et	−2.3	—
	Pr^i	−1.6	(−2.8)[b]
	Bu^t	−0.6	−1.3

[a] Reaction (4) from R. T. McIver and J. H. Silvers, *J. Am. Chem. Soc.* **95**, 8462 (1973); and reactions (5) and (6) from R. T. McIver and J. S. Miller, *J. Am. Chem. Soc.* **96**, 4323 (1974); and J. S. Miller, Ph.D. thesis, University of California, Irvine, 1975.
[b] Experimental result for Pr^n rather than Pr^i.

precisely these or closely related reactions and are thus more reliable than heats of reaction evaluated in the usual way (as the small difference between two large numbers).

The agreement with experiment is good, particularly for the larger phenolic systems. Thus, the correct ordering of substituent effects with respect to the position of the substituent in the ring is obtained in reaction (4), even to the extent of reproducing the small stabilizing effect of an *o*-methyl substituent. It may well be that calculations on large anions will be more successful than those on small anions because of the greater dispersal of the negative charge in the former which may consequently be better described by basis sets constructed essentially for neutral molecules. The gas-phase experimental sequence of acidities of the alcohols, ROH, R = Me < Et < Pr^i < Bu^t is reproduced theoretically. We note that the solution acidities are in the reverse order. The gas-phase ordering of acidities has been rationalized[71] in terms of the calculated charge distributions. For the alkoxide anions RO$^-$, the Bu^t substituent is *electron-withdrawing* relative to Me. The more effective charge dispersal in Bu^{t}O$^-$ leads to stabilization of the anion and hence increased acidity for Bu^{t}OH. Finally, the unusual ordering of acidities indicated by icr studies for the acetylenes RCCH, R = H < Bu^t < Me, is reproduced by the calculations.

A comparison[71] of STO-3G and 4-31G results for reactions (5) and (6) shows that the two basis sets lead to identical qualitative conclusions but there are differences of 1–3 kcal mol^{-1} in the calculated heats of reaction. These are not large errors in an absolute sense but they unfortunately *are* large in relation to the heats of reaction being calculated. The 4-31G heats of reaction are larger in absolute magnitude than the corresponding STO-3G values and surprisingly, in poorer argeement with the experimental results.

4.2.2. Other Acidity Studies

The first *ab initio* comparisons of the acidities of water and small aliphatic alcohols were reported by Hehre and Pople[21] and by Owens, Wolf, and Streitwieser[22] using the STO-3G basis set. Both sets of workers find that methanol is more acidic than water in agreement with gas-phase experimental data. This result implies that substituting a methyl for hydrogen in OH$^-$ has a stabilizing effect. Owens *et al.*[22] also find that ethanol is more acidic than methanol but that the result is sensitive to geometry and exponent optimization. Subsequent studies[71] with standard geometries have yielded good results for the relative acidities of methanol and ethanol with both standard and optimized exponents.

Streitwieser and co-workers[57,75] have also carried out calculations on the acidities of *distorted* methanes, ethanes, and ethylenes. The methane distortion is achieved by reducing one HCH bond angle while keeping the other HCH angle at the tetrahedral value, the purpose being to simulate a small ring system. The result of this distortion is to increase the *s* character in the "exocyclic" bonds and to increase their acidity. Similar results were noted for distortions at the β position in the ethyl and vinyl anions.

4.3. Anion Hydration

Detailed studies have been reported for the hydration of the fluoride[76–80] and hydroxide[23,24] anions. For the fluoride anion, the mono- and polyhydrates have been studied by Diercksen and Kraemer[76,77] and by Kistenmacher, Popkie, and Clementi.[78–80] Both sets of workers used large basis sets which include polarization functions and their results are therefore close to the Hartree–Fock limit.

The monohydrate of F$^-$ is a slightly bent hydrogen-bonded structure (Fig. 3).[78] Calculated binding energies (Table 8) relative to F$^-$ + H$_2$O are in close

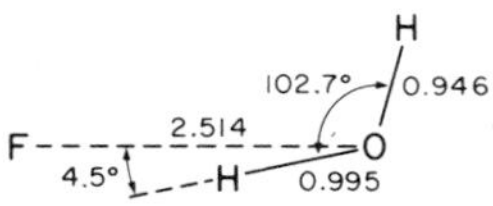

Fig. 3. Optimized structure[78] of the monohydrate (XV) of F$^-$.   XV

Table   8. *Calculated   and   Experimental   Binding   Energies*
(kcal mol^{-1}) for Hydrates of F$^-$ and OH$^-$

Ion	Binding energy[a]	
	Calculated	Experimental
F$^-$·H$_2$O	22.8,[b] 24.1[c]	23.3[e]
F$^-$·2H$_2$O	20.7,[b] 20.8[d]	16.6[e]
F$^-$·3H$_2$O	17.6[b]	13.7[e]
OH$^-$·H$_2$O	24.3,[f] 40.8[g]	22.5,[h] 34.6[i]
OH$^-$·2H$_2$O	30.2[g]	16.4,[h] 23.0[i]
OH$^-$·3H$_2$O	23.0[g]	15.1,[h] 18.5[i]
OH$^-$·4H$_2$O	20.8[g]	14.2[h]

[a] Binding energy quoted relative to next lower hydrate plus H$_2$O.
[b] Reference 80.
[c] Reference 76.
[d] Reference 77.
[e] M. Arshadi, R. Yamdagni, and P. Kebarle, *J. Phys. Chem.* **74**, 1475 (1970).
[f] Reference 24.
[g] Reference 23.
[h] M. Arshadi and P. Kebarle, *J. Phys. Chem.* **74**, 1483 (1970).
[i] M. De Paz, A. G. Guidoni, and L. Friedman, *J. Chem. Phys.* **52**, 687 (1970).

agreement with a recent gas-phase experimental estimate. They are modified by less than 2 kcal mol^{-1} if correlation and zero-point vibrational energy corrections are applied.[79]

The best structure found by Kraemer and Diercksen[77] from a limited geometry search for the dihydrate of the fluoride ion corresponds to a symmetrical geometry (C_2) with the fluoride ion bonded to each of the two water molecules by linear hydrogen bonds. The second water is calculated to be less strongly bound than the first, although the decrease in binding is not as pronounced as is found experimentally.

Kistenmacher, Popkie, and Clementi have examined[80] the dihydrate and higher hydrates (up to five water molecules) of F$^-$ using a pairwise additivity approximation to determine the most favorable structures and then carrying out full SCF calculations on these structures. They also find that successive water molecules are less strongly bound but again the theoretical binding energies decrease more slowly than the experimental values.

The hydration of OH$^-$ has been studied by Newton and Ehrenson (NE)[23] and by Kraemer and Diercksen (KD).[24] Optimum (subject to certain simplifying assumptions such as linear hydrogen bonds in the open forms) structures found by the former workers for OH$^-$·nH$_2$O ($n = 1, 2$, and 3) are shown in Fig. 4 and corresponding binding energies are included in Table 8. Both sets of workers find that the monohydrate of OH$^-$ has a slightly asymmetric hydrogen bond (i.e., O1–H1 not equal to O2–H1 in XVI) but this could be an artifact of the assumed values for the terminal O–H lengths. The calculated energy difference[24] between this asymmetrical structure and a symmetrical (C_2) structure is less than 0.5 kcal mol^{-1}.

There is a large difference between the two theoretical estimates of the binding energies for OH$^-$·H$_2$O. The near-Hartree–Fock calculations of KD predict 24.3 kcal mol^{-1} compared to NE's 4-31G value of 40.8 kcal mol^{-1}. It is difficult to make an absolute judgment as to which value is better on the basis of the experimental results alone, since there is some disagreement in these results as well. In addition, corrections for zero-point vibration and electron correlation of about 2–4 kcal mol^{-1} need to be applied. However, it seems likely that the larger basis-set calculations are the more accurate and that the NE binding energies for the mono and higher hydrates of OH$^-$ are too high.

Newton and Ehrenson[23] find, for the hydrates of OH$^-$, that chain structures with branching where possible are generally favored over cyclic structures, although for the higher hydrates (e.g., H$_9$O$_5^-$) the cyclic structures become increasingly competitive. For electrostatic reasons, the ionic moieties prefer internal rather than outer positions in the complexes. Most of the bridging protons in the chain structures are found to prefer asymmetric equilibrium positions and are associated with potentials characterized by a single asymmetrically located minimum rather than a double-potential well.

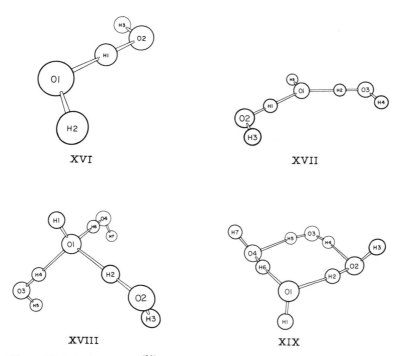

Fig. 4. Optimized structures[23] of the monohydrate (XVI), dihdrate (XVII), and trihydrate (XVIII) of OH$^-$. The cyclic structure (XIX) forms the basis of the optimum structure for the tetrahydrate.

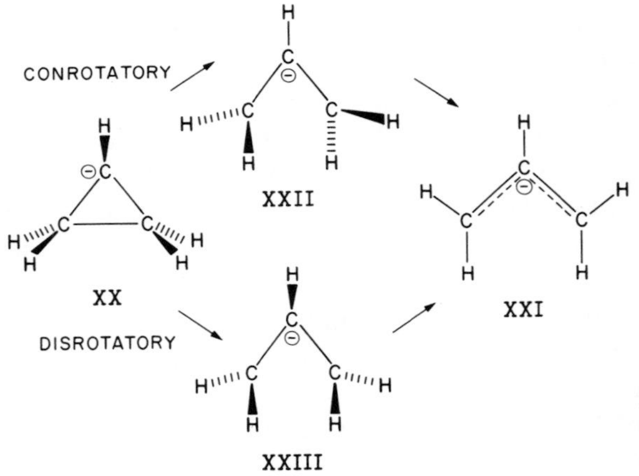

Fig. 5. Electrocyclic transformation of the cyclopropyl (XX) to the allyl (XXI) anion.

The calculated solvation energies associated with the outer solvation shell, although smaller in magnitude than inner-shell energies, are appreciably larger than association energies for neutral water molecules.

5. Mechanistic Studies

5.1. Electrocyclic Transformation of Cyclopropyl to the Allyl Anion

An *ab initio* study of the electrocyclic transformation of the cyclopropyl to the allyl anion has been carried out by Clark and Armstrong.[81] Both conrotatory and disrotatory modes of transformation (Fig. 5) require activation energy (27 and 74 kcal mol^{-1}, respectively), the former being strongly favored, in agreement with the Woodward–Hoffmann predictions.[82] An interesting feature of the conrotatory transformation concerns the movement of the central C—H bond. For the corresponding cation transformation, detailed *ab initio* calculations[83,84] show that this bond, which is in the plane of the carbon atoms in both cyclopropyl and the allyl cations, moves out of plane during the favored disrotatory transformation. In contrast, for the anions the C—H bond which is initially *out* of plane for the cyclopropyl anion, is predicted to move *into* the plane during the conrotatory transformation. It would be of interest to confirm this result with more detailed calculations, i.e., geometry optimization and use of a larger basis set.

5.2. Reaction of the Hydride Ion with Small Molecules

In an interesting series of papers, Ritchie and King[52,85-87] have examined the reaction of the hydride ion with H_2, HF, H_2O, NH_3 and CH_4. For the

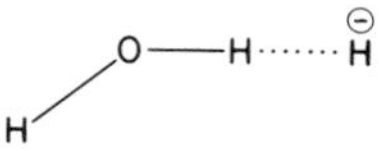

O—H······H⊖ (with H and O bonds shown)

Fig. 6. Hydride ion with linear hydrogen bonding to a neutral molecule.   XXIV

reaction of H^- with H_2, a linear symmetric transition state is found with an energy 15 kcal mol^{-1} higher than that of separated H_2 and H^-. For the other reactions, the energy is found to *decrease* as the hydride ion approaches the neutral molecule, with minimum energy structures having linear hydrogen bonds (see, for example, XXIV, Fig. 6). For H_2F^-, the best structure of this type resembles a complex of hydrogen molecule and fluoride ion. For CH_5^-, NH_4^-, and OH_3^-, the best structures resemble weak complexes of the neutral molecule with hydride ion. These results, however, could have been influenced by the fact that the basis sets used in the study yield the wrong sign for the heats of reactions (7) ($X = CH_3$, NH_2, OH),

$$XH + H^- \rightarrow X^- + H_2 \tag{7}$$

5.3. The S_N2 Reaction

5.3.1. The Reaction of the Hydride Ion with Methane

The simplest model for the S_N2 reaction is the reaction of hydride ion with methane

$$H^- + CH_4 \rightarrow [CH_5^-] \rightarrow CH_4 + H^- \tag{8}$$

For this reaction, the nucleophile is H^-, the leaving group is H^- and the reaction proceeds via a CH_5^- intermediate or transition state. Several studies[88–92] have been carried out on possible structures (XXV–XXVII, Fig. 7) of CH_5^- and the more important results are summarized in Table 9.

The calculations by van der Lugt and Ros[88] and Mulder and Wright[56] both show that the D_{3h} structure (XXV) is substantially (about 50 kcal mol^{-1}) lower in energy than the C_{4v} (XXVI) or C_s (XXVII) structures. This result is consistent with inversion (rather than retention) of configuration in the substitution reaction. Similar values of the C—H bond lengths in the D_{3h} structure are obtained in all studies except that of Mulder and Wright[56] which suffers from a minimal basis set geometry optimization (cf. Section 3.1). A remarkable

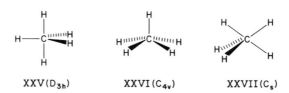

Fig. 7. Structures of CH_5^-.   XXV(D_{3h})   XXVI(C_{4v})   XXVII(C_s)

Table 9. Calculations on CH_5^-

| Reference | Basis set[a] | Relative energies, kcal mol^{-1} | | | D_{3h} bond lengths, Å | | E_a, kcal mol^{-1} for reaction (8) |
		D_{3h}	C_{4v}	C_s	Axial	Equatorial	
88	Extended	0	54	57	1.74	1.07	—
56	Extended	0	55	55	1.629	1.134	69
89	Extended	—	—	—	1.75	1.06	49
90, 91	Extended +polarization	—	—	—	1.737	1.062	60 (55)^c
92	Extended +polarization	—	—	—	1.737	(1.096)^b	70 (55)^c

[a] Basis set specified used to obtain relative energies but not necessarily to obtain optimum geometries.
[b] Assumed value.
[c] Including correlation.

result noted by van der Lugt and Ros[88] is that the charge on C in CH_5^- is more *positive* than in CH_4 while in CH_5^+ the charge on C is more *negative*.

The best single-determinant estimates[90,92] for the activation energy for the S_N2 reaction (8) are in the range 60–70 kcal mol^{-1}. When electron correlation is taken into account, two sets of calculations[91,92] predict that the barrier is reduced to 55 kcal mol^{-1}, although the magnitudes of the corrections are somewhat different.

5.3.2. Detailed Studies of S_N2 Reactions

High-quality calculations (i.e., with geometry optimization and large basis sets including polarization functions) have now been reported by Dedieu and Veillard[90,93] and by Duke and Bader[94,95] for the six S_N2 reactions shown in Table 10.

Table 10. Activation Energies for S_N2 Reactions

Reaction	Activation energy kcal mol^{-1}	Reference
(8) $H^- + CH_4 \rightarrow [HCH_3H]^- \rightarrow CH_4 + H^-$	+60.2	90
(9) $F^- + CH_3F \rightarrow [FCH_3F]^- \rightarrow CH_3F + F^-$	+7.9 +7.3	90 94
(10) $H^- + CH_3F \rightarrow [HCH_2FH]^- \rightarrow CH_3F + H^-$	+60.5	90
(11) $H^- + CH_3F \rightarrow [HCH_3F]^- \rightarrow CH_4 + F^-$	+3.8	90
(12) $F^- + CH_3CN \rightarrow [FCH_3CN]^- \rightarrow CH_3F + CN^-$	+17.3	94
(13) $CN^- + CH_3F \rightarrow [NCCH_3F]^- \rightarrow CH_3CN + F^-$	+22.6	94

An immediate comment that can be made from Table 10 is that there is an activation energy for all reactions. Earlier calculations by Berthier, David, and Veillard[96] had predicted that, for reaction (9), the [FCH$_3$F]$^-$ species was lower in energy than the reactants. Subsequently, this has been shown to be due to the omission of polarization functions from the basis set. The result for (13) is consistent with a recent gas-phase experimental study.[97]

The data in Table 10 also enable us to compare the abilities of the various ions to function as nucleophiles and leaving groups. Comparison of reactions (10) and (11) shows that for a given nucleophile (H$^-$), F$^-$ is displaced much more readily than H$^-$, while reactions (9) and (12) show that F$^-$ displaces F$^-$ more easily than it does CN$^-$. Finally, comparison of (9), (11), and (13) shows that the displacement of F$^-$ in CH$_3$F occurs with comparable ease for the nucleophiles H$^-$ and F$^-$ but is more difficult when CN$^-$ is the nucleophile.

The variation in energy along the reaction coordinate is interesting and typical behavior is shown by reaction (9). There is an initial *decrease* in energy as the nucleophile (F$^-$) approaches the substrate (CH$_3$F), corresponding to a favorable electrostatic interaction. The extent of this energy lowering is quite large (about 12 kcal mol^{-1}), with the energy minimum occurring for a C$\cdots$F distance of about 3 Å, i.e., a weak complex of CH$_3$F and F$^-$. The energy then rises rapidly until the D_{3h} transition state is reached. As the leaving group departs, the energy decreases, reaches a minimum, and then increases to a value corresponding once again to separated CH$_3$F and F$^-$.

A striking feature of all the reactions is the very late departure of the leaving group. Thus, for example, for reaction (8) when the incoming nucleophile (H$^-$) is within 0.15 Å of its position in the transition state, the leaving group still has 0.6 Å to go, having moved only 0.07 Å from its equilibrium position in CH$_4$. Thus, the leaving group remains relatively tightly bound until the nucleophile has approached to a distance close to its transition-state value.[90,94]

Calculations for reactions (11) and (12) yield geometries for the transition states which are in accord with the Hammond postulate,[98] i.e., for an exothermic reaction the geometry of the transition state resembles the reactants rather than the products. Thus, for example, for (12) which is exothermic by about 5 kcal mol^{-1} the bond lengths in the transition state [FCH$_3$CN]$^-$ are closer to the values in CH$_3$CN than in CH$_3$F

$$F^- + CH_3 \overset{1.46}{\text{---}} CN \rightarrow [F \overset{2.25}{\text{---}} CH_3 \overset{1.74}{\text{---}} CN]^- \rightarrow F \overset{1.40}{\text{---}} CH_3 + CN^- \quad (12)$$

The changes in charge distribution during S_N2 reactions have been discussed both by Dedieu and Veillard[90] and by Duke and Bader,[95] the latter workers using the virial partitioning method for analyzing charges. They find that the major change in charge distribution in the transition states for (9) and (13) is a transfer of charge from the nucleophile to the leaving group. For [FCH$_3$F]$^-$ the net positive charge on the CH$_3$ fragment is increased slightly, while in $[NCCH_3F]^-$ the positive charge is decreased through electron donation from CN^-.

Finally, we should note that CI calculations[91] for reaction (9) suggest that the correlation energy remains approximately constant along the reaction path.

6. Conclusions

The main conclusion to be drawn from this review is that useful chemical information *can* be derived from *ab initio* single-determinant molecular orbital calculations on anions. Reliable structural predictions may be made from double-zeta basis-set calculations. In order to obtain good estimates of heats of reaction, near-Hartree–Fock calculations may be required for general reactions such as protonation of anions, which involve closed-shell reactants and closed-shell products. On the other hand, for the particular case of isodesmic proton transfer reactions, good results can be achieved with a minimal basis set. Finally, the ability to describe adequately molecular structure and stability enables the mechanisms of certain chemical reactions to be studied and yields detailed information which is difficult to obtain by other means.

ACKNOWLEDGMENTS

The author is indebted to Dr. Dieter Poppinger and Dr. Peter J. Stiles for their helpful comments on the manuscript.

References

1. D. E. Milligan and M. E. Jacox, *in*: *Molecular Spectroscopy: Modern Research* (K. N. Rao and C. W. Mathews, eds.), pp. 259–286, Academic Press, New York (1972).
2. J. L. Beauchamp, Ion cyclotron resonance spectroscopy, *Ann. Rev. Phys. Chem.* **22**, 527–561 (1971).
3. E. Clementi and A. D. McLean, Atomic negative ions, *Phys. Rev.* **133**, A419–A423 (1964).
4. E. Clementi, A. D. McLean, D. L. Raimondi, and M. Yoshimine, Atomic negative ions. Second period, *Phys. Rev.* **133**, A1274–A1279 (1964).
5. (a) P. E. Cade, Hartree–Fock wavefunctions, potential curves, and molecular properties for OH^- ($^1\Sigma^+$) and SH^- ($^1\Sigma^+$), *J. Chem. Phys.* **47**, 2390–2406 (1967); (b) P. E. Cade, The electron affinities of the diatomic hydrides CH, NH, SiH and PH, *Proc. Phys. Soc., London* **91**, 842–854 (1967).
6. F. Driessler, R. Ahlrichs, V. Staemmler, and W. Kutzelnigg, *Ab initio* calculations on small hydrides including electron correlation. XI. Equilibrium geometries and other properties of CH_3, CH_3^+, and CH_3^-, and inversion barrier of CH_3^-, *Theor. Chim. Acta* **30**, 315–326 (1973).
7. (a) J. Simons and W. D. Smith, Theory of electron affinities of small molecules, *J. Chem. Phys.* **58**, 4899–4907 (1973); (b) J. Kenney and J. Simons, Theoretical studies of molecular ions: BeH^-, *J. Chem. Phys.* **62**, 592–599 (1975).

8. L. C. Snyder, Heats of reaction from Hartree–Fock energies of closed-shell molecules, *J. Chem. Phys.* **46**, 3602–3606 (1967).

9. L. C. Snyder and H. Basch, Heats of reaction from self-consistent field energies of closed-shell molecules, *J. Am. Chem. Soc.* **91**, 2189–2198 (1969).

10. W. J. Hehre, R. Ditchfield, L. Radom, and J. A. Pople, Molecular orbital theory of the electronic structure of organic compounds. V. Molecular theory of bond separation, *J. Am. Chem. Soc.* **92**, 4796–4801 (1970).

11. W. J. Hehre, L. Radom, and J. A. Pople, Molecular orbital theory of the electronic structure of organic compounds. VII. A systematic study of energies, conformations, and bond interactions, *J. Am. Chem. Soc.* **93**, 289–300 (1971).

12. L. Radom, W. J. Hehre, and J. A. Pople, Conformations and heats of formation of organic molecules by use of a minimal Slater type basis, *J. Chem. Soc. A* **1971**, 2299–2303.

13. M. Cohen and A. Dalgarno, Stationary properties of the Hartree–Fock approximation, *Proc. Phys. Soc., London* **77**, 748–750 (1961).

14. K. F. Freed, Geometry and barriers to internal rotation in Hartree–Fock theory, *Chem. Phys. Lett.* **2**, 255–256 (1968).

15. L. Radom and J. A. Pople, *in*: *M.T.P. International Review of Science (Theoretical Chemistry)* (W. Byers Brown, ed.) pp. 71–112, Butterworths, London (1972).

16. H. F. Schaefer, *in*: *Critical Evaluation of Chemical and Physical Structural Information* (D. R. Lide, ed.) pp. 591–602, National Academy of Science, Washington (1974).

17. J. W. Moskowitz and M. C. Harrison, Gaussian wavefunctions for the 10-electron systems. III. OH^-, H_2O, H_3O^+, *J. Chem. Phys.* **43**, 3550–3555 (1965).

18. C. D. Ritchie and H. F. King, Gaussian basis SCF calculations for OH^-, H_2O, NH_3, and CH_4, *J. Chem. Phys.* **47**, 564–570 (1967).

19. A. C. Hopkinson, N. K. Holbrook, K. Yates, and I. G. Csizmadia, Theoretical study on the proton affinity of small molecules using Gaussian basis sets in the LCAO–MO–SCF framework, *J. Chem. Phys.* **49**, 3596–3601 (1968).

20. A. A. Frost, A floating spherical Gaussian orbital model of molecular structure. III. First-row atom hydrides, *J. Phys. Chem.* **72**, 1289–1293 (1968).

21. W. J. Hehre and J. A. Pople, The methyl inductive effect on acid-base strengths, *Tetrahedron Lett.* **1970**, 2959–2962.

22. P. H. Owens, R. A. Wolf, and A. Streitwieser, *Ab initio* calculations of the acidities of some alcohols and hydrocarbons, *Tetrahedron Lett.* **1970**, 3385–3388.

23. M. D. Newton and S. Ehrenson, *Ab initio* studies on the structures and energetics of inner- and outer-shell hydrates of the proton and the hydroxide ion, *J. Am. Chem. Soc.* **93**, 4971–4990 (1971).

24. W. P. Kraemer and G. H. F. Dierckson, SCF MO LCGO studies on hydrogen bonding: the system $(HOHOH)^-$, *Theor. Chim. Acta* **23**, 398–403 (1972).

25. H. Lischka, *Ab initio* calculations on small hydrides including electron correlation. IX. Equilibrium geometries and harmonic force constants of HF, OH^-, H_2F^+ and H_2O and proton affinities of F^-, OH^-, HF and H_2O, *Theor. Chim. Acta* **31**, 39–48 (1973).

26. L. Radom, Structures of simple anions from *ab initio* molecular orbital calculations, *Aust. J. Chem.*, **29**, 1635–1640 (1976).

27. W. J. Hehre, R. F. Stewart, and J. A. Pople, Self-consistent molecular-orbital methods. I. Use of Gaussian expansions of Slater-type atomic orbitals, *J. Chem. Phys.* **51**, 2657–2664 (1969).

28. R. A. Hegstrom, W. E. Palke, and W. N. Lipscomb, Optimized minimum basis set for BH_4^-, *J. Chem. Phys.* **46**, 920–922 (1967).

29. P. H. Owens and A. Streitwieser, *Ab initio* quantum organic chemistry. I. STO-NG calculations of methane and methyl anion, *Tetrahedron* **27**, 4471–4493 (1971).

30. R. Ditchfield, W. J. Hehre, and J. A. Pople, Self-consistent molecular-orbital methods. IX. An extended Gaussian-type basis for molecular-orbital studies of organic molecules, *J. Chem. Phys.* **54**, 724–728 (1971).

31. (a) W. Kutzelnigg, Solution of the two-electron problem in quantum mechanics by direct determination of the natural orbitals. I. Theory, *Theor. Chim. Acta* **1**, 327–342 (1963); (b) M. Jungen and R. Ahlrichs, *Ab initio* calculations on small hydrides including electron

correlation. III. A study of the valence shell intrapair and interpair correlation energy of some first row hydrides, *Theor. Chim. Acta* **17**, 339–347 (1970).

32. P. C. Hariharan and J. A. Pople, Accuracy of AH_n equilibrium geometries by single determinant molecular orbital theory, *Mol. Phys.* **27**, 209–214 (1974).

33. E. L. Albasiny and J. R. A. Cooper, The calculation of electronic properties of BH_4^-, CH_4 and NH_4^+ using one-centre self-consistent field wave functions, *Proc. Phys. Soc., London* **82**, 289–303 (1963).

34. D. M. Bishop, A one-centre treatment of the ammonium and borohydride ions, *Theor. Chim. Acta* **1**, 410–417 (1963).

35. M. Krauss, Calculation of the geometrical structure of some AH_n molecules, *J. Res. Natl. Bur. Stand., Sect. A* **68**, 635–644 (1964).

36. P. Pulay, *Ab initio* calculation of force constants and equilibrium geometries. III. Second-row hydrides, *Mol. Phys.* **21**, 329–339 (1971).

37. J. R. Easterfield and J. W. Linnett, Applications of a simple molecular wavefunction. Part 4. The force fields of BH_4^-, CH_4 and NH_4^+, *J. Chem. Soc., Faraday Trans.* 2 **1974**, 317–326.

38. P. A. Kollman and L. C. Allen, A theory of the strong hydrogen bond. *Ab initio* calculations on HF_2^- and $H_5O_2^+$, *J. Am. Chem. Soc.* **92**, 6101–6107 (1970).

39. P. N. Noble and R. N. Kortzeborn, LCAO–MO–SCF studies of HF_2^- and the related unstable systems HF_2^0 and HeF_2, *J. Chem. Phys.* **52**, 5375–5387 (1970).

40. J. Almlöf, Hydrogen bond studies. 71. *Ab initio* calculation of the vibrational structure and equilibrium geometry in HF_2^- and DF_2^-, *Chem. Phys. Lett.* **17**, 49–52 (1972).

41. T. W. Archibald and J. R. Sabin, Theoretical investigation of the electronic structure and properties of N_3^-, N_3 and N_3^+, *J. Chem. Phys.* **55**, 1821–1829 (1971).

42. P. K. Pearson, H. F. Schaefer, J. H. Richardson, L. M. Stephenson, and J. I. Brauman, Three isomers of the NO_2^- ion, *J. Am. Chem. Soc.* **96**, 6778–6779 (1974).

43. M. D. Newton, W. A. Lathan, W. J. Hehre, and J. A. Pople, Self-consistent molecular orbital methods. V. *Ab initio* calculation of equilibrium geometries and quadratic force constants, *J. Chem. Phys.* **52**, 4064–4072 (1970).

44. W. A. Lathan, W. J. Hehre, L. A. Curtiss, and J. A. Pople, Molecular orbital theory of the electronic structure of organic compounds. X. A systematic study of geometries and energies of AH_n molecules and cations, *J. Am. Chem. Soc.* **93**, 6377–6387 (1971).

45. L. Radom, *Ab initio* molecular orbital calculations on acetyl cations. Relative hyperconjugative abilities of C—X bonds, *Aust. J. Chem.* **27**, 231–239 (1974).

46. L. Radom, P. C. Hariharan, J. A. Pople, and P.v.R. Schleyer, Molecular orbital theory of the electronic structure of organic compounds. XXII. Structures and stabilities of $C_3H_3^+$ and C_3H^+ cations, *J. Am. Chem. Soc.* **98**, 10–14 (1976).

47. P. G. Lykos, R. B. Hermann, J. D. S. Ritter, and R. Moccia, *Ab initio* calculations on simple π-electron systems, *Bull. Am. Phys. Soc.* **9**, 145 (1964).

48. R. N. Rutledge and A. F. Saturno, One-center expansion wavefunctions for CH_3^-, CH_4 and CH_5^+, *J. Chem. Phys.* **43**, 597–602 (1965).

49. B. D. Joshi, Study of CH_3^- and OH_3^+ by one-center expansion self-consistent-field method, *J. Chem. Phys.* **47**, 2793–2798 (1967).

50. W. J. Hehre, R. F. Stewart, and J. A. Pople, Atomic electron populations by molecular orbital theory, *Symp. Faraday Soc.* **2**, 15–22 (1968).

51. P. Millie and G. Berthier, All-electron calculations of open-shell polyatomic molecules. I. SCF wave function in Gaussians for methyl and vinyl radicals, *Int. J. Quantum Chem., Symp.* **2**, 67–73 (1968).

52. C. D. Ritchie and H. F. King, Theoretical studies of proton-transfer reactions. III. The reactions of hydride ion with ammonia and methane, *J. Am. Chem. Soc.* **90**, 838–843 (1968).

53. R. E. Kari and I. G. Csizmadia, Near-molecular Hartree–Fock wavefunction for CH_3^-, *J. Chem. Phys.* **46**, 4585–4590 (1967).

54. R. E. Kari and I. G. Csizmadia, Potential-energy surfaces of CH_3^+ and CH_3^-, *J. Chem. Phys.* **50**, 1443–1448 (1969).

55. R. E. Kari and I. G. Csizmadia, Configuration interaction wavefunctions and computed inversion barriers for NH_3 and CH_3^-, *J. Chem. Phys.* **56**, 4337–4344 (1972).

56. J. J. C. Mulder and J. S. Wright, The electronic structure and stability of CH_5^+ and CH_5^-, *Chem. Phys. Lett.* **5**, 445–449 (1970).

57. A. Streitwieser and P. H. Owens, SCF calculations of acidities of distorted methanes, *Tetrahedron Lett.*, **1973**, 5221–5224.

58. A. J. Duke, A Hartree–Fock study of the methyl anion and its inversion potential surface: use of an augmented basis set for this species, *Chem. Phys. Lett.* **21**, 275–282 (1973).

59. A. Rauk, L. C. Allen, and E. Clementi, Electronic structure and inversion barrier of ammonia, *J. Chem. Phys.* **52**, 4133–4144 (1970).

60. R. M. Stevens, Accurate SCF calculation for ammonia and its inversion motion, *J. Chem. Phys.* **55**, 1725–1729 (1971).

61. P. Dejardin, E. Kochanski, A. Veillard, B. Roos, and P. Siegbahn, MC-SCF and CI calculations for the ammonia molecule, *J. Chem. Phys.* **59**, 5546–5553 (1973).

62. R. M. Stevens, CI calculations for the inversion barrier of ammonia, *J. Chem. Phys.* **61**, 2086–2090 (1974).

63. H. Lischka and V. Dyczmons, The molecular structure of H_3O^+ by the *ab initio* SCF method and with inclusion of correlation energy, *Chem. Phys. Lett.* **23**, 167-172 (1973).

64. D. T. Clark, Non-empirical LCAO–MO–SCF calculations with Gaussian type functions on the aromaticity and anti-aromaticity of cyclopropenyl cation and anion, *Chem. Commun.* **1969**, 637–638.

65. D. T. Clark and D. R. Armstrong, Pseudo-aromaticity and -anti-aromaticity in cyclopropyl cation and anion, *Chem. Commun.* **1969**, 850–851.

66. J. M. Lehn, B. Munsch, and P. Millie, Theoretical conformational analysis. IV. An *ab initio* SCF–LCAO–MO study of methylenimine and of vinyl anion, *Theor. Chim. Acta* **16**, 351–372 (1970).

67. R. Hoffmann, L. Radom, J. A. Pople, P.v.R. Schleyer, W. J. Hehre, and L. Salem, Strong conformational consequences of hyperconjugation, *J. Am. Chem. Soc.* **94**, 6221–6223 (1972).

68. S. Wolfe, L. M. Tel, J. H. Liang, and I. G. Csizmadia, Stereochemical consequences of adjacent electron pairs. A theoretical study of rotation-inversion in ethylene dicarbanion, *J. Am. Chem. Soc.* **94**, 1361–1364 (1972).

69. S. Wolfe, L. M. Tel, and I. G. Csizmadia, The *gauche* effect. A theoretical study of the topomerization (degenerate racemization) and tautomerization of methoxide ion tautomer, *Can. J. Chem.* **51**, 2423–2432 (1973).

70. R. Bonaccorsi, C. Petrongolo, E. Scrocco, and J. Tomasi, SCF wavefunction for the ground state of CN^- and the change of the correlation energy in some simple protonation processes, *Chem. Phys. Lett.* **3**, 473–475 (1969).

71. L. Radom, Effects of alkyl groups on acidities and basicities in the gas phase. An *ab initio* molecular orbital study, *Aust. J. Chem.* **28**, 1–6 (1975).

72. A. C. Hopkinson and I. G. Csizmadia, The proton affinities of the acetylene molecule, and of the acetylide and diacetylide ions, *Chem. Commun.* **1971**, 1291–1292.

73. L. M. Tel, S. Wolfe, and I. G. Csizmadia, Near-molecular Hartree–Fock wavefunctions for CH_3O^-, CH_3OH, and $CH_3OH_2^+$, *J. Chem. Phys.* **59**, 4047–4060 (1973).

74. L. Radom, *Ab initio* molecular orbital calculations on anions. Determination of gas phase acidities, *J. Chem. Soc., Chem. Commun.* **1974**, 403–404.

75. A. Streitwieser, P. H. Owens, R. A. Wolf, and J. E. Williams, *Ab initio* SCF calculations of the acidity of distorted ethanes and ethylenes, *J. Am. Chem. Soc.* **96**, 5448–5451 (1974).

76. G. H. F. Diercksen and W. P. Kraemer, SCF MO LCGO studies on hydrogen bonding. The system $(FHOH)^-$, *Chem. Phys. Lett.* **5**, 570–572 (1970).

77. W. P. Kraemer and G. H. F. Diercksen, SCF LCAO MO studies on the hydration of ions. The system $F^- \cdot 2H_2O$, *Theor. Chim. Acta* **27**, 265–272 (1972).

78. H. Kistenmacher, H. Popkie, and E. Clementi, Study of the structure of molecular complexes. III. Energy surface of a water molecule in the field of a fluorine or chlorine atom, *J. Chem. Phys.* **58**, 5627–5638 (1973).

79. H. Kistenmacher, H. Popkie, and E. Clementi, Study of the structure of molecular complexes. V. Heat of formation for the Li^+, Na^+, K^+, F^- and Cl^- ion complexes with a single water molecule, *J. Chem. Phys.* **59**, 5842–5848 (1973).

80. H. Kistenmacher, H. Popkie, and E. Clementi, Study of the structure of molecular complexes. VIII. Small clusters of water molecules surrounding Li^+, Na^+, K^+, F^-, and Cl^- ions, *J. Chem. Phys.* **61**, 799–815 (1974).

81. D. T. Clark and D. R. Armstrong, Non-empirical LCAO–MO–SCF calculations with Gaussian type functions on the electrocyclic transformation of cyclopropyl to allyl. II. Anion transformation, *Theor. Chim. Acta* **14**, 370–382 (1969).

82. R. B. Woodward and R. Hoffmann, The conservation of orbital symmetry, *Angew Chem., Int. Ed. Engl.* **8**, 781–853 (1969).

83. D. T. Clark and D. R. Armstrong, Non-empirical LCAO–MO–SCF calculations with Gaussian type functions on the electrocyclic transformation of cyclopropyl to allyl. I. Cation transformation, *Theor. Chim. Acta* **13**, 365–380 (1969).

84. L. Radom, P. C. Hariharan, J. A. Pople, and P. v. R. Schleyer, Molecular orbital theory of the electronic structure of organic compounds. XIX. Geometries and energies of $C_3H_5^+$ cations. Energy relationships among allyl, vinyl and cyclopropyl cations, *J. Am. Chem. Soc.* **95**, 6531–6544 (1973).

85. C. D. Ritchie and H. F. King, The absence of a barrier in the theoretical potential energy surface for the reaction of hydride with hydrogen fluoride, *J. Am. Chem. Soc.* **88**, 1069–1070 (1966).

86. C. D. Ritchie and H. F. King, Theoretical studies of proton-transfer reactions. I. Reactions of hydride ion with hydrogen fluoride and hydrogen molecules, *J. Am. Chem. Soc.* **90**, 825–833 (1968).

87. C. D. Ritchie and H. F. King, Theoretical studies of proton-transfer reactions. II. The reaction of water with hydride ion, *J. Am. Chem. Soc.* **90**, 833–838 (1968).

88. W. T. A. M. van der Lugt and P. Ros, Retention and inversion in bimolecular substitution reactions of methane, *Chem. Phys. Lett.* **4**, 389–392 (1969).

89. C. D. Ritchie and G. A. Chappell, An *ab initio* LCGO–MO–SCF calculation of the potential energy surface for an S_N2 reaction, *J. Am. Chem. Soc.* **92**, 1819–1821 (1970).

90. A. Dedieu and A. Veillard, A comparative study of some S_N2 reactions through *ab initio* calculations, *J. Am. Chem. Soc.* **94**, 6730–6738 (1972).

91. A. Dedieu, A. Veillard and B. Roos, in: *Proceedings of the 6th Jerusalem Symposium on Quantum Chemistry and Biochemistry* (E. D. Bergmann and B. Pullman, eds.), pp. 371–377, Israel Academy of Sciences and Humanities, Jerusalem (1974).

92. V. Dyczmons and W. Kutzelnigg, *Ab initio* calculations on small hydrides including electron correlation. XII. The ions CH_5^+ and CH_5^-, *Theor. Chim. Acta* **33**, 239–247 (1974).

93. A. Dedieu and A. Veillard, *Ab initio* calculation of activation energy for an S_N2 reaction, *Chem. Phys. Lett.* **5**, 328–330 (1970).

94. A. J. Duke and R. F. W. Bader, A Hartree–Fock SCF calculation of the activation energies for two S_{N^2} reactions, *Chem. Phys. Lett.* **10**, 631–635 (1971).

95. R. F. W. Bader, A. J. Duke, and R. R. Messer, Interpretation of the charge and energy changes in two nucleophilic displacement reactions, *J. Am. Chem. Soc.* **95**, 7715–7721 (1973).

96. G. Berthier, D. J. David, and A. Veillard, *Ab initio* calculations on a typical S_N2 reaction. Electronic structure of methyl fluoride and of the transition state $(FCH_3F)^-$, *Theor. Chim. Acta* **14**, 329–338 (1969).

97. D. K. Bohme, G. I. Mackay, and J. D. Payzant, Activation energies in nucleophilic displacement reactions measured at 296°K *in vacuo*, *J. Am. Chem. Soc.* **96**, 4027–4028 (1974).

98. G. S. Hammond, A correlation of reaction rates, *J. Am. Chem. Soc.* **77**, 334–338 (1955).

$$9$$

Electron Spectroscopy

Maurice E. Schwartz

1. Introduction

Electron spectroscopy has developed so rapidly during the last ten years or so that an adequate general survey would be far beyond the limits of this chapter. Fortunately there are already several survey books available,[1-3] a new journal devoted to the field has been born,[4] reports of three separate international conferences have been published,[5-7] and two review articles especially relevant to theory have recently appeared.[8-9] Even elementary textbooks[10] and pedagogically oriented journals[11,12] now contain serious discussions of this exciting new field for the beginning student. The reader can consult these various publications for details about earlier work which we cannot discuss here, as well as for many beautiful examples of experimental applications of photoelectron spectroscopy.

The flourishing of *ab initio* quantum chemistry of the sort discussed in the present volume has occurred along with that of photoelectron spectroscopy*: theory and experiment have often successfully gone hand in hand. In this chapter we concentrate on those applications of theory which we believe, up to this time, have been most successful (and useful to experimental workers) and which are best understood. These are the calculations and assignments of

---

*Some quantum chemical background is assumed in this chapter without explicit comment or reference. Relevant chapters in the present series which can be consulted are those by Dunning (basis sets), Bagus (SCF), Wahl and Das (multiconfiguration SCF), Shavitt (CI), Meyer (PNO-CI), Roos and Siegbahn (CI), and Kutzelnigg (pair correlations). We should also point out a few terms which are commonly used in photoelectron spectroscopy. PES is a general acronym. ESCA is often used for x-ray PES (as is XPS). UPS may be used for ultraviolet PES.

Maurice E. Schwartz • Department of Chemistry and Radiation Laboratory, University of Notre Dame, Notre Dame, Indiana

vertical ionization potentials for both core and valence electrons, and the prediction and interpretation of the shifts in core electron binding energies* for a given atom with changing chemical environment.

We shall not be concerned with "secondary" events such as x-ray emission or Auger-electron emission,[1,2] but only with the "primary" ionization process. It has been convenient to distinguish between "core" and "valence" electrons in PES, and we shall continue with these classifications. Core electrons are those with BEs on the order of a few hundred electron volts (for the first two rows of the periodic table, where our discussion will center); they have been often examined in XPS. Valence electrons have IPs in the few electron volts to few tens of electron volts range, though most interesting studies on molecules[3,10–12] have been on those with IPs less than about 21 eV.

Here we consider in detail only the direct ionization energy aspect of PES. Important other topics, especially photoionization cross sections (intensities) and satellite lines ("shake-up"), are much less well understood theoretically, and will be referred to only briefly in the last section of the chapter.

2. Studies of Valence Electrons

2.1. Self-Consistent Field Molecular Orbital Methods

2.1.1. Koopmans' Theorem and Orbital Energies

The basic working language of PES is that of molecular orbital (MO) theory, and indeed the experimental measurements on molecular electron IPs are rightly considered a triumph of MO theory as a general description of molecular electronic structure. Thus, canonical MO theory describes the H_2O molecule by the determinantal wave function

$$|1a_1^2 2a_1^2 1b_2^2 3a_1^2 1b_1^2|$$

and the ESCA spectrum shows five ionizations associated with these MOs, listed in Table 1. These are actually total energy differences between the $^1A_1$ ground state and the various ionic states. The simplest theoretical description rests on Koopmans' theorem.[13] This theorem rigorously identifies the energy difference between the $(2n-1)$-electron ion and the $2n$-electron molecule with the negative of the orbital energy $-\varepsilon_i$, of that SCF-MO ϕ_i which is doubly occupied in the ground state and singly occupied in the ion. When both states are described by determinantal wave functions made from the same SCF-MOs

---

*The equivalent terms ionization potentials (IP) and binding energies (BE) are often used interchangebly; most often IP refers to low-energy (valence) and BE to high-energy (core) ionizations.

Table 1. *Experimental Vertical Ionization Energies for H_2O[a]*

Final ionic state	Energy, eV
$^2A_1$	539.7
$^2A_1$	32.2
$^2B_2$	18.4
$^2A_1$	14.7
$^2B_1$	12.6

[a] From p. 83 of Ref. 2.

determined for the ground state*:

$$\Delta E_i = E^+ - E^0 = -\varepsilon_i. \tag{1}$$

If one chooses to identify $-\varepsilon_i$ with an IP the defects of this model as a quantitative one are well known[14]: it does not allow for relaxation of the remaining electronic structure upon ionization (frozen orbital approximation), nor does it include any accounting for defects inherent in the MO model itself (correlation corrections).† Nevertheless, this simple model is important because it has many useful qualitative and semiquantitative applications itself, and, furthermore, it can serve as a starting point for more rigorous theories.

In the LCAO expansion formulation of SCF-MO theory the basis-set problem also influences results, and in order to test this theoretical model one needs definitive, consistent orbital energies. These could be obtained, for example, from near-Hartree–Fock results with extended basis sets, which unfortunately are not yet generally available for wide classes of molecules. Or one could pick a "reasonably good" basis set and stick with it. This latter approach has been taken by the workers at the Bell Telephone Laboratories, whose *double*-zeta (DZ) quality Gaussian-basis SCF-MO wave functions for 56 different molecules have been summarized in book form[15] and applied to a wide variety of PES problems.[16–19] Table 2 summarizes the three valence orbital energies for H_2O from three different basis-set SCF-MOs: a minimum STO, the DZ, and near-Hartree–Fock. The latter two agree well with one another, but are about 1.1, 1.2, and 1.25 eV too high, respectively, when compared to the experimental values in Table 1. It is a common (but not universal) observation that good quality orbital energies estimate IPs which are too high compared with the true values, because relaxation has not been included. While the orbital energies themselves, as given in Table 2, for example, might be useful in assigning an experimental spectrum,[16–19] they will

---

*In a more general description (e.g., with open-shell initial states), one rigorously identifies ΔE_i with $-\varepsilon_i$ only for an unrestricted SCF-MO wave function. As we shall only consider closed-shell ground states in connection with Koopmans' theorem, the above description is sufficient.

†For deeper lying core electrons, the nonrelativistic wave mechanics itself is an approximation, and relativistic corrections may be required.

Table 2. Orbital Energies for the Three Valence MOs in H_2O, with Different Basis Sets

Molecular orbital	Basis		
	Minimum STO,[a] eV	Double-zeta GTO,[b] eV	Near Hartree–Fock,[c] eV
$1b_2$	16.98	19.52	19.50
$3a_1$	12.68	15.42	15.87
$1b_1$	10.95	13.78	13.85

[a] Ref.. 20.
[b] Ref. 15.
[c] Ref. 21.

not give quantitative estimates of IPs. The Bell workers[16–19] found empirically that reduction of their $DZ - \varepsilon_i$ by 8% often gave good estimates of IPs for valence IPs out to about 30 eV; Table 3 contains a few examples. While this seems a generally suitable correction, we have deliberately included two serious failures in the table; the accidental degeneracy of the $6a_1(\sigma)$ and $2b_2(\pi)$ MOs of ethylene oxide,[16] and the reversal of the levels (compared to experiment) of the $10a'(n_0)$ and $2a''(\pi)$ MOs in formamide.[17] These must be treated with a more sophisticated theory. (One should realize, of course, that the

Table 3. IPs Estimated by the "8% Rule" Compared to Experiment

Molecule	State	8% Rule, eV	Experiment, eV
H_2O [a]	$^2B_2$	17.9	18.4
	$^2A_1$	14.1	14.7
	$^2B_1$	12.7	12.6
CH_4 [b]	$^2A_1$	23.6	23.0
	$^2T_2$	13.6	14.0
H_2CO [a]	$^2B_1$	11.0	10.9
	$^2B_2$	13.4	14.5
	$^2A_1$	16.1	16.0
	$^2B_1$	17.6	16.6
$\overset{O}{\overset{/\backslash}{H_2C-CH_2}}$ [c]	$^2A_1$	21.7	~21.8
	$^2B_2$	11.3	10.6
	$^2A_1$	11.3	11.7
	$^2B_1$	13.5	13.7
	$^2A_2$	13.8	~14.2
	$^2A_1$	16.4	16.4
	$^2B_2$	18.0	17.4
$HCONH_2$ [d]	$^2A'$	10.9	10.1
	$^2A''$	10.4	10.5
	$^2A'$	15.0	13.8
	$^2A''$	14.4	14.7
	$^2A'$	16.3	16.8

[a] Ref. 19.
[b] Ref. 18.
[c] Ref. 16.
[d] Ref. 17.

experimentalists often have other means to deduce at least the type and symmetry of "MO being ionized," such as vibrational spacings, intensities, Jahn–Teller splittings, correlations with similar systems, etc.[2,16–19] Other well-known failures of $-\varepsilon_i$ to give the correct ordering of IPs are the diatomic molecules N_2[22] and F_2.[23] As a final really spectacular example we note that Rohmer and Veillard's SCF-MO study[24] of bis(π-allyl)nickel gave the $9a_g$ MO orbital energy at -18.3 eV, which was the 13th MO down from the highest occupied MO based on ε_i ordering. However, this mostly Ni $3d$ orbital does indeed correspond to the first IP, which occurs at 7.85 eV.

2.1.2. Inclusion of Relaxation by Explicit Ionic State Calculations

Within the SCF-MO model one can account for electronic rearrangement upon ionization by performing a new SCF-MO calculation on the ion, and using

$$\Delta E_{\text{SCF}} = E^+_{\text{SCF}} - E^0_{\text{SCF}} \tag{2}$$

instead of $-\varepsilon_i$ as a measure of IP_i. This, too, depends on basis set, as illustrated in Table 4 for H_2O and CH_4. These are for the so-called restricted Hartree–Fock (RHF) model, which maintains double occupancy of those MOs not losing an electron in the ionization. Unless specifically stated otherwise, we shall assume RHF whenever ΔE_{SCF}, as in Eq. (2), is used. Again the DZ and near-Hartree–Fock results are about the same. Rearrangement energies ($E_R = -\varepsilon_i - \Delta E_{\text{SCF}}$) are 1.91, 2.55, and 2.75 eV for H_2O; 1.39 and 1.18 eV for CH_4. The ΔE_{SCF} differ from the experimental IPs because of omitted correlation energy corrections, being 0.8–1.5 eV too low in H_2O, while in CH_4 the a_1 ionization is about 1.3 eV too high and the t_2 about 0.3 eV too low. The H_2O results are qualitatively in agreement with a very naive model that assigns roughly 1 eV or so of the correlation energy to a doubly occupied orbital in the

Table 4. ΔE_{SCF} for Water and Methane with Different Basis Sets

		Basis		
Molecule	State	Atomic orbital GTO,[a] eV	Double-zeta GTO,[b] eV	Near Hartree–Fock,[c] eV
H_2O	$^2B_2$	18.59	17.71	17.59
	$^2A_1$	13.37	13.30	13.32
	$^2B_1$	11.69	11.01	11.10
CH_4	$^2A_1$	25.31	24.44	24.30
	$^2T_2$	14.51	13.72	13.67

[a] Calculations by the author using the "unsplit" Gaussian AOs from which the DZ basis was derived.[15]
[b] Calculations by the author with the DZ-GTO basis.[15]
[c] H_2O from Ref. 21; CH_4 from Ref. 25.

ground state, which is not present in the ion. The CH_4 results cannot be so simply rationalized. Too much is known nowadays (see other contributions to this series) about correlation energy and pair concepts for one to take this simple model too seriously. Nevertheless, it has sometimes been roughly useful, as, for example, in the calculation of IP as the average[16] of $-\varepsilon_i$ (too high) and ΔE_{SCF} (too low).

For assigning valence IPs, ΔE_{SCF} has itself been useful,[15-19] though sometimes subject to (uncertain) correlation corrections. For example, the correct ordering of the IPs in ethylene oxide[16] (Table 3) is obtained for ΔE_{SCF}. The bis(π-allyl)nickel also seemed well described when ΔE_{SCF} estimates of IP were used,[24] even though an essentially minimum basis set was used (see Chapter 5). Similar results have been obtained for other organometallics studied with similar basis sets,[26] and since these are prodigious calculations to perform for most of us, even in these advanced days of quantum chemistry, they are certainly of great use for understanding experimental PES results.

While the ΔE_{SCF} method does represent the best one-configuration MO model for IPs, and will probably see a great deal of application for larger systems in the near future, it does, nevertheless, present some serious theoretical and practical problems. First of all, when there are many ionic states of the same symmetry in the same energy region (for example, in cases of low symmetry—see $HCONH_2$ in Table 3) this near degeneracy will render a single configuration a poor quantitative description. Indeed it is often difficult, if not practically impossible, to obtain a SCF wave function in such circumstances, because of a tendency toward variational collapse to a lower energy configuration of the same symmetry. We noted this difficulty, for example,[27] in intermediate "hole states" such as $2a_1$ in H_2O and NH_3, and Rohmer and Veillard commented[24] that they also had sometimes noticed it. What will probably be required as a minimal SCF-MO model for ionic states in this category (say, for the first n IPs of a given type) will be a multiconfigurational SCF-MO model with enough configurations to make the number of states desired: the n solutions will be variationally bound by the n lowest true energies of the system (by MacDonald's theorem), and will account for relaxation, but only for the "near-degeneracy" part of correlation.

A second problem arises even when several ionic states are the lowest of their symmetries, but the SCF-MO model contains different correlation defects for the different states. Here ΔE_{SCF} may not even give the correct ordering of IPs. This is illustrated for N_2 in Table 5, where the failure of Koopmans' theorem to order the ionic states is repeated in the ΔE_{SCF} model. Note also that for both σ_g and σ_u ionization, ΔE_{SCF} values are above the true IPs and for π_u ionization both the Koopmans' theorem and ΔE_{SCF} IP estimates are below the true IP. Clearly, such complicated deviations from experiment and naive models must require careful correlation corrections.

Table 5. Koopmans' Theorem, ΔE_{SCF}, and Experimental IPs for the Three Lowest Ions of N_2 [a]

State	$-\varepsilon_i$, eV	ΔE_{SCF}, eV	IP, eV
$^2\Sigma_g^+$	17.28	15.85	15.60
$^2\Pi_u$	16.75	15.50	16.98
$^2\Sigma_u^+$	21.17	19.80	18.78

[a] $-\varepsilon_i$ from Ref. 28; $R = 2.068$ a.u. ΔE_{SCF} deduced (for vertical IPs) from potential energy curves in Ref. 28.

2.2. Methods Including Electron Correlation

2.2.1. Simple Estimates using Pair Correlation Energy Corrections

In their work on small rings[16] and on amides and carboxylic acids,[17] the Bell Laboratories workers used a simple scheme including correlation corrections in which they used ΔE_{SCF} to account for reorganization, and a Mulliken population analysis weighting of Nesbet atomic-pair correlation energies to account for correlation. While this was recognized as somewhat naive, it did help in the resolution of assignments of PES which $-\varepsilon_i$ and ΔE_{SCF} had still left ambiguous. In ethylene oxide, for example,[16] the corrected estimates of the first three IPs were 10.37, 11.73, and 13.6 eV, in fine agreement with experiment (see Table 3). The incorrect ordering of $-\varepsilon_i$ for formamide (Table 3) was reversed, but quantitative agreement was less satisfactory (12.12 and 13.75 eV predicted). I am unaware of any further applications along these lines.

2.2.2. Explicit Configuration Interaction Studies

From my own viewpoint it appears that many photoelectron spectroscopists are just beginning to appreciate the reorganization failure of Koopmans' theorem (correlation corrections will come later), while many quantum chemists are just beginning to appreciate the needs of PES as outlined at the close of Section 2.1.2. There are, thus, few systematic CI studies on correlation energy changes with valence ionization. Meyer has applied his very powerful PNO–CI technique to H_2O[21] and CH_4.[25] Table 6 displays the computed valence IPs for H_2O and CH_4, which agree well with experimental values (Tables 1 and 3). Apparently, careful studies of first row diatomics have also been done,[22,23] and I refer the reader (and myself!) to Meyer's discussion of the PNO–CI technique in Chapter 11, Volume 3 of this Series.

2.2.3. Perturbation Theory Corrections to the Single-Configuration SCF-MO Model

Because many chemically interesting quantities such as IPs are total energy differences between two states, it has long been of interest to try to

Table 6. Meyer's PNO–CI Computed IPs for H_2O and CH_4

Molecule	State	IP, eV
H_2O [a]	$^2A_1$	32.25
	$^2B_2$	18.73
	$^2A_1$	14.54
	$^2B_1$	12.34
CH_4 [b]	$^2A_1$	23.38
	$^2T_2$	14.29

[a] Ref. 21.
[b] CEPA results from Ref. 25.

calculate the difference directly by a perturbation-type calculation on a reference state, rather than by taking a relatively small difference between two separate calculations which are each subject to different kinds and magnitudes of errors. PES has inspired two perturbation approaches to IP calculations via corrections to the Koopmans' theorem ground-state SCF-MO model.

Many-Body Perturbation Scheme. The Munich workers have been developing[22,23,29–32] approaches based on the second-quantization formulation of many-body perturbation theory, in which the IPs are found from the negative real parts of the poles of the one-particle Green's function. With a Hartree–Fock reference state for the perturbation scheme, the zero-order poles correspond to IP $= -\varepsilon_i$ for the occupied MOs, which is just Koopmans' theorem. First-order terms vanish in this formulation, and higher-order terms are analyzed by diagram techniques. The second-order expansion of the self-energy operator contains sums over three orbital indexes; third order, sums over five-index quantities; fourth order, seven indexes, etc. It thus rapidly becomes impossible to evaluate the higher-order terms.[31] Of course, a finite basis LCAO expansion of the MOs also affects the use of this formalism via use of a finite number of MOs and the orbital energies ε_i and electron-repulsion integrals $V_{ij,kl}$ which must be used.

When first applied[30] to H_2CO with inclusion of terms up through second order, and with only the highest five occupied and lowest five unoccupied MOs used, the method gave the first three calculated IPs within about 0.2 eV of experiment. The next two IPs were too high by about 1 and 2 eV. It is not clear whether this error is due to the relatively small number of orbitals used or to the second-order truncation of the expansion. Further studies suggested that both effects are important,[22,23,31] since the second-order expansion failed to give quantitatively acceptable IPs for several other small molecules, and the second-order results became progressively worse as the number of MOs included increased. In fact, the self-energy expansion up through third order can be carried out numerically,[31] but even at this stage the results are not generally satisfying. From numerical results for the most important types[22] of terms in third order, these types were assumed to dominate the higher-order sums. A further assumption that the higher-order contributions constitute a geometric series yielded a direct estimate of all higher-order terms. This "effective interaction" representation of all higher orders was combined with the expansion through third order to give a representation of the complete self-energy operator expansion. IPs calculated from this procedure were combined with the Koopmans' theorem IPs from the MOs actually used in the calculation, to obtain *corrections* to Koopmans' theorem which were then used with more accurate orbital energies from other sources (if different from those used) to compute the final estimate of the IPs. The N_2, F_2, and H_2O results summarized in Table 7 show the accurate results obtained for the valence IPs (the deviation for $2a_1$ ionization in H_2O was discussed by these workers[23]). H_2CO and HCCH were also successfully examined by this approach.

For the systems considered so far, this approach does seem to work, though it is quite elaborate and in a sense "semiempirical" because of the "effective interaction" estimate of the higher-order contributions to the self-energy operator. In a very interesting extension of this work, semiempirical MO theory (e.g., CNDO/2) was used to calculate approximately the defects of Koopmans' theorem, which were combined with *ab initio* orbital energies to estimate IPs. This showed some promise in its first applications. For example, it predicted the correct ordering and values of F_2 IPs and of the ethylene oxide IPs previously mentioned here (Table 3 and related discussion). Also, a rather confused story on the assignment of the first few IPs of BF_3 was set straight. It will be interesting to see whether or not this method can be made generally applicable to PES and easily accessible to the non-many-body theorist.

Rayleigh–Schrödinger Third-Order Energy Corrections. Chong, Herring, and McWilliams[33-35] have been examining the use of straightforward Rayleigh–Schrödinger perturbation theory for calculating IPs. Standard perturbation analysis of a system with a closed-shell SCF-MO reference state ψ^0

Table 7. *Experimental and Computed Valence IPs from Approximate Many-Body Perturbation Scheme*

Molecule	State	Experiment, eV	Computed, eV
N_2	$^2\Sigma_g^+$	15.60	15.50
	$^2\Pi_u$	16.98	16.83
	$^2\Sigma_u^+$	18.78	18.59
F_2	$^2\Pi_g$	15.83	15.69
	$^2\Pi_u$	18.80	18.83
	$^2\Sigma_g^+$	21.0	20.91
H_2O	$^2B_1$	12.6	12.69
	$^2A_1$	14.7	14.91
	$^2B_2$	18.4	18.96
	$^2A_1$	32.2	34.82

[a] N_2 from Ref. 22; F_2 and H_2O from Ref. 23.

with energy E^0 leads to explicit forms of the first-order correction to the wave function ψ^1, from which the energy corrections E^2 and E^3 can be directly calculated by standard perturbation formulas. A similar analysis for the qth ion (spin-orbital ϕ_q now unoccupied in unrelaxed reference state ψ_q^0 with energy $E^0 - \varepsilon_q$) also produces ψ_q^1, E_q^2, and E_q^3 by standard techniques. Five different techniques were first considered[33] for estimating IPs from these wave functions and partial energy quantities.*

1. $\Delta E(3)$, the difference between the usual perturbation theory energies calculated through 3rd order: $\Delta E(3) = -\varepsilon_q + \Delta(E^2) + \Delta(E^3)$.

2. $\Delta \langle H \rangle$, the difference between the expectation values of the Hamiltonians, calculated over the perturbed wave functions through 1st order.

3. $\Delta \mathscr{E}$, as in 2 but using the "scaled perturbation" approach with optimum scale factor η for each state.

4. $\Delta(E^{GA})$, where E^{GA} refers to the energy for each state scaled so that the partial energy sum through 3rd order is stationary with respect to η, leading to a geometric approximation to the energy: $E^{GA} = E^0 + E^2/(1 + E^3/E^2)$, etc.

5. $(\Delta E)^{GA}$, similar to 4, except that the geometric approximation applies to the partial sum differences in 1: $(\Delta E)^{GA} = -\varepsilon_q + \Delta(E^2)/[1 - \Delta(E^3)/\Delta(E^2)]$.

While the first-order wave function corrections ψ^1 and ψ_q^1 are explicit, they do contain very many configurations (all double excitations for ψ^1, all single and double excitations for ψ_q^1). The first simplification used by Chong *et al.*[33] was the omission of all configurations with coefficients less than a threshold 10^{-T} in magnitude. Some test calculations of $\Delta \mathscr{E}$ for several IPs of H_2O, CH_2, and NH_3 with DZ-STO based MOs showed that $\Delta \mathscr{E}$ had essentially converged by $T = 2.6$, which seems to have been the cutoff used everywhere thereafter. Using the DZ-STO MOs, subsequent examination of the formulas 1–5 for valence IPs in H_2O, CH_2, NH_3, H_2CO, and F_2O led to the rejection of $\Delta \langle H \rangle$ and $\Delta \mathscr{E}$ because they generally gave poorer results than the others. (Also, $\Delta \mathscr{E}$ is appropriate only for the lowest ion of each symmetry, because it derives from a variational determination of η.) $\Delta E(3)$, $\Delta(E^{GA})$, and $(\Delta E)^{GA}$ were comparable in accuracy, and no one was preferred; they all gave average absolute errors of about 0.5 eV for all IPs studied. Subsequent examinations of different basis sets were made[34] for the above-mentioned molecules (and CO), and as a "working basis set" based on a balance between accuracy and economy, a "$1\frac{1}{2}$ zeta" STO basis was chosen. This uses a minimum basis for cores and H atoms and double-zeta for valence orbitals. It gave average errors of about 0.3–0.5 eV for valence IPs [depending on whether $\Delta E(3)$, $\Delta(E^{GA})$ or $(\Delta E)^{GA}$ was used], and poorer results (0.6–1.4) for intermediate and core IPs. Subsequent application to molecules containing Si, P, S, and Cl showed similar success.

---

*In their discussion of these, Chong *et al.*[33] sometimes inadvertently wrote down the 1st-order energy correction E^1, which is zero in both neutral and ion as the theory is formulated. We have omitted E^1 in our formulas here.

Table 8. Calculated IPs from the Third-Order Rayleigh–Schrödinger
Perturbation Theory[a]

Molecule	State	$\Delta E(3)$, eV	$\Delta(E^{GA})$, eV
N_2	$^2\Sigma_g^+$	15.18	14.90
	$^2\Pi_u$	16.62	16.67
	$^2\Sigma_u^+$	19.09	18.39
F_2	$^2\Pi_g$	16.39	15.66
	$^2\Pi_u$	19.63	18.69
	$^2\Sigma_g^+$	21.02	20.80
H_2O	$^2B_1$	12.79	12.42
	$^2A_1$	15.07	14.73
	$^2B_2$	19.17	18.97

[a] Ref. 35.

We conclude our discussion here with Table 8, which shows the $\Delta E(3)$ and $\Delta(E^{GA})$ results[35] from MOs in the same flexible Gaussian basis used by the many-body theorists. The H_2O basis is not identical, but is nearly the same (see Ref. 35 for details). The agreement of this work with that summarized in Table 6 (theory and experiment) for the difficult N_2 and F_2 systems and for H_2O is quite impressive. This straightforward approach thus seems very promising as a useful tool for PES, and the reader should consult the original papers for much more detail and for many numerical examples.

3. Studies of Core Electrons

Theoretical studies of core BEs have been much more extensively reviewed than those for valence electrons. The reader should consult the references[1,2,8,9,36,37] for more detailed discussions. The general features of core electron BEs and their shifts (ΔBE) with molecular environment are well understood theoretically. Here we summarize the broad features of this understanding, and point to a few places where some important, interesting problems remain to be solved.

3.1. Use of Orbital Energies

3.1.1. Comparison of Absolute and Relative Values of Core Orbital Energies with Experiment

For a highly localized core MO which is essentially an AO we expect a great deal of "transferability" from system to system. Accurate LCAO–SCF–MO calculations clearly show this.[9,36] Experimentally, XPS shows ionizations in many molecules which are characteristic of core electrons in a given atom, and which can vary over a range of a few electron volts because of varying molecular environment. It is this feature which makes XPS useful for *electron spectroscopy for chemical analysis* (ESCA).

Although by now our discussion will have suggested that core orbital energies will not represent quantitatively the *absolute* values of the core BEs, there is still good reason to hope for some understanding of the *changes* ΔBE in terms of core orbital energy changes $\Delta(-\varepsilon)$. Thus, let

$$BE = -\varepsilon + \delta \tag{3}$$

where δ will include rearrangement, correlation, and relativistic corrections to the core orbital energy $-\varepsilon$. For a given atom in two systems (a and b) we might consider in the expression

$$\Delta BE = BE_a - BE_b$$
$$= -\varepsilon_a - (-\varepsilon_b) + \delta_a - \delta_b$$
$$= \Delta(-\varepsilon) + \Delta(\delta) \tag{4}$$

that *changes* in the *error* $\Delta\delta$ could, as a first approximation, be ignored because of the highly localized invariant nature of the core orbital itself (more later on this important and not completely accurate assumption). Of course, even with this assumption we realize that the basis set used in the LCAO expansion can be important. This basis set problem has already been well discussed.[2,8,9,36,37] Suffice it to say here that about a double-zeta or better basis set is required to get $-\varepsilon$ which are acceptably close to Hartree–Fock, and which, as it turns out, vary over the same energy range as the BEs do. For example,[36] in a systematic study of ΔBE vs $\Delta(-\varepsilon)$ for C 1s, we found that the Pople 3G-STO basis gave $\Delta(-\varepsilon)$ about one-half as large as ΔBE, an unsplit near HF-GTO-AO basis gave $\Delta(-\varepsilon)$ about twice as large as ΔBE, and the DZ-GTO basis gave $\Delta(-\varepsilon)$ which were about the same as ΔBE (examples of these follow shortly). Shirley's review[8] summarizes in tabular and graphic form a large number of ΔBE–$\Delta(\varepsilon)$ comparisons for C, N, and O 1s BEs which show a generally good equality (within 1 eV) between ΔBE and $\Delta(-\varepsilon)$. The 2p BEs in S-containing molecules were similarly well described.[38] For illustrative purposes we summarize in Table 9 a number of C 1s BEs and $-\varepsilon$ recently tabulated by Basch,[37] along with the calculated ΔBE and $\Delta(-\varepsilon)$ values (both measured from CH_4). The δ are of the order of -14 eV, and are roughly the same to within about 7%–8% (1 eV or so). BEs are of the order of about 291 eV, and they vary by about 0%–4% (0–11 eV). The overall trend of BEs is shown by the $-\varepsilon$. However, for small ΔBE the variations, $\Delta(\delta)$ can overwhelm $\Delta(-\varepsilon)$ and cause important errors in equating ΔBE and $\Delta(-\varepsilon)$ (e.g., C_2H_6, C_3H_6, C_2H_4), while the $\Delta(\delta)$ are relatively less important for larger ΔBE (e.g., CH_3OH, CH_3F, etc.), and $\Delta(-\varepsilon)$ measures ΔBE well. Note that for some molecules there are Hartree–Fock

Table 9. Summary of Experimental C 1s BEs and GTO-DZ Orbital
Energies, and Their Shifts[a]

Molecule	BE	$-\varepsilon$, eV	δ, eV	ΔBE, eV	$\Delta(-\varepsilon)$, eV
CH_4	290.7	305.0	−14.3	0	0
		304.9	−14.2		0
C_2H_6	290.5	305.2	−14.7	−0.2	0.2
C_3H_6	290.5	305.5	−15.0	−0.2	0.5
C_2H_4	290.6	305.9	−15.3	−0.1	0.9
CH_3OH	292.2	307.0	−14.8	1.5	2.0
C_2H_2	291.1	306.3	−15.3	0.4	1.3
		305.9	−14.8		1.0
HCN	293.3	307.8	−14.5	2.6	2.8
		306.8	−13.5		1.9
C_2H_4O	292.7	307.4	−14.7	2.0	2.4
CH_3F	293.5	308.0	−14.5	2.8	3.0
CO	296.1	310.5	−14.4	5.4	5.5
		309.1	−13.0		4.1
CH_2F_2	296.2	311.1	−14.9	5.5	6.1
C_2F_4	296.3	312.6	−16.3	5.6	7.6
CO_2	297.5	313.6	−15.8	6.8	8.6
		311.4	−13.9		6.5
CHF_3	299.0	314.4	−15.4	8.3	9.4
CF_4	301.7	317.7	−16.0	11.0	12.7

[a] From Basch.[37] For CH_4, C_2H_2, NCN, CO, and CO_2 the first line contains the DZ values; the second line, near-Hartree–Fock values.

results in this table in addition to the DZ-GTO ones. Although these do give better calculated shifts for C_2H_2 and CO_2, the agreement with experiment is worse for HCN and CO. Obviously, a more rigorous theory than this one based on Koopmans' theorem will be required for a more accurate quantitative description of ΔBEs for core electrons. But since it essentially describes those chemical shifts when BEs are appreciable, it does have the enormous appeal of being a ground-state molecular model which allows correlations of ΔBEs with ground-state molecular properties to be made.

3.1.2. Relationships of Core Orbital Energy Shifts and Other Properties

In their first paper presenting the large set of $\Delta(-\varepsilon)$ from the DZ-GTO calculations, Basch and Snyder[39] noted a nearly linear relationship between the shifts and the charge at the core orbital host atom

$$\Delta(-\varepsilon_A) = k_A Q_A + X \tag{5}$$

where the charge Q_A derived from a Mulliken population analysis of the SCF-MO wave functions (k_A and X are just fitting parameters here). This agreed with an earlier simple model based on a charged spherical atom,[1,2] which has been well reviewed.[8,9] A more attractive and more accurate

Table 10. Core Orbital Energy and Potential Changes[a]

Molecule	$\Delta(-\varepsilon_{1s})$, eV	$\Delta\Phi$, eV	$\Delta\Phi_{\text{val}}$, eV
CH_4	0	0	0
CH_3F	3.0	3.0	3.0
CH_2F_2	6.1	6.2	6.2
CHF_3	9.4	9.6	9.5
CF_4	12.7	13.1	13.0

[a] From Ref. 41.

point-charge model recognized the existence of charges at all atoms, and used

$$\Delta(\text{BE}_A) = k_A Q_A + V_A + l \tag{6}$$

where

$$V_A = \sum_{B \neq A} Q_B / R_{AB} \tag{7}$$

is the potential at A due to all other atoms (k_A and l are still fitting parameters). We have written $\Delta(\text{BE}_A)$ in Eq. (6), since this approach was first used as an empirical relationship[2] with charges from CNDO/2 SCF–MO theory. It has subsequently been used with orbital energies and Mulliken charges from *ab initio* calculations on S $2p$[38] and C $1s$.[40] But it is well known that such charges are not necessarily well-defined physical quantities; for example,[8,9] the same sets of ΔBEs [or $\Delta(-\varepsilon)$ in this case[40]] for C $1s$ are fit equally well by quite different Q, k, and l values, one set from CNDO/2 and one set from *ab initio* Mulliken charges.

A more appealing and unambiguous approach was found independently and simultaneously by Basch[41] and Schwartz,[42] who found that $\Delta(-\varepsilon_A)$ are the same as changes in the potential at the nucleus on which the core orbital resides:

$$\Phi_A = \sum_i -2\langle \phi_i | 1/r_A | \phi_i \rangle + \sum_{B \neq A} Z_A / R_{AB} \tag{8}$$

Schwartz[42] gave an analysis which suggested that this $\Delta(-\varepsilon) - \Delta\Phi$ relationship would be a general one. To illustrate, Table 10 summarizes the results for C $1s$ in the fluoromethanes[41] which were already included in Table 9. Also included are changes in the "valence potential" Φ_{val} obtained from Eq. (8) by omitting from the first sum (the electronic contribution) the core orbital at A and those core orbitals at other centers B, and by using the net *core* charges in the nuclear contribution (the second sum).[9,36,43]

This has been a widely useful result.[9] Basch noted[41] the possible connection with NMR chemical shifts because of the diamagnetic contribution which is essentially Eq. (8); this has been recently reviewed.[9,43] The earlier point-charge potential of the form of Eq. (6), with k_A and l well defined, was found as a direct consequence of using this potential[36,44] with the CNDO approximation. Use of Φ_{val} has also suggested a way to estimate core ΔBEs directly from the potential changes found with *valence* electron semiempirical MO theories.[36] This latter idea has been extensively employed[45] for computing BE shifts from CNDO/2 theory; it has been a successful approach. The reader is referred to our review[9] and a recent paper by Shirley[46] for a survey discussion of these and related topics. Shirley has also shown[47] that the "equivalent-core" thermodynamic model for BE shifts is nearly equivalent to the use of the potential model of Eq. (8).

3.2. Beyond Koopmans' Theorem: More Accurate Theoretical Models

3.1.1. Inclusion of Relaxation in the SCF-MO Model

Direct Hole-State Calculations. By using the aufbau principle and forcing a deep-lying core orbital to be singly occupied,* one can account for orbital relaxation and can determine RHF-SCF-MO wave functions for the core-hole-state ions appropriate for ESCA. Bagus[48] first did this; he obtained near-Hartree–Fock results for various ions of Ne and Ar. Bagus and others[8,9,25,27,49] have discussed the fact that such states are not the lowest of their symmetry, and therefore, strictly speaking, are not protected from variational collapse. They are, however, so far away from other bound states of the ion that this is no real problem; they do not try to "mix," and they are not subject to variational collapse. Later this approach was used for molecular systems,[27] for which some discussion of basis-set requirements has already been published.[27,36] Table 11 summarizes ΔE_{SCF} results for the 10-electron

*For molecules with several symmetry-equivalent sites, it is now well known[8,9] that in the MO model the proper way to treat core ionization is via a localized hole state rather than by delocalized symmetry-adapted core MOs.

Table 11. ΔE_{SCF} and BE for 1s Ionization in First Row 10-Electron Systems

Molecule	DZ-GTO,[a] eV	Extended split[b] s–p GTO, eV	Hartree–Fock,[c] eV	Experiment,[d] eV
CH_4	292.8	291.0	290.8	290.7
NH_3	407.6	405.7	—	405.6
H_2O	541.6	539.4	539.1	539.7
HF	—	693.3	—	—
He	872.2	868.8	868.6	870.2

[a] Our calculations; see Ref. 36.
[b] Ref. 27.
[c] CH_4, Ref. 25; H_2O, Ref. 21; Ne, Ref. 48.
[d] Ref. 2.

systems CH_4, NH_3, H_2O, HF, and Ne for a DZ-GTO basis, for an "extended split" $s-p$ GTO basis, and for Hartree–Fock calculations. Experimental BEs are also included. The DZ-GTO ΔE_{SCF} are about 2 eV too high, while the extended split $s-p$ basis[27] is 0.2 eV too high compared to the Hartree–Fock. This is essentially an atomic error due to core choice, since further (unpublished) test calculations on Ne showed that a further "splitting" of the 4-term s group in the core basis set[27] into one two-term and two one-term basis functions gave the Hartree–Fock result; the published results[27] missed about 0.2 eV of core relaxation. Further work on H_2O (addition of d basis on O, p on H) gave no change of ΔE_{SCF}, until the core basis was further split as in Ne, at which point the near-Hartree–Fock result was obtained. Unfortunately, for the study of the changes ΔBE, this author is not aware of any other published results on closed-shell molecules* in which near Hartree–Fock results have been obtained for ΔE_{SCF}. The extended basis STO results of Gianturco and Guidotti,[49] for which ΔE_{SCF} fell well below BE of $1s$ in CH_4, H_2O, and NH_3, have been shown to be incorrect.[9] The fluoromethane series has been studied at the DZ-GTO level,[18,50] as have a few other small molecules.[36] From these DZ-GTO calculations one readily concludes that the ΔE_{SCF} results are not superior to $\Delta(-\varepsilon)$ when compared to ΔBE. Hillier *et al.*[51,52] have also studied $\Delta(-\varepsilon)$ and $\Delta(\Delta E_{SCF})$ vs $\Delta(BE)$ with a variety of different $s-p$ basis sets. They have done RHF and UHF calculations on C, N, and O $1s$ problems. Here, too, no advantage of $\Delta(\Delta E_{SCF})$ to $\Delta(-\varepsilon)$ was found.

Because $\Delta(\Delta E_{SCF})$ have not yet been studied with quite accurate basis sets it is not yet clear whether small deviations of $\Delta(-\varepsilon)$ from ΔBE are due only to neglect of relaxation, as has been implicit in semiempirical theories,[45] or whether the SCF-MO model itself is incorrect because of differential effects of electron correlation. Some evidence that the latter may also be quantitatively important comes from two recently obtained unpublished sets of near-Hartree–Fock results on CO: one by J. W. Kress and M. E. Schwartz, and the other by P. S. Bagus and H. Schrenk quoted in a recent paper on CO.[53] For C $1s$ ionization, both sets of calculations give $\Delta(-\varepsilon) = 4.0$ eV and $\Delta(\Delta E_{SCF}) = 6.1$ eV when referenced to CH_4, while $\Delta BE = 5.4$ eV (Table 9).

Direct Estimates of Relaxation Energies. The largest defect in Koopmans' theorem for core orbitals is relaxation (e.g., about 14 eV for C $1s$—Tables 9 and 11), and the estimation of E_R without recourse to elaborate hole-state calculations of ΔE_{SCF} would be desirable. This has already been reviewed,[9,41] and our discussion here will be somewhat limited. The most widely used approach up to now is one based on a paper by Hedin and Johansson,[54] who showed that a good approximation for UHF theory is

$$E_R = \tfrac{1}{2}\langle \phi_c | V_p | \phi_c \rangle \qquad (9)$$

---

*The open-shell systems O_2 and NO have been accurately studied—see Section 4.

Table 12. Comparison of Relaxation Energies for 1s Ionization from ΔE_{SCF} and Equivalent Core Relaxation Potential Models

Molecule	$\Delta E_{SCF},^a$ eV	Equiv. core,^{b} Eq. (10), eV
CH_4	13.9	16.9
NH_3	17.1	21.0
H_2O	20.0	25.2
HF	21.9	26.7
Ne	23.2	27.7
Ar	32.2	40.0

^{a} Molecules from Ref. 27. Ne and Ar from Ref. 48.
^{b} Ref. 57.

where ϕ_c is the ground-state core orbital from which ionization occurs, and V_p is the "polarization potential" $V_p = V^* - V$. Here V is the Hartree–Fock potential from all other ground-state orbitals, and V^* is the same quantity for the relaxed orbitals of the ion. To make Eq. (9) usable without the need for explicit determination of the ionic potential V^* from SCF calculations, two approximations have been introduced.[45,55] First the equivalent-cores idea* is used, so that integrals for the $\langle \phi_c | V^* | \phi_c \rangle$ part of Eq. (9) might be taken from the next atom in the isoelectronic sequence. Second, the electron–electron repulsion integrals which would enter Eq. (9) are replaced by the negative of the valence potential integrals as defined in Eq. (8) and its discussion (this is equivalent to assuming the core orbital has a δ-function form). Thus, the working theory becomes the simple formula

$$E_R = -\tfrac{1}{2}(\Phi_{val}^* - \Phi_{val}) \tag{10}$$

To illustrate, calculation of E_R for CH_4 would use Φ_{val}^* from NH_4^+ and Φ_{val} from CH_4. This has been widely used[45,55,56] with CNDO/2 MO theory to examine cases where $\Delta(E_R)$ may be needed to correct $\Delta(-\varepsilon)$ (computed as $\Delta\Phi_{val}$). Comparisons with experimental ΔBEs have suggested that this may sometimes be important. However, it is a long and approximate series of steps from an accurate $\Delta(-\varepsilon)$ to $\Delta\Phi_{val}$ from CNDO, corrected by E_R calculated as in Eq. (10) from CNDO, to the true $\Delta(\Delta E_{SCF})$. For example, we know from *ab initio* DZ-GTO-SCF-MO results in Table 12 that Eq. (10) gives E_R which are about 20% too large when compared to the *ab initio* E_R for the systems studied by ΔE_{SCF} in Table 11. The theory thus has uncertain *ab initio* support, and the CNDO/2 E_R values[55] (in eV) of 15.9 for CH_4, 19.0 for NH_3, and 20.6 for H_2O are also

*The equivalent-core model[8,9] says that core ionization may be treated as equivalent to an increase of the nuclear charge by +1.

worrisome. Perhaps the differential errors are less.[9] Nevertheless, the correlations of *changes* of Φ_{val} and E_R with experimental changes in BEs are so appealing[45,56] that they call for more thorough study for better understanding.[9]

A second approach with some promise is based on the "transition operator" concept widely used in the "MS X_α" technique—but formulated for rigorous *ab initio* SCF theories.[58,59] Here $-\Delta E_{SCF}$ is given directly as the eigenvalue ε of the orbital being ionized, when the SCF orbitals are determined for a spin-orbital occupancy of 1/2—midway between no ionization and complete ionization! This can be formulated within both UHF and RHF theories,[58,59] with excellent *ab initio* results. When used for chemical *shifts*, then $\Delta(\Delta E_{SCF})$ is given by $\Delta(-\varepsilon)$, requiring just one calculation per molecule. First sample applications to 15 carbon-containing molecules used CNDO calculations (formulated to admit fractional core charges corresponding to $1s^{1.5}$), and produced results a bit more in agreement with experimental ΔBEs than the previous $\Delta\Phi_{val}+\Delta E_R$ results.[55] We shall no doubt see many more applications of the transition operator concept with *ab initio* methods.

3.2.2. Calculations Including Correlation Energy

There have been only a few accurate correlated calculations for core-electron ionization, but they are fundamentally important. Table 13 summarizes these, which are all for $1s$ ionization in the first row. We have also included the well-known results for $1s^2$ He-like ions, because it had been

Table 13. Correlation Energy Corrections to $\Delta E_{SCF}{}^a$ for
$1s$ Ionization

System	Configuration	ΔE_{corr}, eV
C^{+4}	$1s^2$	1.23^b
C^0	$1s^2 2s^2 2p^2$	-0.21^c
CH_4	$1s^2 2a_1^2 1t_2^6$	-0.08^d
O^{+6}	$1s^2$	1.24^b
O^0	$1s^2 2s^2 2p^4$	-0.24^c
H_2O	$1s^2 2a_1^2 1b_2^2 3a_1^b 1b_1^2$	0.50^e
Ne^{+8}	$1s^2$	1.24^b
Ne^0	$1s^2 2s^2 2p^6$	0.81^f
		0.60^g

^{a} RHF results.
^{b} Ref. 60; exact $1s^2$ from correlated wave functions.
^{c} Ref. 61, Öksüz–Sinanoğlu theory.
^{d} Ref. 25, PNO-CI.
^{e} Ref. 21, PNO-CI.
^{f} Ref. 62, many-body perturbation theory.
^{g} Ref. 63, Bethe–Goldstone–Nesbet pair theory.

naively assumed[8,9,27] that this nearly constant correction of 1.2 eV could simply be used to correct ΔE_{SCF} for correlation in all systems involving $1s$ ionization. One sees from Table 13 that this obviously is not an acceptable correction; in fact we would have known this from Table 11, where the ΔE_{SCF} are too close to experiment for such a model. The various methods used for these correlation corrections are footnoted in Table 13, and in the original papers much discussion is given to the details of the correlation corrections. We make just a few remarks here. Contraction of the valence orbitals upon core ionization is important, because outside the core the ionization resembles an increase of $+1$ in the nuclear charge Z (the equivalent core concept). Because atomic correlation energy generally increases with nuclear charge, it is tempting to assume[9] that an increase of intravalence-shell correlation contributions tends to go against the loss of correlation in the ionized core to give a net E_{corr} for the ion which is closer to E_{corr} of the neutral than the intraorbital core correlation which was lost. This is too simplistic, however, as an analysis of the $Ne-Ne^+(1s$ hole) results of Moser, Nesbet, and Verhaegen shows[63]: correlation from the valence pairs sums up to -0.31834 a.u. over the Ne^+ orbitals, and to -0.31730 a.u. for the Ne orbitals. The interorbital $1s$–all valence contribution in $1s^1$ Ne^+ is *greater* in magnitude (-0.02933) than is $1s^2$–valence correlation in Ne (-0.02500). Also, there is -0.01253 a.u. of 1-particle valence contribution in the Ne^+ valence orbitals not present in Ne. The net result is that 0.0399 of $1s^2E_{corr}$ is lost in Ne^+ relative to Ne, but it is partly compensated for by a variety of other terms which amount to -0.01787 a.u. Hence, the net $\Delta E_{corr} = 0.0399 - 0.01787 = 0.0220$ a.u. $= 0.60$ eV.

These results make a very important point: failure of $\Delta(-\varepsilon)$ to agree with experimental $\Delta(BE)$ need not be due solely to differential relaxation effects in the SCF-MO model. Our example on CO in Section 3.2.1 showed this clearly when $\Delta(-\varepsilon)$ (4.0 eV) and $\Delta(\Delta E_{SCF})$ (6.1 eV) both disagreed with $\Delta BE = 5.4$ eV. There is a no *a priori* reason to expect that ΔE_{corr} for $1s$ ionization in CH_4 (equivalent core ion NH_4^+) will be the same as that in CO (equivalent core ion NO^+).

We have come a long way without saying anything specific about relativistic corrections to core BEs calculated with nonrelativistic quantum mechanics. We have very little evidence to go on, but at least for those systems which have been accurately studied (CH_4, H_2O, and Ne with both relaxation and correlation accounted for), simple $1s^2$ relativistic corrections are sufficient to bring the theoretical results into such close agreement with experiment (0.1–0.3 eV) that one has to worry about the interpretation of experimental data. Vibrational effects are even of the order of magnitude of the agreement, and the accurate theoretical results on the CH_4 $1s$ hole-state ion[25] have been useful in understanding the core spectra line broadening observed in the very precise measurements recently made in Uppsala.[64]

4. Summary and Prospectus for the Future

By now we have seen that accurate vertical IPs can be found by *ab initio* quantum chemical methods for both core and valence electrons. For specific molecules, one may expect that variations in geometry will be accurately studied for both ground states and ions, and that much of the fine structure, line broadening, etc., which occur in photoelectron spectra will be theoretically accessible.[25,64] While shifts in BEs are fairly well understood, they are not *very*-well understood, and we may expect to see highly accurate $\Delta E_{\rm SCF}$ and correlated calculations done for a given type of core ionization in different molecular systems. We have seen that semiempirical theories derived from *ab initio* ones can be very powerful, and we can expect this sort of progress to continue.

Several topics of some importance have been completely left out of the discussion here. For example, multiplet splitting due to ionization from open-shell systems has not been considered. Bagus and Schaefer have carried out near-Hartree–Fock $\Delta E_{\rm SCF}$ calculations for $1s$ ionization in $NO^{(65)}$ and O_2,[66] and Basch has studied[41] a few small hydride free radicals with DZ-GTO $\Delta E_{\rm SCF}$. Shirley[8] has reviewed other applications, especially to the important cases of transition metals. The intensity problem (photoionization cross sections) has also been omitted. Proceedings of both the Asilomar Conference (pp. 187–288 of Ref. 5) and the Namur Conference (pp. 811–984 of Ref. 7) can be consulted for general surveys. Finally, the appearance of rather low-intensity satellite peaks at lower electron energies than the main photoionization peak has not been considered. These "shake-up" lines provide severe tests of theory, for their understanding will require both accurate energies and intensities for final states which are "excited" with respect to the main core-hole state. For example, *ab initio* UHF-SCF-MO calculations[67] were said to agree well with a previously reported shake-up spectrum associated with $1s$ ionization in N_2, in which three satellite peaks were seen. However, the detailed N_2 $1s$ experimental spectrum obtained with the new Uppsala ESCA spectrometer[68] shows at least five prominent and three less prominent satellite peaks. Two recent papers[69,70] have reformulated the theory and applied it with quantitative success in *ab initio* CI studies of satellite lines associated with F $1s$ ionization in HF, and we can accordingly expect our theoretical understanding of these phenomena to increase in the near future.

In conclusion we must mention three recent papers[71–73] which present some very exciting prospects for future accurate direct calculations of IPs. Smith *et al.*[71,72] have formulated a "transition matrix" from an exact, correlated wave function for a given initial system. Eigenvalues of this matrix (analogous to Koopmans' theorem) correspond directly to IPs calculated, but with correlation and relaxation included. It was shown[72] how this could be used for correlated (but not necessarily exact) initial wave functions, and in particular the approach was related to multiconfiguration SCF theory. Excellent results were found in applications to ionization in 4-electron atoms. Morrell, Parr, and Levy[73] obtained a formally equivalent result via density-matrix analysis. They pointed out the important relations to MC-SCF theory, and presented some accurate results for He. They also noted that the "higher" (more negative) eigenvalues within a given symmetry class could correspond to "shake-up" ionization processes.

ACKNOWLEDGMENTS

The Radiation Laboratory is operated by the University of Notre Dame under contract with the U.S. Energy Research and Development Administration. This is ERDA Document No. COO-38-988.

I would especially like to record my gratitude to the late Professor Charles Coulson, who stimulated my interest on these and other topics, and who would be very pleased with the real chemical importance of quantum chemistry as discussed in this volume.

Note Added in Proof

Impressive application of the many-body Green's function[74] and Rayleigh–Schrödinger[75,76] perturbation schemes have appeared since this chapter was written; see these papers for examples and other recent references. A clear, thorough discussion of the application of the equations of motion method to IP calculations has just appeared.[77] Smith reports[78] excellent IPs from MC SCF formulations of the transition matrix[71] for first row diatomics, including the difficult F_2 and N_2 cases.

References

1. K. Siegbahn, C. Nordling, A. Fahlman, R. Nordberg, K. Hamrin, J. Hedman, G. Johansson, T. Bergmark, S.-E. Karlsson, I. Lindgren, and B. Lindgren, *ESCA: Atomic, Molecular, and Solid State Structure Studied by Means of Electron Spectroscopy*, Almqvist and Wiksells, Uppsala, Sweden (1967).
2. K. Siegbahn, C. Nordling, G. Johansson, J. Hedman, P. F. Hedén, K. Hamrin, U. Gelius, T. Bergmark, L. O. Werme, R. Manne, and Y. Baer, *ESCA Applied to Free Molecules*, North-Holland Publishing Co., Amsterdam (1969).
3. D. W. Turner, C. Baker, A. D. Baker, and C. R. Brundle, *Molecular Photoelectron Spectroscopy*, Wiley–Interscience, London (1970).
4. *Journal of Electron Spectroscopy and Related Phenomena*, Elsevier Scientific Publishing Company, Amsterdam.
5. D. A. Shirley (ed.), *Electron Spectroscopy: Proceedings of an International Conference held at Asilomar, Pacific Grove, California, 7–10 September, 1971*, North-Holland Publishing Co., Amsterdam (1972).
6. J. N. Murrell (ed.), A general discussion on the photoelectron spectroscopy of molecules, 14–16 September, 1972, *Faraday Discuss. Chem. Soc.* **No. 54** (1972).
7. Proceedings of the international conference on electron spectroscopy, Namur, April 16–19, 1974, *J. Electron Spectrosc. Relat. Phenom.* **5** (1974) (entire volume in one issue).

8. D. A. Shirley, ESCA, *in*: *Advances in Chemical Physics* (I. Prigogine and S. A. Rice, eds.), Vol. XXIII, pp. 85–159, Wiley, New York (1973).

9. M. E. Schwartz, ESCA, *in*: *MTP International Review of Science* (*Physical Chemistry Series 2*: *Theoretical Chemistry*) (C. A. Coulson and A. D. Buckingham, eds.) pp. 189–216, Butterworths, London (1975).

10. C. A. Coulson, *The Shape and Structure of Molecules*, Oxford Univ. Press, Oxford (1973).

11. H. Bock and P. D. Mollére, Photoelectron spectra: an experimental approach to teaching molecular orbital models, *J. Chem. Educ.* **51**, 506–514 (1974).

12. H. Bock and B. G. Ramsey, Photoelectron spectra of nonmetal compounds and their interpretation by MO models, *Angew. Chem., Int. Ed. Engl.* **12**, 734–752 (1973).

13. T. Koopmans, Über die Zuordnung von Wellenfunktionen und Eigenwerten zu den Einzelnen Elektronen eines Atoms, *Physica* **1**, 104–113 (1934).

14. W. G. Richards, The use of Koopmans' theorem in the interpretation of photoelectron spectra, *J. Mass Spectrom. Ion Phys.* **2**, 419–424 (1969).

15. L. C. Snyder and H. Basch, *Molecular Wavefunctions and Properties*: *Tabulated from SCF Calculations in a Gaussian Basis Set*, Wiley–Interscience, New York (1972).

16. H. Basch, M. B. Robin, N. A. Kuebler, C. Baker, and D. W. Turner, Optical and photoelectron spectra of small rings. III. The saturated three-membered rings, *J. Chem. Phys.* **51**, 52–66 (1969).

17. C. R. Brundle, D. W. Turner, M. B. Robin, and H. Basch, Photoelectron spectroscopy of simple amides and carboxylic acids, *Chem. Phys. Lett.* **3**, 292–296 (1969).

18. C. R. Brundle, M. B. Robin, and H. Basch, Electronic energies and electronic structures of the fluoromethanes, *J. Chem. Phys.* **53**, 2196–2213 (1970).

19. M. B. Robin, N. A. Kuebler, and C. R. Brundle, Using the perfluoro effect and He(II) intensity effects for identifying photoelectron transitions, *in*: *Electron Spectroscopy* (D. A. Shirley, ed.), pp. 351–378, North-Holland Publishing Co., Amsterdam (1972).

20. S. Aung, R. M. Pitzer, and S. M. Chan, Approximate Hartree–Fock wavefunctions, one-electron properties, and electronic structure of the water molecule, *J. Chem. Phys.* **49**, 2071–2080 (1968).

21. W. Meyer, Ionization energies of water from PNO–CI calculations, *Int. J. Quantum Chem.* **S5**, 341–348 (1971).

22. L. S. Cederbaum, G. Hohlneicher, and W. von Niessen, On the breakdown of Koopmans' theorem for nitrogen, *Chem. Phys. Lett.* **18**, 503–508 (1973).

23. L. S. Cederbaum, G. Hohlneicher, and W. von Niessen, Improved calculations of ionization energies of closed-shell molecules, *Mol. Phys.* **26**, 1405–1424 (1973).

24. M.-M. Rohmer and A. Veillard, Photoelectron spectrum of bis-(Π-allyl) nickel, *J. Chem. Soc. D* **1973**, 250–251.

25. W. Meyer, PNO-CI studies of electron correlation effects. I. Configuration expansion by means of nonorthogonal orbitals and application to the ground state and ionized states of methane, *J. Chem. Phys.* **58**, 1017–1035 (1973).

26. M. F. Guest, I. H. Hillier, B. R. Higginson, and D. R. Lloyd, The electronic structure of transition metal complexes containing organic ligands. II. Low energy photoelectron spectra and *ab initio* SCF MO calculations of dibenzene chromium and benzene chromium tricarbonyl, *Mol. Phys.* **29**, 113–128 (1975).

27. M. E. Schwartz, Direct calculation of binding energies for inner-shell electrons in molecules, *Chem. Phys. Lett.* **5**, 50–52 (1970).

28. P. E. Cade, K. D. Sales, and A. C. Wahl, Electronic structure of diatomic molecules. III. A. Hartree–Fock wavefunctions and energy quantities for $N_2(X^1\Sigma_g^+)$ and $N_2^+(X^2\Sigma_g^+, A^2\Pi_u, B^2\Sigma_u^+)$ molecular ions, *J. Chem. Phys.* **44**, 1973–2003 (1966).

29. G. Hohlneicher, F. Ecker, and L. Cederbaum, Direct calculation of ionization potentials by means of a perturbation method based on the use of Green's functions, *in*: *Electron Spectroscopy* (D. A. Shirley, ed.), pp. 647–659, North-Holland Publishing Co., Amsterdam (1972).

30. L. S. Cederbaum, G. Hohlneicher, and S. Peyerimhoff, Calculation of vertical ionization potentials of formaldehyde by means of perturbation theory, *Chem. Phys. Lett.* **11**, 421–424 (1971).

31. L. S. Cederbaum, Direct calculation of ionization potentials of closed shell atoms and molecules, *Theor. Chim. Acta* **31**, 239–260 (1973).

32. B. Kellerer, L. S. Cederbaum, and G. Hohlneicher, Calculation of Koopmans' defect using semiempirical molecular orbital methods, *J. Electron Spectrosc. Relat. Phenom.* **3**, 107–122 (1974).

33. D. P. Chong, F. G. Herring, and D. McWilliams, Perturbation corrections to Koopmans' theorem. I. Double-zeta slater-type-orbital basis, *J. Chem. Phys.* **61**, 78–84 (1974).

34. D. P. Chong, F. G. Herring, and D. McWilliams, Perturbation corrections to Koopmans' theorem. II. A study of basis set variation, *J. Chem. Phys.* **61**, 958–962 (1974).

35. D. P. Chong, F. G. Herring, and D. McWilliams, Perturbation corrections to Koopmans' theorem. III. Extension to molecules containing Si, P, S, and Cl and comparison with other methods, *J. Chem. Phys.* **61**, 3567–3570 (1974).

36. M. E. Schwartz, J. D. Switalski, and R. E. Stronski, Core-level binding energy shifts from molecular orbital theory, in: *Electron Spectroscopy* (D. A. Shirley, ed.), pp. 605–627, North-Holland Publishing Co., Amsterdam (1972).

37. H. Basch, Theoretical models for the interpretation of ESCA spectra, *J. Electron Spectrosc. Relat. Phenom.* **5**, 463–500 (1974).

38. U. Gelius, B. Roos, and P. Siegbahn, *Ab initio* MO SCF calculations of ESCA shifts in sulphur-containing molecules, *Chem. Phys. Lett.* **8**, 471–475 (1970).

39. H. Basch and L. C. Snyder, ESCA: chemical shifts of K-shell electron binding energies for first row atoms in molecules, *Chem. Phys. Lett.* **3**, 333–336 (1969).

40. U. Gelius, P. F. Hedén, J. Hedman, B. J. Lindberg, R. Manne, R. Nordberg, C. Nordling, and K. Siegbahn, Molecular spectroscopy by means of ESCA. III. Carbon compounds, *Phys. Scr.* **2**, 70–80 (1970).

41. H. Basch, On the interpretation of K-shell electron binding energy shifts in molecules, *Chem. Phys. Lett.* **5**, 337–339 (1970).

42. M. E. Schwartz, Correlation of 1*s* binding energy with the average quantum mechanical potential at a nucleus, *Chem. Phys. Lett.* **6**, 631–635 (1970).

43. B. J. Lindberg, Can we expect any meaningful correlations between NMR and ESCA shifts? *J. Electron Spectrosc. Relat. Phenom.* **5**, 149–166 (1974).

44. M. E. Schwartz, Core level binding energy shifts and the average quantum mechanical potential at a nucleus from CNDO theory, *J. Am. Chem. Soc.* **94**, 6899–6901 (1972).

45. D. W. Davis and D. A. Shirley, The prediction of core-level binding energy shifts from CNDO molecular orbitals, *J. Electron Spectrosc. Relat. Phenom.* **3**, 137–163 (1974).

46. D. A. Shirley, ESCA results vs. other physical and chemical data, *J. Electron Spectrosc. Relat. Phenom.* **5**, 135–148 (1974).

47. D. A. Shirley, Near-equivalence of the quantum mechanical potential model and the thermochemical model of ESCA shifts, *Chem. Phys. Lett.* **15**, 325–330 (1972).

48. P. S. Bagus, Self-consistent-field wave functions for hole states of some Ne-like and Ar-like ions, *Phys. Rev.* **139**, A619–A634 (1965).

49. F. A. Gianturco and C. Guidotti, Some notes on K-electron energies in molecular systems, *Chem. Phys. Lett.* **9**, 539–543 (1971).

50. E. Clementi and A. Routh, Study of the electronic structure of molecules. XV. Comments on the molecular orbital valency state and on the molecular orbital energies, *Int. J. Quantum Chem.* **VI**, 525–539 (1972).

51. I. H. Hillier, V. R. Saunders, and M. H. Wood, On the contribution of orbital relaxation to ESCA chemical shifts, *Chem. Phys. Lett.* **7**, 323–324 (1970).

52. L. J. Aarons, M. F. Guest, M. B. Hall, and I. H. Hillier, Use of Koopmans' theorem to interpret core electron ionization potentials, *J. Chem. Soc., Faraday Trans. 2* **69**, 563–568 (1973).

53. J. Cambray, J. Gasteiger, A. Streitwieser, Jr., and P. S. Bagus, Self-consistent field calculation of hole states of carbon monoxide. Electron density functions by computer graphics, *J. Am. Chem. Soc.* **96**, 5978–5984 (1974).

54. L. Hedin and A. Johansson, Polarization corrections to core levels, *J. Phys. B.* **2**, 1336–1346 (1969).

55. D. W. Davis and D. A. Shirley, A relaxation correction to core-level binding energy shifts in small molecules, *Chem. Phys. Lett.* **15**, 185–190 (1972).

56. D. W. Davis, M. S. Banna, and D. A. Shirley, Core-level binding-energy shifts in small molecules, *J. Chem. Phys.* **60**, 237–245 (1974).

57. M. E. Scnwartz and S. R. Rothenberg, Concerning the calculation of relaxation energies from relaxation potential equivalent core models, to be published (1977).

58. O. Goscinski, B. T. Pickup, and G. Purvis, Direct calculation of ionization energies. Transition operator for the ΔE_{SCF} method, *Chem. Phys. Lett.* **22**, 167–171 (1973).

59. G. Howat and O. Goscinski, Relaxation effects on ESCA chemical shifts by a transition potential model, *Chem. Phys. Lett.* **30**, 87–90 (1975).

60. C. C. J. Roothaan and A. W. Weiss, Correlated orbitals for the ground state of heliumlike systems, *Rev. Mod. Phys.* **32**, 194–205 (1960).

61. S. Corvilain and G. Verhaegen, K-shell binding energies in C and O, *Int. J. Quantum Chem.* **S7**, 69–81 (1973).

62. R. L. Chase, H. P. Kelley, and H. J. Köhler, Correlation energies and auger rates in atoms with inner-shell vacancies, *Phys. Rev. A* **3**, 1550–1557 (1971).

63. C. M. Moser, R. K. Nesbet, and G. Verhaegen, A correlation energy calculation of the 1s hole state in neon, *Chem. Phys. Lett.* **12**, 230–232 (1971).

64. U. Gelius, S. Svensson, H. Siegbahn, E. Basilier, Å. Faxälv, and K. Siegbahn, Vibrational and lifetime line broadenings in ESCA, *Chem. Phys. Lett.* **28**, 1–7 (1974).

65. P. S. Bagus and H. F. Schaefer III, Direct near-Hartree–Fock calculations on the 1s hole states of NO^+, *J. Chem. Phys.* **55**, 1474–1475 (1971).

66. P. S. Bagus and H. F. Schaefer III, Localized and delocalized 1s hole states of the O_2^+ molecular ion, *J. Chem. Phys.* **56**, 224–226 (1972).

67. L. J. Aarons, M. Barber, M. F. Guest, I. H. Hillier, and J. H. McCartney, Satellite peaks in the high energy photoelectron spectra of some small first row molecules. An experimental and theoretical study. *Mol. Phys.* **26**, 1247–1256 (1973).

68. U. Gelius, E. Basilier, S. Svensson, T. Bergmark, and K. Siegbahn, A high resolution ESCA instrument with x-ray monochromator for gases and solids, *J. Electron Spectrosc. Relat. Phenom.* **2**, 405–434 (1974).

69. R. L. Martin and D. A. Shirley, Theory of core-level photoemission spectra, *J. Chem. Phys.* **64**, 3685–3689 (1976).

70. R. L. Martin, B. E. Mills, and D. A. Shirley, Fluorine 1s correlation states in the photoionization of hydrogen fluoride: experiment and theory, *J. Chem. Phys.*, **64**, 3690–3698 (1976).

71. D. W. Smith and O. W. Day, Extension of Koopmans' theorem. I. Derivation, *J. Chem. Phys.* **62**, 113–114 (1975).

72. O. W. Day, D. W. Smith, and R. C. Morrison, Extension of Koopmans' theorem. II. Accurate ionization energies from correlated wavefunctions for closed-shell atoms, *J. Chem. Phys.* **62**, 115–119 (1975).

73. M. M. Morrell, R. G. Parr, and M. Levy, Calculation of ionization potentials from density matrices and natural functions, and the long-range behavior of natural orbitals and electron density, *J. Chem. Phys.* **62**, 549–554 (1975).

74. W. von Niessen, L. S. Cederbaum, and G. H. F. Dierckson, The electronic structure of molecules by a many-body approach. IV. Ionization potentials and one-electron properties of pyrrole and phosphole, *J. Am. Chem. Soc.* **98**, 2066–2073 (1976).

75. D. P. Chong, F. G. Herring, and D. McWilliams, Calculation of vibrational structure in molecular photoelectron spectra, *J. Electron Spectrosc. Relat. Phenom.* **7**, 429–443 (1975).

76. D. P. Chong, F. G. Herring, and D. McWilliams, Theoretical study of vertical ionization potentials of HNO, FNO, O_3, CF_2, and N_2H_2, *J. Electron Spectrosc. Relat. Phenom.* **7**, 445–455 (1975).

77. G. B. Baksay and N. S. Hush, Theoretical study of the N_2^+ molecular ion, *Chem. Phys.* **16**, 219–227 (1976).

78. D. W. Smith, personal communication.

Molecular Fine Structure

Stephen R. Langhoff
and
C. William Kern

1. Introduction

Weak interactions in the complete many-body Hamiltonian are an important part of modern chemical theory and experiment.[1] As we have seen in this volume, these terms are usually neglected in the construction of electronic wave functions because they account for relatively small amounts of energy, typically on the order of cm^{-1}. Nevertheless, they can be measured to very high accuracy, such as in EPR experiments that probe the coupling of the electron spins with themselves and with the angular momenta of their orbital motion.[2,3] These particular interactions, which are the subject of this chapter, are so small that they act as perturbations to split the nonrelativistic electronic states into a "fine structure" of levels. Since the spacings between these spin multiplets are often very sensitive to the details of the charge distribution, they provide a test of the zero-order wave functions that are used to calculate them.* We are, therefore, dealing with weak forces that have conspicious, measurable, and calculable effects.

Closely related to fine structure are other classes of phenomena that have been known for a long time but which have only recently been treated in a

---

*The first successful EPR measurement of zero-field splittings was performed by Hutchison and Mangum[4] on a photoexcited triplet state of naphthalene which was embedded in single crystals of durene. Afterward, Gouterman and Moffitt[5] interpreted the splitting as a manifestation of the pure spin–spin interactions between the unpaired electrons.

Stephen R. Langhoff • Battelle Columbus Laboratories, Columbus, Ohio, and *C. William Kern* • Battelle Columbus Laboratories and The Ohio State University, Department of Chemistry, Columbus, Ohio

satisfactory quantitative manner. The optical spectra of a large number of molecules show phosphorescence from the lowest excited triplet state to the ground singlet state.[6] The lifetime of this triplet depends on weak interaction terms in the Hamiltonian, namely the spin–orbit operators, which govern its radiative unimolecular decay and any predissociation that occurs.

In this chapter, we shall examine molecular fine structure and the effects of phosphorescence and predissociation by applying the methods of *ab initio*[7] quantum chemistry. That is, we shall be concerned with the detailed interpretation of spectra from a knowledge of the number of particles and the values of their masses, charges, and spins. Although many studies have been carried out, relatively few have simultaneously included, without approximation, all electrons in the Breit–Pauli Hamiltonian[8] and all of the interaction matrix elements that arise from the use of *ab initio* wave functions. This is particularly true of molecules, for which a common practice[9] is to neglect completely integrals that are not centered on the same atom. Several authors have pointed out that this is not always a good approximation.[10,11] For the most part, then, we exclude from our discussion studies of molecules whose multicenter and many-electron nature are not fully taken into account. Also, since some rather complete atomic calculations, especially at the Hartree–Fock level, have been extensively documented[12–17] we restrict our attention exclusively to molecules.

In Section 2 the theory of spin–spin and spin–orbit coupling is outlined, using the Breit–Pauli Hamiltonian. After the zero-field splitting parameters are defined and related to the observed effects, orders of magnitude are estimated for states of different symmetry. Since most of the formal theory has been available for about forty years,[18–20] we concentrate in the remaining sections on *ab initio* studies that illustrate recent applications of the theory. Sections 3 and 4 deal successively with matrix element evaluation, and the numerical aspects of zero-field splittings including spin–orbit coupling. Section 5 discusses phosphorescence and predissociation with respective reference to H_2CO and O_2. Our conclusions are given in Section 6.

2. Theory

To describe fine structure it is necessary to extend the Born–Oppenheimer electronic Hamiltonian

$$\mathcal{H}_0 = \frac{1}{2m} \sum_i p_i^2 + \sum_{i<j} \frac{e^2}{r_{ij}} - \sum_{i,\alpha} \frac{Z_\alpha e^2}{r_{i\alpha}} + \sum_{\alpha<\beta} \frac{Z_\alpha Z_\beta e^2}{r_{\alpha\beta}} \tag{1}$$

to include relativistic effects and thereby to incorporate the interactions that are commonly associated with the spin and orbital angular momenta of the electrons (spin–orbit, spin–other–orbit, and spin–spin interactions). If the molecule is allowed to rotate, then terms must also be added which correspond to the coupling of the electronic spin angular momenta with the rotational motion of the nuclei (spin–rotation interaction). All of these interactions remove the degeneracy of the spin levels of an electronic state even in the absence of any external magnetic field and give rise to the so-called fine structure or zero-field splitting (ZFS).

Since these splittings of the spin levels are 10^5–10^6 times smaller than the spacings between electronic states, perturbation theory is ideally suited to characterize the effects of all terms that must be added to $\mathcal{H}_0$. It can be shown by symmetry arguments that the spin–spin term contributes to the ZFS in first order, whereas for orbitally nondegenerate electronic states the leading interactions between the spin and orbital angular momenta begin at second order in the coupling parameter. Nonetheless, the spin–orbit contributions to the ZFS are large enough in this case to be important even for molecules containing only first row atoms. By contrast, the spin–rotation coupling is small and is treated only briefly in this chapter.

2.1. Breit–Pauli Hamiltonian

To obtain the microscopic Hamiltonian for molecular fine structure, we write the Breit equation[21] for the relativistic, stationary-state wave function ψ of N electrons, interacting with each other and with an external electromagnetic field, as[8]

$$\sum_{i<j}^{N}\left[E-\mathcal{H}(i)-\mathcal{H}(j)-\frac{e^2}{r_{ij}}\right]\psi=\frac{-e^2}{2}\sum_{i<j}^{N}\left[\boldsymbol{\alpha}_i\cdot\boldsymbol{\alpha}_j+\frac{(\boldsymbol{\alpha}_i\cdot\mathbf{r}_{ij})(\boldsymbol{\alpha}_j\cdot\mathbf{r}_{ij})}{r_{ij}^2}\right]\psi \tag{2}$$

where

$$\mathcal{H}(i)=-e\varphi(\mathbf{r}_i)+\beta_imc^2+\boldsymbol{\alpha}_i\cdot c\mathbf{p}_i \tag{3}$$

depends on $\mathbf{p}_i=-ih\boldsymbol{\nabla}_i+(e/c)\mathbf{A}_i$, the linear momenta of the ith electron, and on the total scalar $[\varphi(\mathbf{r}_i)]$ and vector potentials $[\mathbf{A}(\mathbf{r}_i)]$ of the external electromagnetic field at the position of electron i. Here $\boldsymbol{\alpha}_i$ and β_i are Dirac matrices which operate on the spinor components of ψ for electron i. These are 4×4 matrices and can be written as

$$\boldsymbol{\alpha}=\begin{pmatrix}0&\boldsymbol{\sigma}^P\\\boldsymbol{\sigma}^P&0\end{pmatrix}, \qquad \beta=\begin{pmatrix}I&0\\0&-I\end{pmatrix} \tag{4}$$

where the three Cartesian components of $\boldsymbol{\sigma}^P$ are 2×2 matrices, called the Pauli spin matrices, and are given explicitly by

$$\sigma_1^P=\begin{pmatrix}0&1\\1&0\end{pmatrix}, \qquad \sigma_2^P=\begin{pmatrix}0&-i\\i&0\end{pmatrix}, \qquad \sigma_3^P=\begin{pmatrix}1&0\\0&-1\end{pmatrix} \tag{5}$$

In a commonly used notation, I is the unit 2×2 matrix, $\mathbf{r}_{ij}$ is the vector distance between electrons i and j, and E is the total energy. The operator $\mathcal{H}(i)$ is identical with the Dirac Hamiltonian and leads to the correct relativistic equation for particles of spin $\frac{1}{2}$. However, since the Breit equation is not fully Lorentz invariant, only an approximation to the true relativistic interaction between electrons is obtained. For the purposes of this chapter, this limitation introduces negligible error.

To simplify the Breit equation, we use the Pauli approximation[22] in which the electrons are assumed to move in a weak external electromagnetic field, or equivalently that the average electronic energy of any bound state is close to the rest mass of N electrons. Since the mathematical details are rather involved* and are given elsewhere,[8] we state the final result in the form of the Breit–Pauli Hamiltonian[8,23]

$$\mathcal{H} - \mathcal{H}_0 = \mathcal{H}_1 + \mathcal{H}_2 + \mathcal{H}_3 + \mathcal{H}_4 + \mathcal{H}_5 + \mathcal{H}_6 \tag{6}$$

where

$$\mathcal{H}_1 = \frac{e^2\hbar}{2m^2c^2}\left\{\sum_{i,\alpha}\frac{Z_\alpha}{r_{i\alpha}^3}(\mathbf{r}_{i\alpha}\times\mathbf{p}_i)\cdot\mathbf{s}_i + \sum_{i,j\neq i}\left[\frac{(2\mathbf{p}_j-\mathbf{p}_i)\times\mathbf{r}_{ij}}{r_{ij}^3}\right]\cdot\mathbf{s}_i\right\} \tag{7}$$

$$\mathcal{H}_2 = \frac{e^2\hbar^2}{m^2c^2}\sum_{i<j}\frac{r_{ij}^2(\mathbf{s}_i\cdot\mathbf{s}_j)-3(\mathbf{r}_{ij}\cdot\mathbf{s}_i)(\mathbf{r}_{ij}\cdot\mathbf{s}_j)}{r_{ij}^5} - \frac{8\pi}{3}\frac{e^2\hbar^2}{m^2c^2}\sum_{i<j}(\mathbf{s}_i\cdot\mathbf{s}_j)\delta(\mathbf{r}_{ij}) \tag{8}$$

$$\mathcal{H}_3 = \frac{-e^2\hbar}{4m^2c^2}\sum_i\sum_\alpha\frac{Z_\alpha}{r_{i\alpha}^3}(\mathbf{r}_{i\alpha}\cdot\mathbf{p}_i) - \sum_{j\neq i}\frac{(\mathbf{r}_{ij}\cdot\mathbf{p}_i)}{r_{ij}^3} \tag{9}$$

$$\mathcal{H}_4 = \frac{-e^2}{2m^2c^2}\sum_{i<j}\left\{\frac{(\mathbf{p}_i\cdot\mathbf{p}_j)}{r_{ij}}+\frac{[\mathbf{r}_{ij}\cdot(\mathbf{r}_{ij}\cdot\mathbf{p}_j)\mathbf{p}_i]}{r_{ij}^3}\right\} \tag{10}$$

$$\mathcal{H}_5 = -\frac{1}{8m^3c^2}\sum_i p_i^4 \tag{11}$$

$$\mathcal{H}_6 = \frac{e\hbar}{2m^2c^2}\sum_i[(\mathbf{E}_i\times\mathbf{p}_i)\cdot\mathbf{s}_i] - \frac{ie\hbar}{4m^2c^2}\sum_i(\mathbf{E}_i\cdot\mathbf{p}_i) + \frac{e\hbar}{mc}\sum_i(\mathbf{H}_i\cdot\mathbf{s}_i) \tag{12}$$

In the last term $\mathbf{E}_i$ and $\mathbf{H}_i$ are the external electric and magnetic field strengths at the position of electron i, respectively.

The successive terms in the Breit–Pauli Hamiltonian have the following significance:

---

*To obtain an equivalent Hamiltonian that can be used with nonrelativistic wave functions, the 16-component Breit equation is transformed into momentum space and the "small components" of ψ are eliminated. This results in an integral equation in momentum space involving only one Pauli spinor.[8] The integrand of this approximate integral equation is then expanded in powers of (p/mc) and (k/mc) and all terms of higher order than $(1/c^2)$ are dropped. After the resulting equation is Fourier transformed back into position space, the eigenvalue equation generated by Eqs. (6)–(12) is obtained.

$\mathcal{H}_1$: The microscopic Hamiltonian describing the spin–orbit and spin–other–orbit interaction.

$\mathcal{H}_2$: The microscopic Hamiltonian describing the spin–spin dipolar interactions. The Dirac delta function $\delta(\mathbf{r}_{ij})$ accounts for the so-called Fermi contact interaction which occurs when electrons i and j coincide spatially. Because this term contributes an equivalent amount to each spin level, it does not affect the level splitting.

$\mathcal{H}_3$: Contains correction terms that Dirac introduced because of electron spin. This term is also present in the Hamiltonian for a single electron in an electric field.

$\mathcal{H}_4$: The Hamiltonian describing the orbit–orbit coupling of the electrons.

$\mathcal{H}_5$: The approximate relativistic correction to the kinetic energy that describes the change of electron mass with velocity.

$\mathcal{H}_6$: Represents the effect of external electric and magnetic fields.

The relativistic perturbations $\mathcal{H}_3$, $\mathcal{H}_4$, and $\mathcal{H}_5$ do not remove the degeneracy of the eigenvalues in the direction of the magnetic field. Since their only effect is to shift each eigenvalue by a small amount, they can be neglected in the level splitting problems that are considered in this chapter, as can the field-dependent term $\mathcal{H}_6$ which vanishes in zero field.

In order to evaluate theoretically the zero-field splitting, the eigenvalues and eigenfunctions of the operator

$$\mathcal{H}_{\text{ZFS}} = \mathcal{H}_0 + \mathcal{H}_1 + \mathcal{H}_2' \tag{13}$$

must be determined. The prime on $\mathcal{H}_2'$ in Eq. (13) indicates that the Dirac-delta function term has been deleted from $\mathcal{H}_2$. It should also be noted that Eq. (13) was derived under the assumption that the nuclei are fixed. As we have indicated, if the molecule is allowed to rotate, or for that matter vibrate, additional small corrections must be added to the ZFS. These terms arise from direct interactions of the spin and orbital angular momenta of the electron with the rotational motion of the nuclei and from the Born–Oppenheimer vibration–rotation Hamiltonian. Although these correction terms are considered briefly in Section 4, we proceed to develop our subject of molecular fine structure by concentrating on $\mathcal{H}_1$ and $\mathcal{H}_2'$ (hereafter referred to as $\mathcal{H}_{\text{so}}$ and $\mathcal{H}_{\text{ss}}$, respectively).

2.2. The ZFS Parameters D and E

In one of several types of experiments involving electron paramagnetic resonance (EPR) spectroscopy, the splitting of the spin levels is characterized by parameters D and E. To establish a connection between these parameters and the theory, it is useful to expand $\mathcal{H}_{\text{so}}$ and $\mathcal{H}_{\text{ss}}$ into tensor components of rank 1 and 2, respectively. For $g = 2$, $\mathcal{H}_{ss}$ can be written as[24]

$$H_{ss} = \frac{-e^2\hbar^2}{m^2c^2} \sum_{i<j}^{N} \frac{3x_{ij}y_{ij}}{r_{ij}^5}(s_{ix}s_{jy} + s_{iy}s_{jx})$$

$$+ \frac{3x_{ij}z_{ij}}{r_{ij}^5}(s_{ix}s_{jz} + s_{iz}s_{jx}) + \frac{3y_{ij}z_{ij}}{r_{ij}^5}(s_{iy}s_{jz} + s_{iz}s_{jy})$$

$$+ \frac{3}{2}\frac{x_{ij}^2 - y_{ij}^2}{r_{ij}^5}(s_{ix}s_{jx} - s_{iy}s_{jy}) + \frac{1}{2}\frac{3z_{ij}^2 - r_{ij}^2}{r_{ij}^5}(3s_{iz}s_{jz} - \mathbf{s}_i \cdot \mathbf{s}_j) \qquad (14)$$

Within a triplet manifold, $\mathcal{H}_{ss}$ has the same matrix elements as the operator

$$\mathbf{S} \cdot \mathbf{D} \cdot \mathbf{S} \qquad (15)$$

where $\mathbf{S} = \hat{\mathbf{i}}S_x + \hat{\mathbf{j}}S_y + \hat{\mathbf{k}}S_z$ is the total spin operator and $\mathbf{D}$ is the symmetric ZFS tensor.[25,26] Its components D_{xx}, D_{xy}, etc., involve integrals over the operators $\sum_{i<j}(r_{ij}^2 - 3x_{ij}^2/r_{ij}^5)$ and $\sum_{i<j}(-3x_{ij}y_{ij}/r_{ij}^5)$, respectively. If x, y, and z are the principal ZFS axes, then

$$\langle \mathcal{H}_{ss} \rangle = -D_{xx}S_x^2 - D_{yy}S_y^2 - D_{zz}S_z^2 \qquad (16)$$

However, the tensor is traceless so that Eq. (17) can be rewritten in terms of just two independent constants, which are commonly called D and E; namely

$$\langle \mathcal{H}_{ss} \rangle = D(S_z^2 - \tfrac{1}{3}\mathbf{S}^2) + E(S_x^2 - S_y^2) \qquad (17)$$

where

$$D = -\tfrac{3}{2}D_{zz}, \qquad E = \tfrac{1}{2}(D_{yy} - D_{xx}) \qquad (18)$$

Returning to the spin–orbit contribution $\mathcal{H}_{so}$, we note that it can be cast in the same form as Eq. (15); namely[25]

$$\mathbf{S} \cdot \mathbf{D}' \cdot \mathbf{S} \qquad (19)$$

This is easily seen if the microscopic spin–orbit Hamiltonian given in Eq. (7) is written in the form

$$\mathcal{H}_{so} = \sum_i \mathbf{A}_i \cdot \mathbf{s}_i \qquad (20)$$

where $\mathbf{A}$ and $\mathbf{s}$ denote the spatial and spin part of $\mathcal{H}_{so}$, respectively. The elements of the tensor $\mathbf{D}'$ are then

$$D'_{lm} = \sum_{n=1}^{\infty} \frac{\langle \Psi_0 | \sum_i A_{il}s_{il} | \Psi_n \rangle \langle \Psi_n | \sum_i A_{im}s_{im} | \Psi_0 \rangle}{E_n - E_0}, \qquad l, m = x, y, z \qquad (21)$$

This derivation clearly shows how second-order spin–orbit effects generate an effective spin–spin coupling mechanism. If the first-order spin–orbit contribution does not vanish, there can be an additional contribution to the diagonal elements of the tensor [cf. Eq. (29)].

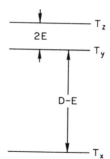

Fig. 1. The triplet (T) level splitting pattern for a typical molecule. The identification of x, y, and z with spatial direction depends upon both the molecule and its electronic state.

The parameters D and E are related to the level splitting pattern shown in Fig. 1, where the axes are chosen to satisfy $|D| \geq 3|E|$. The quantity that is experimentally determined is the energy difference between the inner and outer levels, the so-called $\Delta M_s = \pm 1$ transition, or alternately the energy between the two outer levels, the $\Delta M_s = \pm 2$ transition. The first successful measurements of D and E were carried out by Hutchison and Mangum[4] on a system of oriented naphthalene molecules substituted into a single crystal of durene. They performed a conventional EPR experiment and observed the $\Delta M_s = \pm 1$ transition. More recently, it has become possible to study both the $\Delta M_s = \pm 1$ and $\Delta M_s = \pm 2$ transitions in a variety of systems, including low-temperature glasses and systems of randomly oriented molecules. It is also possible to measure fine-structure parameters by such techniques as optical detection of magnetic resonance,[27] molecular-beam resonance experiments,[28] phosphorescence (optical) microwave double resonance,[29] and microwave optical magnetic resonance induced by electrons.[30]

Use of Perturbation Theory. To evaluate D and E theoretically, we use the fact that $\mathscr{H}_{ss}$ and $\mathscr{H}_{so}$ are small perturbations to $\mathscr{H}_0$, the unperturbed Hamiltonian. Changes in the corresponding unperturbed eigenvalues E_k and eigenfunctions Ψ_k can therefore be described by some form of perturbation theory. Since the properties under study depend explicitly on electron spin, we need to exercise care that the zero-order starting functions Ψ_k are eigenfunctions of the usual spin angular momentum operators $\mathbf{S}^2$ and S_z. Noting that $\mathscr{H}_0$ commutes with both $\mathbf{S}^2$ and S_z, we can find simultaneous eigenfunctions of these operators as characterized by their respective quantum numbers k, $\mathscr{S}$, and $M_s = m$. Thus,

$$\mathscr{H}_0 \, {}^{2\mathscr{S}+1}\Psi_{k,m} = E_k \, {}^{2\mathscr{S}+1}\Psi_{k,m} \tag{22}$$

$$\mathbf{S}^2 \, {}^{2\mathscr{S}+1}\Psi_{k,m} = \mathscr{S}(\mathscr{S}+1)\hbar^2 \, {}^{2\mathscr{S}+1}\Psi_{k,m} \tag{23}$$

$$S_z \, {}^{2\mathscr{S}+1}\Psi_{k,m} = M_s \hbar \, {}^{2\mathscr{S}+1}\Psi_{k,m} \tag{24}$$

The only other condition (Pauli exclusion principle) imposed on the ${}^{2\mathscr{S}+1}\Psi_{k,m}$ is that they be antisymmetric with respect to permutations of the electron coordinates. In most cases, the ${}^{2\mathscr{S}+1}\Psi_{k,m}$ are either self-consistent field (SCF) or configuration-interaction (CI) wave functions for the specific electronic states of interest (cf. Section 3).

To illustrate how $\mathcal{H}_{ss}$ and $\mathcal{H}_{so}$ remove the degeneracy of unperturbed states with different M_s quantum numbers, consider the three components of the kth triplet state ${}^3\Psi_{k,1}$, ${}^3\Psi_{k,0}$, and ${}^3\Psi_{k,-1}$. The energy of these states in the absence of spin–spin and spin–orbit effects is ${}^3E_k$. In the presence of the interactions, the energies become[31,32]

$$ {}^3E_{k,i} = {}^3E_k + \lambda_i \tag{25} $$

where the λ_i are obtained as roots of the secular equation

$$ \begin{vmatrix} a_{1,1} & a_{1,0} & a_{1,-1} \\ a_{0,1} & a_{0,0} & a_{0,-1} \\ a_{-1,1} & a_{-1,0} & a_{-1,-1} \end{vmatrix} = 0 \tag{26} $$

with elements

$$ a_{ij} = \langle {}^3\Psi_{k,i}|\mathcal{H}_{ss}|{}^3\Psi_{k,j}\rangle - \sum_{m \neq k}^{\infty} \frac{\langle {}^3\Psi_{k,i}|\mathcal{H}_{so}|\Psi_m\rangle\langle\Psi_m|\mathcal{H}_{so}|{}^3\Psi_{k,j}\rangle}{E_m - {}^3E_k} \tag{27} $$

In Eq. (27) we assume implicitly that the ${}^3\Psi_k$ state is orbitally nondegenerate; otherwise, there would be a first-order contribution from $\mathcal{H}_{so}$. The infinite sum in the second term goes over all states Ψ_m which couple with the ${}^3\Psi_k$ state via *any* component of the vector operator $\mathcal{H}_{so}$. The matrix corresponding to Eq. (26) is symmetrical about its diagonal and can generally be simplified using the Wigner–Eckart theorem.[33] In Section 3, we outline how the a_{ij}, and subsequently the splitting parameters D and E, are calculated.

2.3. Magnitude of Fine-Structure Contributions

It is instructive to estimate the orders of magnitudes for the spin–spin, spin–orbit, and spin–rotation interactions in various classes of systems. In all molecules considered in this chapter, the spin–rotation interaction is much smaller than either the spin–spin or spin–orbit interaction. The relative sizes of these last two terms depend on factors such as the nuclear charges of the constituent atoms, the symmetry, the internuclear distance, the orbital degeneracy of the triplet state ${}^3\Psi_{k,i}$, and the detailed nature and energy distribution of other states Ψ_m which mix via Eq. (27).

2.3.1. Spin–Spin Interaction

The leading spin–spin contribution to the ZFS always connects states with the same total spin (**S**) and orbital angular momentum (**L**) but with m_s quantum numbers that differ by 0, $\hbar$, or $2\hbar$. As a result, it generally dominates the ZFS in light molecules with orbitally nondegenerate triplet states (e.g., He_2).[34] The magnitude of the spin–spin term tends to decrease sharply with an increase in the distance between the unpaired electrons. Specifically, it tends to be larger in systems (e.g., NH,[35] CH_2[36]) where the unpaired electrons are mostly localized on the same atom, than where the unpaired spin distribution is highly delocalized over the entire molecule (e.g., C_6H_6,[37] porphyrin,[38] hemin[39]).

2.3.2. Spin–Orbit Interaction

By considering the selection rules[40] for $\mathcal{H}_{so}$, it is possible to enumerate those states (Ψ_m) in Eq. (27) that contribute to the ZFS through the spin–orbit mechanism. In general, the integrand of a nonvanishing matrix element must contain the symmetric representation of the point group. For diatomic molecules, $\mathcal{H}_{so}$ has nonvanishing matrix elements in the spaces Λ, S, and S_z when their respective quantum numbers differ by 0 or $\pm\hbar$, with the restriction that ($\Lambda + S_z$) must be conserved. Here Λ is the projection of the orbital angular momentum along the internuclear axis (z axis). In addition, this operator, which includes both one- and two-electron parts, conserves parity and only connects states that differ in symmetry with respect to reflection in the plane of the nuclei. Thus for example, $\Sigma^{\pm}$ and $\Sigma^{\mp}$ states mix and guarantee that no first-order contribution to the ZFS is present. For $\Lambda \neq 0$, however, the $\Psi(\Lambda = 1)$ state is degenerate with the $\Psi(\Lambda = -1)$ state, and only the linear combinations

$$\Psi^{\pm} = (1/2^{1/2})[\Psi(\Lambda = 1) \pm \Psi(\Lambda = -1)] \qquad (28)$$

satisfy this symmetry requirement. There is, therefore, a nonvanishing matrix element,

$$A = \langle \Psi^{\pm} | \mathcal{H}_{so} | \Psi^{\pm} \rangle \qquad (29)$$

which is referred to as *spin-orbit* coupling (SOC) constant A. Examples of some coupling constants are given in Table 1. The sensitivity of A to molecular state means that spin–orbit coupling can supplement the usual energy criteria for assigning the electronic states.[9]

Even for triplet states with vanishing first-order contributions, the spin–orbit interactions in second order can be larger than the first-order spin–spin term. In contrast to the spin–spin interaction, which is essentially independent of nuclear charge, the spin–orbit parameter increases approximately as the fourth power of the effective nuclear charge. Thus, even for the $^3\Sigma_g^-$ ground state of the oxygen molecule, the second-order spin–orbit term accounts for $\frac{2}{3}$ of the ZFS (cf. Section 4). The ZFS in heavy atoms and transition metal complexes is almost wholly attributable to spin–orbit coupling.[41]

Table 1. Spin–Orbit Coupling Constants and Zero-Field Splitting Parameters for Selected Diatomic and Polyatomic Molecules[a]

Molecule	Electronic state	Theoretical value, cm^{-1}	Experimental value,[b] cm^{-1}
		Spin–orbit coupling constants	
OH	$X\,^2\Pi$	-141.4,[9] -141.24[61]	-139.235[c]
NO	$X\,^2\Pi$	146,[102] 105.3,[95] 94.8[119]	122.2[94]
	$B\,^2\Pi$	31[102]	33[d]
PO	$X\,^2\Pi$	190.1,[95] 208[118]	224[94]
CH	$X\,^2\Pi$	30.4,[9] 27.8,[119] 28.44[92]	27.95[94]
BH$^+$	$X\,^2\Pi$	15.9,[9] 13.15[92]	14.00[94]
CO	$A\,^3\Pi$	39.0,[95] 39.5[127]	41.51[e]
	$A\,^3\Delta$	-16.3[127]	-16.0[e]
BO	$A\,^2\Pi$	122.7[95]	126.7[94]
		Zero-field splitting parameters[f]	
NH	$^3\Sigma^-$	$D = 2.0$[g] $D = 2.091$[35,86]	$D = 1.856 \pm 0.014$[h]
O$_2$	$^3\Sigma_g^-$	$D = 3.846$[61,62] $D = 3.856$[91]	$D = 3.965$[132,133]
CH$_2$	$^3B_1$	$D = 0.807$[36] $E = 0.049$[36]	$D = 0.76 \pm 0.02$[134] $E = 0.052 \pm 0.017$[134]
CH$_2$O	$^3A_2$	$D = 0.32$[145] $E = 0.04$[145]	$D = 0.42$[144] $E = 0.04$[144]
C$_6$H$_6$	$^3B_{1u}$	$D = 0.207$[152]	$D = 0.1580 \pm 0.0003$[45] $E = -0.0064 \pm 0.0003$[45]

[a] Numerical superscripts refer to reference numbers within the text.
[b] Superscript letters refer to references below that are not cited in text.
[c] M. Mizushima, *Phys. Rev.* **A5**, 143 (1972).
[d] R. F. Barrow and E. Miescher, *Proc. Phys. Soc., London* **70**, 219 (1957).
[e] R. W. Field, Ph.D. thesis, Harvard University, 1971.
[f] $E = 0$ if not explicitly given.
[g] M. Horani, J. Rostas, and H. Lefebvre-Brion, *J. Can. Phys.* **45**, 3319 (1967).
[h] R. N. Dixon, *Can. J. Chem.* **37**, 1171 (1959).

Another factor that is important in determining the magnitude of the second-order spin–orbit contribution is the distribution of electronic states. For example, states that are nearly degenerate with the triplet state of interest are generally large contributors because of the very small energy denominators in Eq. (27). Thus substituent groups and hosts can affect the ZFS of aromatic carbonyls and benzaldehydes through changes in the energy separation of the nearly degenerate $n\pi^*$ and $\pi\pi^*$ triplet levels.[42]

Finally, the spatial symmetry of the molecule can also affect the magnitude of the spin–orbit contribution. The classic example is planar benzene, where McClure[43] has argued that the spin–orbit contribution to the ZFS in the $^3B_{1u}$ state is unusually small because the one- and two-center integrals vanish by symmetry, leaving only the small three- and four-center two-electron integrals. Thus, the spin–orbit contribution for this state is estimated[44] to be 0.0001 cm^{-1}, whereas the experimental value[45] of D is 0.1580 cm^{-1}.

2.3.3. Spin–Rotation Interaction

Molecular rotation couples with the spin angular momenta of the electrons to give an added contribution to the ZFS. The appropriate microscopic spin–rotation Hamiltonian[20,46] is given by

$$\mathcal{H}_{sr} = \frac{-g\beta}{c} \sum_{i,\alpha} \frac{Z_\alpha e}{r_{i\alpha}^3} (\mathbf{r}_{i\alpha} \times \mathbf{v}_\alpha) \cdot \mathbf{s}_i \cdot \tag{30}$$

where β is the Bohr magneton and $\mathbf{v}_\alpha$ is the velocity of nucleus α. This Hamiltonian has been used by Tinkham and Strandberg[24] and by Kayama and Baird[40] in computing the first-order contribution to the splitting of the $^3\Sigma_g^-$ ground state of O_2. Their treatments are not rigorous, however, because they neglected the influence of the nuclear rotation on the electronic velocities.

To illustrate the relatively small importance of $\mathcal{H}_{sr}$, consider the $c^3\Pi_u(1s, 2p)$ state of H_2 which has been examined by Jette and Miller[47] using Hund's case (b) coupling. They find that the spin–rotation constant of this state is about 26 MHz, to be compared with the spin–orbit coupling constant of -3741 MHz. Although the contribution from $\mathcal{H}_{sr}$ is negligible in this case, it does account nicely for differences in fine structure between *ortho* and *para* hydrogen.

In addition to this first-order contribution from $\mathcal{H}_{sr}$, there is a second-order interaction that arises from the product of matrix elements between the full spin–orbit operator $\mathcal{H}_{so}$ and the orbit–rotation operator $\mathcal{H}_{or}$ which couples the nuclear to the electronic orbital angular momenta. As with $\mathcal{H}_{so}$ itself, it is observed for Σ states that this second-order effect can be greater than the first-order $\mathcal{H}_{sr}$ contribution. For example, this is observed to be the case in O_2[24,40] and even in He_2[34,48] where the spin–orbit matrix elements are small. Also Dixon[49] has shown that second-order spin–rotation effects for bent AH_2 molecules are particularly important because of mixing of low-lying $^2\Pi$ states into the $^2\Sigma$ ground states. On the other hand, first-order effects dominate the spin–rotation coupling constant of the $^2\Sigma$ ground state of the H_2^+ molecular ion.[50]

It is clear that additional calculations are needed to obtain a fundamental understanding of spin–rotation interactions in molecules. No second-order *ab initio* calculations exist at the present time for molecules with two or more electrons; estimates can be made by subtracting the first-order contribution from the experimental value. Therefore, a complete treatment for this interaction in a system such as O_2 would be very worthwhile.

3. Computational Aspects

Before the theory described in Section 2 can be quantitatively implemented, it is necessary to evaluate matrix elements of $\mathcal{H}_{ss}$ and $\mathcal{H}_{so}$, given the many-electron wave functions. We begin to simplify these matrix elements by using the Wigner–Eckart theorem which allows us to separate all dependence on the M_s quantum number into a $3j$ symbol. This factorization reflects the tensorial nature of the operators and leaves the so-called "reduced matrix element." An explicit form for the wave function as a linear combination of Slater determinants is then introduced and the reduced matrix elements are decomposed further to expansions in the elementary integrals over one-electron basis functions. Section 3.2 takes up the evaluation of these elementary integrals for both Slater- and Gaussian-type orbitals.

3.1. Matrix Element Reduction

For illustration, we assume that the state functions $\Psi(\gamma \Lambda S M_s)$ are characterized by the operators L_z, S^2, and S_z with corresponding eigenvalues of $\hbar\Lambda$, $\hbar S$ $[S = \mathcal{S}(\mathcal{S}+1)]$ and $\hbar M_s$. All other quantum numbers are denoted by γ. This classification corresponds to Hund's coupling case (b) which is valid when the separation between rotational levels is greater than the spin–orbit interaction, as in all Σ states and other states of light molecules. Thus, the matrix elements we wish to evaluate can be written as

$$\langle \mathcal{H}' \rangle = \langle \gamma' \Lambda' S' M_s' | \mathcal{H}' | \gamma \Lambda S M_s \rangle \tag{31}$$

where $\mathcal{H}'$ is one of the fine-structure operators and where Ψ has been suppressed for notational simplicity.

Many authors[51-57] have developed techniques to evaluate $\langle \mathcal{H}' \rangle$ for spin-dependent operators. Following McWeeny,[56] we write the one- and two-body operators in tensor form

$$\sum_{im} (-1)^m F_{l,-m}(i) \sigma_{l,m}(i) \tag{32}$$

and

$$\sum_{ij}' \sum_m (-1)^m F_{l,-m}(i,j) \sigma_{l,m}(i,j) \tag{33}$$

where the component of an irreducible tensor of rank l is designated by m. Application of the Wigner–Eckart theorem allows $\langle \mathcal{H}' \rangle$ to be written as

$$\langle \mathcal{H}' \rangle = (-1)^{S-M_s} \begin{pmatrix} S & l & S' \\ -M_s & m & M_s' \end{pmatrix} \langle \gamma' \Lambda' S' \| \mathcal{H}' \| \gamma \Lambda S \rangle . \tag{34}$$

where the M_s quantum number is now contained only in a Wigner $3j$ symbol which vanishes unless $m = M_s - M_s'$. The quantity $\langle \gamma' \Lambda' S' \| \mathcal{H}' \| \gamma \Lambda S \rangle$ is the so-called reduced matrix element. We then make a convenient choice of M_s and

M'_s, usually S and S', respectively, and express all other matrix elements as

$$\langle \mathcal{H}' \rangle = C_l(S'M'_s; SM_s)\langle \gamma', \Lambda', S' = M'_s|\mathcal{H}'|\gamma, \Lambda, S = M_s \rangle \tag{35}$$

where

$$C_l(S'M'_s; SM_s) = (-1)^{S-M_s}\begin{pmatrix} S & l & S' \\ -M_s & M_s - M'_s & M'_s \end{pmatrix}\begin{pmatrix} S & l & S' \\ -S & S - S' & S' \end{pmatrix}^{-1} \tag{36}$$

The coefficients C_1 and C_2 are tabulated in Refs. 52 and 56. Thus, it is only necessary to evaluate $\langle \gamma', \Lambda', S' = M'_s|\mathcal{H}'|\gamma, \Lambda, S = M_s \rangle$ for each of the spatial components of $\mathcal{H}'$, using in each case $\sigma_{l,o}(i)$ and $\sigma_{l,o}(i, j)$ for the spin operators. This allows us to construct the entire zero-field-splitting tensor from wave functions with $M_s = S$, thereby circumventing time-consuming calculations for the other components of Ψ.

To proceed further, we need to introduce an explicit form for the wave function. Most studies take a configuration-interaction wave function which can be written as

$$\Psi(\gamma \Lambda SM_s) = \sum_{a=1} c_a X_a(\gamma \Lambda SM_s) \tag{37}$$

with

$$X_a = \sum_k c_{ka} D_k \tag{38}$$

where D_k is a Slater determinant (with $M_s = S$) and the c_{ka} are chosen to make X_a an eigenfunction of S^2. Thus, since

$$\langle \gamma', \Lambda', S' = M'_s|\mathcal{H}'|\gamma, \Lambda, S = M_s \rangle = \sum_a \sum_b \sum_k \sum_l c_a c_b c_{ka} c_{lb} \langle D_k|\mathcal{H}'|D_l \rangle \tag{39}$$

the problem reduces to a determination of matrix elements between determinants.

Bottcher and Browne[53] have given general formulas for evaluating matrix elements between determinants without the restriction to an orthonormal basis. Although these expressions are sometimes useful, for example, to evaluate certain off-diagonal spin–orbit matrix elements (cf. Section 4), the orbitals in D_k are usually orthogonal to those in D_l, in which case quite simple rules can be given for $\langle D_k|\mathcal{H}'|D_l \rangle$.

To illustrate how this reduction to elementary integrals over one-electron basis functions occurs, we consider a diagonal element of the spin–spin operator $\mathcal{H}_{ss}$, for which Eq. (39) becomes[58]

$$\langle \gamma, \Lambda, S = M_s|\mathcal{H}_{ss}|\gamma, \Lambda, S = M_s \rangle = \sum_{a \leq b} \sum_k \sum_l (2 - \delta_{ab}) c_a c_b c_{ka} c_{lb} \langle D_k|\mathcal{H}_{ss}|D_l \rangle \tag{40}$$

Now let R and G represent, respectively, any of the irreducible spatial components of $\mathcal{H}_{ss}$ and the $m = 0$ component of the spin operator $\sigma_{l,o}(i, j) = 2s_{zi}s_{zj} - s_{xi}s_{xj} - s_{yi}s_{yj}$. If the determinants D_k and D_l differ by more than two spin-orbitals, the matrix element is zero since R and G are both two-electron operators. Thus, there are three unique nonzero cases that must be considered explicitly, namely when D_k and D_l differ by 2, 1, and 0 spin-orbitals.

Case I: Determinants Differing in Two Spin-Orbitals. Let D_k and D_l differ by the replacement of the two spin-orbitals $\Phi_i\chi_i$ and $\Phi_j\chi_j$ by $\Phi_i'\chi_i'$ and $\Phi_j'\chi_j'$, where Φ_i and χ_i represent the spatial and spin (α or β) parts of the spin–orbital respectively. For this case the matrix element becomes

$$\langle D_k|\mathcal{H}_{ss}|D_l\rangle = [\Phi_i^*\Phi_i'|R|\Phi_j^*\Phi_j'][\chi_i^*\chi_i'|G|\chi_j^*\chi_j']$$
$$- [\Phi_i^*\Phi_j'|R|\Phi_j^*\Phi_i'][\chi_i^*\chi_j'|G|\chi_j^*\chi_i'] \tag{41}$$

where the spin elements $[\chi_i^*(1)\chi_i'(1)|G|\chi_j^*(2)\chi_j'(2)]$ are given by the matrix

	$|\alpha\alpha]$	$|\beta\beta]$	$|\alpha\beta]$	$|\beta\alpha]$
$[\alpha\alpha|$	$\frac{1}{2}$	$-\frac{1}{2}$	0	0
$[\beta\beta|$	$-\frac{1}{2}$	$\frac{1}{2}$	0	0
$[\alpha\beta|$	0	0	0	$-\frac{1}{2}$
$[\beta\alpha|$	0	0	$-\frac{1}{2}$	0

$$\tag{42}$$

Case II: Determinants Differing in One Spin–Orbital. Let D_k and D_l be identical except that the spin-orbital $\Phi_j\chi_j$ in D_k is replaced by the $\Phi_j'\chi_j'$ in D_l. For this case the matrix element becomes

$$\langle D_k|\mathcal{H}_{ss}|D_l\rangle = \sum_{i\neq j} [\Phi_i^*\Phi_i|R|\Phi_j^*\Phi_j'][\chi_i^*\chi_i|G|\chi_j^*\chi_j']$$
$$- [\Phi_i^*\Phi_j'|R|\Phi_j^*\Phi_i][\chi_i^*\chi_j'|G|\chi_j^*\chi_i] \tag{43}$$

However, if Φ_i is doubly occupied, there is no net contribution since from Eq. (42), $[\alpha\alpha|G|\chi_j^*\chi_j'] = -[\beta\beta|G|\chi_j^*\chi_j']$ and $[\alpha\chi_j'|G|\chi_j^*\alpha] = -[\beta\chi_j'|G|\chi_j^*\beta]$. Therefore, the sum in Eq. (43) can be restricted to only those Φ_i that are singly occupied in either configuration.

Case III: Determinants Are Identical. Following the same argument for Case II, we find that when the determinants are identical, the matrix element becomes

$$\langle D_k|\mathcal{H}_{ss}|D_l\rangle = \sum_{i<j} [\Phi_i^*\Phi_i|R|\Phi_j^*\Phi_j][\chi_i^*\chi_i|G|\chi_j^*\chi_j]$$
$$- [\Phi_i^*\Phi_j|R|\Phi_j^*\Phi_i][\chi_i^*\chi_j|G|\chi_j^*\chi_i] \tag{44}$$

where the summation extends only over orbitals that are singly occupied in either configuration.

It is apparent from Eqs. (42)–(44) that the spin–spin interaction is a purely valence–valence interaction,[59] since the closed-shell part makes no contribution to the matrix element. That this is not the case for the spin–orbit interaction can be shown in a similar way. For $\mathcal{H}_{so}$, the sums in Eqs. (43) and (44) *do* include closed-shell terms and physically represent a two-electron shielding by "core" electrons.

3.2. Integral Evaluation

To proceed further, we must evaluate the matrix elements $[\Phi_i^*\Phi_j|R|\Phi_k^*\Phi_l]$, where R signifies the components of $\mathcal{H}_{ss}$ and $\mathcal{H}_{so}$. The MO basis functions Φ_i are usually taken as linear combinations of either Slater

$$\varphi_{nlm} = N_{nlm}r^{n-1}Y_{l,m}(\hat{r})\exp(-\zeta r) \tag{45}$$

or Gaussian-type atomic orbitals

$$\varphi_{ijk} = N_{ijk}x^iy^jz^k\exp(-\zeta r^2) \tag{46}$$

where $\mathbf{r}$ represents the radial vector from a suitably chosen origin. The functions $Y_{l,m}(\hat{r})$ are spherical harmonics, and N is a normalizing constant that depends on the indices (nlm) or (ijk) and on the effective nuclear charge ζ (orbital exponent). Frequently, linear combinations of the φ are chosen as the basis set, with weighting factors and indices judiciously selected to optimize accuracy and to facilitate computation (cf. Volume 3, Chapter 1).

Since it is desirable that the molecular wave functions and fine-structure parameters be determined with compatible methodologies, the detailed evaluation of fine-structure integrals over basis functions generally varies from one programming system to another. We shall consider the principal features of two approaches in common use; the first arises in the application of *Slater-type orbitals* (STO) to linear molecules and the other involves a multicenter technique in which integrals over *Gaussian-type orbitals* (GTO) enter at an intermediate stage in the treatment of the four-center STO case.

3.2.1. Diatomic Molecules

During a period of nearly fifteen years, many important analytical and computational techniques for diatomic molecules were developed at the Laboratory of Molecular Structure and Spectra at the University of Chicago. These are described in a series of papers[60] entitled Study of Two-Center Integrals Useful in Calculations on Molecular Structure. Building on the analysis and computer programs[61] developed in these studies, Kern and co-workers[62] have constructed a package for *ab initio* diatomic fine-structure calculations that employs the STO in Eq. (45).

To illustrate what is involved, we consider the $m = 0$ component of the spin–spin interaction operator in Eq. (33) which we write as

$$F^{ss}_{2,0}(1, 2) = \frac{\partial}{\partial z_1} \frac{\partial}{\partial z_2} \frac{1}{r_{12}} \tag{47}$$

Four types of integrals are then needed, namely one-center integrals and two-center Coulomb, hybrid, and exchange integrals.

The procedure for evaluating the one-center (designated A), two-electron spin–spin integral

$$\langle F^{ss}_{2,0}\rangle_O = \iint \varphi^*_A(1)\varphi''^*_A(2)\left(\frac{\partial}{\partial z_1} \frac{\partial}{\partial z_2} \frac{1}{r_{12}}\right) \varphi'_A(1)\varphi'''_A(2) \, dV_1 \, dV_2 \tag{48}$$

makes use of an expansion of the interelectron potential

$$\frac{1}{r_{12}} = 4\pi \sum_{k=0}^{\infty} \sum_{q=-k}^{k} \frac{(-1)^q}{(2k+1)} Y_{k,-q}(\hat{r}_1) Y_{k,q}(\hat{r}_2) \frac{r^k_<}{r^{k+1}_>} \tag{49}$$

where $r_<$ and $r_>$ refer, respectively, to the lesser and greater of r_1 and r_2. Essentially, when Eqs. (45) and (49) are substituted into Eq. (48) and the operators $\partial/\partial z_1$ and $\partial/\partial z_2$ allowed to act on r_{12}^{-1}, the three dimensions of integration over spherical polar coordinates for each electron iterate to yield an analytical expression for $\langle F^{ss}_{2,0}\rangle_O$ which depends upon the quantum numbers and orbital exponents of each STO. The general expression for $\langle F^{ss}_{2,0}\rangle_O$ is given by Matcha, Kern, and Schrader (MKS);[63] special cases[64,65] are also given in the literature for the one-center spin–spin and spin–orbit integrals.

The MKS procedures for evaluating the two-center Coulomb and hybrid integrals are similar and can be discussed together. As an example, we consider here the spin–spin Coulomb case which we write as

$$\langle F^{ss}_{2,0}\rangle_C = \iint \varphi^*_A(1)\varphi'^*_B(2)\left(\frac{\partial}{\partial z_1} \frac{\partial}{\partial z_2} \frac{1}{r_{12}}\right) \varphi'_A(1)\varphi_B(2) \, dV_1 \, dV_2 \tag{50}$$

$$= \int dV_2 \Omega^*_B(2) \int dV_1 \varphi^*_A(1)\left(\frac{\partial}{\partial z_1} \frac{\partial}{\partial z_2} \frac{1}{r_{12}}\right) \varphi'_A(1) \tag{51}$$

$$= \int \partial V_2 \Omega^*_B(2) \mathcal{U}_A(2) \tag{52}$$

where

$$\mathcal{U}_A(2) = \int_0^{2\pi} d\phi_1 \int_0^{\pi} d\theta_{A1} \sin \theta_{A1} \int_0^{\infty} dr_{A1} r^2_{A1} \Omega_A(1)\left(\frac{\partial}{\partial z_1} \frac{\partial}{\partial z_2} \frac{1}{r_{12}}\right) \tag{53}$$

is a center-A potential in the coordinates of electron 2 and where

$$\Omega_\alpha(i) = \varphi^*_\alpha(i)\varphi'_\alpha(i) \tag{54}$$

is an α-centered ($\alpha = A$ or B), one-electron ($i = 1$ or 2) distribution. We therefore see, in analogy with the one-center integral, that $\mathcal{U}_A(2)$ can be evaluated analytically by direct integration. The resulting potential times $\Omega_B(2)$ can then be tabulated and integrated numerically [cf. Eq. (52)]. In particular, we transform to prolate spheroidal coordinates, integrate over the azimuthal angle ϕ_2 analytically, and finally use Gauss–Legendre quadrature formulas to integrate over the remaining two coordinates. A detailed discussion of the computational requirements for this double numerical integration is given in Refs. 60 and 66.

The two-center hybrid integrals can also be evaluated by a scheme similar to that developed in Eqs. (50)–(54). The principal change is that $\Omega_B^*(2)$ in Eq. (52) is replaced by a two-center distribution function

$$\Omega_{AB}(2) = \varphi_A''^* \varphi_B(2) \tag{55}$$

The corresponding Coulomb and hybrid integrals for the spin–orbit operator can be treated in a similar fashion.

Following Christoffersen and Ruedenberg,[67] Matcha *et al.*[68] have presented an alternate scheme for evaluating hybrid integrals which replaces the double numerical integration over the product of charge distributions and potential functions by a one-dimensional integration over products of spherical Bessel functions and potential functions. A significant saving of computational effort results because values of the potential function are required at only N grid points rather than N^2 as in the previous treatment. Implementation of this scheme for fine-structure integrals has not been attempted but would be very worthwhile.

The two-center exchange integrals are evaluated by partial integration rather than by using expansions[69] in prolate spheroidal coordinates for these operators. This alternative was first proposed by Schrader[70] for $F_{l,-m}^{ss}$ and by Hall and Hardisson[71] for $F_{l,-m}^{so}$. The spin–spin and two-electron spin–orbit exchange integrals are thus transformed into repulsion-like integrals with modified charge distributions.

Again, choosing the spin–spin case as an illustration, we have

$$\langle F_{2,0}^{ss} \rangle_E = \int\int \varphi_A^*(1)\varphi_B^{*\prime}(2)\left(\frac{\partial}{\partial z_1}\frac{\partial}{\partial z_2}\frac{1}{r_{12}}\right)\varphi_B(1)\varphi_A'(2)\, dV_1\, dV_2 \tag{56}$$

as the integral to be evaluated. Integration twice by parts gives a delta function term [cf. Eq. (75) *et. seq.*], plus

$$\langle F_{2,0}^{ss} \rangle_E = \int\int \frac{1}{r_{12}}\tilde{\Omega}_{AB}(1)\tilde{\Omega}_{BA}(2)\, dV_1\, dV_2 \tag{57}$$

where the modified charge distributions are given by

$$\tilde{\Omega}_{AB}(1) = \frac{\partial}{\partial z_1}\varphi_A^*(1)\varphi_B(1) \tag{58}$$

and

$$\tilde{\Omega}_{BA}(2) = \frac{\partial}{\partial z_2} \varphi_B'^*(2)\varphi_A'(2) \tag{59}$$

Equation (57) is then evaluated by following Ruedenberg's analysis[72] of electron repulsion integrals; that is, the prolate spheroidal expansion for $1/r_{12}$ is introduced, the azimuthal angles are integrated out analytically, and the remaining variables are integrated over using either numerical quadrature or analytically by means of suitable recursion formulas.

An alternative to this integration-by-parts procedure is to introduce Neumann-type expansions[69] for the two-electron fine-structure operators directly into Eq. (56). The details and merits of such an approach have not been reported.

We also remark, in completing this subsection on diatomic molecules, that the one-electron part of the spin–orbit operator leads to integrals that are simpler than those specifically discussed here. The reader is referred to Ref. 73 for details and a summary of the literature on this subject.

3.2.2. Polyatomic Molecules

When a molecule has four or more atoms located off a single axis, the STO methods described so far must be extended to include multicenter integrals. For example, let us consider the space part of the two-electron spin–orbit operator in Eq. (7), for which

$$\langle F_{1,0}^{so}\rangle_{ACBD} = \iint \varphi_A^*(1)\varphi_C(2)\left(\frac{\partial}{\partial z_1} \frac{1}{r_{12}} \frac{\partial}{\partial z_1}\right)\varphi_B^*(1)\varphi_D(2)\, dV_1\, dV_2 \tag{60}$$

Matcha and Kern[74] have pointed out that the MKS results for diatomic fine-structure integrals can be incorporated into numerical integration schemes such as those developed by McLean,[75] by Magnusson and Zauli,[76] and by Wahl and Land[77] for multicenter electron repulsion integrals. The essential idea is to write Eq. (60) as a vector product of a modified two-center charge distribution times a two-center potential; that is,

$$\langle F_{1,0}^{so}\rangle_{ACBD} = \tilde{\Omega}_{CD}(\mathbf{p}) \cdot \mathcal{U}_{AB}^{so}(\mathbf{p}) \tag{61}$$

where the set of points $\mathbf{p}$ is determined by dividing space into a set of rectangular parallelepipeds and then using Gaussian quadrature in each of 3 directions such that the integration grid is concentrated about the nuclei.

An entirely different approach[78,79] for STOs involves the Laplace transform

$$\exp(-\zeta r) = \frac{\zeta}{2\pi^{1/2}} \int_0^\infty s^{-3/2} \exp\left(-\frac{\zeta^2}{4s} - sr^2\right) ds \tag{62}$$

in which each of the four exponential orbitals in Eq. (60) is replaced by the right-hand side of Eq. (62). By interchanging the electronic and transformation variables, the dimensionality of $\langle F^{so}_{1,0}\rangle_{ACBD}$ is reduced from six to four. This reduction in the number of integration variables occurs because four-center two-electron matrix elements over GTOs can be evaluated in closed form.[80] The basic step here is to take the product of two GTOs referred to different centers, let us say A and B, and express it as a constant times a Gaussian function centered at a third point (P) on line segment AB between centers A and B.[80,81]

To illustrate why this simplification has made GTOs a viable basis for evaluating multicenter fine-structure integrals, consider the evaluation of

$$\langle F^{ss}_{2,0}\rangle_{ACBD} = \iint \varphi_A(1)\varphi_B(1)\left(\frac{\partial}{\partial z_1}\frac{\partial}{\partial z_2}\frac{1}{r_{12}}\right)\varphi_C(2)\varphi_D(2)\, dV_1\, dV_2 \quad (63)$$

when the φ are Gaussian-lobe type orbitals or equivalently unnormalized $1s$ GTOs. Since the charge distribution $\varphi_A(1)\varphi_B(1)$ can be expressed as

$$\exp(-\alpha_A|\mathbf{r}-\mathbf{R}_A|^2)\exp(-\alpha_B|\mathbf{r}-\mathbf{R}_B|^2) = E_{AB}\exp(-\alpha_P|\mathbf{r}-\mathbf{R}_P|^2) \quad (64)$$

where

$$\alpha_P = \alpha_A + \alpha_B \quad (65)$$

$$\mathbf{R}_P = (\alpha_A\mathbf{R}_A + \alpha_B\mathbf{R}_B)\alpha_P^{-1} \quad (66)$$

and

$$E_{AB} = \exp(-\alpha_A\alpha_B\alpha_P^{-1}|\mathbf{R}_A - \mathbf{R}_B|^2) \quad (67)$$

with a similar expression for $\varphi_C(2)\varphi_D(2)$ in terms of a point Q on the line segment CD, the integral in Eq. (63) can be simply written as the two-center integral

$$\langle F^{ss}_{2,0}\rangle_{ACBD} = E_{AB}E_{CD}\int \varphi_P(1)\left(\frac{\partial}{\partial z_1}\frac{\partial}{\partial z_2}\frac{1}{r_{12}}\right)\varphi_Q(2)\, dV_1\, dV_2 \quad (68)$$

This may be simplified further by the change of variable $\mathbf{r} = \frac{1}{2}(\mathbf{r}_1+\mathbf{r}_2)$, $\mathbf{r}_{12} = \mathbf{r}_1 - \mathbf{r}_2$ followed by integration over $\mathbf{r}$ to give

$$\langle F^{ss}_{2,0}\rangle_{ACBD} = E_{AB}E_{CD}\left(\frac{\pi}{\alpha_P+\alpha_Q}\right)^{3/2}\int \varphi_T(\mathbf{r}_{12})\left(\frac{\partial}{\partial z_1}\frac{\partial}{\partial z_2}\frac{1}{r_{12}}\right)dV_{12} \quad (69)$$

where

$$\mathbf{R}_T = \mathbf{R}_P - \mathbf{R}_Q \quad (70)$$

and

$$\alpha_T = \alpha_P\alpha_Q(\alpha_P+\alpha_Q)^{-1} \quad (71)$$

The final integration is easily completed to give

$$\langle F_{2,0}^{ss}\rangle_{ACBD} = \frac{8\pi}{3} E_{AB} E_{CD} \left(\frac{\pi}{\alpha_P + \alpha_Q}\right)^{3/2} \alpha_T (R_T^2 - 3Z_T^2) F_2(\alpha_T R_T^2) \qquad (72)$$

where

$$F_2(t) = \int_0^1 u^4 \exp(-tu^2)\, du \qquad (73)$$

Although Eq. (72) requires the evaluation of F_2, the second derivative of the error function, this generally is much less time consuming than the numerical integrations required for STOs. Thus, although more GTOs than STOs are required to describe the space, the increased speed with which the integrals can be evaluated make GTOs attractive for determining the fine structure of polyatomic molecules (cf. Volume 3, Chapter 1).

As is suggested by Eq. (57), it is instructive to emphasize the close relationship between the fine structure and the corresponding electron repulsion integrals.[10,82] Noting that

$$\frac{\partial}{\partial z_i}[\varphi_K(i)\varphi_L(i)] = -\left(\frac{\partial}{\partial Z_K} + \frac{\partial}{\partial Z_L}\right)[\varphi_K(i)\varphi_L(i)] \qquad (74)$$

where the K and L subscripts on $\partial/\partial Z$ denote differentiations with respect to the nuclear coordinates, we can integrate Eq. (63) twice by parts to obtain

$$\langle F_{2,0}^{ss}\rangle_{ACBD} = \left(\frac{\partial}{\partial Z_A} + \frac{\partial}{\partial Z_B}\right)\left(\frac{\partial}{\partial Z_C} + \frac{\partial}{\partial Z_D}\right)\left\langle \frac{1}{r_{12}}\right\rangle_{ACBD} - \frac{4\pi}{3}\langle\delta(\mathbf{r}_{12})\rangle_{ACBD} \qquad (75)$$

It is, therefore, possible to generate formulas for spin–spin integrals directly from those for $1/r_{12}$, though care must be exercised in interchanging the order of integration and differentiation. In general, the one-, two-, and three-center integrals generated in this way can only be obtained from the four-center integrals by allowing the appropriate nuclei to coalesce. The results are in agreement with those derived by Geller and Griffith[83] with a Fourier transform technique.[84–87]

Before concluding this section, we also point out that the spin–spin and two-electron spin–orbit matrix elements are related as well. Matcha and Kern[88] have shown that the identity

$$\langle F_{2,0}^{ss}\rangle_{ACBD} = -\langle F_{1,0}^{so}\rangle_{ACBD} - \langle F_{1,0}^{so}\rangle_{BCAD} - \frac{4\pi}{3}\langle\delta(\mathbf{r}_{12})\rangle_{ACBD} \qquad (76)$$

holds for arbitrary atomic or molecular orbital basis functions. Equation (76) provides a powerful tool for generating matrix elements and also for checking purposes, since it is a necessary condition that all integrals must satisfy.

The delta-function terms in Eqs. (75) and (76) represent the contribution to the spin–spin integrals in the neighborhood of $\mathbf{r}_{12} = 0$ and are needed because these integrals are conditionally convergent.[89] Although this term has a form similar to the Fermi contact interaction [Eq. (8)], the two are not the same, as is indicated by their derivation and by the fact that the latter is independent of the coordinate system.

3.2.3. Integral Values

Construction of automated computer codes for the evaluation of the atomic integrals has been a relatively recent development. Although it has been possible to calculate individual fine-structure integrals for a number of years, it is necessary in large-scale calculations to develop highly efficient programs that eliminate redundancy and that rapidly transform the AO integrals to a MO basis. Codes must optimize arithmetic and "administrative" operations associated with mass storage and data handling. Exhaustive tests of these computer routines corroborate their correctness for all types of spin–spin and spin–other–orbit interactions over both STO and GTO basis sets.[90,91]

Early workers were apprehensive that spin–orbit effects could not be represented accurately using GTOs because of their incorrect cusp behavior near the nucleus. Recent calculations[91–93] indicate that this is not the case, at least for molecules containing atoms of the first row of the periodic table. Table 2 compares the MO integrals over GTO and STO bases which are required to evaluate the $\langle {}^3\Sigma_g^- | \mathcal{H}_{so} | {}^1\Sigma_g^+ \rangle$ matrix element that is principally responsible for determining the spin–orbit contribution to the zero-field splitting in O_2 (cf. Section 4). A comparison of columns 1 and 2 or especially 3 and 4 of this table strongly indicates that Gaussians represent a viable alternative to using Slater functions.

It has been shown earlier in this section that the spin–spin contribution is predominantly a valence–valence type interaction. By contrast, we see from Table 2 that the core contributes significantly to spin–orbit matrix elements. To emphasize this point, the individual Coulomb and exchange integrals involving the sigma orbitals are given explicitly. By far the largest contribution to the two-electron blocks come from the integrals involving the $1\sigma_g$ and $1\sigma_u$ core orbitals. Also, the exchange integrals are almost as large as the corresponding Coulomb integrals. Physically, the one-electron term expresses the interaction between the nuclear field and the electron spins, whereas the two-electron terms of opposite sign represent a screening of the nuclear field.[40] It would be very interesting to identify a situation where the latter effect were antishielding but so far none has been found.

The apparatus described in this section has been implemented for a number of molecules and states, some of which are discussed in the remainder of this chapter.

Table 2. Molecular Orbital Integrals Needed to Evaluate $\langle{}^3\Sigma_g^-|H_{so}|^1\Sigma_g^+\rangle$ *for the Oxygen Molecule* $(cm^{-1})^a$

Integral contributions by block type	Minimum GTO[b]	Minimum STO[c]	DZ-GTO[b]	DZ-STO[c]	DZP-GTO[b]
$\langle\pi|h_{so}|\pi\rangle^d$	−192.4348	−209.5336	−275.4766	−286.0457	−271.9074
$[\pi\pi|H_{so}|\sigma\sigma]^d$	48.5118	50.3173	57.5948	59.2855	56.9427
$[\pi\sigma|H_{so}|\pi\sigma]$	28.9904	31.3394	43.0869	44.3257	42.2756
$[\pi\pi|H_{so}|\pi\pi]^e$	−2.0215	−2.1070	−2.1992	−2.4509	−2.0893
		Selected integrals within blocks[f]			
$[\pi\pi|H_{so}|\sigma\sigma]$Block					
$[1\pi_{gx}1\pi_{gy}|H_{so}|1\sigma_g1\sigma_g]$	19.3556	20.2371	23.7336	23.8668	23.4016
$[1\pi_{gx}1\pi_{gy}|H_{so}|1\sigma_u1\sigma_u]$	19.3580	20.2192	23.7359	23.8692	23.4040
$[1\pi_{gx}1\pi_{gy}|H_{so}|2\sigma_g2\sigma_g]$	2.9334	2.7849	3.2138	3.0117	3.2372
$[1\pi_{gx}1\pi_{gy}|H_{so}|2\sigma_u2\sigma_u]$	2.9965	2.9644	2.8983	2.9882	2.9357
$[1\pi_{gx}1\pi_{gy}|H_{so}|3\sigma_g3\sigma_g]$	3.8683	4.1116	4.0132	4.0324	3.9642
$[\pi\sigma|H_{so}|\pi\sigma]$Block					
$[1\pi_{gx}1\sigma_g|H_{so}|1\pi_{gy}1\sigma_g]$	13.3534	14.4412	20.9334	20.8674	20.4520
$[1\pi_{gx}1\sigma_u|H_{so}|1\pi_{gy}1\sigma_u]$	13.5473	14.6718	21.1765	21.1115	20.6880
$[1\pi_{gx}2\sigma_g|H_{so}|1\pi_{gy}2\sigma_g]$	−0.2804	−0.2243	−0.7176	−0.7775	−0.5936
$[1\pi_{gx}2\sigma_u|H_{so}|1\pi_{gy}2\sigma_u]$	0.3553	0.3693	−0.4301	−0.4461	−0.3015
$[1\pi_{gx}3\sigma_g|H_{so}|1\pi_{gy}3\sigma_g]$	2.0148	2.0867	2.1247	2.1181	2.0307

[a]The electronic configuration of both the $^3\Sigma_g^-$ and $^1\Sigma_g^+$ states of O_2 is $1\sigma_g^2 1\sigma_u^2 2\sigma_g^2 2\sigma_u^2 3\sigma_g^2 1\pi_{ux}^2 1\pi_{uy}^2 1\pi_{gx}^1 1\pi_{gy}^1$, respectively.

[b]Reference 91.

[c]References 61 and 62.

[d]h_{so} and H_{so} are the one-and two-electron parts of the spin–orbit operator [Eq. (7)], respectively.

[e]Includes contributions from $[\pi\pi|H_{so}|\bar{\pi}\bar{\pi}]$ integral types where the overbar denotes complex conjugate.

[f]Only the individual integrals for the two large contributing blocks are given explicitly here. The individual integrals composing the $[\pi\pi|H_{so}|\pi\pi]$ block are given in Ref. 91 for the Gaussian bases and are observed to be in good agreement with the comparable STO basis result.

4. Numerical Studies of Fine Structure

The formalism described in Sections 2 and 3 has been employed by several authors to determine the zero-field splitting parameters D and E and the spin–orbit coupling constants A of small to medium size molecules. In this section, we illustrate the range of effects and problems encountered, and the present limitations of *ab initio* theory as applied to these interactions. Assorted molecules that have been investigated extensively are listed* in Table 1. Although the overall agreement between theory and experiment is seen to be rather satisfactory, we shall find upon examining the details that large basis sets and configuration lists are often required for consistently accurate results. Because of the relative difficulty in carrying out these calculations, especially for heavier molecules, a large number of studies have introduced approximations. Some of these are discussed in the context of the larger diatomic and polyatomic molecules considered in this section.

4.1. Diatomic Molecules

Because of the difficulty in evaluating multicenter spin–spin and spin–orbit integrals, much of the work on fine structure has been confined to diatomic molecules. Progress in this area has recently accelerated by the steamlining and automation of computer codes[61] that evaluate all the diatomic fine-structure integrals accurately over a Slater-type basis set.

4.1.1. Hydrogen Molecule

The most extensive calculations have probably been carried out for the $2p\ ^3\Pi_u$ state of H_2. If we neglect the spin–rotation interaction (cf. Section 2), the fine-structure constants for this state are[97,98]

$$\begin{pmatrix} A_1 \\ A_2 \\ B_0 \end{pmatrix} = \left\langle {}^3\Pi_u, M_L = M_S = 1 \left| \begin{pmatrix} h_{so} \\ H_{so} \\ -2\mathcal{H}_{ss} \end{pmatrix} \right| {}^3\Pi_u, M_L = M_S = 1 \right\rangle \qquad (77)$$

and

$$B_2 = -(2/6^{1/2})\langle {}^3\Pi_u, M_L = -1, M_S = 1 | \mathcal{H}_{ss} | {}^3\Pi_u, M_L = 1, M_S = -1 \rangle \qquad (78)$$

where h_{so} and H_{so} are the one- and two-electron axial parts of the spin–orbit Hamiltonian, respectively, and $\mathcal{H}_{ss}$ is the appropriate component of the spin–spin Hamiltonian in Eq. (33).

---

*Additional lists of fine-structure parameters can be obtained from Refs. 94–96.

A summary of the results obtained so far is given in Table 3. For the spin–spin constants B_0 and B_2, the agreement between theory and experiment improves as the calculations are refined. For example, the early work of Chiu[98] used a simple linear combination of Heitler–London and ionic functions which led to B_0 and B_2 values that differed from experiment by about 15%. In the more recent and elaborate study of Pritchard et al.,[11] who used a 50-term CI wave function constructed from the Zemke, Lykos, and Wahl (ZLW) extended Slater basis,[99] and the study of Lombardi,[100] who used the 50 configuration elliptic coordinate natural spin–orbital wave function of Rothenberg and Davidson (RD),[101] agreement within about 3% is obtained for these constants. By contrast, the spin–orbit coupling constant $(A_1 + A_2)$ seems to be more sensitive to the details of the electronic distribution. The results of Pritchard, Sink, and Kern (PSK)[11] are a disappointing 30% too high, with most of the error probably residing in the one-electron part A_1. Since A_1 is more strongly dependent on basis set than CI expansion length, we infer that the RD elliptical basis is of superior quality to that of ZLW which was optimized for the lowest singlet state of Π_u symmetry. This sensitivity of the one-electron spin–orbit contribution to the quality of the basis has also been clearly demonstrated[62] for the $X\,^3\Sigma_g^-$ state of O_2 (see below). We remark that absurd results are obtained for all the H_2 interaction constants if two-center integrals are omitted.

In comparing theory with experiment, it can be important to compensate for zero-point vibrational motion. For any property P, we need to examine the slope (m) and curvature (c) of P vs internuclear distance in order to establish qualitative trends (cf. Appendix[102–114]). We anticipate, other factors being

Table 3. *Comparison of Vibrationally Averaged Fine-Structure Constants with Previous Calculations and Experiment*: $2p\,^3\Pi_u\,H_2\,(MHz)$

Case	A_1	A_2	$A_1 + A_2$	B_0	B_2
A^a	5737	−9800	−4100 (10)^g	−1330 (16)	−3840 (16)
B^b	8424	−12644	−4220 (13)	−1651 (5)	−4681 (3)
C^c	8433	−12360	−3927 (5)	−1725 (9)	—
D^d	7433	−12350	−4917 (31)	−1530 (3)	−4456 (2)
E^e	7413	−12308	−4895 (31)	−1522 (3)	−4439 (2)
Experimental^f	—	—	−3741	−1577	−4547

^{a}Published results of Chiu, $R = 1.9608$ bohr, Ref. 98.
^{b}Published results of Lombardi, Ref. 100.
^{c}Lombardi's values vibrationally averaged by Pritchard and Kern; those not shown became divergent.
^{d}Published results of Pritchard et al.,[11] ZLW basis, full CI, vibrationally averaged using the (experimental) Morse potential.
^{e}Published results of Pritchard et al.,[11] ZLW basis, full CI, vibrationally averaged using the (ab initio) Dunham potential.
^{f}Recent experimental results quoted by A. N. Jette, J. Chem. Phys. **61**, 816 (1974).
^{g}Numbers in parentheses are percent differences with experiment.

Table 4. Vibrationally Averaged Fine-Structure Constants[a]:
$2p\ ^3\Pi_u\ H_2\ (MHz)$

v	A_1+A_2	B_0	B_2
$-\frac{1}{2}$[b]	-5036	-1571	-4542
0	-4917	-1530	-4456
	-4895	-1522	-4439
1	-4597	-1415	-4212
	-4668	-1443	-4271
2	-4316	-1315	-4001
	-4435	-1362	-4099

[a] Obtained by Pritchard *et al.* using the Slater basis of Zemke, Lykos, and Wahl and a full CI (Ref. 11). Upper values were calculated using (*ab initio*) Dunham potential; lower values were calculated using (experimental) Morse potential.
[b] Value at R_e without vibrational averaging.

equal, that the zero-point correction is large (small) when m and c are appreciable and have equal (opposite) signs. It is apparent from the results of Table 4, where we list vibrationally averaged fine-structure constants for $v = 0$, 1, 2, 3, that (A_1+A_2) is indeed relatively sensitive to these effects but that they are still insufficient to explain all of the discrepancy. The best agreement (5%) is obtained after vibrationally averaging the results of Lombardi.

From these studies, we conclude that the following conditions must be satisfied in order to determine quantitatively the very small fine-structure splitting parameters of the $2p\ ^3\Pi_u$ state of H_2: (i) the use of an extensive well optimized basis set; (ii) the use of the large CI expansions; (iii) the inclusion of all multicenter integrals; and (iv) the proper use of vibrational averaging. It is also possible that the second-order spin–orbit contributions must be included for extremely high accuracy.

4.1.2. Diagonal Spin–Orbit Coupling (SOC) in Larger Diatomics

In comparison to H_2, the accurate evaluation of spin–orbit parameters for Π states of heavier diatomics is achieved without the need for high sophistication. Qualitative, and often quantitative, agreement with experiment can be obtained from SCF wave functions, even ignoring all two-center integrals. The relative ease of calculating reliable spin–orbit constants and their large variation from state to state have made them valuable aids for assigning the Π states of diatomic molecules.[115–118] In this way, for example, Roche and Lefebvre-Brion[118] have concluded that the observed vibrational levels of the $D'\ ^2\Pi$ state of PO actually correspond to the high vibrational levels of the $B'\ ^2\Pi$ state.

An approach that has been quite successful in correlating the main effects of spin–orbit coupling, is that of replacing the Breit–Pauli spin–orbit operator by the expression

$$\mathcal{H}'_{so} = \sum_{\alpha,i} \xi_\alpha(\mathbf{r}_i)[(\mathbf{r}_{\alpha i} \times \mathbf{p}_i) \cdot \mathbf{s}_i] \tag{79}$$

where $\xi_\alpha(\mathbf{r}_i)$ is an empirically determined function that depends on the electronic configuration of atom α and on the fourth power of its nuclear charge. For an atom, this operator is rigorously equivalent to the complete form in Eq. (7). For molecules, Moores and McWeeny[119] have noted that the reliability of Eq. (79) depends on the accuracy with which (i) the spin density can be resolved into a superposition of one-center contributions, and (ii) the Breit–Pauli two-electron one-center matrix elements can be consistently generated from the equivalent operator in Eq. (79). These qualifications reflect the fact that when the one-center interactions dominate, the spin–orbit parameter A as defined in Eq. (29) is controlled by the strong inverse dependence of ξ on $\mathbf{r}_i$ [cf. Eq. (7)].

Ishiguro and Kobori[120] have used Eq. (79) to obtain SOC constants for a large number of diatomic molecules. They derived values of $\xi_\alpha(\mathbf{r}_i)$ from Moore's table[121] and from the application of Slater's rules. Considering the simplicity of their model, the agreement with experiment (cf. Table 5) is remarkably good. Since no attempt was made to include the effects of electron correlation in the model, the largest disagreement with experiment occurs for electronic states that are poorly represented by one configuration. This is

Table 5. Spin–Orbit Coupling Constants A in Diatomic Molecules by Various Methods

Molecular state	Walker and Richards[a]			Ishiguro and Kobori[b]	Experiment[c]
	1	2	3		
BO $A\,^2\Pi$	−121.2	—	−122.7	−116.7	−126.7
CO $A\,^3\Pi$	38.2	—	39.0	43.0	41.5
NO $X\,^2\Pi$	98.6	—	105.3	129.2	122.2
CO$^+$ $A\,^2\Pi$	−121.4	—	−121.7	−112.4	−117.5
O$_2^+$ $^2\Pi_u$	—	—	—	48.0	8.2
OH $X\,^2\Pi$	−141.4	—	—	−151	−139.7
BeF $A\,^2\Pi$	23.7	—	24.3	—	21.8
MgH $A\,^2\Pi$	26.8	36.4	—	40.5	35
AlH$^+$ $A\,^2\Pi$	92.9	110.9	—	99.8. 124.9[d]	108
SiH $X\,^2\Pi$	129.5	155.1	—	148.9	142
SH $X\,^2\Pi$	−355.7	−375.7	—	−382.4	−382
PO $X\,^2\Pi$	153.7	190.1	—	—	224

[a] Columns: 1, inclusion of only one-center integrals; 2, same as 1 with the inclusion of the "core polarization" corrections; 3, all integrals retained (the "exchange hybrid" integrals are evaluated using Mulliken's approximation).
[b] Results using the model Hamiltonian [Eq. (79)], Ref. 120.
[c] Reference 94.
[d] Values obtained using $[\xi_{3,1}(\mathrm{Al}) + \xi_{3,1}(\mathrm{Al}^+)]/2$ and $\xi_{3,1}(\mathrm{Al}^+)$, respectively.

manifested by the fact that the larger the electronegativity difference between the two atoms, the larger is the discrepancy with experiment. A specific, dramatic example of a CI effect occurs for the $^2\Pi_u$ state of O_2^+, where even using a three-term CI wave function the calculated value is about six times too large. In most cases, however, this model approach to SOC leads to consistently satisfactory results. It is perhaps most useful for estimating orders of magnitude and for understanding the origin of the principal effects and trends. The phenomenological Hamiltonian $A\,\mathbf{L}\cdot\mathbf{S}$ is of limited utility, as we shall see below for O_2, because it can give zero expectation value for states with nonzero coupling. Hence, it should be used with special caution for studying second-order spin–orbit effects.*

Richards and co-workers[9,93,123,124] and H. Lefebvre-Brion and co-workers[115,118,125–127] systematized a related device to obtain SOC constants. In their method, the correct Breit–Pauli spin–orbit Hamiltonian is employed but only one-center integrals are retained in the calculation. This approximation is particularly good for diatomic hydrides, since the contributions from the hydrogen atom are relatively small. From a comparison of columns 1 and 3 of Table 5, we see that even for nonhydride molecules where the electron distribution is spread more uniformly over the molecule, the two-center contributions are relatively small due largely to a cancellation between the two-center one- and two-electron integrals. This cancellation is consistent with Eq. (79), in that the SOC parameter is assumed to depend on a simple sum of atomic contributions.

Another point is that restricted Hartree–Fock calculations on second row diatomics all give SOC constants smaller than those experimentally observed. This is ascribed to "core polarization" and can be accounted for within the unrestricted Hartree–Fock formalism by allowing the radial basis functions to change with the quantum numbers M_l and M_s. This relaxation increases A because the closed shells now contribute through the one-electron coupling.[124] The results in column 2 of Table 5 are obtained from those in column 1 by adding an approximate atomic correction for this effect.

Both of these approximate approaches yield SOC constants in qualitative agreement with experiment. Since these methods are computationally simpler than completely *ab initio* treatments, it is likely that they will continue to be used as reasonable, quick aids to molecular fine structure, particularly diagonal spin–orbit interactions. A more inclusive systematic approach that includes the spin–spin Hamiltonian and that explicitly takes account of core–core, core–valence, and valence–valence couplings would be particularly worthwhile in future work.

---

*For diatomic molecules, Veseth[122] has argued that for diagonal as well as off-diagonal elements Eq. (7) leads to the same results as the operator $A\,\mathbf{L}\cdot\mathbf{S}$ if $\Delta S = 0$ and the constant A is chosen in a prescribed way.

4.1.2. Off-Diagonal Spin–Orbit Coupling in C_2

Although diagonal spin–orbit interactions are reasonably well described by SCF wave functions, off-diagonal matrix elements are often sensitive to correlation mixing. This type of situation is exemplified by the coupling between the $X\ ^1\Sigma_g^+$ and the $b\ ^3\Sigma_g^-$ $(1\pi^2 \to 3\sigma_g^2)$ states of C_2 expressed by,*

$$A = \langle\, ^3\Sigma_g^- | \mathcal{H}_{\text{so}} | ^1\Sigma_g^+ \rangle \tag{80}$$

Small perturbations in these two states were discovered by Ballik and Ramsay[128] in 1963 to have equal and opposite sign. From these observations, they deduced that the ground state of C_2 was the $^1\Sigma_g^+$ state instead of the $^3\Pi_u$, as formerly believed. Given the magnitude of these perturbations and an explicit form for the variation of A with internuclear distance (R), it is possible to derive[129] a value of A at the $X\ ^1\Sigma_g^+ - b\ ^3\Sigma_g^-$ crossing point near 2.68 bohr. Since the variation of A with R can be determined by the *ab initio* methods described in this chapter, an excellent opportunity is provided for the interplay of theory and experiment.

Using a minimal Slater basis and the $^1\Sigma_g^+$ SCF orbitals for each state, the single configuration result yields $A = 0.069\ \text{cm}^{-1}$, which is two orders of magnitude too small.[61,129] Analysis shows that isoconfigurational $^3\Sigma_g^-$ $(^1\Sigma_g^+)$ functions, or those singly excited with respect to the $^3\Sigma_g^-$ $(^1\Sigma_g^+)$ state, are entirely absent at the Hartree–Fock level of approximation. The weighting coefficients of these terms may be relatively small, but their spin–orbit coupling to the dominant root configurations project out large valence and core–valence integrals whose weighted sum is significant. Although the $b\ ^3\Sigma_g^-$ state is well represented by a single configuration over the range of R near the crossing point, the $X\ ^1\Sigma_g^+$ state needs several configurations to be described correctly. Of these configurations, however, only the one that is isoconfigurational with the dominant configuration of the $b\ ^3\Sigma_g^-$ state has a large spin–orbit matrix element. Hence, the value of A is controlled by the coefficient of the (core) $3\sigma_g^2 1\pi_u^2$ configuration in the singlet wave function that is dominated by the (core) $1\pi_u^4$ configuration.

With this background, CI calculations were performed[129] using a DZP basis of Gaussian-lobe functions at R values from 2.28 to 3.08 bohr in increments of 0.2 bohr. The $b\ ^3\Sigma_g^-$ state was described by all singly excited and a subset of doubly excited configurations (based on a perturbation theory estimate of their energy contributions) away from the (core) $3\sigma_g^2 1\pi_u^2$ config-uration. The wave function for the $X\ ^1\Sigma_g^+$ state was determined using the orbitals for the triplet state in the same manner except that a combination of ten configurations was used as the starting point for selection. This procedure insures that all important triply and quadruply excited configurations away from the (core) $1\pi_u^4$ configuration are included. These configurations are

---

*In this subsection, A refers to the off-diagonal spin–orbit matrix element defined in Eq. (80).

precisely the ones required to insure the proper weighting of the (core) $3\sigma_g^2 1\pi_u^2$ configuration in the singlet wave function. Using these wave functions, we obtain values for A of 1.0, 2.2, 5.0, 9.7, and 17.2 cm^{-1} at R values of 2.28, 2.48, 2.68, 2.88, and 3.08 bohr, respectively. A plot of A vs R shows significant nonlinearity and further parallels the behavior of the coefficient of the (core) $3\sigma_g^2 1\pi_u^2$ configuration in the singlet wave function that is needed for proper dissociation of the molecule into C atoms. Assuming no spin–rotation interaction, this curvature leads[129] to a value of 5.6 ± 0.5 cm^{-1} for A at 2.68 bohr, in good agreement with the purely *ab initio* value of 5.0 cm^{-1}.

One of the shortcomings of the present approach is the need to use one set of orbitals to describe both states. This was not a serious limitation in the present calculation since a large number of configurations was used, and since the SCF orbitals for the two states are similar. When this latter condition is not satisfied, such as often occurs for the crossing between a repulsive and bound state (see Section 5), it may be better to reformulate the problem as outlined by Hall *et al.*[127] In their notation

$$A = \sum_{a,b} \rho_1(^1\Sigma_g^+; {}^3\Sigma_g^-|b; a)\langle\chi_a|h_{so}|\chi_b\rangle$$

$$+ \sum_{\substack{a,b, \\ c,d}} \rho_2(^1\Sigma_g^+; {}^3\Sigma_g^-|bd; ac)\langle\chi_a\chi_b|H_{so}|\chi_c\chi_d\rangle \qquad (81)$$

where ρ_1 and ρ_2 are one- and two-electron transition density matrices in some basis of functions $\{\chi\}$, and h_{so} and H_{so} are, respectively, the one- and two-electron spin–orbit operators. In their procedure, Hartree–Fock (HF) MOs appropriate to each state are used and nonorthogonality between them is incorporated directly into ρ_1 and ρ_2. Although this general problem of orbital nonorthogonality is troublesome no matter how it is treated, simplifications occur in Eq. (81) when the wave function is a linear combination of single Slater determinants. Hall *et al.*[127] also conclude from their work on CO that a wide disparity can arise in A if the set of HF orbitals of either state is taken to describe both states, and furthermore, that the results using nonorthogonal orbitals can be quite different from the average value between these two canonical approximations.

Two practical approaches to this difficulty, therefore, are either to obtain an optimum set of MOs for describing both states simultaneously and perform an extensive CI, or to use separate MOs for each state and perform a more modest CI calculation, explicitly taking into account the nonorthogonality.

4.1.3. Oxygen Molecule

The first complete theoretical and experimental treatment of the $X\,^3\Sigma_g^-$ fine structure was reported by Tinkham and Strandberg (TS)[24] in 1955. This system is especially interesting because it combines most of the difficulties encountered so far. To proceed chronologically, TS measured the fine-structure constant $D = D_{so} + D_{ss}$ as 3.962 cm^{-1} from microwave spectra. These data were interpreted with the phenomenological operator $A\mathbf{L}\cdot\mathbf{S}$ instead of the true microscopic spin–orbit Hamiltonian in Eq. (7). Since $\mathbf{L}\cdot\mathbf{S}$ mixes only states of $^3\Pi_g$ symmetry into the $X^3\Sigma_g^-$ state in second-order perturbation theory,[40] TS obtained a very small value for the spin–orbit contribution $D_{so} = 0.02$ cm^{-1} and concluded that the spin–spin splitting D_{ss} was the dominant interaction. Subsequent studies by Kayama[130] and particularly by Pritchard *et al.*[131] established, however, that $D_{ss} = 1.54 \pm 0.04$ cm^{-1} for all single-configuration wave functions constructed from minimum to near-Hartree–Fock Slater basis sets. Inclusion of correlation effects by Pritchard *et al.*[62] indicated that this value should be somewhat smaller, about $D_{ss} = 1.42$ cm^{-1}. Since these results pointed to serious discrepancy between theory and experiment, later work concentrated on justifying a much larger spin–orbit contribution of $D_{so} \cong 2.6$ cm^{-1}.

Two independent calculations[61,91] have recently been carried out for the second-order spin–orbit splitting parameter with the correct microscopic spin–orbit Hamiltonian. The first of these studies was performed at the SCF level using single- and double-zeta bases of Slater functions, and the second with CI wave functions and a polarized double-zeta basis of contracted Gaussian-lobe functions. Table 6 compares the contributions to D_{so} from ten electronic states in the second-order sum [Eq. (27)].

Table 6. Second-Order Spin–Orbit Contributions to D_{so} for the $X^3\Sigma_g^-$ State of O_2[a]

State	Excitation	SZ Slater basis,[b] D_{so}^{SCF}	DZ Slater basis,[b] D_{so}^{SCF}	DZP Gaussian-lobe basis[c] D_{so}^{SCF}	D_{so}^{CI}
$^1\Sigma_g^+$	—	1.276	2.404	2.316	2.400
$^3\Sigma_g^{+d}$	$3\sigma_g\pi_u \to 3\sigma_u\pi_g$	-3.2×10^{-8}	-4.0×10^{-8}	-4.4×10^{-5}	-2.8×10^{-4}
$^5\Sigma_g^+$	$3\sigma_g\pi_u \to 3\sigma_u\pi_g$	1.8×10^{-5}	1.3×10^{-5}	1.0×10^{-5}	2.4×10^{-5}
$^3\Pi_g$	$3\sigma_g \to \pi_g$	0.074	0.138	0.094	0.110
$^1\Pi_g$	$3\sigma_g \to \pi_g$	-0.062	-0.120	-0.080	-0.096
$^5\Pi_g$	$\pi_u \to 3\sigma_u$	-0.014	-0.020	-0.028	-0.034
$^3\Pi_g^{d,e}$	$\pi_u \to 3\sigma_u$	—	—	0.040	0.046
$^1\Pi_g^{d,e}$	$\pi_u \to 3\sigma_u$	—	—	-0.020	-0.024
$^3\Pi_g^{e}$	$\pi_u\bar\pi_g \to 3\sigma_u\pi_g$	—	—	8.2×10^{-4}	3.6×10^{-4}
$^1\Pi_g^{e}$	$\pi_u\bar\pi_g \to 3\sigma_u\pi_g$	—	—	-5.0×10^{-4}	-4.8×10^{-5}
Net D_{so}	—	1.274	2.402	2.322	2.402
Net D_{ss}	—	—	1.444	1.545	1.453
$D_{so}+D_{ss}^f$	—	—	3.846	3.867	3.855

[a] The experimental term values were used for all the lower-lying excited states.
[b] References 61 and 62.
[c] Reference 91.
[d] Corresponds to the contribution of the 3(2) lowest states of this symmetry for the triplet (singlet) state.
[e] State not considered in the Slater basis calculations.
[f] Experimental value is 3.965 cm^{-1} (Ref. 133).

The following conclusions emerge from the table: (i) the discrepancy between experiment and theory is resolved, meaning that the second-order spin–orbit contribution accounts for about $\frac{2}{3}$ of the zero-field splitting; (ii) only states that are singly excited with respect to the $X\,^3\Sigma_g^-$ state couple significantly; (iii) the overall effect of electron correlation on D_{so} is less than 10%; (iv) a basis set of at least double-zeta quality must be used since the spin–orbit integrals are very sensitive to the charge density distribution (see Table 2); and (v) the low-lying $^1\Sigma_g^+$ state generates virtually 100% of the total splitting parameter, with self-cancelling contributions from the higher states. This last observation lends credence to arguments that the second-order sum has converged. Adding in the best results for D_{ss}, we obtain $D = 3.846 \text{ cm}^{-1}$ and $D = 3.856 \text{ cm}^{-1}$ for the two most complete *ab initio* studies. This is in quite satisfactory agreement with a $D = 3.965 \text{ cm}^{-1}$ from recent measurements.[132,133] We see again in this example that very elaborate calculations are sometimes required to identify the origin of molecular fine structure.

4.2. Polyatomic Molecules

The *ab initio* investigation of zero-field splittings in polyatomic molecules has been accelerated by the recent development of Gaussian-lobe integral programs which evaluate all fine-structure multicenter integrals accurately and rapidly through closed-form analytical formulas of the type derived in Eq. (72). In this section, we discuss some classic systems to which these programs have been applied, namely the $^3B_1$ ground state of methylene, the $^3A_2$ $(n \to \pi^*)$ state of formaldehyde, and the lowest $^3B_{1u}$ state of benzene.

4.2.1 Methylene

One of the simplest nonlinear molecules with a triplet ground state is CH_2. The currently accepted experimental values[134,135] of the zero-field parameters for its $X\,^3B_1$ state are $D = 0.76 \pm 0.02 \text{ cm}^{-1}$ and $E = 0.052 \pm 0.017 \text{ cm}^{-1}$. These values refer to the isolated free molecule, having been extracted from raw data appropriately corrected for motional averaging of CH_2 and CD_2 frozen in either octafluorocyclobutane or SF_6 at 4°K. Rotational motion of the methylene within the host matrix results in observed values of D between 15% and 20% less than that for the free molecule, depending on the solvent. Nearly free rotation about the long axis of the molecule also makes E much smaller.

To analyze the zero-field effect, several authors[36,136–142] have calculated the spin–spin and spin–orbit contributions to D and E. The first *ab initio* study of the spin–spin interaction was carried out by Harrison[137] with a Gaussian-lobe basis. He used the SCF orbitals from the $^1A_1$ state and included limited CI.

Values of $D_{ss} = 0.71$ cm^{-1} and $E_{ss} = 0.05$ cm^{-1} were obtained at the computed equilibrium geometry of $R_{CH} = 1.985$ a.u. and $\theta = 132.5°$. A later and more extensive calculation by Langhoff and Davidson (LD)[138] using a polarized double-zeta of a similar type, the $^3B_1$ SCF orbitals, and a more extensive CI expansion led to $D_{ss} = 0.781$ cm^{-1} and $E_{ss} = 0.050$ cm^{-1} at $R_{CH} = 2.0714a_0$ and $\theta = 132°$. Since two independent and technically different studies produced almost identical results, it is certain that, unlike the situation in O_2, D and E are dominated by the first-order spin–dipolar interaction.

Until very recently, no fully satisfactory *ab initio* treatment existed for the second-order spin–orbit contributions. Following LD, Langhoff performed a calculation in which the two lowest states of $^1A_1$ symmetry as well as the lowest states of $^{1,3}A_2$ and $^{1,3}B_2$ symmetry were incorporated into the second-order perturbation sum.[36] Two HCH bond angles, 135° and 180°, and extensive CI expansions were used at a CH bond length of $2.0714a_0$. A spin–orbit contribution to D of 0.023 cm^{-1} was obtained at the 135° bond angle which, when combined with D_{ss}, gave a value of $D = D_{ss} + D_{so} = 0.807$ cm^{-1}, somewhat larger than the experimental value of $D = 0.76 \pm 0.02$ cm^{-1}. The dominant contribution to the splitting comes from the two low-lying $^1A_1$ states. The total contribution to D_{so} from the higher energy $^{1,3}A_2$ and $^{1,3}B_2$ states is only about 0.0005 cm^{-1} or about 2% at the 135° bond angle. Also, electron correlation reduces D_{so} by 50% at the equilibrium angle. This large effect occurs primarily because the $^1A_1$ states need two configurations to be properly represented. The parameter E_{so} was found to be zero to three significant figures.

Although the origin of the zero-field splitting parameters in methylene appears to be reasonably well understood, it is again instructive to consider the effects of zero-point motion. For more than one degree of vibrational freedom, we must be concerned with the additive and interactive effects of all the modes. The purely symmetric and bending vibrations can be treated in much the same way as with H_2 (see above). However, since interactions between the normal modes can be large, the overall sign and magnitude of the zero-point corrections to D and E are determined by a complex mixture of factors (cf. Appendix). Furthermore, the experimental data refer to CD_2 so that a proper treatment of vibration is essential for a strict comparison between theory and experiment. Employing the full Watson[114] Hamiltonian and proceeding by analogy to the Hartree–Fock and CI variational methods[107–113] discussed in Volume 3 of this series, we obtain the results[143] shown in Table 7. Clearly, vibrational motion is unimportant at the SCF level of approximation, though a similarly rigorous treatment including electron correlation has not yet been carried out. This small correction to D_{ss} and E_{ss} is due to the facts that their dependence on geometry is almost linear in the vicinity of equilibrium and that the potential energy surface itself contains relatively little anharmonic character in this region. While the vibrational corrections to D_{so} and E_{so} may be larger percentagewise, their absolute values are sufficiently small that nuclear effects can be safely neglected except in the most accurate work.

Table 7. Vibrationally Averaged SCF Results for the Zero-Field Splitting Parameters D and E of Methylene (cm^{-1})

Vibrational state[a]			CH$_2$ molecule results[b,c]		CD$_2$ molecule results[b,c]	
n_1	n_2	n_3	ΔD	ΔE	ΔD	ΔE
0	0	0	0.0003	−0.0006	0.0001	−0.0003
0	1	0	0.0043	−0.0051	0.0028	−0.0033
1	0	0	−0.0007	−0.0006	−0.0006	−0.0003
0	0	1	−0.0009	0.0012	−0.0010	0.0015

[a] n_1, n_2, and n_3 refer to the number of quanta in the symmetric stretching, bending, and asymmetric stretching modes, respectively.
[b] The averaged D and E parameters are given relative to the equilibrium values: $\Delta D = D - D_{eq}$ and $\Delta E = E - E_{eq}$, where D_{eq} and E_{eq} are 0.76233 and 0.06961 cm^{-1} at the equilibrium values of $\theta_{eq} = 130.12°$ and $R_{eq} = 2.0232$ bohr.
[c] The parameters were vibrationally averaged using the 165-term CI vibration wave functions obtained for CH$_2$ and CD$_2$ using the 5th degree polynomial fit to the SCF potential surface.

It, therefore, emerges that the spin–spin contribution accounts for about 97% of the zero-field splitting at both the 135° and 180° bond angles in methylene. That this is a special case and not a general rule is reemphasized in the following section on formaldehyde, where it is again found that the second-order spin–orbit effects are significant.

4.2.2. Formaldehyde

The first experimental values reported for the zero-field splitting in the $^3A_2(n \to \pi^*)$ state of H$_2$CO are those of Raynes[144] who obtained $D = 0.42$ cm^{-1} and $E = 0.04$ cm^{-1} from a rotational analysis of the $1^+ \leftarrow 0$ band of the $^3A_2 \leftarrow {}^1A_1$ transition.* Experiments are hampered by the short lifetime of the $^3A_2$ state and the tendency of formaldeyde to polymerize.

A CI study of D_{ss} for this state was reported by Langhoff *et al.*[139] who found $D_{ss} = 0.539$ cm^{-1} and $E_{ss} = 0.031$ cm^{-1} with $^3A_2$ wave functions of the general type described for CH$_2$. Comparing the theoretical and experimental values of D_{ss} and D, we infer that spin–orbit coupling diminishes D_{ss} by 25%.

A direct calculation of D_{so} has been recently completed by Langhoff *et al.*[145] A set of diffuse functions was added to the polarized double-zeta basis described in Ref. 139 so that excited Rydberg states could be described in a realistic way. The spin–orbit parameter D_{so} was determined from the relation

$$D_{so} = \sum_n^\infty \left[\frac{|\langle {}^3A_2|\mathcal{H}_{so}|{}^1A_1, n\rangle|^2}{E({}^1A_1, n) - E({}^3A_2)} - \frac{|\langle {}^3A_2|\mathcal{H}_{so}|{}^3A_1, n\rangle|^2}{E({}^3A_1, n) - E({}^3A_2)} \right.$$
$$- \frac{\frac{1}{2}|\langle {}^3A_2|\mathcal{H}_{so}|{}^1B_2, n\rangle|^2}{E({}^1B_2, n) - E({}^3A_2)} + \frac{\frac{1}{2}|\langle {}^3A_2|\mathcal{H}_{so}|{}^3B_2, n\rangle|^2}{E({}^3B_2, n) - E({}^3A_2)}$$
$$\left. - \frac{\frac{1}{2}|\langle {}^3A_2|\mathcal{H}_{so}|{}^1B_1, n\rangle|^2}{E({}^1B_1, n) - E({}^3A_2)} + \frac{\frac{1}{2}|\langle {}^3A_2|\mathcal{H}_{so}|{}^3B_1, n\rangle|^2}{E({}^3B_1, n) - E({}^3A_2)} \right] \qquad (82)$$

*More recent experimental values can be found in Ref. 139.

The signs and coefficients of each term reflect the fact that states which lower the $T_z(T_x$ or $T_y)$ level increase (decrease) the splitting. The contribution of each excited state to the sublevels of the $^3A_2(n \to \pi^*)$ state is presented in Table 8. Several important conclusions emerge: (i) the $^3A_1(\pi \to \pi^*)$ and $X\,^1A_1$ states both reduce D, the amount of lowering being quite sensitive to the energy separations [especially the separation of the $^3A_2$ and nearby $^3A_1(\pi \to \pi^*)$ states]; (ii) the Rydberg states contribute a negligible amount to D_{so}; (iii) the states that contribute significantly to D_{so} are not necessarily the ones that determine the phosphorescent lifetimes of the $^3A_2$ sublevels (cf. Section 5); and (iv) the nonnegligible spin–orbit contribution to E comes almost entirely from the $^{1,3}B_1(\sigma \to \pi^*)$ states.

Combining $D_{so} = -0.224$ cm^{-1} and $E_{so} = 0.009$ cm^{-1} from Table 8 with the spin–spin results, we find $D = 0.315$ cm^{-1} and $E = 0.04$ cm^{-1}. If the transition energies reported by Peyerimhoff *et al.*[146] are used, the values $D = 0.304$ cm^{-1} and $E = 0.041$ cm^{-1} are obtained. Although spin–orbit coupling acts qualitatively in the expected ways, that is to decrease (increase) D (E), the complexity of the second-order calculation still makes D_{so} and E_{so} the largest uncertainties. The following three approximations probably have the most significant effect on the overall reliability: (i) the use of $X\,^1A_1$ state geometry; (ii) the truncation of the infinite sums in Eq. (82); and (iii) the use of the $^3A_2$ SCF orbitals to construct the CI wave functions for all states. Only the first of these factors is relevant to the spin–spin calculation.

4.2.3. Benzene

Another polyatomic molecule of considerable interest, and the largest one that we consider in this chapter, is C_6H_6. For the $^3B_{1u}$ state, the zero-field parameters $D = 0.1580 \pm 0.0003$ cm^{-1} and $E = -0.0064 \pm 0.0003$ cm^{-1} have been determined from electron spin resonance (ESR) spectra of C_6H_6 in C_6D_6 host crystals.[45] The nonzero value of E clearly demonstrates that triplet benzene, at least in the host crystal, has less than D_{6h} symmetry. Both of these experiments and recent hyperfine measurements[147] indicate that the molecule, at least *on the average*, favors an elongated form as opposed to a quinoid (compressed) structure.

McClure[43] has shown that for planar aromatic molecules the class of potentially large one- and two-center spin–orbit integrals vanishes and causes the second-order spin–orbit interaction to be extremely small ($D_{so} \cong 0.0001$ cm^{-1}). It appears reasonable, therefore, to interpret the experimental results in terms of the spin–spin contribution alone.

Theoretical calculations[37,44,64,148–153] of D_{ss} have suffered from the many approximations that were needed to make them tractable. The first published treatment[64] expanded the triplet-state wave functions for the π electrons as single Slater functions centered on each carbon atom and also neglected all

Table 8. Formaldehyde Spin–Orbit Matrix Elements

Matrix elements with the $^3A_2(n\to\pi^*)$ state

State Y	Type of state	$\langle{}^3A_2\|\mathcal{H}_{so}\|Y\rangle i$,[a] cm^{-1}	$E(Y)-E(^3A_2)$,[b] eV (cm^{-1})	Transition[c] moment, a.u. $\langle X\,^1A_1\|\mathbf{r}\|Y\rangle$	Fine-structure contributions, cm^{-1} T_x	T_y	T_z
$X\,^1A_1$[d]	Valence	-61.97	$-3.43\ (-27674)$	17.263	—	—	0.139
$^1A_1(\pi\to\pi^*)$	Valence	-41.32	$7.19\ (58011)$	-0.884	—	—	-0.029
$^1A_1(n^2\to\pi^{*2})$	Valence	-53.07	$7.90\ (63740)$	-0.047	—	—	-0.044
$^1A_1(2b_2\to3b_2)$	Rydberg	-0.14	$5.06\ (40826)$	0.042	—	—	-4.9×10^{-7}
$^3A_1(2b_2\to3b_2)$	Rydberg	0.33	$5.06\ (40826)$	—	-2.7×10^{-6}	-2.7×10^{-6}	—
$^3A_1(\pi\to\pi^*)$	Valence	55.27	$2.27\ (18315)$	—	-0.167	-0.167	—
$^1B_2(2b_2\to6a_1)$	Rydberg	2.40	$4.02\ (32435)$	-0.451	-1.8×10^{-4}	—	—
$^3B_2(2b_2\to6a_1)$	Rydberg	-2.45	$4.01\ (32354)$	—	—	-1.9×10^{-4}	-1.9×10^{-4}
$^1B_1(\sigma\to\pi^*)$	Valence	-52.59	$5.89\ (47522)$	0.080	—	-0.058	—
$^1B_1(1b_1\to6a_1)$	Rydberg	0.14	$7.58\ (61158)$	0.150	—	-3.2×10^{-7}	—
$^3B_1(\sigma\to\pi^*)$	Valence	55.15	$5.02\ (40503)$	—	-0.075	—	-0.075
$^3B_1(1b_1\to6a_1)$	Rydberg	0.22	$7.57\ (61077)$	—	-7.9×10^{-7}	—	-7.9×10^{-7}
Totals					-0.242	-0.225	-0.009

Matrix elements with the $^1A_1$ ground state

State Y	Type of state	$\langle Y\|\mathcal{H}_{so}\|^1A_1\rangle i$, cm^{-1}	$E(Y)-E(^1A_1)$, eV (cm^{-1})	Transition[c] moment, a.u. $\langle{}^3A_2\|\mathbf{r}\|Y\rangle$
$^3A_2(1b_1\to3b_2)$	Rydberg	-0.87	$12.07\ (97384)$	0.001
$^3B_2(2b_2\to6a_1)$	Rydberg	3.77	$7.44\ (60028)$	0.297
$^3B_1(\sigma\to\pi^*)$	Valence	-57.79	$8.45\ (68177)$	0.035
$^3B_1(1b_1\to6a_1)$	Rydberg	1.61	$11.00\ (88751)$	-0.008

[a] The spin component associated with the $^3A_2$ is $M_s = 0$ and $M_s = S$ for Y a singlet and triplet state, respectively.

[b] The energies reported here do not agree quite as well with experiment as say those of Peyerimhoff *et al.* (Ref. 146). This is because the $^3A_2(n\to\pi^*)$ canonical orbitals are used to describe each state and the method of configuration selection biases some states. Only small changes occur in the fine structure and phosphorescent lifetimes, however, when the experimental energies are used instead of the ones reported here.

[c] When Y is $X\,^1A_1$, this represents the electronic part of the dipole moment.

[d] The ground state electronic configuration is $1a_1^2 2a_1^2 3a_1^2 4a_1^2 5a_1^2 1b_1^2 1b_2^2 2b_2^2$.

three- and four-center integrals. Refinements beyond this simple approach include the introduction of multiterm Gaussians[37] to fit the atomic exponential orbitals and the accurate evaluation of all three- and four-center Slater integrals via the Gaussian transform technique.[151] Although there are a number of uncertainties in the quantitative comparison of theory with experiment (e.g., the uncertainty in molecular geometry), all semiempirical studies along these lines reported agreement with experiment to about 10%. To test the sensitivity of D_{ss} to the form of the wave function and particularly to assess the reliability of Hückel wave functions, *ab initio* CI calculations were carried out at the D_{6h} ground-state geometry with a double-zeta basis of contracted Gaussian-lobe functions.[152] The results are summarized in Table 9.

At the SCF level of approximation, $D_{ss} = 0.1133$ cm^{-1}. To understand better the electron-correlation effect, the selection of configurations was resolved into five distinct stages, each corresponding to unique classes of excitation. It is observed that excitations of the types $\sigma^2 \to \pi^2$ and $\pi^2 \to \sigma^2$ have little effect on D_{ss}, whereas types $\pi^2 \to \pi'^2$, $\pi \to \pi'$, and $\sigma\pi \to \sigma'\pi'$ increase D and types $\sigma^2 \to \sigma'^2$ and $\sigma \to \sigma'$ decrease D.† Furthermore, the sum of the absolute values of the CI effects for these five classes of excitations is greater than the value of D_{ss} itself. A CI wave function with almost 8400 symmetry-adapted configurations,‡ including all singles and the most important doubles from the three important classes of excitation, yielded $D_{ss} = 0.1823$ cm^{-1}.

Before interpreting these results, it is instructive to consider the most accurate semiempirical calculation of Godfrey *et al.*[151] They use the three-term wave functions of de Groot and van der Waals[153] for the hexagonal and quinoid forms. Simple Hückel orbitals, ignoring overlap, were employed and all Slater integrals were accurately evaluated as described in Section 3. With these wave functions and the MO integrals given in their tables, one obtains $D^* = (D^2 + 3E^2)^{1/2}$ to be 0.1719 cm^{-1} and 0.1520 cm^{-1} for the hexagonal and quinoid forms, respectively.§ If one instead uses extended Hückel orbitals with overlap included, the wave functions of de Groot and van der Waals lead to a substantially larger splitting parameter of $D^* = 0.1995$ cm^{-1} and 0.1769 cm^{-1} for these respective structures. It would seem, therefore, that the rather good agreement with experiment was somewhat fortuitous, especially since the *ab initio* results in Table 9 clearly indicate that a three-term wave function does not include all the configurations which make significant contributions to D^*.

One difficulty in comparing theory and experiment is that not enough configurations could be included in the wave function to account adequately for

---

†The notation $\sigma^2 \to \sigma'^2$ is meant to imply $\sigma\sigma' \to \sigma''\sigma'''$, etc.
‡The number of configurations refers to the number generated with a program using D_{2h} symmetry. The number would be somewhat less if D_{6h} symmetry were used instead.
§The value of 0.1671 cm^{-1} quoted in Ref. 151 is inconsistent with their tables. The corrected values quoted here were obtained by using their Table III for the hexagonal form and their Table IX for the quinoid form.

Table 9. Results of the Ab Initio Spin–Spin Calculations for the $^3B_{1u}$ State of Benzene

Description of calculation	Size of Hamiltonian matrix[a]	Leading coefficient in wave function[b]	Correlation energy, a.u.	D, cm^{-1}
SCF	2	0.7071	—	0.1133
SCF+single, double, and triple π excitations	2131	0.6590	0.1004	0.1511
SCF+325 most important[c] $\pi^2 \to \sigma^2$ excitations	1097	0.7059	0.0051	0.1128
SCF+443 most important $\sigma^2 \to \pi^2$ excitations	2069	0.7056	0.0073	0.1103
SCF+779 most important $\sigma\pi \to \sigma'\pi'$ excitations	3265	0.6887	0.0898	0.1769
SCF+all single $\sigma \to \sigma'$ excitations+688 most important $\sigma^2 \to \sigma'^2$ double excitations	6036	0.6955	0.0629	0.0593
SCF[d] $+173\pi^2 \to \pi'^2$+all single $\sigma \to \sigma'$ $+498\sigma^2 \to \sigma'^2 +779\sigma\pi \to \sigma'\pi'$	8376	0.6587	0.1956	0.1823

[a] The configurations are generated using a program written for D_{2h} symmetry; however, hexagonal geometry is assumed.

[b] The SCF wave function has the form $2^{-1/2}[|\ldots a_{2u}^2 e_{1gx}^2 e_{1gy}^1 e_{2uxy}^1| + |\ldots a_{2u}^2 e_{1gx}^2 e_{1gy}^1 e_{2u_{x^2-y^2}}^1|]$, hence the coefficient can at most be $2^{-1/2}$.

[c] "Most important" implies those configurations having the largest coefficients in the wave functions formed from a large number of excitations of the particular class under consideration.

[d] The configurations included are again the most important in the sense defined in footnote c.

the triplet spin density. Although the effect of the $\pi^2 \to \pi'^2$, $\pi \to \pi'$, $\pi^2 \to \sigma^2$, and $\sigma^2 \to \pi^2$ excitations was fairly well accounted for, only a relatively small number of the potentially important $\sigma^2 \to \sigma'^2$ and $\sigma\pi \to \sigma'\pi'$ excitations could be included. Even if the configurations that were excluded do not make a direct contribution to D^*, they do so indirectly by affecting the coefficients for the configurations which are included. Another shortcoming of the results in Table 9 is that only a basis set of double-zeta quality was used. Although it may be that the SCF result for D^* is not sensitive to the quality of the basis set,[62] the better description accorded by a larger basis set may be quite important when the calculation is extended to the CI level.[91] In future work, it would be especially important to examine the effect of $3d\pi$ orbitals.

The question of what molecular geometry to use for the $^3B_{1u}$ state is very important since the spin–spin interals are inversely proportional to the cube of the distance between the electrons with unpaired spin (cf. Section 2). To examine how changes in molecular conformation affect the ZFS, CI calculations of similar quality were carried out for three additional geometrical forms[153]; a hexagonal form with C–C bond lengths appropriate to the triplet state (1.427 Å), a quinoid (D_{2h}) form with four long bonds (1.448 Å) and two short bonds (1.381 Å), and an elongated (D_{2h}) form with four short bonds (1.405 Å) and two long bonds (1.468 Å). Table 10 compares the ZFS for these conformations and the hexagonal (1.395 Å) form with the ESR results of de Groot et al.[45] Best agreement is obtained for the hexagonal form (1.427 Å) that predicts D to be 0.1676 cm^{-1}, larger than experiment by about 0.01 cm^{-1}. This value of D is 0.015 cm^{-1} less than that obtained using 1.395 Å C–C bond lengths and is similar to that expected by simply scaling the results by the ratio of the bond lengths cubed.

The ZFS patterns that are predicted for the quinoid and elongated forms are qualitatively different from experiment. For the quinoid form, the T_x and T_y levels are in opposite order to experiment and the overall spacing between the levels is substantially less. This latter effect is a consequence of the unpaired

Table 10. Fine-Structure Results for Various Geometrical Forms of Benzene

Geometrical form	Fine-structure levels,[a] cm^{-1}		
	T_x	T_y	T_z
Elongated	0.1299	−0.0677	−0.0623
Quinoid	0.0006	0.0422	−0.0428
Hexagonal, 1.395 Å bond lengths	0.0608	0.0608	−0.1216
Hexagonal, 1.427 Å bond lengths	0.0559	0.0559	−0.1118
Experiment[b]	0.0590	0.0462	−0.1054

[a] Compression or elongation is made along the x axis which passes through the two para carbon atoms. The z direction is perpendicular to the plane of the molecule.
[b] ESR results of M. S. de Groot, I. A. M. Hesselman, and J. H. van der Waals.

electrons becoming localized on opposite carbons, which increases their average separation. This effect is, in turn, due to double bond character introduced into the short bonds upon compression. For the elongated form, the T_x and T_y levels are in the same order as experiment, but the splitting between T_x and T_y is so large that the T_y level is now predicted to lie below the T_z level. It follows, therefore, that the conformation experimentally observed, *on the average*, in crystals of durene is nearly hexagonal with only a slight elongation. The distortions must be much smaller than those considered in the present study (2%–3%) and certainly less than 1%. The phrase "on the average" must be included to allow for the possibility that the spectra result from an interconversion of equivalent distorted quinoid conformations.[153] If this process occurs, then the rate of interconversion must be relatively rapid compared to the characteristic time resolution of the resonance experiment. In fact, any distortion that is present may be due to no more than the crystal field effect of the host crystal.

Calculations of the hyperfine splittings for the various conformations support the conclusions drawn on the basis of the ZFS, and provide additional evidence that the spin density in the molecule can change dramatically from small changes in molecular conformation.

We see, therefore, that the ZFS can be very sensitive to the geometric parameters implicit in a molecular wave function. It should be possible, therefore, to use the ZFS as a means of distinguishing different conformations of a molecule, just as spin–orbit coupling constants can be used to distinguish between electronic states of a molecule.[115–118]

5. Phenomena Related to Fine Structure

So far in this chapter we have examined spin–spin and spin–orbit terms in the context of molecular fine structure. One of these terms, the spin–orbit Hamiltonian, is responsible for other classes of phenomena that are very important to biology and photochemistry. We consider two of them in this section, namely phosphorescence and predissociation.

5.1. Phosphorescence

Phosphorescence is the emissive passage of the same molecule between two states which are of different multiplicity. According to Kasha's rule,[154] this process occurs most often between the lowest triplet and the singlet ground state. Although many workers have studied phosphorescence, perhaps the most complete theoretical treatment is that of Goodman and Laurenzi,[155] who employed relativistic quantum mechanics. To illustrate what information

Table 11. *Radiative Lifetimes of the* $^3A_2(n \to \pi^*)$ *State of Formaldehyde*

Component of triplet (Γ_r)	Radiative lifetime (τ_{Γ_r}), sec	
	Present calculation	Results of ESJ[a]
Γ_z	0.0057	0.010
Γ_x	5.8	36.7
Γ_y	3.1	0.030
High-temperature limit[b]	0.017	0.025

[a] Reference 156.
[b] Computed using Eq. (87).

can be obtained from *ab initio* quantum chemistry, we begin with their results and apply them to calculate phosphorescent lifetimes of the sublevels of the $^3A_2(n \to \pi^*)$ state of formaldehyde.

In the absence of spin–orbit coupling, S is a good quantum number and singlet–triplet transitions are strictly forbidden. When spin–orbit mixing is included in the Hamiltonian, however, each state is perturbed to a small extent by the others. Thus, the singlet ground state and first-excited triplet state wave functions can be written as

$$S_0 = {}^1\Psi_0 + \sum_t \sum_{\Gamma_r} [\langle{}^3\Psi_t^{\Gamma_r}|\mathscr{H}_{so}|{}^1\Psi_0\rangle/({}^1E_0 - {}^3E_t)]{}^3\Psi_t^{\Gamma_r} \tag{83}$$

and

$$T_1^{\Gamma_r} = {}^3\Psi_1^{\Gamma_r} + \sum_s [\langle{}^1\Psi_s|\mathscr{H}_{so}|{}^3\Psi_1^{\Gamma_r}\rangle/({}^3E_1 - {}^1E_s)]{}^1\Psi_s \tag{84}$$

where the Ψ and E are nonrelativistic eigenfunctions and eigenvalues appropriately summed over all the triplet (t index) and singlet (s index) states of the molecule. The electric dipole transition moment between the ground and various components (Γ_r) of the lowest triplet state is then given by

$$M(S_0, T_1^{\Gamma_r}) = \sum_s \{[\langle{}^1\Psi_s|\mathscr{H}_{so}|{}^3\Psi_1^{\Gamma_r}\rangle^*/({}^3E_1 - {}^1E_s)]M({}^1\Psi_0, {}^1\Psi_s)\}$$

$$+ \sum_t \{[\langle{}^3\Psi_t^{\Gamma_r}|\mathscr{H}_{so}|{}^1\Psi_0\rangle/({}^1E_0 - {}^3E_t)]M({}^3\Psi_1^{\Gamma_r}, {}^3\Psi_t^{\Gamma_r})\} \tag{85}$$

The natural radiative lifetime (τ) of the various triplet components is proportional to the square of the transition moment[156,157]

$$1/\tau_{\Gamma_r} = [64\pi^4\nu^3(\Gamma_r)/3hc^3]|M(S_0, T_1^{\Gamma_r})|^2 \tag{86}$$

where $\nu(\Gamma_r)$ is the frequency of the singlet–triplet transition.

At high temperatures, where it can be assumed that the sublevels are equally populated, the phosphorescence will appear as a single first-order decay with a lifetime given by[158]

$$1/\tau = (g_0/g_1) \sum_{\Gamma_r} 1/\tau_{\Gamma_r} \tag{87}$$

where g_0 and g_1 are the degeneracies of the single and triplet states with values of 1 and 3, respectively. The summation extends over all three components of the triplet.

Formaldehyde. This molecule contains the prototype of the carbonyl group and can be accurately treated by *ab initio* methods. Since the spatial symmetry of the lowest triplet state is a_2 with spin functions belonging to the representations a_2, b_1, and b_2 in C_{2v}, the symmetries of the three triplet sublevels are a_1, b_2, and b_1. Electronic transitions between the ground-state and all three components of the triplet state are therefore allowed by symmetry.

There have been several calculations of phosphorescent lifetimes for the $^3A_2(n \to \pi^*)$ reported in the literature.[156,158,159] Probably the most ambitious treatment so far is that of Ellis, Squire, and Jaffe (ESJ)[156] who used CNDO/S wave functions with all singly excited configurations. Although an approximate one-electron spin–orbit Hamiltonian (Eq. 79) was used and two-center integrals were neglected, all possible electronic states (24 of each multiplicity) were included in the summations [Eq. (27)].

To assess the quality of these semiempirical calculations, a fully *ab initio* treatment was performed using the ground-state geometry. The calculation employed a polarized double-zeta basis of contracted Gaussian-lobe functions augmented with a set of diffuse functions. The correct microscopic Hamiltonian in Eq. (7) was used and all integrals were accurately evaluated. Configuration-interaction wave functions for each of the states were obtained by accepting all singly excited configurations and all doubly excited configurations contributing at least 0.0003 hartree to the second-order Rayleigh–Schrödinger perturbation energy, with the restriction that the MOs corresponding to $1s$ orbitals on carbon and oxygen remained frozen. The $^3A_2(n \to \pi^*)$ MOs were used throughout to construct the wave functions. The spin–orbit matrix elements and transition energies and moments are given in Table 8. The phosphorescent lifetimes of the three $^3A_2$ components are compared to those of ESJ in Table 11.

To learn which states are primarily responsible for phosphorescence, it is useful to examine Eqs. (85) and (86) for the shortest lifetime component, namely the sublevel Γ_z of a_1 symmetry. The transition moment between $X\,^1A_1$ and this $^3A_2$ component is

$$M(X\,^1A_1, {}^3A_2^{\Gamma_z}) = \sum_s^\infty \left[\frac{\langle ^1A_1, s | \mathcal{H}_{so} | ^3A_2^{\Gamma_z} \rangle^*}{E(^3A_2) - E(^1A_1, s)} \right] M(X\,^1A_1, {}^1A_1, s)$$

$$+ \sum_t^\infty \left[\frac{\langle ^3A_2^{\Gamma_z}, t | \mathcal{H}_{so} | X\,^1A_1 \rangle}{E(X\,^1A_1) - E(^3A_2, t)} \right] M(^3A_2^{\Gamma_z}, {}^3A_2^{\Gamma_z}, t) \qquad (88)$$

Lohr[160] has shown that the infinite sums over s and t should contain the $X\,^1A_1$ and $^3A_2(n \to \pi^*)$ states, respectively, and that the transition moments should be computed using the length formulation. A major contribution to the lifetime should then come from the spin–orbit matrix element between these states times the difference in their dipole moments. The other sizable contribution to the Γ_z component comes from the $^1A_1(\pi \to \pi^*)$ state, in agreement with other authors.[156,158] This calculation predicts the excitation energy for the $^1A_1(\pi \to \pi^*)$ state to be 10.6 eV, although more elaborate calculations[161] place it in the range 11.2–11.5 eV and, hence, in the continuum since the first ionization potential is 10.87 eV.[162] In fact, the calculation of ESJ indicates that substantial contributions to the lifetimes may actually come from valence-like states embedded in the continuum.

Two additional conclusions emerge from these results. First, Rydberg states have small matrix elements between the $X\,^1A_1$ and $^3A_2(n \to \pi^*)$ states and, hence, contribute negligibly to the phosphorescent lifetime. The second observation, made previously by ESJ, is that for an accurate calculation excited triplet states must be allowed to mix with the singlet ground state in addition to allowing excited singlet states to mix with the excited triplet.

Turning now to the in-plane components, we note that the lifetime of the Γ_x level is determined by mixing of $^1B_2$ states into $^3A_2(n \to \pi^*)$ and by coupling between $^3B_1$ states and the $^1A_1$ ground state. The $^1B_2(2b_2 \to 6a_1)$ Rydberg state and the $^3B_1(\sigma \to \pi^*)$ valence state are the primary perturbation of each symmetry type but neither state contributes appreciably because of small spin–orbit coupling and a weak transition moment with $^3A_2$. The result is a computed lifetime of 5.8 sec, to be compared with the much shorter lifetime of 0.0057 sec for the Γ_z level.

The lifetime of the Γ_y component is determined by the mixing of $^1B_1$ states with $^3A_2(n \to \pi^*)$ and of $^3B_2$ states with $X\,^1A_1$. The only state having a significant perturbation in this case is $^1B_1(\sigma \to \pi^*)$. Although the spin–orbit matrix element between $^1B_1$ and $^3A_2$ is very large, the small oscillator strength of 0.0015 for $^1B_1$ makes the overall contribution relatively small and leads to a lifetime of 3.1 sec. The calculation of ESJ predicts a lifetime for this component of 0.03 sec which is about a factor of 100 less than the *ab initio* calculation. Their short lifetime was obtained for two reasons. First, ESJ obtained a much larger oscillator strength of 0.144 for the $^1B_1$ state. The small oscillator strength obtained in the *ab initio* work is, however, more consistent with the value of 0.002 obtained by Yeager and McKoy[163] from the equations-of-motion method and with experiment, in the sense that this state has not been observed. The second reason is that ESJ obtain a significant contribution from $^3B_2$ states of quite high energy. This is consistent with the earlier contention that valence-like states embedded in the continuum make substantial contributions to phosphorescent lifetimes. Inclusion of these states in the *ab initio* study should reduce the 3.1 sec lifetimes to a value in closer agreement with ESJ.

Both of the calculations discussed here predict that the Γ_z component should have the shortest lifetime and that the emitted light should therefore be polarized primarily along the CO bond direction. This agrees with the conclusions reached by Raynes[144] from an analysis of the rotational bands of the $^3A_2 \leftarrow {}^1A_1$ transition. The calculations also predict a high-temperature phosphorescent lifetime of 0.017 sec (*ab initio*) and 0.025 sec (ESJ). Although the lifetime has not been experimentally determined, it is estimated to be about 0.01 sec on the basis of a comparison with similar molecules.

One of the approximations in these calculations is the use of the formaldehyde ground-state geometry. The $^3A_2$ state is known to be nonplanar (out-of-plane angle for the hydrogens is 36°) and to have a CO bond length about 10% longer than in the ground state.[144] Since the $n \rightarrow \pi^*$ transition takes place almost entirely in the CO bond region, it probably is a good approximation to use C_{2v} geometry, but the approximation of a shortened CO bond length in the excited state is of uncertain validity.[164]*

We have seen in this section that although it is possible to identify the principal perturbing states and to determine the polarization of the emitted light, the quantitative validity of the numerical results is limited by several approximations, including the truncation of summations in Eq. (27) and molecular geometry. The reliable *a priori* theoretical determination of phosphorescent lifetimes within a factor of three is still difficult, even for molecules of only moderate size.

5.2. Molecular Predissociation

Predissociation is the radiationless process by which an atom or molecule passes from a metastable discrete state to an energetically degenerate continuum state. Of the competing mechanisms that mix Born–Oppenheimer states of different symmetry, and thus give rise to molecular predissociation, either spin–orbit or orbit–rotation coupling is usually rate determining. Although no completely satisfactory *ab initio* calculation of molecular predissociation has been reported, the principal perturbing states and the dominent mechanism of predissociation have been identified in several diatomic molecules.[165–167] To examine some of the computational difficulties, we consider the predissociation in the Schumann–Runge bands ($B\,{}^3\Sigma_u^- - X\,{}^3\Sigma_g^-$) of O_2. Since predissociation broadens the discrete vibrational spectrum of radiation transmitted through the 1750–1900 Å window of the upper atmosphere,

---

*Recently, Bendazzoli and Palmieri[164] have reported a calculation to determine the lifetime of the $^3A_2$ state using STO-3G and STO-4.31G bases with limited valence-shell CI. Their study, which included contributions from the first few singlet and triplet excited states, was carried out at both the planar ground state geometry and at the experimental geometry of the $^1A_2$ state. They found that the mean radiative lifetime of the state was reduced from 0.0266 to 0.0182 sec by changing from the planar ground state geometry to the pyramidal conformation of the $^1A_2$ state. In agreement with ESJ and the present study, they observe that the $X\,{}^1A_1$ and $^1A_1(\pi \rightarrow \pi^*)$ states are principally responsible for the phosphorescence, but, in contrast, they find the contributions of these states are in opposite directions. The source of this discrepancy is not clear to us.

the Schumann–Runge system is of considerable practical importance and has been studied extensively.[165,168–173]

The *full width* at *half-maximum* (FWHM) associated with a discrete vibrational state is[172,173]

$$\tau_v = 2\pi \sum_r |V_{vr}(E_v)|^2 \qquad (89)$$

where V_{vr} represents the interaction of v with the continuous vibrational states of the repulsive state r. The quantity $(\hbar/\tau_v)$ is the average lifetime of the discrete state before it predissociates. In addition to line broadening, predissociation also shifts the bound level v from its zero-order position, E_v^0, by[172,173]

$$S_v = E_v - E_v^0 = P \int \frac{\sum_r |V_{vr}(E)|^2}{E_v - E} \, dE \qquad (90)$$

where P denotes the Cauchy principal value integral for levels that are sufficiently far separated that they do not interact with one another through the continuum.

If we assume that the electronic and nuclear contributions to the width enter multiplicatively and further that the electronic interaction at the crossing point R_X dominates, then

$$\tau_v = 2\pi |V(R_X)|^2 \sum_r |\langle v|r \rangle|^2 \qquad (91)$$

Equation (91) holds when the electronic matrix element V_{vr} varies slowly with internuclear separation (R) and when the R-centroid of the bound continuum overlap is close to R_X. In this case, the linewidths depend on the spin–orbit and orbit–rotation matrix elements at the crossing point and on the vibrational overlap or Franck–Condon factors $\langle v|r \rangle$.

Predissociation in the Schumann–Runge Bands of O_2. Any of the states $c\ ^1\Sigma_u^-$, $^3\Sigma_u^+$ (2 states), $^5\Sigma_u^-$, $^1\Pi_u$, $^3\Pi_u$, and $^5\Pi_u$, that dissociate into two $O(^3P)$ atoms are candidates for predissociating the $B\ ^3\Sigma_u^-$ state of O_2. However, only the $^{1,3,5}\Pi_u$ and $^3\Sigma_u^+$ states have nonvanishing first-order spin–orbit matrix elements with $B\ ^3\Sigma_u^-$. Moreover, only $^3\Pi_u$ can couple via the orbit–rotation interaction, since this operator only connects states for which $\Delta S = 0$ and $\Delta \Lambda = \pm 1$. Figure 2 shows the potential energy curves for the RKR and Schaefer–Miller (SM) $B\ ^3\Sigma_u^-$ states, the $^1\Pi_u$ and $^3\Pi_u$ repulsive states of SM, and the $^5\Pi_u$ repulsive state of Julienne and Krauss. In the notation of Herzberg,[174] as modified by Mulliken,[175] the $(^3\Pi_u \rightarrow B\ ^3\Sigma_u^-)$ intersection corresponds to c^- predissociation, whereas the $(^1\Pi_u \rightarrow B\ ^3\Sigma_u^-)$ and $(^5\Pi_u \rightarrow B\ ^3\Sigma_u^-)$ intersections correspond to c^+. A central question is which Π state is primarily responsible for the line broadening of the $B\ ^3\Sigma_u^-$ state.

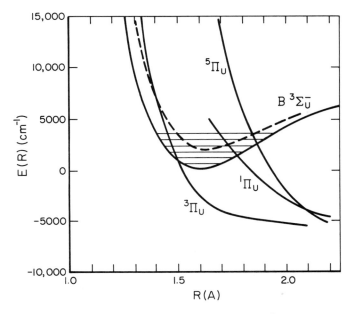

Fig. 2. Potential energy curves for O_2, showing the $B\ ^3\Sigma_u^-$ RKR curve (solid) and that calculated by Schaefer and Miller (dotted). The $^3\Pi_u$ and $^1\Pi_u$ repulsive curves are those of Schaefer and Miller (Ref. 165) and the $^5\Pi_u$ repulsive curve is that of Julienne and Krauss (Ref. 173).

Two recent experimental studies[168,171] on the linewidth of the Schumann–Runge bands (Fig. 3) indicate that the $v = 4$ level is the most strongly predissociated, with subsidiary maxima occuring at $v = 8$ and $v = 11$. There appears to be little predissociation below $v = 3$ or above $v = 12$.

Murrell and Taylor[170] (MT) and Riess and Ben-Aryeh[169] have recently applied the Franck–Condon principle to predissociation in these bands, with the result that an intersection on the *outer* limb of the $B\ ^3\Sigma_u^-$ state by a single repulsive curve can account for the experimental linewidths. By contrast, and in contradiction to Fig. 3, an intersection on the repulsive *inner* branch of the $B\ ^3\Sigma_u^-$ potential results in Franck–Condon factors with fairly uniform values over all vibrational levels above the crossing point. Assuming a single repulsive curve (assigned $^3\Pi_u$), MT obtained the best agreement with experiment for a crossing of the outer limb of the B state at about $v = 4$.

From the Schaefer and Miller (SM) calculations on the $B\ ^3\Sigma_u^-$, $^3\Pi_u$, and $^1\Pi_u$ states of O_2, the repulsive $^3\Pi_u$ and $^1\Pi_u$ states cross the inner and outer limbs of $B\ ^3\Sigma_u^-$, respectively, which on the basis of the previous argument rules out $^3\Pi_u$ as the principal perturbing state. At the same time, SM note that the predissociating state at $v = 4$ must be either $^5\Pi_u$ or $^5\Sigma_u^-$, since $^1\Pi_u$ appears to cross too low on the outer branch, thereby suggesting that several repulsive states may predissociate $B\ ^3\Sigma_u^-$ to about the same degree.

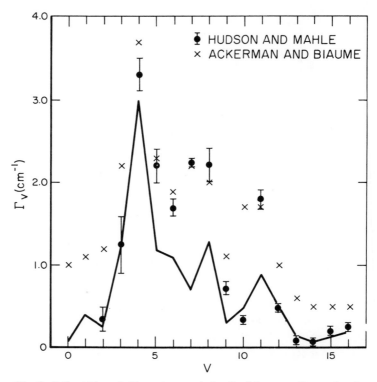

Fig. 3. Full width at half-maximum of the O_2 Schumann–Runge bands, showing the experimental data of Hudson and Mahle (Ref. 171) and of Ackerman and Biaume (Ref. 168). The calculated widths of Julienne and Krauss (Ref. 173), including the contributions from the $^1\Pi_u$, $^3\Pi_u$, and $^5\Pi_u$ repulsive curves, are shown by the solid line.

Another important point established by SM is that spin–orbit coupling is the dominant predissociating mechanism. The orbit–rotation interaction (which couples only $^3\Pi_u$ to the $B\ ^3\Sigma_u^-$ state) is estimated to be about 1.2 J cm^{-1}, or at room temperature about 10 cm^{-1}, which is nearly an order of magnitude too small to explain the observed linewidths.

The most recent and extensive treatment of predissociation in the Schumann–Runge bands is the study of Julienne and Krauss (JK).[172,173] They first identify the principal perturbing states by calculating the $^3\Sigma_u^-$ and $^{1,3,5}\Pi_u$ potential curves and spin–orbit matrix elements which couple them. Multiconfiguration self-consistent-field wave functions are employed with an approximate spin–orbit Hamiltonian of the form

$$\mathscr{H}'_{so} = A \sum_i \mathbf{l}_i \cdot \mathbf{s}_i \tag{92}$$

Potential curves similar to those of SM are obtained for the $^1\Pi_u$ and $^3\Pi_u$ repulsive states. These curves cross the $^3\Sigma_u^-$ potential at 3.25 and 2.7a_0, respectively. Significantly, their $^5\Pi_u$ curve crosses the outer limb around the $v = 4$ level ($3.5a_0$) and, hence, near the model curve of Murrell and Taylor (cf. Fig. 2). Using the atomic value for A and neglecting two-center integrals, they obtain spin–orbit matrix elements of 63, 47, and 36 cm^{-1} for the coupling between the $^3\Sigma_u^-$ and $^{5,3,1}\Pi_u$ states. Based on both the position of the crossing points and the spin–orbit matrix elements, it is concluded that the $^5\Pi_u$ is principally responsible for the predissociation.

Julienne and Krauss then took this overall qualitative picture and modeled the position and form of the $^5\Pi_u$ repulsive curve to obtain the best agreement with the experimental line shifts [cf. Eq. (90)]. For the $^5\Pi_u$ repulsive state, an exponential form with variable position and slope at the crossing point was fit to the experimental data for the second difference in the level shifts. The best deperturbation of these shifts was obtained with a fitted potential for the dominant $^5\Pi_u$ state which was close to the *ab initio* curve with a spin–orbit matrix element of 65 cm^{-1}, in excellent agreement with the calculated value of 63 cm^{-1}. Using the $B\,^3\Sigma_u^-$ curve determined from the RKR method, the *ab initio* curves for the $^{1,3}\Pi$ states, and the modeled curve for the $^5\Pi$ state, JK evaluated the linewidths for each of the vibrational levels. The calculated widths are shown by the solid line in Fig. 3. Quantitative agreement is obtained for all vibrational levels except the intermediate ones from $v = 5$ to 11. Part of this discrepancy may result from blending and from unresolved triplet splittings of the lines. It is also possible that the inclusion of the $2\,^3\Sigma_u^+$ state in the final treatment may improve the results for these intermediate vibrational levels.[176]

Although these results offer concrete hope that the O_2 Schumann–Runge predissociation bands are now understood, it would be very worthwhile in future work to determine the spin–orbit matrix elements between the relevant states with the full Breit–Pauli spin–orbit Hamiltonian, and thereby confirm the conclusions of JK that the $^5\Pi_u$ crossing of the outer limb of the $B\,^3\Sigma_u^-$ state at $v \cong 4$ accounts for most of the observed effects.

6. Conclusions

In this chapter, we have sought to illustrate the role of quantum chemistry in interpreting molecular fine structure and the related effects of phosphorescence and predissociation which depend on spin–orbit coupling. We have seen that *ab initio* methods are often required to obtain correct results, even for qualitative understanding. For quantitative comparisons with experiment, the use of extensive basis sets, large CI expansions, and proper vibrational averaging techniques are generally required. Although much progress has been made in identifying the forces responsible for these effects and their relationship to the electron distribution, much remains to be learned, particularly in synthesizing the results of the calculations as they become available on different molecules in their ground and excited states. Computational developments in this field depend upon concomitant advances in the subject matter described in Volume 3 of this Series. In the future, we hope that reliable rules of thumb can be developed for purposes of correlation and prediction.

Appendix. Vibration–Rotation Corrections to the ZFS Parameters

The standard treatment[102] of nuclear motion near potential energy minima hinges on the assumption that the harmonic oscillator and rigid-rotor models provide good approximations to the vibrational and rotational motions, respectively, and that corrections for anharmonicity, Coriolis, and centrifugal distortion effects can be introduced as perturbations. This approach has proved useful for low-lying states of vibration in which displacements from equilibrium are small.

Briefly, this perturbation method[103,104] involves the following steps. First, the nonrelativistic Schrödinger equation for the electrons, $\mathcal{H}_0 \Psi = E\Psi$, is solved for its eigenvalues E and eigenfunctions Ψ over a grid of judiciously chosen nuclear configurations. This is followed by the computation of ZFS parameters $P = D_{ij}$ or $E_{ij}(i, j = x, y, z)$ at each point on this grid. The resulting surfaces for E and P are then fit in n normal coordinates (q) to expansions (cf. Chapter 4 for an alternate technique) of the form

$$P = P_e + \sum_r^n \alpha_r q_r + \sum_{r,s}^n \beta_{rs} q_r q_s + \sum_{r,s,t}^n \gamma_{rst} q_r q_s q_t + \cdots \qquad (A1)$$

Using vibration–rotation wave functions developed to the appropriate order of perturbation theory, P is finally averaged over the nuclear motion. For n modes of pure vibration corresponding to the set of quantum numbers $v = (v_1, v_2, \ldots, v_n)$, the result[105] is

$$\langle P \rangle_v = P_0 + \sum_i^n A_i(v_i + \tfrac{1}{2}) + \sum_{i \le j}^n B_{ij}(v_i + \tfrac{1}{2})(v_j + \tfrac{1}{2}) + \cdots \qquad (A2)$$

where $P_0, A_i, B_{ij}, \ldots$ are certain constants that depend on the zero-field expansion parameters in Eq. (A1), and on the harmonic, cubic, quartic, and higher force constants in the corresponding expansion of the Born–Oppenheimer potential energy E. These constants characterize the behavior of $\langle P \rangle_v$ with vibrational state. For example, in one dimension, they develop as $P_0 = P_e - \tfrac{7}{8}r\gamma + \cdots$, $A = \beta - 3r\alpha + \cdots$, $B = \tfrac{3}{2}(\delta - 5r\gamma + \cdots)$, where P_e is the value of P at $R = R_e$ and where r is the dimensionless ratio of the cubic force constant to the harmonic frequency. We see that each constant in the expansion depends on the curvatures of both E and P. Although this perturbation analysis is instructive and provides insight, it is severely limited in a quantitative sense by asymptotic expansion behavior[106] and by computational intractability,[102] especially when carried to high order for polyatomic molecules.

An alternative formulation[107–113] that avoids these difficulties is based on the variational principle. Employing the full Watson[114] Hamiltonian for vibration and rotation, one proceeds (without necessarily expanding into normal coordinates) by analogy to the Hartree–Fock and configuration-interaction methods discussed in Volume 3 of this Series. The trial wave function is constructed as a superposition of products of either model harmonic or Morse oscillator basis functions times rigid-rotor functions. Solution of a secular equation then yields the expansion coefficients and energy levels in the customary manner. Although the details are affected by the type and number of oscillator basis sets, whether an SCF step is included, how the matrix elements are evaluated and so on, a vibration–rotation wave function is ultimately obtained and used to average expansions like Eq. (A1) over the nuclear motion. Recently, one variant of this approach has been applied[113,143] to about a dozen, nonlinear polyatomic molecules including methylene, water, and ozone. The reader is referred to Refs. 107–113 for details which are beyond the scope of this chapter.

ACKNOWLEDGMENTS

Many colleagues have helped us in significant ways to compile this review. Of special importance are our collaborators on much of the research summarized here: Dr. J. D. Allen, Dr. C. F. Bender, Dr. E. R. Davidson, Dr. M. Godfrey, Dr. M. Gouterman, Dr. M. Karplus, Dr. D. J. Kouri, Dr. R. L. Matcha, Dr. R. H. Pritchard, Dr. D. M. Schrader, and Dr. M. L. Sink. This chapter would probably not have been written without their interest in, and contributions to this subject. The support of this research by the Battelle Institute (BI) Program is also gratefully acknowledged.

We would also like to thank Dr. P. Julienne and Dr. M. Krauss for sending us preprints of their work concerning the predissociation in the Schumann–Runge bands of the oxygen molecule. Figures 2 and 3 of this chapter are essentially taken from their work.

Finally, we would like to thank Dr. R. M. Pitzer for a careful reading of the manuscript and Dr. P. R. Certain for helpful comments concerning spin–rotation interactions.

References

1. M. Karplus, Weak interactions in molecular quantum mechanics, *Rev. Mod. Phys.* **32**, 455–460 (1960).

2. A. B. Zahlan (ed.), *The Triplet State*, Proceedings of an International Symposium held at the American University of Beirut, Lebanon, Cambridge Univ. Press, Cambridge (1967).

3. A. Carrington, D. H. Levy, and T. A. Miller, *in: Advances in Chemical Physics* (I. Prigogine and S. A. Rice, eds.), Vol. 18, pp. 149–248, Wiley–Interscience, New York (1970).

4. C. A. Hutchison and B. W. Mangum, Paramagnetic resonance absorption in naphthalene in its phosphorescent state, *J. Chem. Phys.* **29**, 952–953 (1958); **34**, 908–922 (1961).

5. M. Gouterman and W. Moffitt, Origin of zero-field splittings in triplet states of aromatic hydrocarbons, *J. Chem. Phys.* **30**, 1107–1108 (1959).

6. G. N. Lewis and M. Kasha, Phosphorescence and the triplet state, *J. Am. Chem. Soc.* **66**, 2100–2116 (1944); Phosphorescence in fluid media and the reverse process of singlet–triplet absorption, *J. Am. Chem. Soc.* **67**, 994–1003 (1945).

7. L. C. Allen and A. M. Karo, Basis functions for *ab initio* calculations, *Rev. Mod. Phys.* **32**, 275–285 (1960).

8. H. A. Bethe and E. E. Salpeter, *Quantum Mechanics of One- and Two- Electron Atoms*, Springer-Verlag, Berlin (1957).

9. T. E. H. Walker and W. G. Richards, *Ab initio* computation of spin–orbit coupling constants in diatomic molecules, *Symp. Faraday Soc.* **2**, 64–68 (1968).

10. H. Ito and Y. J. I'haya, Evaluation of molecular spin–orbit integrals by a Gaussian expansion method, *Mol. Phys.* **24**, 1103–1115 (1972).

11. R. H. Pritchard, M. L. Sink, and C. W. Kern, Theoretical study of the fine-structure coupling constants in the $2p\ ^3\Pi_u$ state of H_2, *Mol. Phys.* **30**, 1273–1282 (1975).

·12. S. Fraga and K. M. S. Saxena, Electronic Structure of Atoms, Division of Theoretical Chemistry, University of Alberta, Technical Report TC-AS-I-72, 1972; S. Fraga and J. Karwowski, Electronic Structure of Atoms, Division of Theoretical Chemistry, University of Alberta, Technical Report TC-AS-II-73, 1973.

13. G. Malli, Spin–other–orbit interaction in many-electron atoms, *J. Chem. Phys.* **48**, 1088–1091 (1968).

14. G. Malli, Spin–spin interaction in many-electron atoms, *J. Chem. Phys.* **48**, 1092–1094 (1968).

15. K. M. S. Saxena and G. Malli, Spin–orbit and spin–other–orbit interactions for f^4 electron configuration, *Can. J. Phys.* **47**, 1829–1862 (1969).

16. S. Fraga, K. M. S. Saxena, and B. W. N. Lo, Hartree–Fock values of energies, interaction constants, and atomic properties for the ground states of the negative ions, neutral atoms, and first four positive ions from helium to krypton, *At. Data* **3**, 323–361 (1971); Hartree–Fock values of energies, interaction constants, and atomic properties for excited states with p^N configurations of the negative ions, neutral atoms, and first positive ions from boron to bromine, *At. Data* **4**, 255–267 (1972); S. Fraga and K. M. S. Saxena, Hartree–Fock values of energies, interaction constants, and atomic properties for excited states with $3d^N4s^0$ and $3d^N4s^2$ configurations of the negative ions, neutral atoms, and first four positive ions of the transition elements, *At. Data* **4**, 269–287 (1972).

17. S. Fraga and G. Malli, *Many-Electron Systems: Properties and Interactions*, W. B. Saunders Company, Philadelphia (1968).

18. P. A. M. Dirac, Quantum theory of the electron, *Proc. R. Soc. London, Ser. A* **117**, 610–624 (1928).

19. H. A. Kramers, Structure of the multiplet S-states in molecules with two atoms. Parts I and II, *Z. Phys.* **53**, 422–438 (1929).

20. J. H. van Vleck, The coupling of angular momentum vectors in molecules, *Rev. Mod. Phys.* **23**, 213–227 (1951).

21. G. Breit, Effect of retardation on the interaction of two electrons, *Phys. Rev.* **34**, 553–573 (1929).

22. W. Pauli, Quantum mechanics of the magnetic electron, *Z. Phys.* **43**, 601–623 (1927).

23. J. O. Hirschfelder, C. F. Curtiss, and R. B. Bird, *Molecular Theory of Gases and Liquids*, pp. 1044–1046, John Wiley and Sons, New York (1954).

24. M. Tinkham and M. W. P. Strandberg, Theory of the fine structure of the molecular oxygen ground state, *Phys. Rev.* **97**, 937–951 (1955).

25. A. Carrington and A. D. McLachlan, *Introduction to Magnetic Resonance*, Harper and Row, New York (1967); A. D. McLachlan, Spin–spin coupling Hamiltonian in spin multiplets, *Mol. Phys.* **6**, 441–444 (1963).

26. R. D. Sharma, Spin–spin interaction in methylene, *J. Chem. Phys.* **38**, 2350–2352 (1963); erratum *J. Chem. Phys.* **41**, 3259 (1964).

27. A. L. Kwiram, in *M.T.P. International Review of Science* (C. A. McDowell, ed.), Vol. 4, pp. 271–315, Butterworths, London (1972).

28. D. De Santis, A. Lurio, T. A. Miller, and R. S. Freund, Radio-frequency spectrum of metastable $N_2(A\ ^3\Sigma_u^+)$. II. Fine structure, magnetic hyperfine structure, and electric quadrupole constants in the lowest 13 vibrational levels, *J. Chem. Phys.* **58**, 4625–4665 (1973); see also references contained therein.

29. M. A. El-Sayed, *M.T.P. International Review of Science* (A. D. Buckingham and D. A. Ramsay, eds.), Vol. 3, pp. 119–153, Butterworths, London (1972).

30. R. S. Freund and T. A. Miller, Microwave optical magnetic resonance induced by electrons (MOMRIE) in $H_2\ G(3d\ ^1\Sigma_g^+)$, *J. Chem. Phys.* **56**, 2211–2219 (1972).

31. H. F. Hameka, *Advanced Quantum Chemistry*, Addison-Wesley, Reading, Massachusetts (1965).

32. O. Zamani-Khamiri and H. F. Hameka, Spin–orbit contribution to the zero-field splitting of the oxygen molecule, *J. Chem. Phys.* **55**, 2191–2196 (1971); R. H. Pritchard, C. W. Kern, O. Zamani-Khamiri, and H. F. Hameka, Comment on the spin–orbit contribution to the zero-field splitting of the oxygen molecule, *J. Chem. Phys.* **56**, 5744–5745 (1972).

33. A. Messiah, *Quantum Mechanics*, Appendix C, North-Holland, Amsterdam (1962).

34. D. R. Beck, C. A. Nicolaides, and J. I. Musher, Calculation on the fine structure of the $a\ ^3\Sigma_u^+$ state of molecular helium, *Phys. Rev. A* **10**, 1522–1527 (1974).

35. J. B. Lounsbury, Calculation of zero-field splitting in NH. II. One-center representation of triplet states, *J. Chem. Phys.* **46**, 2193–2200 (1967); I. One-center minimal basis and atomic orbital representations of the ground state, *J. Chem. Phys.* **42**, 1549–1554 (1965).

36. S. R. Langhoff, Spin–orbit contribution to the zero-field splitting in CH_2, *J. Chem. Phys.* **61**, 3881–3885 (1974).

37. S. A. Boorstein and M. Gouterman, Zero-field splittings. IV. Gaussian approximation of integrals, *J. Chem. Phys.* **41**, 2776–2781 (1964).

38. S. R. Langhoff, E. R. Davidson, M. Gouterman, W. R. Leenstra, and A. L. Kwiram, Zero-field splitting of the triplet state of porphyrins. II, *J. Chem. Phys.* **62**, 169–176 (1975).

39. P. S. Han, T. P. Das, and M. F. Rettig, Calculation of the spin–spin and spin–orbit contribution to the zero-field splitting in hemin, *J. Chem. Phys.* **56**, 3861–3873 (1972).

40. K. Kayama and J. C. Baird, Spin–orbit effects and the fine structure in the $^3\Sigma_g^-$ ground state of O_2, *J. Chem. Phys.* **46**, 2604–2618 (1967).

41. J. S. Griffith, *The Theory of Transition-Metal Ions*, Cambridge Univ. Press, Cambridge (1961).

42. H. Hayashi and S. Nagakura, Correlation of the zero-field splittings with the $n\pi^*$ and $\pi\pi^*$ triplet levels of benzaldehydes, *Chem. Phys. Lett.* **18**, 63–66 (1973); The lowest $n\pi^*$ and $\pi\pi^*$ triplet levels of benzaldehydes and their correlation with the zero-field splittings, *Mol. Phys.* **27**, 969–979 (1974).

43. D. S. McClure, Spin–orbit interaction in aromatic molecules, *J. Chem. Phys.* **20**, 682–686 (1952).

44. S. A. Boorstein and M. Gouterman, Theory for zero-field splittings in aromatic hydrocarbons, *J. Chem. Phys.* **39**, 2443–2452 (1963).

45. M. S. de Groot, I. A. M. Hesselmann, and J. H. van der Waals, Paramagnetic resonance in phosphorescent aromatic hydrocarbons, *Mol. Phys.* **16**, 45–60 (1969).

46. R. F. Curl, The relationship between electron spin–rotation coupling constants and g-tensor components, *Mol. Phys.* **9**, 585–597 (1965).

47. A. N. Jette, Fine-structure of the metastable, $c\ ^3\Pi_u$ (1s, 2p), state of molecular hydrogen, *Chem. Phys. Lett.* **25**, 590–592 (1974); A. N. Jette and T. A. Miller, Fine structure in Rydberg states of the H_2 molecule, *Chem. Phys. Lett.* **29**, 547–550 (1974).

48. W. Lichten, M. V. McCusker, and T. L. Vierima, Fine structure of the metastable $a\ ^3\Sigma_u^+$ state of the helium molecule, *J. Chem. Phys.* **61**, 2200–2212 (1974).

49. R. N. Dixon, Spin–rotation interaction constants for bent AH_2 molecules in doublet electronic states, *Mol. Phys.* **10**, 1–6 (1966).

50. S. K. Luke, The radio-frequency spectrum of H_2^+, *Astrophys. J.* **156**, 761–769 (1969). See also P. M. Kalaghan and A. Dalgarno, Hyperfine structure of the molecular ion H_2^+, *Phys. Lett.* **38A**, 485–486 (1972).

51. F. E. Harris, Matrix elements of spin-interaction operators, *J. Chem. Phys.* **47**, 1047–1061 (1967).

52. I. L. Cooper and J. I. Musher, Evaluation of matrix elements of spin-dependent operators for *N*-electron systems. I. One-body operators, *J. Chem. Phys.* **57**, 1333–1342 (1972); II. Two-body operators, *J. Chem. Phys.* **59**, 929–938 (1973).

53. C. Bottcher and J. C. Browne, Matrix elements of spin-dependent operators over total molecular wavefunctions, *J. Chem. Phys.* **52**, 3197–3201 (1970); C. Bottcher, Calculations on the Small Terms in the Hamiltonian of a Diatomic Molecule (Part I) and Variational Principles for the Study of Resonances (Part II), Ph.D. thesis, The Queen's University of Belfast, 1968.

54. Y.-N. Chiu, On singlet–triplet transitions induced by exchange with paramagnetic molecules and the intermolecular coupling of spin angular momenta, *J. Chem. Phys.* **56**, 4882–4898 (1972).

55. L. C. Chiu, Electron magnetic perturbation in diatomic molecules of Hund's case (b), *J. Chem. Phys.* **40**, 2276–2285 (1964).

56. R. McWeeny, On the origin of spin–Hamiltonian parameters, *J. Chem. Phys.* **42**, 1717–1725 (1965).

57. K. F. Freed, Theory of the hyperfine structure of molecules: Application to $^3\Pi$ states of diatomic molecules intermediate between Hund's cases (a) and (b), *J. Chem. Phys.* **45**, 4214–4241 (1966).

58. S. R. Langhoff, Spin Dipole–Dipole Contribution to the Zero-Field Splitting in Methylene and Formaldehyde, Ph.D. thesis, University of Washington (1973).

59. H. M. McConnell, A theorem on zero-field splittings, *Proc. Natl. Acad. Sci. USA* **45**, 172–174 (1959).

60. A. C. Wahl, P. E. Cade, and C. C. J. Roothaan, Study of two-center integrals useful in calculations on molecular structure. V. General methods for diatomic integrals applicable to digital computers, *J. Chem. Phys.* **41**, 2578–2599 (1964).

61. R. H. Pritchard, A Theoretical Study of Fine Structure Interactions in Diatomic Molecules, Ph.D. thesis, The Ohio State University, 1974.

62. R. H. Pritchard, M. L. Sink, J. D. Allen, and C. W. Kern, Theoretical studies of fine structure in the ground state of O_2, *Chem. Phys. Lett.* **17**, 157–159 (1972).

63. R. L. Matcha, C. W. Kern, and D. M. Schrader, Fine-structure studies of diatomic molecules: Two-electron spin–spin and spin–orbit integrals, *J. Chem. Phys.* **51**, 2152–2170 (1969); erratum, *J. Chem. Phys.* **57**, 2598 (1972).

64. R. M. Pitzer and H. F. Hameka, Evaluation of the spin–spin interaction in benzene, *J. Chem. Phys.* **37**, 2725 (1962); H. F. Hameka, Theory of the electron spin resonance of benzene in the triplet state, *J. Chem. Phys.* **31**, 315–321 (1959).

65. M. Blume and R. E. Watson, Theory of spin–orbit coupling in atoms. I. Derivation of the spin–orbit coupling constant, *Proc. R. Soc. London, Ser. A* **270**, 127–143 (1962); Theory of spin–orbit coupling in atoms. II. Comparison of theory with experiment, *Proc. R. Soc. London, Ser. A* **271**, 565–578 (1963).

66. Z. Kopal, *Numerical Analysis*, John Wiley and Sons, New York (1961).

67. R. E. Christoffersen and K. Ruedenberg, Hybrid integrals over Slater-type atomic orbitals, *J. Chem. Phys.* **49**, 4285–4292 (1968).

68. R. L. Matcha, G. Malli, and M. B. Milleur, Two-center two-electron spin–spin and spin–orbit hybrid integrals, *J. Chem. Phys.* **56**, 5982–5989 (1972); G. Malli, M. B. Milleur, and R. L. Matcha, Two-center hybrid integrals, *J. Chem. Phys.* **54**, 4964–4965 (1971).

69. R. L. Matcha, R. H. Pritchard, and C. W. Kern, Prolate-spheroidal expansions of the spin–orbit, spin–spin, and orbit–orbit operators, *J. Math. Phys.* **12**, 1155–1159 (1971) and references therein.

70. D. M. Schrader, Calculation of spin–spin interaction integrals, *J. Chem. Phys.* **41**, 3266–3267 (1964).
71. G. G. Hall and A. Hardisson, The anistropy of the *g*-factor for polycyclic hydrocarbons, *Proc. R. Soc. London, Ser. A* **278**, 129–136 (1964).
72. K. Rüdenberg, A study of two-center integrals useful in calculations on molecular structure. II. The two-center exchange integrals, *J. Chem. Phys.* **19**, 1459–1477 (1951).
73. R. L. Matcha, D. J. Kouri, and C. W. Kern, Relativistic effects in diatomic molecules: Evaluation of one-electron integrals, *J. Chem. Phys.* **53**, 1052–1059 (1970).
74. R. L. Matcha and C. W. Kern, Evaluation of three- and four-center integrals for operators appearing in the Breit–Pauli Hamiltonian, *J. Chem. Phys.* **55**, 469 (1971).
75. A. D. McLean, LCAO–MO–SCF ground state calculations on C_2H_2 and CO_2, *J. Chem. Phys.* **32**, 1595–1597 (1960).
76. E. A. Magnusson and C. Zauli, Evaluation of molecular integrals by a numerical method, *Proc. Phys. Soc., London* **78**, 53–64 (1961).
77. A. C. Wahl and R. H. Land, The evaluation of multicenter integrals by polished brute force techniques I. Analysis, numerical methods, and computational design of the potential-charge distribution scheme, *Int. J. Quantum Chem.* **IS**, 375–401 (1967).
78. I. Shavitt and M. Karplus, Gaussian-transform method for molecular integrals. I. Formulation for energy integrals, *J. Chem. Phys.* **43**, 398–414 (1965).
79. C. W. Kern and M. Karplus, Gaussian-transform method for molecular integrals. II. Evaluation of molecular properties, *J. Chem. Phys.* **43**, 415–429 (1965).
80. S. F. Boys, Electronic wavefunctions. I. A general method of calculation for the stationary states of any molecular system, *Proc. R. Soc. London, Ser. A* **200**, 542–554 (1950).
81. I. Shavitt, *in*: *Methods in Computational Physics* (B. Alder, S. Fernbach, and M. Rotenberg, eds.) vol. 2, pp. 1–45, Academic Press, New York (1963).
82. R. L. Matcha and C. W. Kern, Relationships between spin–spin and electron repulsion integrals, *J. Phys. B* **4**, 1102–1108 (1971).
83. M. Geller and R. W. Griffith, Zero-field splitting, one- and two-center Coulomb-type integrals, *J. Chem. Phys.* **40**, 2309–2325 (1964); erratum, *J. Chem. Phys.* **40**, 2309–2310 (1964).
84. F. E. Harris and H. H. Michels, *in*: *Advances in Chemical Physics* (I. Prigogine, ed.), Vol. 13, pp. 205–266, Interscience, New York (1967).
85. J. B. Lounsbury and G. W. Barry, General solution for one-center zero-field splitting integrals, *J. Chem. Phys.* **44**, 4367–4372 (1966).
86. J. W. McIver, Jr. and H. F. Hameka, Effect of spin–orbit interactions on the zero-field splitting of the NH radical, *J. Chem. Phys.* **45**, 767–773 (1966).
87. F. P. Prosser and C. H. Blanchard, On the evaluation of two-center integrals, *J. Chem. Phys.* **36**, 1112 (1962).
88. R. L. Matcha and C. W. Kern, Identities relating spin–spin and orbit–orbit to spin–orbit interactions, *Phys. Rev. Lett.* **25**, 981–982 (1970).
89. R. M. Pitzer, C. W. Kern, and W. N. Lipscomb, Evaluation of molecular integrals by solid spherical harmonic expansions, *J. Chem. Phys.* **37**, 267–274 (1962).
90. R. H. Pritchard and C. W. Kern, Spin–spin and spin–other–orbit integrals for diatomic molecules, *J. Chem. Phys.* **57**, 2590–2591 (1972); erratum, *J. Chem. Phys.* **61**, 754 (1974).
91. S. R. Langhoff, *Ab initio* evaluation of the fine structure of the oxygen molecule, *J. Chem. Phys.* **61**, 1708–1716 (1974).
92. P. W. Abegg and T-K. Ha, *Ab initio* calculation of the spin–orbit coupling constant from gaussian lobe SCF molecular wavefunctions, *Mol. Phys.* **27**, 763–767 (1974).
93. R. K. Hinkley, T. E. H. Walker, and W. G. Richards, Spin–orbit coupling constants from Gaussian wavefunctions, *J. Chem. Phys.* **52**, 5975–5976 (1970).
94. G. Herzberg, *Spectra of Diatomic Molecules*, 2nd ed., D. Van Nostrand, Princeton (1950).
95. K. P. Huber, Constants of diatomic molecules, *in*: *American Institute of Physics Handbook*, McGraw-Hill, New York (1972).
96. S. P. McGlynn, T. Azumi, and M. Kinoshita, *Molecular Spectroscopy of the Triplet State*, Prentice-Hall, Englewood Cliffs, New Jersey (1969).

97. P. R. Fontana, Spin–orbit and spin–spin interactions in diatomic molecules. I. Fine structure of H_2, *Phys. Rev.* **125**, 220–228 (1962).

98. L. C. Chiu, Fine structure constants of metastable H_2 in the $c\ ^3\Pi_u$ state, *Phys. Rev.* **137**, A384–A387 (1965); Fine-structure constants of the metastable $c\ ^3\Pi_u$-state hydrogen molecule, *J. Chem. Phys.* **41**, 2197–2198 (1964).

99. W. T. Zemke and P. G. Lykos, Double configuration self-consistent field study of the $^1\Pi_u$, $^3\Pi_u$, $^1\Pi_g$, and $^3\Pi_g$ states of H_2, *J. Chem. Phys.* **51**, 5635–5650 (1969).

100. M. Lombardi, Fine and hyperfine structure of the $2p$ and $3p\ ^3\Pi_u$ states of H_2, *J. Chem. Phys.* **58**, 797–802 (1973).

101. S. Rothenberg and E. R. Davidson, Natural orbitals for hydrogen-molecule excited states, *J. Chem. Phys.* **45**, 2560–2576 (1966).

102. H. H. Nielsen, The vibration–rotation energies of molecules, *Rev. Mod. Phys.* **23**, 90–136 (1951).

103. C. W. Kern and R. L. Matcha, Nuclear corrections to electronic expectation values: Zero-point vibrational effects in the water molecule, *J. Chem. Phys.* **49**, 2081–2091 (1968).

104. W. C. Ermler and C. W. Kern, Zero-point vibrational corrections to one-electron properties of the water molecule in the near Hartree–Fock limit, *J. Chem. Phys.* **55**, 4851–4860 (1971).

105. B. J. Krohn, W. C. Ermler, and C. W. Kern, Nuclear corrections to molecular properties. IV. Theory for low-lying vibrational states of polyatomic molecules with application to the water molecule near the Hartree–Fock limit, *J. Chem. Phys.* **60**, 22–33 (1974).

106. L. L. Sprandel and C. W. Kern, A test of perturbation theory for determining anharmonic vibrational corrections to properties of diatomic molecules, *Mol. Phys.* **24**, 1383–1389 (1972).

107. G. D. Carney and R. N. Porter, Abstracts of the 1972 Molecular Structure and Spectroscopy Symposium held at The Ohio State University; G. D. Carney, *Ab-initio* Calculation of Vibration–Rotation Properties for the Ground Electronic State of the H_3^+ Molecular Ion, Ph.D. thesis, University of Arkansas, 1973.

108. L. L. Sprandel, Quantum Mechanical Studies of Molecular Vibrations, Part II—A Study of the Use of the Self-Consistent Field and Configuration Interaction Methods for Solving the Vibrational Schrodinger Equation of a Bent AB_2 Molecule With Application to Water, Ph.D. thesis, The Ohio State University 1974.

109. M. G. Bucknell, N. C. Handy, and S. F. Boys, Vibration–rotation wave-functions and energies for any molecule obtained by a variational method, *Mol. Phys.* **28**, 759–776 (1974).

110. M. G. Bucknell and N. C. Handy, Vibration–rotation wavefunctions and energies for the ground electronic state of the water molecule by a vibrational method, *Mol. Phys.* **28**, 777–792 (1974).

111. S. A. Gribov and G. V. Khovrin, Determination of the potential surface and analysis of the anharmonic vibrations of the water molecule, *Opt. Spectrosc.* **36**, 274–279 (1974).

112. E. K. Lai, M. S. thesis, Department of Chemistry, Indiana University, 1975.

113. G. D. Carney and C. W. Kern, Vibration-rotation analysis of some nonlinear molecules by a variational method, *Int. J. Quantum Chem. Symp.* **9**, 317–323 (1975).

114. J. K. G. Watson, Simplification of the molecular vibration–rotation Hamiltonian, *Mol. Phys.* **15**, 479–490 (1968).

115. H. Lefebvre-Brion and C. M. Moser, Calculation of valence states of NO and NO^+, *J. Chem. Phys.* **44**. 2951–2954 (1966).

116. T. E. H. Walker and W. G. Richards, The nature of the first excited electronic state in BeF, *Proc. Phys. Soc., London* **92**, 285–290 (1967).

117. T. E. H. Walker and W. G. Richards, The nature of the first excited electronic state in MgF, *Proc. Phys. Soc., London* **1** (Ser. 2), 1061–1065 (1968).

118. A. L. Roche and H. Lefebvre-Brion, Valence-shell states of PO: An example of the variation of the spin–orbit coupling constants with internuclear distance, *J. Chem. Phys.* **59**, 1914–1921 (1973).

119. W. H. Moores and R. McWeeny, The calculation of spin–orbit splitting and g tensors for small molecules and radicals, *Proc. R. Soc. London, Ser. A* **332**, 365–384 (1973).

120. E. Ishiguro and M. Kobori, Spin–orbit coupling constants in simple diatomic molecules, *J. Phys. Soc. Japan* **22**, 263–270 (1967).

121. C. E. Moore, Atomic Energy Levels, I., II., and III. National Bureau of Science, Circular No. 467, U.S. Government Printing Office, Washington, D.C. (1949).

122. L. Veseth, Spin–orbit and spin–other–orbit interaction in diatomic molecules, *Theor. Chim. Acta*, **18**, 368–384 (1970).

123. T. E. H. Walker and W. G. Richards, Calculation of spin–orbit coupling constants in diatomic molecules from Hartree–Fock wavefunctions, *Phys. Rev.* **177**, 100–101 (1969).

124. T. E. H. Walker and W. G. Richards, Molecular spin–orbit coupling constants. The role of core polarization, *J. Chem. Phys.* **52**, 1311–1314 (1970).

125. H. Lefebvre-Brion and C. M. Moser, On the calculation of spin–orbit interaction in diatomic molecules, *J. Chem. Phys.* **46**, 819–820 (1967).

126. H. Lefebvre-Brion and N. Bessis, Spin–orbit splitting in $^2\Delta$ states of diatomic molecules, *Can. J. Phys.* **47**, 2727–2730 (1969).

127. J. A. Hall, J. Schamps, J. M. Robbe, and H. Lefebvre-Brion, A theoretical study of the perturbation parameters in the a $^3\Pi$ and A $^1\Pi$ states of CO. *J. Chem. Phys.* **62**, 1802–1805 (1975).

128. E. A. Ballik and D. A. Ramsay, The $A'\,^3\Sigma_g^- - X'\,^3\Pi_u$ band system of the C_2 molecule, *Astrophys. J.* **137**, 61–83 (1963); An extension of the Phillips system of C_2 and a survey of C_2 states, *Astrophys. J.* **137**, 84–101 (1963); G. Herzberg, A. Lagerqvist, and C. Malmberg, New electronic transitions of the C_2 molecule in absorption in the vacuum ultravoilet region, *Can. J. Phys.* **47**, 2735–2743 (1969).

129. S. R. Langhoff, M. L. Sink, R. H. Pritchard, C. W. Kern, S. J. Strickler, and M. J. Boyd, *Ab initio* study of perturbations between the $X\,^1\Sigma_g^+$ and b $^3\Sigma_g^-$ states of the C_2 molecule, *J. Chem. Phys.* (submitted for publication).

130. K. Kayama, Spin dipole–dipole interaction in O_2, *J. Chem. Phys.* **42**, 622–630 (1965).

131. R. H. Pritchard, C. F. Bender, and C. W. Kern, Fine-structure interactions in the ground state of O_2, *Chem. Phys. Lett.* **5**, 529–532 (1970).

132. T. J. Cook, B. R. Zegarski, W. H. Breckenridge, and T. A. Miller, Gas phase EPR of vibrationally excited O_2, *J. Chem. Phys.* **58**, 1548–1552 (1972) and references therein.

133. T. Amano and E. Hirota, Microwave spectrum of the molecular oxygen in the excited vibrational state, *J. Mol. Spectrosc.* **53**, 346–363 (1974).

134. E. Wasserman, R. S. Hutton, V. J. Kuck, and W. A. Yager, Zero-field parameters of "free" CH_2: Spin–orbit contributions in xenon, *J. Chem. Phys.* **55**, 2593–2594 (1971); Electron paramagnetic resonance of CD_2 and CHD isotope effects, motion and geometry of methylene, *J. Am. Chem. Soc.* **92**, 7491–7493 (1970); E. Wasserman, V. J. Kuck, R. S. Hutton, E. D. Anderson, and W. A. Yager, ^{13}C hyperfine interactions and geometry of methylene, *J. Chem. Phys.* **54**, 4120–4121 (1971).

135. R. A. Bernheim, H. W. Bernard, P. S. Wang, L. S. Wood, and P. S. Skell, Electron paramagnetic resonance of triplet CH_2, *J. Chem. Phys.* **53**, 1280–1281 (1970); ^{13}C hyperfine interaction in CD_2, *J. Chem. Phys.* **53**, 1280–1281 (1970).

136. J. Higuchi, On the effect of the bond angle on the electron spin–spin interaction. Methylene derivatives, *J. Chem. Phys.* **39**, 1339–1341 (1963); Zero-field splittings in molecular multiplets. Spin–spin interaction of methylene derivatives, *J. Chem. Phys.* **38**, 1237–1245 (1963).

137. J. F. Harrison, An *ab initio* study of the zero field splitting parameters of $^3B_1$ methylene, *J. Chem. Phys.* **54**, 5413–5417 (1971).

138. S. R. Langhoff and E. R. Davidson, An *ab initio* calculation of the spin dipole–dipole parameters for methylene, *Int. J. Quantum Chem.* **7**, 759–777 (1973).

139. S. R. Langhoff, S. T. Elbert, and E. R. Davidson, A configuration interaction study of the spin dipole–dipole parameters for formaldehyde and methylene, *Int. J. Quantum Chem.* **7**, 999–1019 (1973).

140. S. H. Glarum, Spin–orbit interactions in molecular radicals, *J. Chem. Phys.* **39**, 3141–3144 (1963).

141. S. J. Fogel and H. F. Hameka, Spin–orbit interactions and their effect on the zero-field splitting of the methylene radical, *J. Chem. Phys.* **42**, 132–136 (1965).
142. W. R. Hall and H. F. Hameka, Second-order effect of spin–orbit coupling on the angular dependence of the zero-field splitting in CH_2, *J. Chem. Phys.* **58**, 226–231 (1973). See also erratum, *J. Chem. Phys.* **60**, 4104 (1974).
143. S. R. Langhoff and G. D. Carney, to be published.
144. W. T. Raynes, Rotational analysis of some bands of the triplet←singlet transition in formaldehyde, *J. Chem. Phys.* **44**, 2755–2777 (1966); Spin splittings and rotational structure of nonlinear molecules in doublet and triplet electronic states, *J. Chem. Phys.* **41**, 3020–3032 (1964).
145. S. R. Langhoff and E. R. Davidson, *Ab initio* evaluation of the fine structure and radiative lifetime of the $^3A_2(n \to \pi^*)$ state of formaldehyde, *J. Chem. Phys.* **64**, 4699–4710 (1976).
146. S. D. Peyerimhoff, R. J. Buenker, W. E. Kammer, and H. Hsu, Calculation of the electronic spectrum of formaldehyde, *Chem. Phys. Lett.* **8**, 129–135 (1971).
147. A. M. Ponte Goncalves and C. A. Hutchinson, Jr., Electron nuclear double resonance in photoexcited triplet-state benzene-h_6 molecules in benzene-d_6 single crystals, *J. Chem. Phys.* **49**, 4235–4236 (1968).
148. Y-N. Chiu, Zero-field splittings in some triplet-state aromatic molecules, *J. Chem. Phys.* **39**, 2736–2748 (1963).
149. M. Geller, Two-electron, one- and two-center integrals, *J. Chem. Phys.* **39**, 853–854 (1963).
150. J. H. van der Waals and G. ter Maten, Zero-field splitting of the lowest triplet state of some aromatic hydrocarbons: Calculation and comparison with experiment, *Mol. Phys.* **8**, 301–318 (1964).
151. M. Godfrey, C. W. Kern, and M. Karplus, Studies of zero-field splittings in aromatic molecules, *J. Chem. Phys.* **44**, 4459–4469 (1966).
152. S. R. Langhoff, E. R. Davidson, and C. W. Kern, *Ab initio* study of the zero-field splitting parameters of $^3B_{1u}$ benzene, *J. Chem. Phys.* **63**, 4800–4807 (1975).
153. M. S. de Groot and J. H. van der Waals, Paramagnetic resonance in phosphorescent aromatic hydrocarbons. III. Conformational isomerism in benzene and triptycene, *Mol. Phys.* **6**, 545–562 (1963).
154. M. Kasha, Characterization of electronic transitions in complex molecules, *Discuss. Faraday Soc.* **9**, 14–19 (1950).
155. L. Goodman and B. J. Laurenzi, in: *Advances in Quantum Chemistry* (P-O. Lowdin, ed.), Vol. 4, pp. 153–169, Academic Press, New York (1968).
156. R. L. Ellis, R. Squire, and H. H. Jaffe, Use of the CNDO method in spectroscopy. V. Spin–orbit coupling, *J. Chem. Phys.* **55**, 3499–3505 (1971).
157. G. Herzberg, *Electronic Spectra of Polyatomic Molecules*, pp. 417–419 Van Nostrand, Princeton, New Jersey (1966), and references therein.
158. J. W. Sidman, Spin–orbit coupling in the $^3A_2-^1A_1$ transition of formaldehyde, *J. Chem. Phys.* **29**, 644–652 (1958).
159. T. Yonezawa, H. Kato, and H. Kato, Oscillator strength of singlet–triplet transition in formaldehyde, *J. Mol. Spectrosc.* **24**, 500–503 (1967).
160. L. L. Lohr, Spin-forbidden electric-dipole transition moments, *J. Chem. Phys.* **45**, 1362–1363 (1966).
161. S. R. Langhoff, S. T. Elbert, C. F. Jackels, and E. R. Davidson, *Chem. Phys. Lett.* **29**, 247–249 (1974).
162. J. E. Mentall, E. P. Gentieu, M. Krauss, and D. Neumann, Photoionization and absorption spectrum of formaldehyde in the vacuum ultraviolet, *J. Chem. Phys.* **55**, 5471–5479 (1971).
163. D. L. Yeager and V. McKoy, Equations of motion method: Excitation energies and intensities in formaldehyde, *J. Chem. Phys.* **60**, 2714–2716 (1974).
164. G. L. Bendazzoli and P. Palmieri, Spin–orbit interaction in polyatomic molecules: *Ab initio* computations with gaussian orbitals, *Int. J. Quantum Chem.* **8**, 941–950 (1974).
165. H. F. Schaefer III and W. H. Miller, Curve crossing of the $B\ ^3\Sigma_u^-$ and $^3\Pi_u$ states of O_2 and its relation to predissociation in the Schumann–Runge bands, *J. Chem. Phys.* **55**, 4107–4115 (1971).

166. J. A. Hall and W. G. Richards, A theoretical study of the spectroscopic states of the CF molecule, *Mol. Phys.* **23**, 331–343 (1972).

167. P. S. Julienne, M. Krauss, and B. Donn, Formation of OH through inverse predissociation, *Astrophys. J.* **170**, 65–70 (1971).

168. M. Ackerman and F. Biaume, Structure of the Schumann–Runge bands from the 0–0 to the 13–0 band, *J. Mol. Spectrosc.* **35**, 73–82 (1970).

169. I. Riess and Y. Ben-Aryeh, Application of the quantum Franck–Condon principle to predissociation in oxygen, *J. Quant. Spectrosc. Radiat. Transfer* **9**, 1463–1468 (1969).

170. J. N. Murrell and J. M. Taylor, Predissociation in diatomic spectra with special reference to the Schumann-Runge bands of O_2, *Mol. Phys.* **16**, 609–621 (1969).

171. R. D. Hudson and S. H. Mahle, Photodissociation rates of molecular oxygen in the mesophere and lower thermosphere, *J. Geophys. Res.* **77**, 2902–2914 (1972).

172. P. S. Julienne and M. Krauss, Predissociation of the Schumann–Runge Bands of O_2, NRL Memorandum Report 2900, Naval Research Laboratory, Washington, D.C. (1974).

173. P. S. Julienne and M. Krauss, Predissociation of the Schumann–Runge bands of O_2, *J. Mol. Spectrosc.* **56**, 270–308 (1975).

174. G. Herzberg, Predissociation and similar phenomenon, *Ergeb. Exakten. Naturwiss.* **10**, 207–284 (1931).

175. R. S. Mulliken, Some neglected subcases of predissociation in diatomic molecules, *J. Chem. Phys.* **33**, 247–252 (1960).

176. P. S. Julienne, private communication.

Author Index

Boldface page numbers indicate a chapter in this volume.

Subject Index

453

Date Due

FORM 109